THE MEASUREMENT OF AIRBORNE PARTICLES
Richard D. Cadle

ANALYSIS OF AIR POLLUTANTS
Peter O. Warner

ENVIRONMENTAL INDICES
Herbert Inhaber

URBAN COSTS OF CLIMATE MODIFICATION
Terry A. Ferrar, Editor

CHEMICAL CONTROL OF INSECT BEHAVIOR: THEORY AND APPLICATION
H. H. Shorey and John J. McKelvey, Jr.

MERCURY CONTAMINATION: A HUMAN TRAGEDY
Patricia A. D'Itri and Frank M. D'Itri

SULFUR IN THE ENVIRONMENT, Parts I and II
Jerome O. Nriagu

DATE DUE

MAY 19 1986			
OCT 21 1987			
APR 11 1988			
MAY 02 1988			
SEP 26 1989			
NOV 26 '89			
DEC 15 '89			
NOV 21 '92			
APR 25 1996			
MAY 03 2000			

GAYLORD PRINTED IN U.S.A.

SULFUR IN THE ENVIRONMENT

Part II: Ecological Impacts

SULFUR IN THE ENVIRONMENT

Part II: Ecological Impacts

Edited by

JEROME O. NRIAGU
Canada Centre for Inland Waters
Burlington, Ontario, Canada

A WILEY-INTERSCIENCE PUBLICATION
JOHN WILEY & SONS
New York • Chichester • Brisbane • Toronto

Library of Congress Cataloging in Publication Data:

Main entry under title:

Sulfur in the environment.

(Environmental science and technology)
"A Wiley-Interscience publication."
Includes bibliographical references and index.
1. Sulfur—Environmental aspects. I. Nriagu, Jerome O.

TD 196.S95S84 363 78-6807
ISBN 0-471-02942-4 (pt. 1)
ISBN 0-471-04255-2 (pt. 2)

Printed in the United States of America

10 9 8 7 6 5 4 3 2 1

SERIES PREFACE

Environmental Science and Technology

The Environmental Science and Technology Series of Monographs, Textbooks, and Advances is devoted to the study of the quality of the environment and to the technology of its conservation. Environmental science therefore relates to the chemical, physical, and biological changes in the environment through contamination or modification, to the physical nature and biological behavior of air, water, soil, food, and waste as they are affected by man's agricultural, industrial, and social activities, and to the application of science and technology to the control and improvement of environmental quality.

The deterioration of environmental quality, which began when man first collected into villages and utilized fire, has existed as a serious problem under the ever-increasing impacts of exponentially increasing population and of industrializing society. Environmental contamination of air, water, soil, and food has become a threat to the continued existence of many plant and animal communities of the ecosystem and may ultimately threaten the very survival of the human race.

It seems clear that if we are to preserve for future generations some semblance of the biological order of the world of the past and hope to improve on the deteriorating standards of urban public health, environmental science and technology must quickly come to play a dominant role in designing our social and industrial structure for tomorrow. Scientifically rigorous criteria of environmental quality must be developed. Based in part on these criteria, realistic standards must be established and our technological progress must be tailored to meet them. It is obvious that civilization will continue to require increasing amounts of fuel, transportation, industrial chemicals, fertilizers, pesticides, and countless other products; and that it will continue to produce waste products of all descriptions. What is urgently needed is a total systems approach to modern civilization through which the pooled talents of scientists and engineers, in cooperation with social scientists and the medical profession, can be focused on the development of order and equilibrium in the presently disparate segments of the human environment. Most of the skills and tools that are needed are already in existence. We surely have a right to hope a technology that has created such manifold environmental problems is also capable of solving them. It is our hope that this Series in Environmental Sciences and Technology will not only serve to make this challenge more explicit to the established professionals, but that it also will help to stimulate the student toward the career opportunities in this vital area.

Robert L. Metcalf
James N. Pitts, Jr.
Werner Stumm

PREFACE

The energy crisis has fostered heightened interest in the effects of sulfur pollution on the environment. As the skills and techniques of many scientific disciplines are brought to bear on the problem, the difficulty of obtaining a broad picture of the various research activities has greatly increased. *Sulfur in the Environment* was conceived as a systematic endeavor to interface the biological, chemical, geological, and clinical studies on pollutant sulfur, and represents perhaps the first attempt to bring most aspects of environmental sulfur pollution together in a single work. The chapters have been contributed by experts from many scientific disciplines; indeed, the literature on sulfur pollution has become so vast that no single scientist can present a detailed account of all the recent developments.

A comprehensive coverage of the various aspects of sulfur pollution must entail a large number of pages, as indeed is the case here. Part I includes papers on the sources (Chapters 1–3), behavior (Chapters 8 and 9), and transport (Chapters 5–7) of sulfur in the atmosphere. Chapter 4 on the costs and benefits of sulfur emission controls and Chapter 10 on ambient air monitoring for pollutant sulfur compounds add further dimension to the volume. Part II contains reports on the biological (Chapters 12–14), ecological (Chapters 11, 16–21), and health (Chapters 12 and 13) significance of sulfur pollution. The division of chapters into Parts I and II is quite arbitrary in view of the supply and effect relationships for pollutant sulfur in the environment.

Inevitably, several important topics have been slighted or even omitted. For example, we have not emphasized the metabolism and homeostasis of (pollutant) sulfur in mammalian systems because there are several good volumes and review papers devoted to this topic. We have also shied away from considering specific sources of sulfur and the control technologies; to do so would have more than doubled the size of the present work. The primary focus has clearly been on the processes of change and the ecological stresses stemming from environmental sulfur pollution.

Sulfur in the Environment is basically the result of the combined efforts of our distinguished group of contributors. Acknowledgment is also due to Drs. A. L. W. Kemp and P. G. Sly for their generous advice and counsel.

JEROME O. NRIAGU

Burlington, Ontario
April 1978

CONTENTS

CONTENTS
PART I

1

DETERIORATIVE EFFECTS OF SULFUR POLLUTION ON MATERIALS

Jerome O. Nriagu

Canada Centre for Inland Waters, Burlington, Ontario, Canada

INTRODUCTION

The deleterious effects of sulfur pollution on both health and the functions of the various life-support systems for living things are now widely recognized. In this volume the effects of sulfur pollution on the biota, flora, and climatic conditions are discussed in several other chapters. The present chapter focuses on the deteriorative effects of pollutant sulfur on cultural artifacts. The environmental impact of pollutant sulfur oxides shown in Figure 1 may be viewed against the available data (Table 1) on the exposure of materials and the general

Table 1. Percentage of Total U.S. Population (and Materials) Exposed to Different Minimum Average SO_2 Levels between 1967 and 1972[a]

Minimum Annual Average SO_2 ($\mu g/m^3$)	Exposure by Year (% of total material) population × 10^6					
	1967	1968	1969	1970	1971	1972
>100	26.1	20.5	19.3	18.5		
	(13.2)	(10.3)	(8.6)	(9.1)		
>80	33.2	31.5	21.3	25.3		
	(16.8)	(15.8)	(10.1)	(12.4)		
>60	44.8	48.8	37.9	31.8	9.3	2.5
	(22.7)	(24.5)	(18.8)	(15.6)	(4.5)	(1.2)
>40	66.2	62.4	55.6	46.1	35.7	37.9
	(33.5)	(31.3)	(27.6)	(26.6)	(17.3)	(18.2)
>20	85.1	83.9	78.8	75.8	68.2	67.0
	(43.1)	(42.1)	(39.1)	(37.1)	(33.1)	(32.2)
>0	197.5	199.4	201.4	203.8	206.2	208.0
	(100)	(100)	(100)	(100)	(100)	(100)

[a] Data from Gillette (1975).

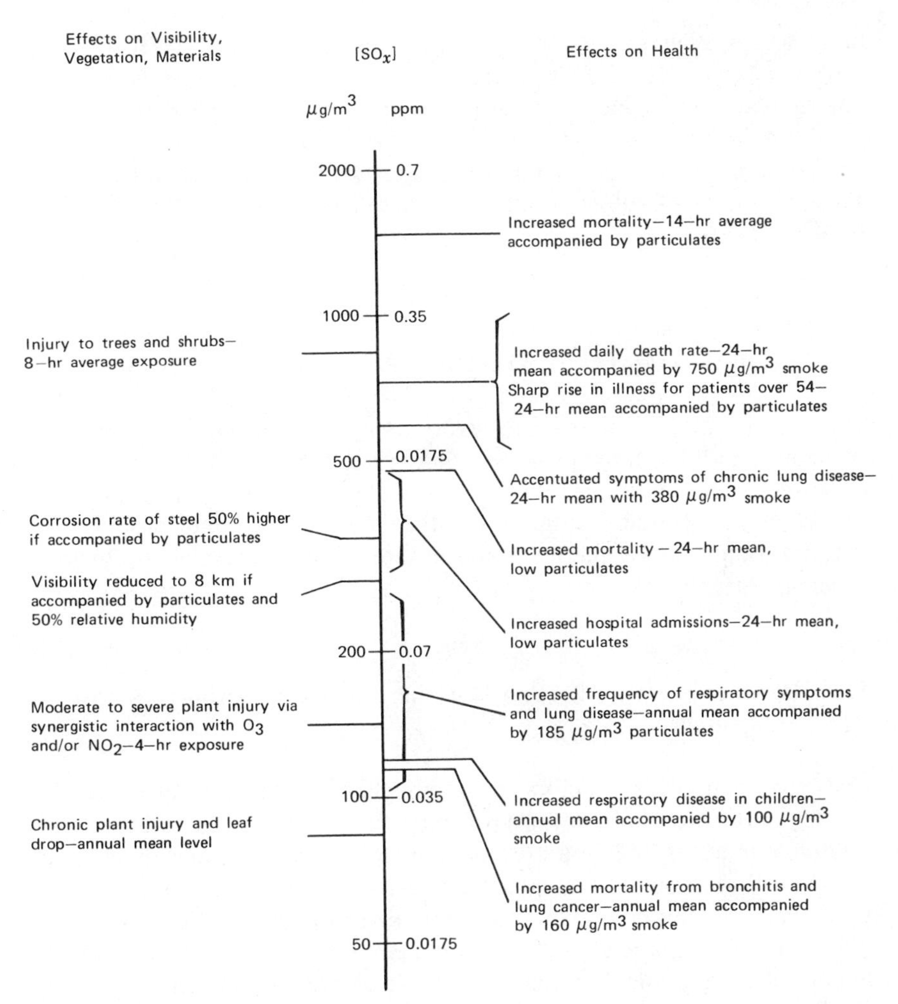

Figure 1. Air quality criteria for sulfur oxides. (Data from *Air Quality Criteria for Sulfur Oxides,* 1969, pp. 161–162.

population in the United States to the various levels of sulfur dioxide between 1967 and 1972.

Material damage by sulfur pollution represents a major burden on the economy of industrialized societies. The annual cost of corrosion to the world has been estimated at over $\$18 \times 10^9$ (Hendrik, 1964). Babcock (Part I) estimated the *total* (including social and indirect) damage due to pollutant SO_2 in the United States to be $\$6.75 \times 10^9$ in 1970; the total damage in the Chicago area alone was estimated to be $\$525 \times 10^6$ in 1966 and $\$228 \times 10^6$ in 1971. Gillette (1975) estimated that the damage to building materials caused by ambient sulfur dioxide was $900 × 10^6 in 1968 and $100 × 10^6 in 1972. Kucera (1976) estimated the loss in Sweden due to material damage by sulfur pollution to be about $35 × 10^6 in 1970. No doubt ambient sulfur compounds contribute greatly to the $280 × 10^6 annual damage suffered by building materials in Canada (Sereda, 1977). The economics of sulfur pollution damage has been treated in detail by Babcock (Part I).

1. MATERIAL DAMAGE BY ATMOSPHERIC SULFUR COMPOUNDS

The severity of damage done to materials by airborne sulfur compounds is determined by four principal factors: (*a*) the weather and other environmental conditions; (*b*) the physical and chemical characteristics of the material under attack; (*c*) the mechanisms and products of deterioration; and (*d*) the level and nature of the aggressive sulfur (and other) pollutants.

1.1. Weather Factors Affecting Material Damage by Airborne Sulfur Compounds

The most important factors in atmospheric corrosion of materials include moisture (in the form of rain, fog, dew, condensation, and relative humidity), temperature, air movement, sunlight, and the general "climate." It should be emphasized at the outset that weather conditions exercise unpredictable and conflicting effects on materials; minor differences in climatic conditions have been known to result in relatively major differences in material damage (e.g., see Black and Lherbier, 1968).

Moisture

Moisture is the critical controlling factor in atmospheric corrosion. In the absence of moisture, even high levels of atmospheric sulfur compounds have little, if any, corrosive effect. Vernon (1935) was the first to demonstrate that there is a "critical humidity" above which the corrosion rate increases sharply. Subsequent work have estimated the critical humidities for many materials in the presence of SO_2; the data for metallic surfaces may be summarized as follows (see Rosenfeld, 1972;

Yocom and McCaldin 1968):

Metal	Critical Humidity (%)
Aluminum	75–80
Brass	60–65
Copper	65–70
Iron	55–70
Magnesium	85–90
Nickel	65–70
Nickel–copper 70/30	65–75
Nickel–copper 80/20	75–80
Nickel–chrome 80/20	65–75
Steel	60–75
Zinc	70–75

The critical humidity is a function of both the hygroscopicity of the corrosion products and of the corrodents present. It is probably conditioned by the minimum number of layers of water molecules required for the corrosion products to accumulate. For example, zinc surfaces accommodate 15 molecular layers of water at 55% relative humidity (RH), 17 layers at 93% RH, and 92 layers at 100% RH (Spedding, 1977). The exact value of the critical humidity apparently is determined by the electrochemical processes in these molecular layers.

Basically, atmospheric corrosion is a discontinuous process with the result that the total corrosion damage is determined primarily by the percentage of time during which the critical humidity is exceeded (Sereda, 1960; Barton, 1973). A method for estimating this duration, known as time of wetness, has been developed by Sereda (1960), and the data collected subsequently by Guttman (1968) and Guttman and Sereda (1968) show that the rate of atmospheric corrosion is related in a logarithmic manner to the time of wetness. Their corrosion data were described quite well by equations of the form

$$r = aX^b(Y + c)$$

where r is the corrosion rate, X is time of wetness, Y is SO_2 concentration, and a, b, and c are empirical constants. Sereda (1974) estimated that the fractions of the total time during which the RH exceeded 70, 80, 90, and 95% were about 0.6, 0.4, 0.17, and 0.07, respectively, for continental Canadian locations; the corresponding fractional values for coastal areas were 0.8, 0.6, 0.34, and 0.2, respectively. In Britain the relative humidity exceeds 90%, on the average, for approximately 146 days each year (Stanners, 1974). By studying zinc corrosion over a 4-year period, Guttman (1968) found that the time of wetness as determined by Sereda's dew meter corresponded to the time during which the RH exceeded 86%. The implication is thus clear that for a substantial fraction (0.2 to 0.6) of a given year, exposed materials are subjected to conditions that promote corrosion by airborne pollutant sulfur.

Several studies (Sydberger and Vannerberg, 1972; Duncan and Spedding, 1973a, 1973b) have used $^{35}SO_2$ as a tracer to study the influence of relative humidity on the absorption of SO_2 on metal surfaces. As expected, there was an increase in $^{35}SO_2$ absorbed as the RH was increased. The data for zinc are portrayed in Figure 2; representative

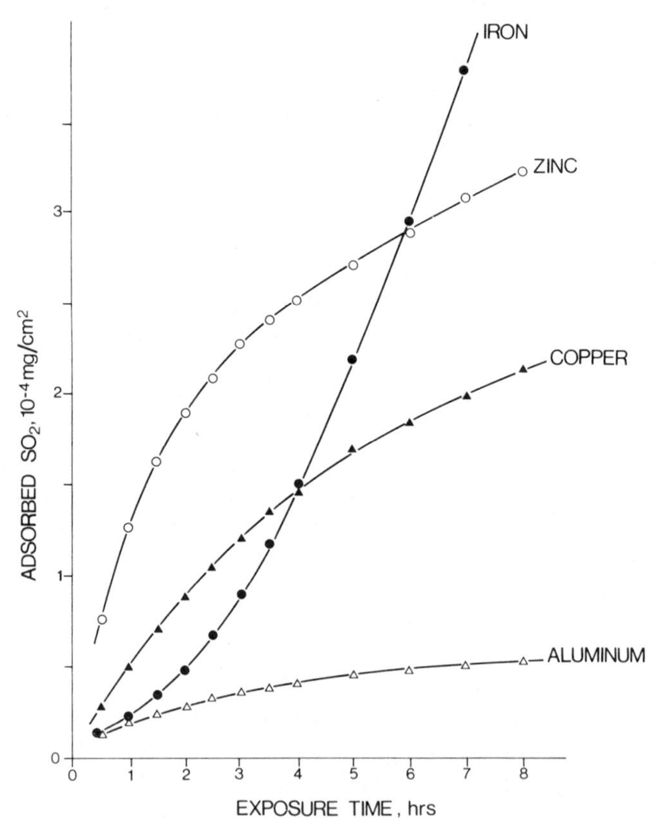

Figure 2. Adsorption of sulfur dioxide on polished metal surfaces at 90% relative humidity. (After Sydberger and Vannerberg, 1972.)

Table 2. Typical Values for Deposition of SO_2 onto Metal Surfaces[a]

Metal	Relative Humidity (%)	SO_2 Concentration ($\mu g/m^3$)	$10^5(SO_2)$ Sorbed ($\mu g/mm^2$)	$10^2 V^b$ (mm/sec)
Copper	90	285	63	42
Aluminum	90	285	15	10
	95	1100	75	30
Zinc	81	93	52	10
	90	285	75	49
	89	145	64	64
	89	157	64	61
Iron	90	285	23	15
	90	65	3	10
	90	100	5	10
	87	67	2	8

[a] Data from Duncan and Spedding (1974a, 1974b).
[b] V_g, the velocity of deposition, is defined as (mass of SO_2 deposited per unit area)/(mean SO_2 concentration in air × time of exposure).

data for other metals are shown in Table 2. In general, it has been found that sample preparation and sample treatment before exposure have considerable effects on the SO_2 uptake behavior. Furthermore, desorption of adsorbed SO_2 from zinc and aluminum surfaces was reported with pseudo-first-order half-lives for the reaction of about 2 hr and 0.1 to 4 hr, respectively (Duncan and Spedding, 1974). Unquantifiable desorption from some iron samples was noted (Duncan and Spedding, 1973a), but desorption from lead surfaces was not detected (Duncan and Spedding 1973b). It is also significant that the uptake rate is strongly dependent on the mean atmospheric SO_2 concentration. The adsorption-desorption of SO_2 has been explained in terms of specific interaction at sites on the surface oxide-hydroxide layers or the underlying metal (Duncan and Spedding, 1974).

The absorption of low concentrations of $H_2{}^{35}S$ from air by lead and zinc foils has also been investigated (Spedding, 1977). In both cases the H_2S uptake rate was markedly influenced by the RH, being very low below 65% RH but gradually increasing above 70% to a maximum at 100% RH. Between RH values of 40 to 65% the hydrogen sulfide uptake by zinc was less than that by lead, but from 75 to 100% RH the H_2S absorption by the two metals were fairly comparable (Spedding, 1977).

Fogs and *dew* generally intensify material corrosion, because they cause the formation of surface films of water that can absorb aggressive pollutants from the air. The effect of *rainfall* is complex. In the short term it reduces the corrosion rate by diluting and washing away the corrosive material. Over a long period of time, however, it can enhance the corrosion rate by constantly stripping away the protective coatings from material surfaces.

Temperature

Insofar as corrosion entails chemical reaction, it is affected by temperature changes according to the Arrhenius equation:

$$r = k \exp(-E/RT)$$

where r is corrosion rate, E is activation energy, T is temperature, and k and R are constants. For most locations the average annual variation in temperature is less than 50°C, implying that the influence of temperature on the electrochemical reaction rate should be minor. Indeed, the application of multiple linear regression and nonlinear curve fitting techniques to corrosion data and the corresponding weather factors generally shows insignificant correlations with temperature (see Guttman, 1968; Haynie and Upham, 1974).

Temperature affects the solubility and availability of oxygen in the corroding solution, however. An increase in temperature decreases the solubility of oxygen in aqueous solution. On the other hand, the diffusion rate for oxygen in aqueous systems increases with temperature. It is therefore commonly found that the corrosion rate increases to a maximum value and then decreases to a lower value at the boiling point of water. Changes in temperature also may (*a*) affect the solubility of the protective corrosion products (lead chloride is a notable example), (*b*) effect changes in the physical nature or chemical composition of the corrosion products, and (*c*) influence the pH of the corrosives. These changes may result in a substantial modification of the corrosion rate.

The influence of temperature may also be manifested in other ways. It determines the rate of drying and the length of time that a surface remains moist. Relative humidity is also strongly dependent on temperature; at high RH a small temperature drop may cause the dew point to be exceeded, thereby promoting corrosion. Because of their heat capacities, metal surfaces may be cooled relative to ambient air to below dewpoint and become liable to severe attack by the corrosive sulfur oxides dissolved in the condensed moisture. The temperature lag has been known to produce severe damage on metal objects in unheated warehouses and metal tools stored in plastic bags.

Air Movement

Air movement exercises a considerable influence on the adsorption-desorption of SO_2 from condensed moisture on material surfaces, and affects the deposition of particulates and water droplets on materials. Near point sources, wind direction is particularly important in determining the supply of SO_2 to material surfaces. Wind movement often causes rapid temperature and humidity fluctuations and, under exceptional circumstances, may promote abrasion-induced corrosion.

The General "Climate"

Under this heading may be grouped such variables as the height of an object above the ground; the orientation of the object; special heating or cooling of the object (e.g., that due to nearby power plants, power lines, or industrial installations); topography; indoor versus outdoor exposure; protective measures; and aeration and concentration differences. In localized situations, these factors may exercise a major influence on the extent of material damage by pollutant sulfur compounds.

1.2. Effects of Acid Precipitation

Sulfur pollution is commonly the dominant factor in the acidity of rainwater (see Kramer, Part I), so that the effect of acid precipitation is difficult to isolate from changes induced by sulfur pollution. It is well known that high acidity promotes corrosion because the hydrogen ions act as a sink for the electrons liberated by the critical corrosion processes:

$$\mathrm{Me} \rightarrow \mathrm{Me}^{n+} + n\mathrm{e}$$

$$n\mathrm{H}^+ + n\mathrm{e} \rightarrow n[\mathrm{H}] \rightarrow \frac{n}{2}\mathrm{H}_2$$

Corrosion of iron has been shown to be drastically increased below the critical pH of 3 to 4 (Rosenfeld, 1972); such pH values are not uncommon for rainwaters in parts of Scandinavia and North America. Acidic waters will also promote corrosion by dissolving away any oxide-hydroxide and carbonate coatings on metal surfaces.

There can be no doubt that the greater corrosion of materials in some urban and industrial centers as compared to rural areas is related to the higher acidity of moisture and rainfall in the former locations. Acid rains exercise a dramatic influence on the corrosion of buildings and works of art made of limestones and marble (Riederer, 1974). It has been suggested that, when nylon deteriorates in SO_2-polluted atmospheres, the damage may have been caused by tiny droplets of sulfuric acid which formed on the material (Yocom, 1959). This underscores the synergistic effects of sulfur pollution and high acidity on the corrosion of materials.

High rainfall and moisture acidity may result in the *hydrogen damage* of materials. As noted above, atomic hydrogen is a common cathodic product during corrosion. Under certain situations the hydrogen may diffuse into the material, accumulate near dislocation sites and micro-voids, and ultimately result in the embrittlement of the material. Hydrogen damage is a major problem only at high temperatures.

Acid rains may also enhance the leaching of ions and mineral nutrients, and consequently may reduce soil fertility and biological productivity (Nyborg, this volume). Another consequence of the high acidity in rainfall is the lowering of the pH of rivers and lakes, which may result in severe corrosive damage to hydroelectric power plants and concrete structures.

1.3. Effects of Metal Types

Metals and their alloys can be divided into four broad groups on the basis of their susceptibility to attack by sulfur oxides:

I. Iron and ferroalloys, which are essentially nonresistant to attack by pollutant sulfur oxides. They are usually not recommended for outside bold exposure or for use in enclosed space without supplementary anticorrosion protection.

II. Metals resistant to attack in atmospheres with low levels of sulfur oxides. In the absence of very aggressive corrodents, they may become immune to corrosion through the formation of insoluble salts by various anodic reactions. This group includes such nonferrous metals as cadmium, copper, magnesium, nickel, and zinc. Lead is more resistant to atmospheric corrosion than the other metals because its corrosion products with SO_2-SO_4^{2-} and Cl^- are only slightly soluble in water.

III. Metals that readily attain a passive state and can withstand exposure to moderate levels of atmospheric sulfur pollution. This group includes chromium and its alloys, aluminum and some of its alloys, titanium, zirconium, and, to a limited extent, atimony, molybdenum, silver, and tungsten. On exposure to high levels of

SO_2, depassivation of the metals occurs, resulting in grain-bound attack and pitting corrosion of varying intensities.

IV. The noble metals, which are essentially inert to pollutant sulfur oxides. Notable members of this group include gold, palladium, and platinum.

The actual mechanism and extent of corrosion of a given metal depends on its present physical characteristics and its past history. Metallurgical attributes that influence the attack by pollutant sulfur oxides include grain boundaries, grain orientation, impurities, surface roughness, differential stress and strain, differential grain size, differential thermal or pressure treatment, differential aeration and illumination, differential erosion, other differential preexposure, contact with other metals, and the form of anticorrosion protection. A detailed discussion of the importance of these factors is outside the principal objectives of this chapter; interested readers may consult Evans (1960), Kaesche (1966), Fontana and Green (1967), Leidheiser (1971), Rosenfeld (1972), Uhlig (1967), Speller (1951), and Sculley (1966).

1.4. Rates of Atmospheric Corrosion by Pollutant Sulfur

The rates of metal corrosion in the atmosphere are determined by a multitude of factors pertaining to the metal type, environmental variables, corrosion products, and time. As would be expected, considerable research effort has gone into elucidating the mechanisms and rates of atmospheric metal corrosion by the pervasive and aggressive sulfur oxides. Excellent phenomenological reviews are given by Evans and Taylor (1972), Rosenfeld (1972), and Barton (1973); a brief summary is included here as an essential introduction to the subsequent sections.

Before considering the rates of corrosion, a word on corrosion mechanism is in order. It is now generally recognized that the corrosion of metals in aqueous environments is caused by electrochemical processes and that the great diversity of corrosion reactions cannot be ascribed to any single set of mechanisms. Most often the reaction is diffusion controlled, and for it to proceed electrochemically three necessary conditions must be met simultaneously: (*a*) a potential difference must exist at points on the corroding surface; (*b*) there must be a mechanism for charge transfer between the electrolytic and the electronic conductors; and (*c*) a continuous conduction path must be available between the cathodic and anodic reaction centers. Detailed treatment of corrosion mechanisms may also be found in the references cited above.

The primary time-dependent step has to be the SO_2 uptake from the atmosphere. It may be recalled that the sorption rate is dependent on relative humidity, temperature, and the metal surface. Two models have been proposed to explain the initial reaction of SO_2 after adsorption onto metal surfaces: (*a*) the surface electrolyte solution reaction (SESR) model, which assumes the presence of a homogeneous layer of electrolyte (on metal surfaces) that resembles a bulk solution in its properties; and (*b*) the specific adsorption site (SAS) model, which holds that SO_2 adsorption occurs at a particular number of sites on metal surfaces exposed to the atmosphere. Duncan and Spedding (1974) considered the two models in detail and concluded that the available experimental data may be explained satisfactorily by either model. Using an autoradiographic technique, Duncan and Spedding (1973a, 1973b) showed that for a RH of 46 to 90% and SO_2 concentrations of 500 to 1000 $\mu g/m^3$ the SO_2 uptake occurred at discrete areas on unrusted iron, stainless steel, aluminum, lead, and copper sheets, whereas there was a fairly homogeneous deposition on zinc and nickel samples.

Increasing evidence suggests that, once corrosion is initiated, the subsequent reaction progress is largely controlled by the sulfate ions formed from oxidation of the adsorbed SO_2. The actual mechanism for the oxidation of SO_2 (and its hydrolytic products) at metal interfaces is essentially unknown (see Duncan and Spedding, 1973a). A schematic oxidation reaction may be written (see Barton, 1973) as

$$SO_2 + O_2 + 2e \rightarrow SO_4^{2-}$$

or

$$HSO_3^- + 1.5O_2 + 2e \rightarrow 25O_4^{2-} + H_2O$$

with the electron coming from the metal oxidation

$$M \rightarrow M^{n+} + ne$$

Using an electrophoretic method, Duncan and Spedding (1974a) found the rates of sulfate formations on iron and zinc surfaces to be similar, and the pseudo-first-order half-life was determined to be about 24 hr. Other workers (e.g., Karraker, 1963; Yoshimura et al., 1964), however, have reported considerably shorter oxidation rates (half-life, 10 to 100 mn) in bulk solutions using Fe(III) catalysts.

Rust on iron and steel, which is first restricted to localized sites or "nests," later spreads across the entire exposed surface. At an SO_2 concentration of about 260 $\mu g/m^3$ obvious corrosion products were observed on iron surfaces after 6 to 8 weeks (Duncan and Spedding, 1973a), whereas at SO_2 levels of 4×10^5 $\mu g/m^3$ corrosion products became evident after only a few hours (Heimler and Vannerberg, 1972). Barton (1973) showed that the critical concentrations for active SO_2 corrosion to become kinetically significant were 6 to 10 μg $SO_2/m^2 \cdot$year for steel and 18 to 20 μg $SO_2/m^2 \cdot$year for zinc and copper. The formation of rust drastically promotes the adsorption rate for SO_2. For example, at a SO_2 concentration of >0.001% and a RH of >96%, virtually all the SO_2 molecules that come into contact with the rusting surface are taken up (Barton, 1973). From the distribution of sulfate in the rust profile (Figures 3 and 4) and the relationship between rust layering and sulfate contents (Table 3), it is clear that the redistribution of sulfate via the rust layer is a very important process. The semiconducting properties of the corrosion products are probably also important in the transport of electrons from the phase boundaries.

Barton and Bartonova (1969) investigated the kinetics of atmospheric corrosion of steel and interpreted their data in terms of the following reaction scheme:

$$Fe + H_2O \rightarrow Fe(OH^-)_{ads} + H^+ \quad (1)$$

$$Fe(OH^-)_{ads} \rightarrow Fe(OH)_{ads} + e \quad (2)$$

$$Fe(OH)_{ads} + SO_4^{2-} \rightarrow FeSO_4 + OH^- + e \quad (3)$$

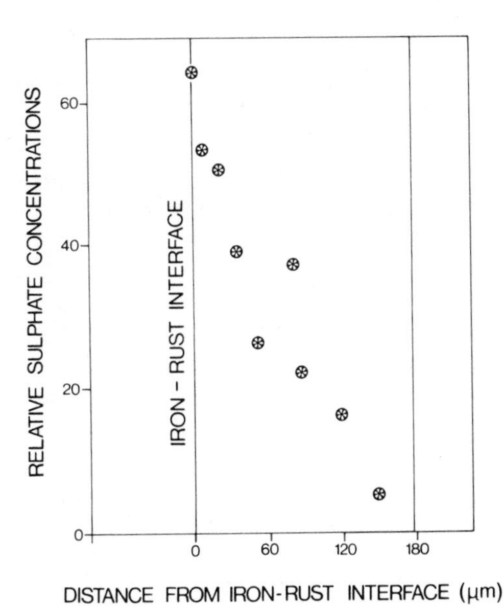

Figure 3. Sulfate profile in a rusted iron surface. (After Barton, 1973.)

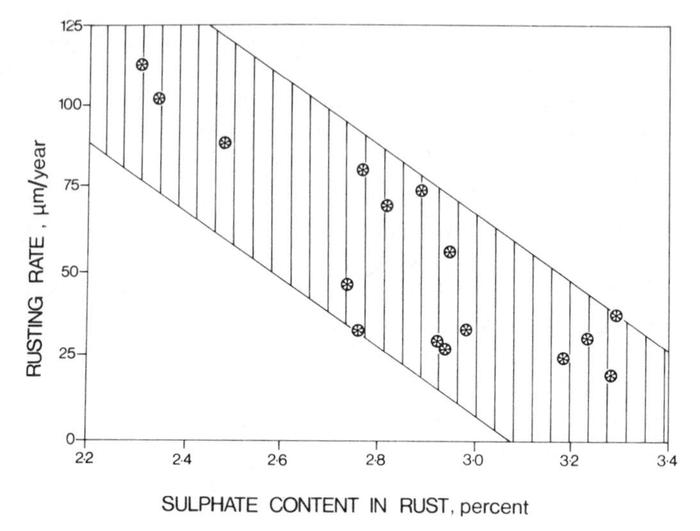

Figure 4. The dependence of the sulfate concentration in rust on the rate of rust formation. (After Barton, 1973.)

Oxidative hydrolysis of $FeSO_4$ will then lead to the formation of ferric hydroxides and the regeneration of the sulfate ions, which in turn will initiate another chain reaction:

$$FeSO_4 + 2H_2O \rightarrow FeOOH + SO_4^{2-} + 3H^+ + e \quad (4)$$

It was suggested that reactions 3 and 4 are the rate-limiting steps,

Table 3. Composition of Rust Layers Accumulated after 34 Months' Exposure in an Industrial Atmosphere[a]

Constituent	Layer[b] I	II	III
Rust (mg/cm^2)	7.0	40	29
Iron content of rust (%)	61	64	67
Fe^{2+} (%)	0	8.7	61
SO_4^{2-} ($\mu g/cm^2$)	281	1128	240
SO_4^{2-}, soluble ($\mu g/cm^2$)	55	87	40
Cl^-, total ($\mu g/cm^2$)	32	186	130

[a] From Barton (1973).
[b] Layer I: removable by brushing or scraping; layer II: "sprung off" by bending the test specimen; layer III: removable by chemical methods.

implying that the corrosion process is second order in relative humidity, but, above a certain critical value, first order in SO_4^{2-} concentration. Barton and Bartonova showed that, below the stationary corrosion rate of 15 to 25 mg/m^2·hr, the reaction progress becomes independent of the sulfate concentration.

Barton's (1973) reaction scheme is just one of the several that have been used to explain the SO_2-catalyzed, atmospheric rusting of iron and steel (e.g., see Evans and Taylor, 1972). In general, the validity of each model is determined by how well it explains the two basic features of atmospheric rusting, which are as follows.

1. Ferrous sulfate is formed before any insoluble rust appears, with the SO_2 needed as the source of sulfate in the soluble compound. Once the ferrous sulfate is formed, however, rusting can continue even when there is no SO_2 in the gas phase (see Evans and Taylor for experimental documentation).
2. A small amount of SO_2 can produce a relatively large amount of rust; in one set of tests (Schiber, 1963, 1967), each sulfur oxide molecule was found to generate between 15 and 40 molecules of rust, depending on the season. Evidently, the sulfate ions play a catalytic role and must be continuously regenerated during the rusting process.

Long-Term Rate of Corrosion

There is now a large body of data on the long-term corrosion of metals in SO_2-polluted environments in many parts of the world (e.g., see ASTM Spec Tech. Publs. 438 and 558). The huge bank of data reveals that atmospheric corrosion is most rapid in places where the SO_2 level is very high, and also that in a given locality rusting proceeds must faster during seasons of the year when the SO_2 concentration is highest. Considerable attention is now being devoted to developing phenomenological models that relate the long-term corrosion rates to the controlling and promoting environmental variables discussed above (see Guttman and Sereda, 1968; Haynie and Upham, 1974; Knotkova-Cermakova et al., 1974; Legault and Preban, 1975).

Figure 5 typifies the graphs of time-dependent corrosion of iron, steel, and type II metals in SO_2-polluted atmospheres. In general, these curvilinear graphs can be described by empirical equations of the form (e.g., see Legault and Preban, 1975)

$$r = kt^n$$

where r is average corrosion rate, t is exposure time, and k is a constant. Although the corrosion graphs are usually shown as smooth curves, it

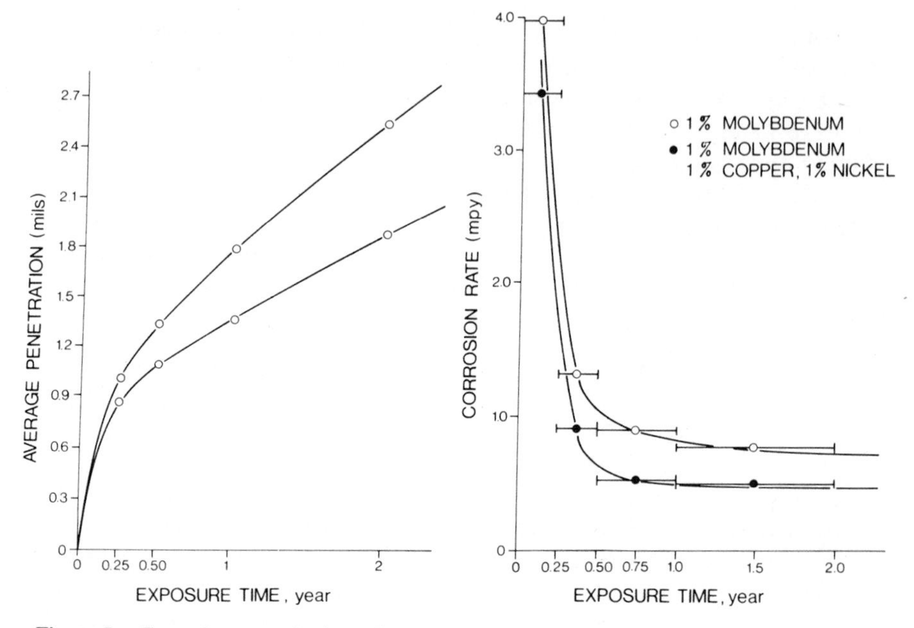

Figure 5. Corrosion rates for low-alloy steels in an industrial atmosphere. (After Legault and Preban, 1975.)

must be remembered that the corrosion process itself is discontinuous, because of the variable rates of absorption of SO_2 and water vapor under fluctuating atmospheric conditions. Indeed, several attempts have been made to develop quantized models of atmospheric corrosion, taking into account the time of wetness, atmospheric SO_2 levels, and so forth (see Barton, 1973; Knotkova-Cermakova et al., 1974).

Barton (1973) made a detailed analysis of the changes in corrosion rate with time during the corrosion process and showed that graphs of time versus corrosion generally consist of three segments:

1. The induction period, when the metal surfaces become covered with inherent oxide layer. It represents anything between a few months and several years, depending on the aggressiveness of the atmosphere.
2. The transition period between the induction period and the establishment of stationary conditions. During this phase of decreasing corrosion rate, the protective rust layer probably becomes properly adjusted to the atmospheric conditions.
3. The stationary phase, when the corrosion is linearly dependent on time. Barton and Bartonova (1969) estimated the stationary corrosion

rates for iron, copper, and zinc in several parts of Czechoslovakia during 1965–1966. Their data are shown in Table 4. Since the corrosion rates are for identical meteorological conditions, the observed disparities in the data must be due mainly to the effect of the SO_2 content. Additional discussions of the stochastic relationships between the corrosion rate and SO_2 concentrations are given in Section 1.5 for individual metals.

Effects of Airborne Particulate Materials

Sulfates and sulfites constitute a significant fraction of the suspended particulates in urban and industrial atmospheres (Moss, Part I). Compounds identified in airborne particulates include $(NH_4)_2SO_4$, NH_4HSO_4, $(NH_4)_3(SO_4)_2$, metal sulfates, and metal sulfites (Urone and Schroeder, Part I). These and other nonsulfate solid particles can promote atmospheric corrosion (*a*) through direct chemical action on the metal or its corrosion products, (*b*) by providing active sites for moisture condensation on the metal surface, and (*c*) by acting as a retaining nest for aggressive air pollutants. Even rusts, which usually contain elevated amounts of sulfate, promote the corrosion of fresh metal surfaces (Singhania et al., 1974).

Since the classic work of Vernon (1935), other investigators (see Singhania et al., 1972; Barton, 1973; Haynie and Upham, 1974) have confirmed that sulfate particulates (particularly ammonium sulfates) accelerate the corrosion of metals by reducing the critical humidity and by providing the sulfate ions that deactivate the protective oxide film. Figure 6 shows the effects of $(NH_4)_2SO_4$ and other solid particles on the corrosion of iron in the presence and the absence of SO_2. The marked deteriorative effect of charcoal dust particles in the presence of SO_2 has been widely reported (see Vernon, 1953; Rosenfeld, 1972) and is mainly

Table 4. Average Stationary Corrosion Rates for Zinc, Copper, and Iron during Wet Periods[a]

Location (in Czechoslovakia)	Stationary Corrosion Rate ($mg/m^2 \cdot hr$)		
	Iron	Zinc	Copper
Hurbanovo	26	3	3
Letnany	33	3.3	3.7
Usti nach Labem	111	11.2	10.7

[a] Tests were conducted during 1965 and 1966. Data from Barton (1973).

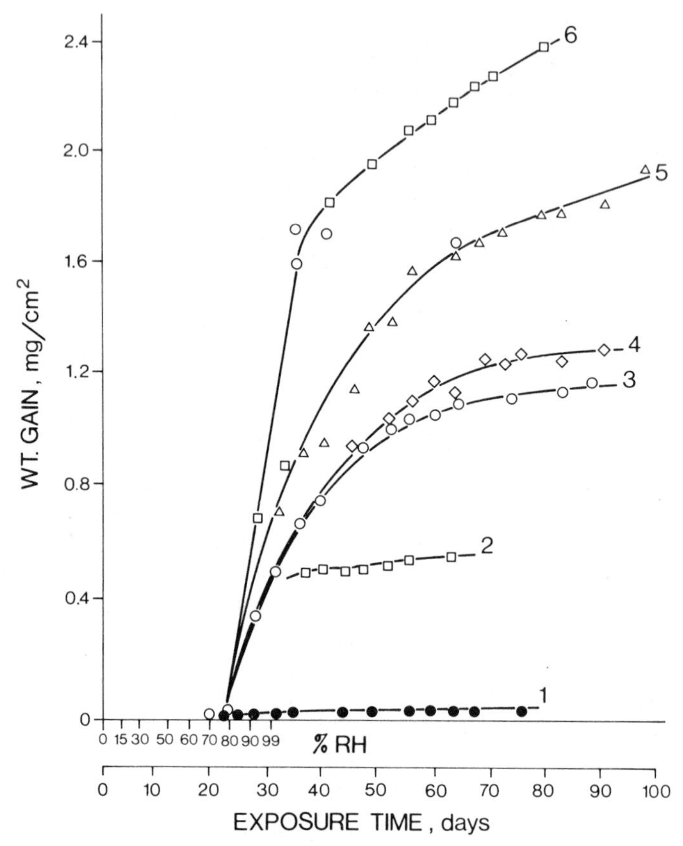

Figure 6. Effects of various particles on the rate of corrosion of iron. 1, Charcoal alone; 2, ammonium sulfate alone; 3, 0.01% SO_2; 4, charcoal + 0.01% SO_2; 5, ammonium sulfate + 0.01% SO_2; 6, charcoal + 0.01% SO_2. (After Rosenfeld, 1973.)

due to the great capacity of charcoal to cull SO_2 from the atmosphere. Barton (1973), however, has emphasized that dusts are particularly influential only during the initial stages of atmospheric corrosion and that their effects are lessened considerably in the presence of a thick layer of corrosion products.

Field tests generally show a very weak correlation between dustfall and corrosion rate (Upham, 1967). Haynie and Upham (1974), however, found a strong covariance between SO_2 and the sulfate in suspended particulates, probably because of chemical interactions either in the air or on the sampler filter. Although it is possible for particulate sulfates to promote the corrosion of steel, SO_2 is more likely to be the causative factor. The sulfate in particulates may, however, provide a good proxy variable for SO_2 at some locations, as emphasized by the results of Haynie and Upham (1974).

1.5. Damage of Specific Materials by Sulfur Pollution

Metals

The corrosiveness of individual sulfur compounds toward the important industrial metals and alloys is given in Table 5. The following discussions will focus on the combined impacts of these compounds on only a selected group of metals exposed to the natural atmosphere.

Iron and Steel. Ferroalloys are the backbone of modern technology, and a large volume of field and laboratory data has been accumulated on air pollutants, especially sulfur compounds, as determinants of the service lives of various iron products. It would be jejune to discuss the various studies in any detail; some highlights have already been noted above.

In the absence of aggressive corrodents, the rate of rusting is usually quite small because of the formation of a protective film of iron oxyhydroxides (e.g., see Rosenfeld, 1972, and Barton, 1973, for in-depth treatments). In the presence of H_2S, SO_2, or even organosulfur compounds, the corrosion rate is greatly accelerated because of the destruction of the protective rust film and the formation of more soluble corrosion products. Thus the corrosion of ferroalloy near point sources of SO_2 has been shown to decrease with the distance from the source (Figure 7). Data from numerous authors clearly indicate that the mean SO_2 level in a city constitutes a measure of the corrosivity of the city's atmosphere (e.g., see Schikorr, 1941; Schikorr and Schikorr, 1943; Uhlig, 1967; Upham, 1967; Guttman and Sereda, 1968; Manweiler, 1968; Thomas and Alderson, 1968; Stanners, 1974; Haynie and Upham, 1974). This fact is best illustrated by Figure 8, which relates the SO_2 corrosion data to urban populations during the early and middle 1960s.

The corrosion rates of ferroalloys exposed to widely different outdoor and sheltered conditions and to synthetic atmospheres have been determined. In most instances a significant correlation has been established between the SO_2 level and the corrosion rate. Table 6 summarizes the numerous empirical equations that have been developed to relate the corrosion to environmental factors. Notice that sulfur pollution is widely recognized as a major causative factor. In general, the SO_2 concentration accounts for over 70% of the variations in the curves of corrosion versus time.

Table 5. Corrosion of Metals by Various Sulfur-Bearing Compounds[a]

Corrosive	Corrosion Rate (g/cm·year)								
	Gray Cast Iron	Ni Cast Iron	Mild Steel	Copper	Brass	Nickel	Aluminum	Lead	Tantalum
Ammonium bisulfite	>10,500	>10,000	>10,000	<4,000	>10,000	<4,000		<600	
Ammonium sulfate	4000–10,000	<4,000	<4,000	4000–10,000	4000–10,000	<4,000	>3,000	<5,000	
Calcium sulfate	<4,000	<400	<4,000	>10,000	>10,000	—	<1,400	<5,000	
Calcium sulfite	<4,000	<4,000	<4,000	>10,000	>10,000	—	<1,400	<5,000	<800
Copper sulfate	>1,000	<400	>10,000	>4,000	>4,000	<4,000	>3,000	<600	<800
Ferric sulfate	>10,000	>10,000	>10,000	>10,000	>10,000	<400	<140	<600	<800
Ferrous sulfate	>10,000	—	>10,000	400–10,000	>10,000	>10,000	<140	500–6,000	<800
									<800
Hydrogen sulfide, dry[b]	<4,000	<4,000	<400	<4,000	<4,000	<4,000	<130	<6,000	<800
Hydrogen sulfide, wet	<4,000	<4,000	<4,000	<4,000	4000–10,000	>10,000	<1,400	6000–14,000	<800

Lead sulfate	>10,000[b]	<4,000	<4,000	—	<4,000	<4,000	>3,000	<600	<8,400
Magnesium sulfate	<400	<400	<400	<400	<400	<400	<140	<600	<800
Magnesium sulfite	>10,000	>10,000	>10,000	>10,000	>10,000	>10,000	<1,400	<6,000	—
Marcaptans[b]	—	<400	<400	>10,000	>10,000	>10,000	<140	—	—
Potassium sulfate	<400	<4,000	<4,000	<400	4000–10,000	<4,000	<140	<600	<800
Sodium sulfite	>10,000	<4,000	400–10,000	<400	>10,000	<4,000	<140	<600	<800
Sulfur[b]	>10,000	<400	>10,000	>10,000	4000–10,000	<400	<140	<6,000	800
Sulfite liquor (with 10% SO_2)	>10,000	>10,000	>10,000	4000–10,000	4000–10,000	4000–10,000	>3,000	<6,000	<800
Sulfur dioxide, dry[b]	>10,000	>10,000	>10,000	>10,000	>10,000	>10,000	>3,000	<600	<800
Sulfuric acid, aerated	>10,000	>10,000	>10,000	<4,000	>10,000	<4,000	—	<6,000	—
Sulfuric acid, no air	>10,000	>10,000	>10,000	<400	>10,000	<4,000	>3,000	<6,000	<800
Sulfurous acid	>10,000	4000–10,000	>10,000	>10,000	>10,000	>10,000	<14,000	<600	<800
Sulfur trioxide[b]	<4,000	<4,000	<4,000	<4,000	<4,000	<4,000	>3,000	<6,000	>21,000
Zinc sulfate	>10,000	<400	<4,000	<400	4000–10,000	<4,000	<1,400	—	—

[a] Based on the compilation by Hamner (1974).

[b] Measurements pertain to pure compounds or very concentrated solutions. All other data refer to dilute solutions (<10 wt %) at 25°C.

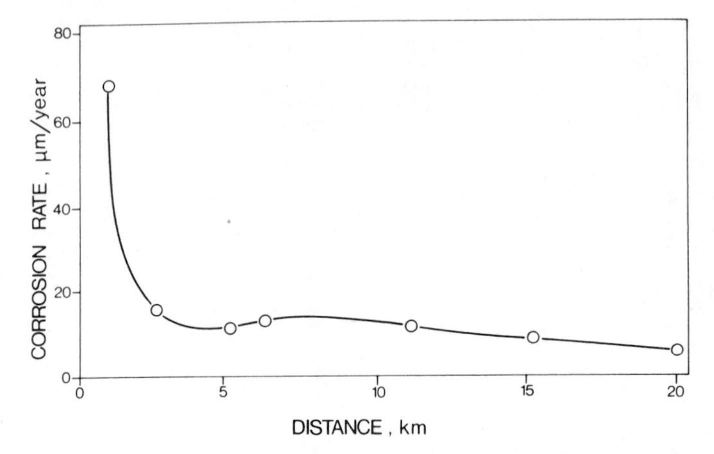

Figure 7. Effect of distance from point source of SO_2 emission (at Kvarntorp, Sweden) on the atmospheric corrosion rate of carbon steel during 1950–1960. (After Mattsson and Kucera, 1971.)

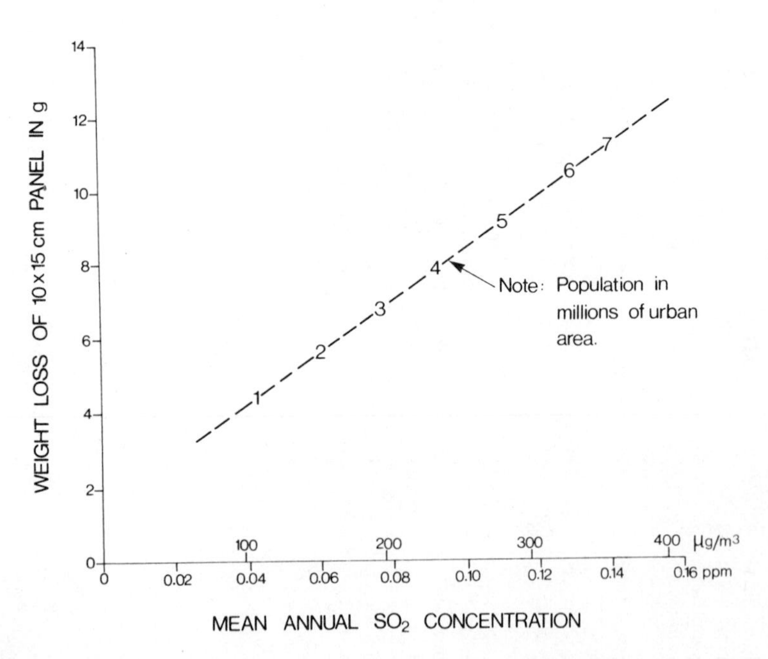

Figure 8. Relationships between corrosion rate of mild steel, mean annual SO_2 concentration, and population of urban area. Data pertain to 16-month exposure tests during the early and middle 1960s. (After Stern et al., 1973.)

Table 6. Some Empirical Expressions for Corrosion of Ferroalloys in Air

Locality	Empirical Equation[a]	Reference
Tokyo (1 month)	$r = (0.083S_1 + 0.066R + 0.028H - 1.63)t$	Ona et al. (1965)
Sheffield (1 year)	$r = (0.70S + 0.035)t$	Chandler and Kilcullen (1968)
Moscow (2–12 months)	$r = (0.025S_2 + 0.031TS_2 + 0.015T + 0.062)kWtP$	Stanners (1972)
Ottawa (1–18 months)	$r = -0.647(S - 0.679)(t_w)^{0.677}$	Guttman and Sereda (1968)
NASN network (1–2 years)	$r = 183.5t^{1/2} \exp(0.0642S_3 - 163.2/H)$	Haynie and Upham (1974)
NASN network (1–2 years)	$r = 325t^{1/2} \exp(0.00275S - 163.2/H)$	Haynie and Upham (1974)
Czechoslovakia (9–15 months)	$r = k_1 (H_2)^{k_2} \exp(k_3T + k_4S)$	Knotkova-Cermakova et al. (1974)

[a] r = corrosion rate; H = relative humidity; H_2 = frequency of occurrence of relative humidity exceeding 80%; R = rainfall; t = duration of exposure; t_w = time of wetness; T = temperature; S = SO_2 content of air; S_1 = SO_3 collected by the lead candle method; S_2 = SO_2 in moisture film; S_3 = sulfate content of suspended particulates; P = rust; W = (time of wetness)/(elapsed time); k, k_1, k_2, k_3, k_4 = empirical constants.

The kinetics of the atmospheric corrosion of iron was outlined above. It was stressed that the initial corrosion during the induction period (of a month or so) is very different, in both mechanism and rate, from the situation when the stationary phase is established. One aspect of atmospheric corrosion that needs to be emphasized is the considerable influence of the physical and chemical properties of the rust layer on the corrosion progress (see Fyfe et al., 1972; Singhania et al., 1974; Satake and Moroishi, 1974; Barton, 1973).

Steel surfaces that are boldly exposed to the environment are normally protected from corrosion by using such measures as nickel-, zinc-, aluminum-, or chrome-plating or antirust painting. As expected, both the maintenance and the replacement of steel products are much more frequent in SO_2-polluted areas than in clean, remote locations. For example, the life of zinc galvanizing is >25 years in rural areas, ~10 years in industrial/urban areas, and <5 years in severely polluted districts (see Gillette, 1975). Paint life is generally about 12 years in rural areas and <6 years in urban and industrial locations. Haynie and Upham

(1974) showed that maintenance frequency against steel corrosion increases exponentially with increasing SO_2 concentrations. This would imply that a small reduction in ambient SO_2 concentrations could result in substantial savings in the costs for maintenance and replacement of steel products.

Copper Materials. The extensive use of copper as an architectural material stems from its good corrosion resistance in pristine environments and its superior ageing characteristics. Many European cities derive a portion of their charm and character from their copper cupolas, domes, spires, railings, and roofs, the beauty of which is generally enhanced by the green corrosion products or patina (Leidheiser, 1971). The corrosion of copper in the atmosphere is also of great interest to electrical and other industrial concerns.

Hydrogen sulfide is a strong corrodent for copper, corrosion rates of up to 140 μm/years having been measured near sources of H_2S (Barton, 1973). The strong influence of RH on H_2S attack can be seen by comparing Figure 9 with the data in Table 5. Copper that has been exposed to H_2S-free air for a significant period of time, however, becomes resistant to subsequent H_2S attack. This is good news for copper artifacts near coastal areas where the emission of H_2S from waste effluents has been on the increase (e.g., see Friend, 1973). The corrosive effect of thioorganics on copper in such environments is unknown, however.

At low relative humidity, SO_2 is only a very mild corrodent for copper. Above the critical RH, however, SO_2 attacks copper materials aggressively. Figure 10 shows Vernon's (1931) data on the corrosion of copper at different SO_2 concentrations. At a RH $>75\%$ and a SO_2 concentration $>1\%$, the corrosion rate is directly proportional to the SO_2 concentration. The anomalous behavior at low SO_2 levels is related to the fact that at a SO_2 concentration of 1% the principal corrosion product is normal copper sulfate, but above 1% the dominant product is the basic sulfate. Although the exact corrosion mechanism is not clear in all cases, the following reaction scheme, proposed by Barton (1973), seems to be generally applicable to copper and other type II metals. The principal initial reaction (at the metal surface) is the direct anodic oxidation of the metal to the oxides and hydroxides:

$$Me + H_2O \rightarrow MeO + 2H^+ + 2e$$

$$Me + 2H_2O \rightarrow Me(OH)_2 + 2H^+ + 2e$$

Subsequent formation of the metal sulfates involves these oxides and

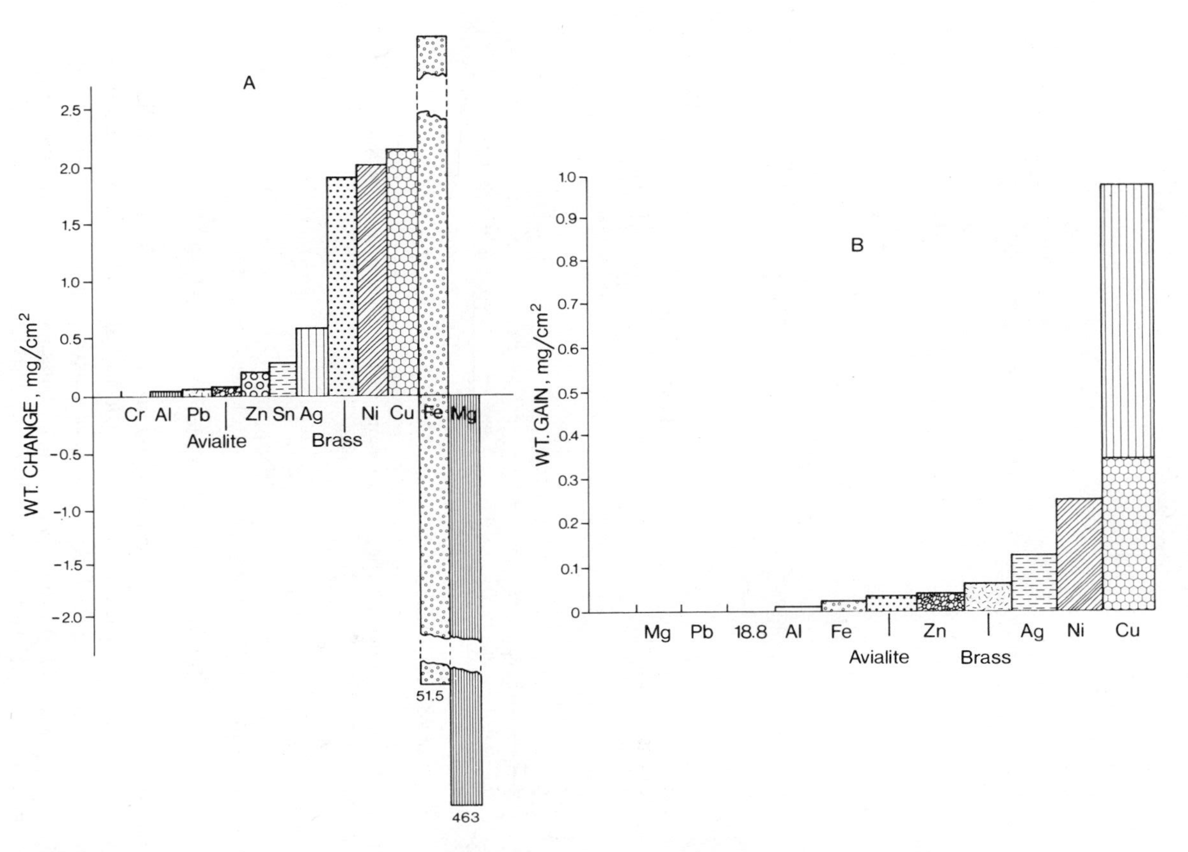

Figure 9. Effects of H_2S on the corrosion of metals. *A*. Roughly 100% relative humidity after 60 days. *B*. Dry H_2S after 180 days. (After Rosenfeld, 1973.)

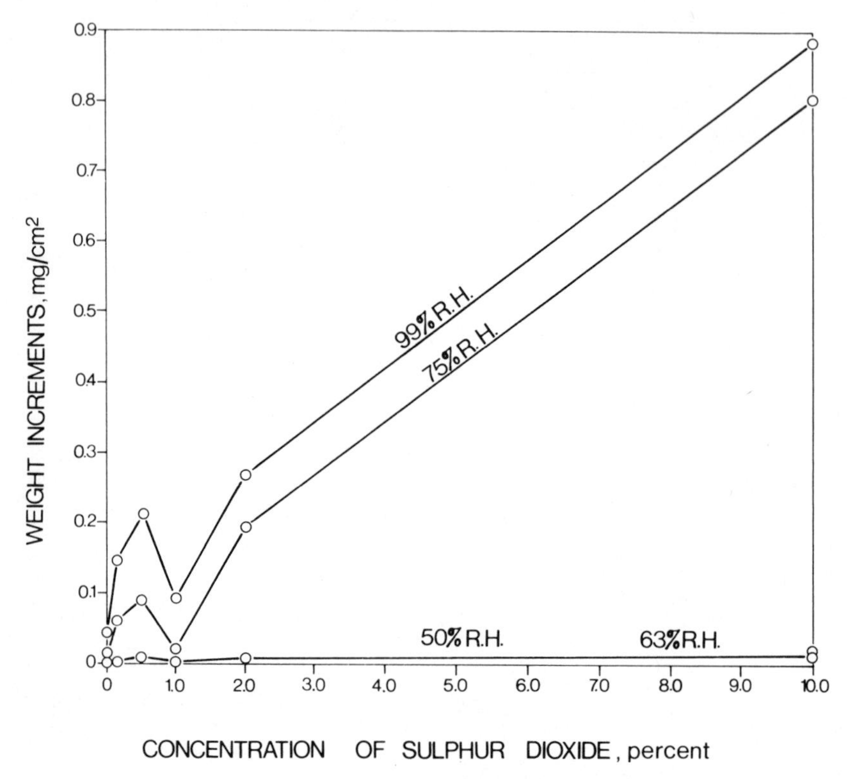

Figure 10. The dependence of copper corrosion on the concentration of SO_2 in the atmosphere. (From Vernon, 1935.)

hydroxides and occurs mostly at the outermost weathered layer:

$$Me(OH)_2 + SO_4^{2-} \rightarrow MeSO_4 + 2OH^-$$

Since the sulfates of the type II metals are very soluble, they are liable to be washed away during rainfall, and the initial oxide-hydroxide layer must be reestablished before the next coating of metal sulfates is formed. Analyses of patinas (see below) suggest, however, that complete conversion of the oxides and hydroxides to sulfates is not common.

Considerable insight into the corrosion of copper by ambient sulfur pollutants has been derived from various outdoor exposure tests (Vernon, 1929; Vernon and Whitby, 1930; Tracy, 1951; Guttman and Sereda, 1968; Thompson, 1968; Mattson and Holm, 1968). The general conclusion from these studies is that the corrosion rate is higher in industrial/

urban atmospheres (0.9 to 2.2 μm/year) than in marine (0.6 to 1.1 μm/year) and rural (0.1 to 0.6 μm/year) atmospheres. Guttman and Sereda (1968) related the corrosion of copper to atmospheric factors and found a strong dependence on ambient SO_2 concentration. Their short-term corrosion data were closely fit by the empirical expression

$$r = -0.0372(B - 0.246)A^{0.832}$$

where r is corrosion rate, A is time of wetness (days), and B is SO_2 concentration (ppm). For long-term exposures, however, they found that the corrosion rates in inland sites were controlled mainly by factors related to the changing surface conditions resulting from accumulation of corrosion products and foreign objects.

The composition of the corrosion products, the so-called green patinas, has been extensively investigated. In a pioneering study, Vernon (1929, 1931) and Vernon and Whitby (1930) demonstrated that the major components of patinas were dependent on local atmospheric conditions. They found that in polluted urban areas the patina consisted of basic sulfate with small amounts of basic carbonates; in marine atmospheres the dominant phase was basic chloride. The compounds they identified included $CuSO_4 \cdot 3Cu(OH)_2$—bronchantile, $CuCO_3 \cdot 3Cu(OH)_3$—malachite, and $CuCl_2 \cdot 3Cu(OH)_2$—atacamite. Aoyama (1961) found identical compounds in corrosion products formed on the 1.5-kV high-tension wires of an electrical railway in Japan. He also showed that the sulfate was the main constituent in industrial regions, chloride in coastal areas, and the basic nitrate, $Cu(NO_3)_2 \cdot 3Cu(OH)_2$, in mountain areas. Osborn (1963) confirmed that the patina on the Statue of Liberty in New York Harbor consisted largely of basic copper sulfate with a basic chloride content of less than 5%. Collins (1930) examined the weathered layers from a number of 1000- to 4000-year-old Chinese bronzes and found mostly malachite, $2CuCO_3 \cdot Cu(OH)_2$—azurite, and Cu_2O—cuprite; sulfur compounds were less prevalent, as would be expected from the fact that ancient Chinese atmospheres were probably free of sulfur pollution.

The rate of patination of copper materials is variable and environment specific (Herman and Castillo, 1974). In atmospheres that contain sulfur compounds, copper first tarnishes quickly to a brown color, which then turns black gradually. The green patina begins to appear after about 6 to 7 years (Herman and Castillo, 1974; Mattson and Holm, 1968). On the other hand, the Australia House in London was still coated black 12 years after it was in place (Vernon and Whitby, 1930). Usually, though, houses 30 or more years old show evidence of patination (Leidheiser, 1971). The present-day shorter induction period for patina to form (compared to the situation before the turn of the century) may be evidence that the copper corrosion rate has recently increased.

Old copper roofs and other copper artifacts have yielded useful insight into the long-term deterioration of copper in the atmosphere. Tracy (1951) estimated the corrosion rate for the roof of Christ Church, Philadelphia, during its 213-years exposure, to be about 0.09 mm/100 years. Copper roofs in parts of Copenhagen, Denmark, exposed for 200 years or more indicate penetration rates of the order of 0.1 to 0.2 mm/100 years (Leidheiser, 1971). Riederer (1974) reported extensive damage to bronze sculptures and other works of art exposed to the atmosphere in Germany, corrosion rates of 0.1 mm/300 years or more being observed in some instances.

Zinc and Galvanized Materials. Zinc exposed to the outdoor environment is usually in the form of galvanized products, such as building accessories, gutters, cables, wires, fencing, fence posts and pole-line hardware. Zinc is used to protect steel as it offers good atmospheric corrosion resistance and is also "anodic" to the steel. This means that, if zinc and steel are in contact in an electrolyte solution, the electrolytic cell formed provides current to protect the steel from corrosion at the expense of the zinc. In polluted atmospheres with high levels of SO_2, however, the protective effect of the zinc coating is severely curtailed (Table 7).

It has been shown that both dry H_2S at 50% RH and moisture-saturated H_2S cause only limited corrosion of zinc, because of the

Table 7. Service Lives of Zinc Coatings in Various Atmospheres[a]

Site	Type of Atmosphere	Corrosion Rate (μm/year)		Estimated Life of 610-g/m² Zinc Coating (years)
		Zinc	Iron Ingot	
Motherwell, Scotland	Industrial	5	61	17
Woolwich, Kent	Industrial	4	69	19
Sheffield (University)	Industrial	5	86	15
Sheffield (Attercliffe)	Industrial	15	119	5
Llanwrtyd Wells	Rural	2	56	34
Calshot, Hants	Maine	3	114	23
Apapa, Nigeria	Marine tropical	0.8	20	100
Basrah, Iran	Dry, subtropical	0.3	8	300
Congella, Durban	Marine industrial	5	76	17

[a] These data (from Fleetwood, 1975) are based on 5-year-exposure tests.

formation of a coherent protective sulfide film (Spedding, 1977). By contrast, the strong correlation between atmospheric SO_2 concentrations and both zinc corrosion and absorption of SO_2 by zinc surfaces is dramatically illustrated in Fig. 11. It is now widely accepted that SO_2 level is the prime factor stimulating atmospheric zinc corrosion. A classic illustration is the fourfold reduction in the rate of zinc corrosion in Pittsburgh associated with a three-fold reduction in ambient SO_2 levels and a twofold reduction in dustfall between 1926 and 1960 (Tice, 1962). Many authors have used empirical equations to closely relate the zinc corrosion rates to just two weather factors: average SO_2 concentration and the time of wetness. For example, Haynie and Upham (1970) expressed the damage function for zinc or galvanized coatings exposed to ambient SO_2 as

$$r = X(Y - 48.8)1.028 \times 10^{-3}$$

where r is corrosion rate (μm/year), X is average SO_2 concentration (μg/m^3), and Y is annual average relative humidity.

Guttman (1968) related the corrosion of zinc to the time of wetness (Y) and the average atmospheric SO_2 level during the time the surfaces are wet (X) as follows:

$$r = Y^{0.8152}(X + 0.02889)5.461 \times 10^{-3}$$

The equation closely fit the data obtained during exposure periods of up to 256 weeks.

Schikorr (1965) made an excellent compilation on the atmospheric corrosion of zinc in which he referred to the works of several investigators who dealt with the effects of atmospheric SO_2. Zinc is a type II metal, and its mechanism of corrosion generally parallels that enunciated for copper. According to Anderson (1955), four major properties of zinc control its response to corrodents: (*a*) zinc has a high hydrogen overvoltage; (*b*) zinc is electronegative to its common metal impurities, to hydrogen, and to most surface contaminants; (*c*) zinc readily forms insoluble basic salts; and (*d*) zinc is attacked by acids at rates that decrease as the pH increases. Anderson (1955) postulated that the zinc surface, when in contact with acid rains, dissolves rapidly and in the process increases the pH of the attacking solvent until a basic salt precipitates. The film of basic salts provides a line of defense since it must first be dissolved before a pH sufficiently low to cause rapid attack can be attained.

The results of several long-term atmospheric exposure test programs involving zinc and its alloys have been reported (e.g., see Guttman, 1968; Knotkova-Cermakova, 1974; Atteraas and Hagerup, 1975). These

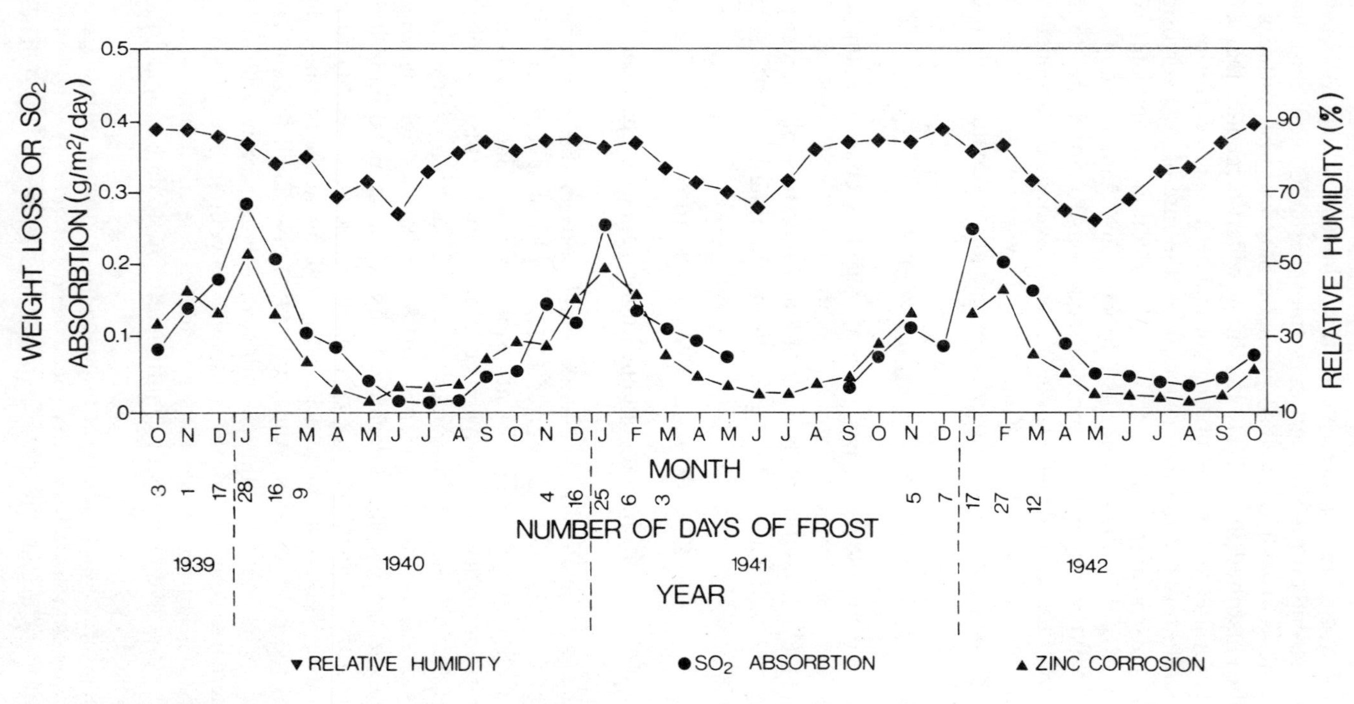

Figure 11. The correlation between the ambient SO_2 levels and the corrosion of zinc in Berlin. (From Kucera, 1976.)

studies generally show that after the induction period of several months the corrosion rate attains a steady-state value which may vary between 0.5 and 5 μm/year, depending on the aggressiveness of the airborne pollutants.

Aluminum. Aluminum and its alloys are used extensively for structural and architectural purposes because of their strong resistance to corrosion in mildly aggressive atmospheres. Even in urban/industrial atmospheres, most aluminum alloys have good corrosion resistance, although the presence of substantial levels of SO_2 usually produces discoloration, pitting, surface roughening, and gradual loss of strength.

Few studies on dose-response and dose-effect relationships involving low SO_2 levels have been reported. At very high SO_2 concentration (7.3×10^5 $\mu g/m^3$), Aziz and Godard (1959) found that, above the critical RH of 72%, aluminum corrodes rather rapidly. The white powdery corrosion product was found to be $Al_2(SO_4) \cdot 18H_2O$. There is no doubt, however, that ambient SO_2 levels do intensify the corrosion of aluminum in the atmosphere. Table 8 compares the corrosion rates for aluminum, copper, lead, tin, and zinc during 20 years's exposure in various locations in the United States. The data clearly show the relatively high corrosion rate for zinc in industrial atmospheres; the increased corrosion rate for zinc in Altoona between 10 and 20 years was attributed to an increase in industrial activity during the test period (Godard, 1967). The very low corrosion rate for aluminum in rural atmospheres is equally noteworthy. Apparently, the 20-fold or so increase in rate of corrosion in industrial areas compared to rural locations is related to the higher concentrations of pollutant sulfur compounds.

Magnesium. In unpolluted atmospheres, magnesium and its alloys show good surface stability and hence have found use in such commercial items as dockboards, foundry flasks, shipping containers, conveyors, ladders, automobile engine parts, and hand trucks. In polluted atmospheres, on the other hand, surface protection is needed before magnesium and its alloys can be boldly exposed even indoors.

In moist open air, magnesium gradually darkens and finally attains a dark gray color. Often the overall reaction may be described by the equation (Bothwell, 1967)

$$Mg + 2H_2O \rightarrow Mg(OH)_2 + H_2$$

In neutral and alkaline environments, the film of magnesium hydroxide deposit may provide considerable protection to the metal. Most often the magnesium hydroxide absorbs CO_2 and SO_2 and is converted to the soluble carbonates, sulfites, and sulfates, which are more likely to be

Table 8. **Corrosion Rates for Four Metals, Based on Weight Loss in ASTM Tests**[a]

Location	Atmosphere	Corrosion Rate (mils/year) for (0.9 mm-Thick Sheet Panels)							
		Aluminum		Copper		Lead		Zinc	
		10 years	20 years	10 years	20 years	10 years	20 years	10 years	20 years
Phoenix, Ariz.	Desert	0.000	0.003	0.005	0.005	0.009	0.014	0.010	0.007
State College, Pa.	Rural	0.001	0.003	0.023	0.017	0.019	0.012	0.042	0.043
Key West, Fla.	Marine	0.004	—	0.020	0.022	0.022	—	0.021	0.026
Sandy Hook, N.J.	Marine	0.008	0.001	0.026	—	—	—	0.055	—
La Jolla, Calif.	Marine	0.0028	0.025	0.052	0.050	0.016	0.021	0.068	0.068
New York, N.Y.	Industrial	0.031	0.029	0.047	0.054	0.017	0.015	0.19	0.22
Altoona, Pa.	Industrial	0.025	—	0.046	0.055	0.027	—	0.19	0.27

[a] Aluminum AA-1100 (99.2% Al, 0.01% Cu); copper, tough pitch (99.9% Cu); lead, commercial (99.2% Pb); zinc, prime western (98.8% Zn). From Walton and King (1955).

washed away by rain. The strong influence of SO_2 and rainfall on the weathering of magnesium alloys is clearly illustrated by the data in Table 9. The attack by the airborne corrodents usually results in a reduction in the tensile strength of the exposed material (Brandt and Adam, 1968).

Building Materials

Building materials and sculptures made of natural stone have not been spared from the devastating effects of sulfur pollution. The deterioration is most pronounced in materials that contain carbonates, and the decay in noncalcareous materials may be influenced by decomposition products from the calcareous materials. There is now undeniable evidence that the rate of deterioration has increased in recent times because of the higher ambient levels of pollutant sulfur compounds.

The deposition of airborne SO_2 on building stone has been studied by Henley (1967), Spedding (1969a, 1969b), Braun and Wilson (1970), and Sengupta and De Gast (1972). Spedding (1969b) reported that at a RH > 80%, the surfaces of oolitic limestone became rapidly (within 10 mn) saturated with SO_2. In an earlier study, Spedding (1969a) found that, when radioactive $^{35}SO_2$ was accidentally released in the laboratory, two of the most strongly contaminated materials were concrete and cement-sand. Braun and Wilson (1970) found that the weathered surfaces of limestone blocks from buildings 100 to 500 years old adsorbed as much SO_2 as a fresh surface cut from the same block; at SO_2 concentration of

Table 9. Composition of Corrosion Products on Magnesium Exposed to Indoor and Outdoor Atmospheres[a]

	Amount (wt %)		
		Indoor[c]	
Compound	Outdoor[b]	A	B
$MgCO_3$		46.5	46.6
$MgCO_3 \cdot 3H_2O$	61.5		
$MgSO_4$		9.8	9.7
$MgSO_4 \cdot 7H_2O$	26.7		
$Mg(OH)_2$	6.4	16.3	15.5
Carbonaceous matter	2.5		
$Fe_2O_3 + Al_2O_3$	2.9		

[a] From Aziz and Godard (1959).
[b] After 217 days' exposure.
[c] After 412 days' (A) and 300 days' (B) exposure.

1 to 5 mg/m^3 the deposition rate was found to be 3 to 19 mg/m^2·hr. This result, as well as the fact that the maximum sulfur concentration was found not at the surface (of the block in place) but some distance below it, led Braun and Wilson (1970) to conclude that the surface reactivity of these blocks has been maintained by the removal of the soluble calcium sulfate by rainfall.

In the presence of moisture, carbonates are attacked rapidly by SO_2:

$$CaCO_3 + SO_2 + \tfrac{1}{2}O_2 \rightarrow CaSO_4 + CO_2$$

$$CaCO_3 \cdot MgCO_3 + 2SO_2 + O_2 \rightarrow CaSO_4 + MgSO_4 + 2CO_2$$

The CO_2 liberated may react further with the calcium carbonate:

$$CaCO_3 + H_2O + CO_2 \rightarrow Ca(HCO_3)_2$$

The end products of these reactions are the relatively soluble calcium sulfate, magnesium sulfate, and calcium bicarbonate. Although the chemical reactions are simple in principle, the mechanism of extensive disruption associated with them is linked in a complex way to the physical properties of the rock such as porosity, permeability, water retention ability, and calcite:dolomite ratio (Sengupta and De Gast, 1972).

Actually, the chemical weathering associated with the removal of sulfate in solution is generally minor. Much of the damage to building stone is attributable to the formation of hard, impenetrable ''skins'' on the surface (Table 10). These skins, aided by alternate freezing and thawing, cause blistering, scalding, exfoliation, and loss of cohesion of the surfaces, and may induce similar effects in neighboring materials not themselves susceptible to direct attack. The stone may also grow

Table 10. Analysis of Surface Skins on Limestone Cylindrical Blocks after Stimulated Chemical Decay Tests[a]

Constituent	Weight (%)
Silica	0.40
Oxides of iron and aluminum	0.80
Calcium carbonate	37.5
Magnesium carbonate	19.6
Calcium sulfate	31.5
Magnesium sulfate	1.05
Water and organic matter	11.0

[a] From Sengupta and De Gast (1972).

leprous because the conversion of carbonates to sulfates results in an increase in volume of crystallization of 1.5 to 4.2 times. Schaffer (1932) referred to the weathering action associated with volume difference between reactant and product as *efflorescence* or *crystallization spalling*. Baines (1934) reported that the crystallization of calcium and magnesium salts along the cleavage planes of building stone in English cities slowly opened up the old sealed vents where they could be attacked by airborne acids, and that the efflorescence spalling lifted stone fragments several tons in weight.

Granites, gneisses, sandstones, and slates which do not contain calcareous cements are less susceptible to damage by sulfur pollution. Sengupta and De Gast (1972), however, reported that Potsdam sandstone developed an ugly discoloration after a short exposure to the natural atmosphere, because of the oxidation of the pyrites in these rocks. Portland cement mortar and concrete are readily attacked by sulfate, the hydrated calcium aluminate and/or calcium hydroxide component of the hardened cement paste being converted into ettringite (calcium sulfoaluminate hydrate) and gypsum ($CaSO_4 \cdot 2H_2O$). The stresses associated with the substantial increase in the volume of crystallization may lead to a breakdown of the paste and ultimately of the mortar or concrete (Sereda, 1977). The deterioration of concrete by reaction with sulfate ions is particularly marked in subaqueous environments and in moist soils (Swenson, 1971; also see below).

The staining of stone facades in major cities is all too familiar and is generally attributed to the action of solid particulates. In many instances, however, sulfur pollution is instrumental in the roughening and the erosion of the exposed calcareous surface, and in the formation of deposits on it. Indeed, Henley (1967) strongly implicated sulfur pollution as the causative factor in the deterioration (estimated rate, 16 kg/m^2·century) of the balustrade of the roof of Covell House, built in 1890 at Charring Cross, London.

Works of art and frescoes are equally liable to SO_2 damage by the same mechanism. Fresco, a pigmented lime plaster and marble, is particularly vulnerable to SO_2 attack. In Rome, the Coliseum, the Arch of Titus, and many sculptures made of fresco and marble have recently come under severe SO_2 attack; the situation in Florence, Italy has been described as disastrous. Cleopatra's Needle, the large stone obelisk moved from Alexandria, Egypt, to London, has suffered more deterioration in the damp, smoky, and acid atmosphere of London in the last 70 to 80 years than in the preceding 3000 years or more of its history (*Air Quality Criteria for Sulfur Oxides,* 1969). The same obelisk developed extensive efflorescence spalling within just 2 years of exposure in

Central Park, New York (Winkler, 1966). The frieze on the Parthenon in Athens was essentially undamaged during its first 2240 years but decayed to an unrecognizable form during the subsequent 136 years of the industrial age (see Yocom and McCaldin, 1968). In Southern Bavaria, a red nodular limestone intercalated with thin layers of limestone was preferred for portals and tombstones. Riederer (1974) showed that airborne acids penetrated through the clay layers deep into the stone, where the resulting efflorescence spalling caused rapid decomposition. Examples abound of other works of art in stone which have been either damaged or seriously threatened by sulfur pollution. York Minster, with one of the largest concentrations of medieval glass in England, recently underwent extensive and expensive rescue and restoration work on the two million or so pieces of painted glass (Calami, 1977). Likewise, many other famous churches in England, including Canterbury Cathedral and St. Paul's Cathedral, have mounted appeals for public donations to help stem the ravages of sulfur pollution on these important national monuments (Calami, 1977).

Paints

The blackening of lead-based paint in the presence of H_2S has been known for a long time. In *Il Libro dell'Arte,* Cennino Cennini (1398) talked about the "rosso chiamato minio," the "giallo chiamato orphimento," and the "bianco archimiato di piombo il quale si chiama biacca," which "divengono neri quando vendono l'aria," as happened in the Cimabue's frescoes in the Upper Church of San Francisco in Assini (Paleni and Curri, 1972). The severity of discoloration is related in a general way to the lead content of the paint, the level of H_2S in the air, the surface texture of the painted surface, the relative humidity at time of exposure, and the duration of exposure (Ward, 1956; Wohlers and Felstein, 1966). Little damage may be experienced if the air and the painted surface are both very dry. It is worth mentioning that sulfide staining also afflicts paints that contain significant quantities of other heavy-metal salts, such as phenylmercury compounds (Hoffman, 1964).

Sulfur dioxide affects paints in two principal ways: (*a*) by increasing the drying time of the film, and (*b*) by attacking the mature dry film, causing loss of gloss. These effects have only been studied in the laboratory at very high concentrations of SO_2. Thus Holbrow (1962) found that 2620 to 5240 μg SO_2/m^3 increased the drying times of several oil-based paints by 50 to 100%. It was also reported that the touch- and hard-dry times of oleoresinous and alkyd paints with titanium oxide pigments were increased substantially. From these observations, Holbrow (1962) concluded that the SO_2 levels likely to be encountered in fogs, in some acid rains, or near industrial sites can increase the drying time and hardening time of some paintwork.

The effects of sulfur pollution on latex paints are essentially unknown. It is conceivable that airborne sulfurous acid can cause instability in the protective colloid around latex particles, resulting in flocculation (Sereda, 1977). The deteriorative effect of acidic aerosols on such paint ingredients as pigments, fungicides, and binders may also be substantial. In particular, some pigments are liable to be bleached by acids of sulfur. A good example is the discoloration of Brunswick green paint, which was noted when the fresh paint was exposed to 520 $\mu g\ SO_2/m^3$ and then to moisture and warmer temperatures (Holbrow, 1962). Under these conditions the lead chromate in the pigment was bleached and became blue.

In the presence of SO_2 (260 $\mu g/m^3$ or more), moisture, and ammonia, a mist of fine-grained ammonium sulfate particles may be formed on paint and varnish surfaces. The troublesome defect called crystalline bloom is readily removed by wiping, but if left undisturbed for months may result in permanent damage to the paintwork. Soot and other particulate matter deposited on painted surfaces will form focal points for SO_2 absorption and for attack on the paint film. Chipping and abrasion associated with the cleansing of these particulates may allow the airborne acids to gain access to the substrate specially if made of steel and masonry.

There is no doubt that sulfur pollution, acting in consort with other air pollutants, affects paint life. Beaver (1954), Fink et al. (1971), and Gillette (1975) have indicated that houses in polluted areas require much more frequent painting than those in remote areas. Other workers also have demonstrated a strong correlation between the frequency of exterior painting and the suspended particulate concentration (see Spence and Haynie, 1972; Stern et al., 1973).

Riederer (1974) made a detailed study of the air pollution damage of wall paintings and painted glasses in Germany. He found that on early wall paintings, where a plaster of lime was used, the reaction to form gypsum either destroyed the paint layer or led to crumbling of the surface. The deterioration of painted glass windows was so rapid that a great many early European windows have already been lost. The sequence in the decay of a painted window from 1166 at St. Patrokli, Soest, was carefully documented. Riederer (1974) found that in the initial stages of decay the glass became more and more opaque as potassium, a major constituent of glass, was dissolved, leaving the silica behind. At the same time, a layer of "weather-stone" was formed on the surface which consisted of dust baked together with sulfates, the sulfates being the principal corrosion product on the glass. Below the surficial skin, little corrosion nests appeared in the glass and the impurities, and from these, networks of cracks spread all over the glass until it was completely shattered. Obviously, similar deterioration must be occurring in other parts of the world.

Textiles and Dyes

The impact of sulfur pollution on textiles may entail both chemical attack (from compounds present in the gaseous state or dissolved in water droplets) and soiling by particulate matter. Quite often, the effects of sulfur pollution are reinforced by the action of sunlight.

The reports of various witnesses concerning damage to nylon (especially nylon stockings) resulting from SO_2-laden soot have been compiled by Travnicek (1966). These phenomenological observations have been collaborated by several outdoor weathering tests. Little and Parsons (1967) subjected cotton, nylon, and Terylene (polyester) fabrics to unprotected outdoor exposure during 1961–1962 at eight sites in Great Britain. Using loss of strength as a measure of degradation, they found that cotton fabric was more resistant than either nylon or polyester in rural areas. In polluted urban areas, however, cotton degraded more rapidly than Terylene and, to a lesser extent, nylon. From a study of the weathering of cotton fabric at several locations in the St. Louis area, Brysson et al. (1968) demonstrated quite clearly that heavy air pollution substantially reduced the service life of the exposed fabric. They found a significant correlation between the degradation of cotton duck and printed fabric and the ambient concentration of SO_2. Cotton, being a cellulosic fiber (linen, hemp, jute, rayon, and synthetic nylons are other examples), is highly susceptible to attack by the acids of sulfur (Petrie, 1948).

The extent of chemical damage from sulfur pollution depends on the level of pollutants, the duration of exposure, the type and composition of fabric, and the relative humidity (Gregory and Manganelli, 1970). In this respect, curtains may be regarded as the guinea pigs in polluted areas because they hang at open windows and to some extent serve as filters for airborne particulates and acids. Considerable shortening of the life of fabric used as curtain material has been reported in many polluted areas (Yocom and McCaldin, 1968).

Laboratory studies addressed to the effects of sulfur pollution on textiles have been limited. Fye et al. (1969) examined the effect of air (RH, 73 to 76%) contaminated with 5240 $\mu g\ SO_2/m^3$ on fabrics made of cotton, nylon, polyester, rayon, wool, and 65-35 polyester-cotton blend. They reported that nylon experienced a small (3.5%) but significant loss in strength, whereas the losses of other fabrics were statistically insignificant. Their findings are somewhat inconsistent with those of other workers who showed that the photodegradation of cotton and rayon fabrics increased substantially if the air contained SO_2 (see Little, 1964; Zeronian et al., 1973). Zeronian et al. (1973) studied the reaction of fabrics made of modacrylic, nylon, and polyester in air containing 520 μg SO_2/m^3. They showed that modacrylic and polyester fabrics were not affected by SO_2. By contrast, the presence of SO_2 during the light exposure reduced the strength of the nylon yarn by 80% and its toughness (work of rupture) by over 90%.

The mechanism of SO_2 attack on textiles is a matter of considerable conjecture, but is probably basically an oxidation process. It is generally believed that during chemical attack either long molecular chains are degraded or the cross-links are broken, leading to the observed loss of strength in the material. Zeronian et al. (1973) showed that the exposure of nylon to SO_2 resulted in a decrease in the relative viscosity and the amine end groups, while the number of carboxyl end groups increased markedly. These observations suggest that SO_2 attack on nylon could occur in the absence of particulate matter and that the degradation was not (entirely) due to acid hydrolysis following conversion of SO_2 to sulfuric acid, as proposed by Yocom and McCaldin (1968).

Dyes are equally susceptible to damage by sulfur pollution. Severe fading was noted for some dyes in fabric exposed in Chicago and Los Angeles, where the annual average SO_2 concentrations are usually high (Slavin, 1963; Ajax et al., 1967). The major difficulty with these practical tests is the proper apportionment of the observed degradation between the various likely contributory causes, many of which act synergistically. For example, controlled laboratory experiments by Ajax et al. (1967) detected no fading in selected dye-fabric combinations exposed either to automobile exhausts or to clean air with 2620 μg SO_2/m^3; auto exhausts irradiated with SO_2 (2620 $\mu g/m^3$), however, produced pronounced fading of the same dyes.

Paper and Wood

Damage of such cellulosic materials as paper, wood, and cardboard by sulfur pollution generally takes the form of discoloration and premature loss of mechanical strength. Kimberley (1932) observed that exposure of books and writing paper to SO_2 concentrations of 2620 to 23,580 $\mu g/m^3$ for 10 days caused embrittlement and reduced their folding resistance. Hudson et al. (1967) showed that between SO_2 concentrations of 2620 and 13×10^6 $\mu g/m^3$ the uptake of SO_2 by paper was linear for 4 days. Over a period of a few decades, the level of sulfuric acid in the paper can rise to as much as 5 wt % (Ede, 1967). The deposition of SO_2 on a given paper depends on its surface finish, pattern design, trace metal content, and the handling it receives, which may result in sweat deposits (which have a great affinity for SO_2) being left on the paper.

Studies with radioactive $^{35}SO_2$ have revealed a preferential deposition of the SO_2 about the metallic impurities left in the paper from the manufacturing process (Spedding, 1974). It is presumed that the SO_2 is oxidized to sulfuric acid at these points and that the basic degradation process is one of acid hydrolysis of the cellulose. In addition, it is thought that the formation of lignosulfonic acids by the reaction between the SO_2 in the surface moisture and the lignins in the paper may also be significant (Spedding, 1974).

Paper objects generally have a short service life, and it is therefore doubtful whether air pollution causes a noticeable depreciation in the value or usefulness of these materials in practice. Problems associated with such attack are, however, of particular concern to museum officials and antique dealers who are involved in the preservation of ancient paperworks. In this respect it is interesting that papers made before 1750, when chemical methods for papermaking were introduced, are not amenable to SO_2 damage (Yocom and McCaldin, 1968).

Unpainted timber and other wooden artifacts may deteriorate because of the reaction of pollutant sulfite with the lignins, analogous to the sulfite-pulping process for the production of cellulose from wood. Damage by other weathering processes, however, will generally override the effect of SO_2.

Leather

Leather has a strong affinity for SO_2 and may accumulate as much as 7% of sulfur acids by weight if left exposed to an atmosphere containing this pollutant (Yocom, 1959). Leather impregnated with sulfuric acid loses its flexibility and mechanical strength and ultimately may disintegrate. Faraday observed, as early as 1843, that the rotting of leather upholstery in his London club was the direct result of atmospheric sulfur pollution (Parker, 1955). Sheepskin valves used to pump air into large electro-pneumatic organs, which formerly wore out in 20 to 30 years, now hold up for only about 5 years because of elevated levels of pollutant sulfur (Stern et al., 1973). Leather safety belts, leather coats, leather boots, and many other leather objects now have their durability and service lives curtailed considerably by sulfur pollution. Likewise, leather-bound books deteriorate in polluted air, and the storage of such books in large city libraries is obviously a serious problem.

The uptake of SO_2 by leather has been shown to be almost completely controlled by the rate of SO_2 diffusion to the surface (Spedding, 1974). The formation of sulfuric acid presumably causes acid hydrolysis of the proteins at the surface and in the pores of the leather, leading to a breakdown of the fibrous structure.

2. SULFUR POLLUTION AND SUBAQUEOUS CORROSION OF MATERIALS

Modifications of the sulfur contents of aquatic ecosystems exercise a considerable influence, both direct and indirect, on the subaqueous corrosion of materials. As a major dissolved ion, SO_4^{2-} is a main factor in the galvanic and electrochemical corrosion of metals, especially in soils. Sulfur compounds are also involved in the speciation and distribution of trace metals (Nriagu and Hem, this volume). Furthermore, the cycling of sulfur generally controls the pH, redox potential, and alkalinity of aquatic ecosystems; these intensive variables chiefly determine the conditions of immunity, corrosion, and passivation of metals in a given aqueous system (see above). To keep this chapter within a reasonable length, the discussions will be confined to direct effects of sulfur pollution, where sulfur species are actually involved in the corrosion reaction.

The sulfur pollution of natural waters has been considered by Kramer (Part I), Dickson and Almer (this volume), and Nriagu and Hem (this volume). Among the principal sources of pollutant sulfur in natural waters are sanitary and industrial sewage and wastewaters, urban runoff, drainage from mining operations, fertilizers used in agriculture, and atmospheric precipitation. In addition, severe contamination of groundwater may stem from the leaching of landfills and dumps of domestic and agricultural refuse and slags, and of waste piles from industrial plants.

2.1. Cement and Concrete

The main areas of application of mass concrete include dams, dry docks, retaining walls, sluices, quays, embankment walls, sewer pipes, foundation walls, canals, and water and wastewater treatment facilities. The use of cement as an adhesive filler between bricks and stone in the construction of buildings and engineering works actually dates back to early Egyptian and Greek times. The early cements (or lime mortars) were essentially mixtures of quicklime and mortar and owed their long-term durability to both the initial slaking reaction:

$$CaO \text{ (quicklime)} + H_2O \rightarrow Ca(OH)_2 \text{ (portlandite)}$$

and the subsequent conversion of the colloidal $Ca(OH)_2$ precipitate to calcium carbonate by reaction with atmospheric CO_2. The Romans made the added discovery that the addition of reactive siliceous material (such as crushed tiles or volcanic ash) produced a cement that developed superior strength and water resistance. Widespread use of the Roman cement continued until the end of the eighteenth century. Portland cement, first formulated around 1824, is basically a mixture of calcium silicates in typical weight proportions of about 50% $3CaO \cdot SiO_2$, 25% $2CaO \cdot SiO_2$, 10% $3CaO \cdot Al_2O_3$, 10% $4CaO \cdot Al_2O_3 \cdot Fe_2O_3$, and about 5% of other oxides. Portland cement and its predecessors are highly susceptible to sulfate attack (see below). High-alumina cement (containing mostly $CaO \cdot Al_2O_3$), which was specifically designed to resist sulfates, came on the commercial market after World War I.

Of all the dissolved constituents, sulfate ions are the principal agent in the subaqueous deterioration of cement (Biczoc, 1964). The sulfate attack takes several forms, which are predicated upon the physicochemical conditions of the concrete and the aquatic system (see Biczoc, 1964, for a detailed discussion). Both laboratory and field evidence, in general, indicates that the disruption of mortars and concrete by sulfate involves three basic interactions:

1. The conversion of the free lime to calcium sulfates (gypsum and/or anhydrite):

   $$Ca(OH)_2 + H_2SO_4 \rightarrow CaSO_4 \cdot 2H_2O$$

   Gypsum is fairly soluble and is liable to be leached away by the aqueous matrix. The large increase in solid volume accompanying the reaction may lead to severe efflorescence in the mortar. More recently, it has been demonstrated that the presence of gypsum also retards the hydration of the $4CaO \cdot Fe_2O_3 \cdot Al_2O_3$ in the cement (Jawed et al., 1976).
2. The formation of insoluble calcium sulfoaluminate and sulfosilicates, notably ettringite or Michaelis-Candlot-Deval salt ($3CaO \cdot Al_2O_3 \cdot 3CaSO_4 \cdot 31H_2O$) and thaumasite [$Ca_6Si_2(OH)_{12}(CO_3)_2(SO_4)_2$]. The formation of ettringite may be conceptualized as follows:

   $$3CaSO_4 + 3CaO \cdot Al_2O_3 + 31H_2O \rightarrow 3CAO \cdot Al_2O_3 \cdot 3CaSO_4 \cdot 31H_2O$$

   Spallation due to the increase in solid volume, retarded caulking of fresh portland cement paste, high early strength of supersulfated cements, and development of prestress in restrained concretes made with expansive cements are some of the phenomena associated with ettringite formation (Van Aardt and Visser, 1975; Lukas, 1976; Mehta, 1976).

3. In magnesium-rich solutions the conversion of the free lime, the hydrated calcium silicates, and calcium sulfoaluminate to gypsum and magnesium hydroxide:

$$MgSO_4 + Ca(OH)_2 \rightarrow CaSO_4 + Mg(OH)_2$$

$$3MgSO_4 + 3CaO \cdot Al_2O_3 \cdot 6H_2O \rightarrow 2CaSO_4 + 2Al(OH)_3 + 3Mg(OH)_2$$

The deposition of $Mg(OH)_2$ in the pores of the corroded cement product usually results in the formation of a hard, glassy skin which tends to hinder further penetration of the solution into the cement or mortar. This effect may be contrasted with the attack by magnesium-poor sulfate solutions, where the corroded cement product tends to be soft and incoherent. As noted previously, surface skins are most susceptible to weathering by spallation.

Because of its high sulfate and high magnesium contents, seawater is noted as being a strong concrete corrodent. The corrosion mechanism, however, is obscure. For example, the rate of concrete sulfidation in seawater tends to be mitigated by the high chloride ion concentration (Addleson, 1972). The chemical action of seawater presumably involves extensive leaching action which removes lime and gypsum from the concrete. The formation of ettringite or thaumasite probably facilitates the seawater attack by opening up the concrete.

Highly acidic waters, commonly linked to the sulfur cycle, also attack concrete aggressively. The deterioration of concrete in acidic media can be attributed to three basic reactions: (*a*) the dissolution (or removal) of $Ca(OH)_2$, a major binding agent in cements; (*b*) the hydrolysis of the various cement components, leading ultimately to the formation of a residue consisting of silica gel, aluminum hydroxide gel, and ferrioxide gel; and (*c*) the conversion of the cement components to the more soluble gypsum or anhydrite. Those reactions substantially reduce the strength and rocklike consistency of the cement and concrete. It is interesting that the wastes of sulfuric acid factories have been known to destroy concrete sewers in a very short time (Biczok, 1964). Many concrete structures surrounded by slags have collapsed because of the adverse effects of sulfur acids leached out of the slag pile (Biczok, 1964). In both Australia and the island of Helgoland (North Sea), rapid corrosion of concrete structures standing in seawater occurred as a result of the oxidation of hydrogen sulfide (derived during the decomposition of seaweeds) to sulfuric acid (Grun, 1937). The portland cement concrete used in the construction of the pier at Marseille harbor deteriorated during a period of 50 years, primarily because of sulfate attack (Biszok, 1964). It is not surprising that many countries have established detailed guidelines for the protection of concrete structures exposed to sulfate concentrations exceeding 200 ppm.

Domestic sewage and wastewaters are usually harmless to concrete, the total SO_4^{2-} concentration being generally less than 300 ppm. When the sewage becomes anaerobic, however, the formation of even low levels (2 ppm or less) of H_2S has a marked deleterious effect on the concrete sewer (Thistlethwayte, 1972; Boon and Lister, 1975). The corrosion ensuing from the escaping H_2S is generally concentrated above the sewage or wastewater surface. Grun (1937) analyzed the severely damaged municipal sewer in the town of Gelsenkirchen, Germany, and found that the undamaged concrete contained only 0.3% SO_3, whereas the destroyed concrete had 1.5% SO_3.

The weak resistance of concrete to some industrial wastes has not always been recognized when industrial sewer installations have been designed, and sewage failures with remarkable consequences inevitably have resulted. The classic example has to be the steel-sheet-rolling mill in Hungary where the sulfuric acid wastes infiltrated for several years into the clay soils around the deteriorated sewers under the pickling plant building. The swelling associated with the reaction between the sulfuric acid wastes and the clayey soils resulted in such a large-scale heaving and tilting of the pickling plant building that (the structure) was abandoned for safety reasons (Biczok, 1964).

2.2. Ferroalloys and Microbial Corrosion

Corrosion under immersed conditions is strictly electrochemical, in the sense of Faraday's law. Unlike corrosion in the atmosphere, which is primarily controlled by the time of wetness, corrosion of metals in totally immersed conditions is controlled mainly by the availability of oxygen. Thus the corrosion mechanism and rate will vary depending on whether the conditions are aerobic or anaerobic. The composition of the water is important insofar as it affects the electrolytic processes at the corroding surface; occasionally the precipitation of carbonates, silicates, and sulfides on a metal surface can have a protective value. Movement of the water may increase the rate of corrosion.

The rusting of iron and other metals immersed in a *sterile* and oxygenated aqueous medium is analogous to the situation in the atmosphere: a surface film of oxyhydroxides is quickly formed which stifles further attack. By and large, the most serious corrosion of buried or underwater metallic pipes, cables, and containers is biologically mediated. The deterioration of a metal by corrosion processes which results directly or indirectly from the activity of any group of organisms is generally referred to as *biological corrosion*. The process of biological corrosion is now well established in the literature (e.g., see Von Wolzogen Kuhr and Van der Vlugt, 1934; Parker, 1955; Starkey, 1956; Postgate, 1959; Booth and Tiller, 1960, 1962; Iverson, 1968; Senez, 1969; Sharpley, 1973). Metal corrosion by sulfur cycle bacteria can be brought about in various ways: (*a*) by creating corrosive conditions, for instance, by producing H_2S, H_2SO_4, and organic acids; (*b*) by producing deposits or modifying the existing protective surface films; and (*c*) by directly modifying the anodic and cathonic reactions. These effects may occur singly or in combination, depending on the environment and the organisms involved.

Sulfate reducers, prevalent under anaerobic conditions, play a dual role in causing corrosion. First, they produce the corrosive H_2S, and, second, they act as cathodic depolarizers. The mechanism of microbial corrosion of ferroalloys can be depicted in the following schematic reactions:

$$2(CH_2O) + SO_4^{2-} \rightarrow H_2S + 2HCO_3^- \quad \text{(production of } H_2S\text{)}$$

$$\left.\begin{array}{l} H_2O = H^+ + OH^- \\ Fe \rightarrow Fe^{2+} + 2e \end{array}\right\} \quad \text{(anodic dissolution of iron)}$$

$$SO_4^{2-} + 8H^+ \xrightarrow{\text{(enzymes)}} S^{2-} + 4H_2O \quad \text{(depolarization)}$$

$$\left.\begin{array}{l} Fe^{2+} + H_2S \rightarrow FeS + 2H^+ \\ Fe^{2+} + 2OH^- \rightarrow Fe(OH)_2 \end{array}\right\} \quad \text{(corrosion products)}$$

Under most conditions the most pronounced effect is the dissolution or corrosion of the ferroalloy structures.

The effect of low levels of dissolved H_2S on the corrosion behavior of iron group metals has been reviewed by Sury (1976). Adsorption of HS^- or H_2S was emphasized as a dominant factor in determining the corrosion rates of these metals. In aerated solutions the oxygenation of the chemisorbed sulfide enhances both the depolarization of the hydrogen evolution reaction and the anodic dissolution; the corrosion rate is thus accelerated. In deoxygenated waters, however, the sulfide ions act as a catalytic poison, inhibiting the recombination of cathodic hydrogen atoms with H_2 molecules. As a consequence, surface coverage with hydrogen atoms approaches saturation, and the diffusion into the bulk material may result in the well-known hydrogen embrittlement of steel in the presence of H_2S (Prowse and Wayman, 1974). The exact mechanism of hydrogen absorption into steel is uncertain, however (see Kawashima et al., 1976). The reaction mechanisms pertaining to the catalytic effect of sulfide ions on anodic dissolution are unknown, but may entail the formation of $FeSH^-$ as a surface catalyst (Iofa et al., 1964). The formation of corrosion cells between the iron (anode) and the precipitated iron sulfide (cathode) has also been postulated (Stumper, 1923). The pitting corrosion of steel in H_2S solutions has been attributed to the differential response of the nonmetallic inhomogeneities at the steel surface to H_2S attack (see Sury, 1976; Dvoracek, 1976). It is also interesting that the inhibition of ferroalloy corrosion by many sulfur-containing organic compounds is largely due to the partial decomposition of these compounds to H_2S at the metallic surface (Iofa and Tomashova, 1960). Microorganisms that cause the precipitated iron sulfide to be removed will, of course, accelerate the corrosion.

When slime or mat of biomass accumulates on the metal, the aerobic organisms on the surface of the slime mass use oxygen and release by-products which may create anaerobic conditions at the metal interface. Because of the commensalism between aerobic and anaerobic microbiota, anaerobic corrosion may also occur in well-aerated systems (Oppenheimer, 1958; Olson and Szybalski, 1960).

Corrosion under aerobic conditions involves chiefly the genus *Thiobacillus,* which produces various types of acid as the corrosive metabolic product. The production of sulfuric acid from elemental sulfur and other sulfur compounds (thiosulfates, polythionates, etc.) is a familiar example. The H_2SO_4 formed, which can lower the pH to 0.7 in cultures (Kempner, 1966), will damage most materials, including concrete. One group of these organisms, *T. ferrooxidans,* oxidizes the pyrites (FeS_2) in coals and sulfide deposits and is chiefly responsible for the acidity of mine drainage waters (see Barton, this volume). Such waters are highly corrosive to the pumping machinery and mining installations (Butlin and Postgate, 1954). Hoey and Dingley (1971) have also made a detailed survey of the corrosion problems in Canadian sulfide ore mines and mills. The corrosion of mild steel in synthetic acid mine waters has been studied by Subrahmanyam and Hoey (1975).

Bacterial and fungal metabolisms produce a large number of sulfur-bearing organic compounds, such as mercaptans, thiocyanates, thiourea derivatives, sulfoxides, and sulfonamides, and the corrosion of ferroalloys by these thioorganics has been reported on several occasions (Ehlert, 1967; Makrides and Hackerman, 1955). Allen (1948) reported corrosion of iron by this mechanism in beet sugar mills; a similar phenomenon was observed in Cuban cane mills (Sharpley, 1973). Corrosion of metals by pollutant thioorganic compounds can also be expected.

2.3. Copper and Its Alloys

Copper and its alloys have been widely used as surface condenser materials for shipboard and seacoast service where once-through seawater is used for cooling. The pollution of waters by sulfides and thioorganics has now been shown to cause extensive deterioration of the condensers. The corrosion is particularly rapid in polluted harbors and estuaries, where the ebb and flow of tides cause diurnal changes from anoxic sulfide-rich brackish waters to relatively fresh, aerated seawater (Bates and Popplewell, 1975). Severe sulfide attacks on other buried copper objects have also been recorded (Gilbert, 1947).

The influence of sulfides, polysulfides, and thioorganic compounds such as mercaptans, cystine, and thiourea on the tarnishing and corrosion of copper and its alloys in natural waters has been studied extensively (Llopsis et al., 1959; Tanabe, 1964; Rowlands, 1965; Leidheiser, 1971; Wolynec and Gabe, 1972; Mor and Beccaria, 1974). Highly localized corrosion is occasionally encountered during the deterioration of condenser tubes. This type of corrosion is associated with hot spots and is usually exacerbated by H_2S (Bem and Campbell, 1961). Table 11 shows representative data for the hot-spot corrosion of copper alloys in seawater containing H_2S. Bates and Popplewell (1975) also studied the corrosion of condenser tube alloys in sulfide-contaminated brines and found that alternate treatment with sulfide and with air was

Table 11. Hot-Spot Corrosion (Temperature, 100°C) of Copper Alloys in Seawater Containing H_2S[a]

Alloy	Depth of Attack (mm in 14 days)
Copper	0.37–0.44
Aluminum brass	0.05–0.55
Admiralty brass	0.09–0.27
70/30 Cu-Ni	0.19–0.82
90/10 Cu-Ni	0.35–0.55

[a] Data from Bem and Campbell (1961).

much more aggressive than sulfide alone or air alone. Polysulfide solutions also attack copper aggressively. After the usual induction period, the sulfidization of copper in polysulfide solutions has been shown to be proportional to the polysulfide concentration (Hoar and Tucker, 1952; Wolynec and Gabe, 1972). In the presence of sulfide ions, the usual corrosion products consist of Cu_2S, CuS, Cu_2O, and small amounts of basic oxides and chlorides (Mor and Beccaria, 1974).

Microbial corrosion of copper is fairly common, contrary to what one would expect from the lethal effects of this element on microbiota. In particular, the Krebs cycle acids are commonly corrosive toward copper so that the deterioration of copper and its alloys occurs in environments where these acids are produced. Jones and Snow (1965) showed that most amino acids are sufficiently corrosive to copper to cause color changes in the solution or to form visible corrosion products on the metal surface. Abnormally rapid pitting of condenser tubes has been associated with bacteria that produce pigments from which melanin has been isolated (Leidheiser, 1971). Rapid pit corrosion of brass has also been reported in samples of seawater containing organic sulfur compounds (Leidheiser, 1971). Invariably, the tarnishing of copper in the presence of thioorganic compounds is characterized by the formation of sulfide film, the catalytic effect of the film on copper dissolution being essentially unknown. In aqueous solutions of thiourea, the growth rate for the tarnish film has been shown to be proportional to the thiourea concentration during the stationary growth phase (Wolycec and Gabe, 1972).

3. SULFUR AND HIGH-TEMPERATURE CORROSION OF METALS

During a multitude of uses, hot metals are exposed to sulfur and its compounds. Interest in hot sulfur corrosion of metals has risen in recent years because of the increased effort to solve the energy and air pollution problems. Hot sulfur corrosion has been recognized as the cause of fireside deterioration in boilers (De Santis, 1973), and as the major factor in the failure of gas turbine blading (Hart and Cutler, 1973; Reising, 1975) and coal-gasification equipment (Natesan, 1976). Hot sulfur corrosion also occurs in furnaces and chimney linings, jet engines, refineries, and geothermal energy systems, and has been implicated in the "burning" of automobile valves (Radcliff and Stinger, 1974). In view of the widespread applications of metals in high-temperature situations, a detailed assessment of the role of sulfur in the deterioration of materials at elevated temperatures would be a Herculean task. Only a bird's-eye view of the general principles involved can be given in this chapter.

The corroding atmosphere during the combustion of a fossil fuel is usually a complex mixture of gaseous, molten, and/or solid components. During complete combustion the flue gases consist mostly of CO_2, N_2, SO_2, and water vapor; in addition, H_2, NO_x, H_2S, CO, and hydrocarbons are formed in appreciable amounts during conditions of incomplete combustion. Materials exposed to the combustion atmospheres may thus undergo interactions leading to oxidation, sulfidation, hot corrosion, carburization, and metal dusting. The extent to which each interaction occurs depends on the temperature, the alloy composition, and its usage history, as well as the gas composition. Although each process is now fairly well understood, little progress has so far been made in the quantitative assessment of the combined reactions in gas mixtures (Rahmel, 1969, 1973; Natesan, 1976).

At lower temperatures (less than about 500°C), most of the corrosion is associated with fly-ash deposits entrained in the flue gas (De Santis, 1973). Table 12 shows the average compositions of the various fuel ashes. The point that needs to be emphasized is that only the pyrosulfates and ferric chlorides are molten in the temperature range of 350 to 500°C, to be expected in steam generating tubes. The rapid corrosion of carbon steel boiler tubes in the temperature range 400 to 500°C by

Table 12. Average Compositions of Fossil Fuel Ashes[a]

	Concentration (wt %)[b]			
Component	British Coals	Fuel Peat	Fuel Oil	Crude Oils
SiO_2	25–50	20–50	2–20	0.8–53
Al_2O_3	20–40	7.15	2–20	
Fe_2O_3	0–30	0–20	1–60	13–98
TIO_2	0–3	Tr	Tr	
CaO	1–10	5–30	0–10	0.7–13
MgO	0.5–5	5–25	Tr	0.2–9
Na_2O	1–6	0–6	1–30	0.1–31
V_2O_5	Tr	Tr	5–60	0–38
SO_3	1–12	5–20	7.30	1–42
Cl_2	Tr	0–1	Tr	0–5
P_2O_5	Tr	1–3	Tr	Undet

[a] From the compilations by De Santis (1973).
[b] Tr = trace amounts; Undet = undetermined.

$KHSO_4$ and $K_2S_2O°$ was demonstrated by Miller et al. (1972); the corrosive reaction had been proposed many years ago by Corey et al. (1949):

$$3K_2S_2O^0 + 2FeO_3 + 1.50_2 \rightarrow 2K_3Fe(SO_4)_3$$

These investigators showed that, if the deposits had the composition of glaserite ($3K_2SO_4 \cdot Na_2SO_4$), the pyrosulfate would be molten at temperatures as low as 280°C if the SO_3 concentration was higher than 7 ppm. Analysis of superheater steel tubes that failed because of molten alkali ash attack showed the major constituents actually to be α-Fe_2O_3, FeO, Na_2SO_3, Na_2SO_4, $KHSO_4$, FeS, and V_2O_5 (De Santis, 1973). The synergistic action of chloride and sulfur oxides in the rapid corrosion of furnace walls is also well documented (Miller et al., 1972; De Santis, 1973).

Molten sulfur corrodes metals aggressively. The wastage of ferroalloys in contact with molten sulfur usually exceeds 0.1 mm/year; in the presence of an aqueous phase and sulfurous gases, the rate may exceed 5 mm/year (Zubakin and Salei, 1976). Lichter and Wagner (1960) studied the corrosion of Cu-Ni alloys by molten sulfur at 400°C and obtained a parabolic rate law. The corrosion of metals by molten sulfur is of particular concern in the production of sulfur.

Metal corrosion by sulfur oxides at high temperatures (>600°C) is generally referred to as hot corrosion and is of particular concern in fossil fuel combustion chambers and in marine gas turbines. The presence of a liquid phase rich in sulfates, notably Na_2SO_4, is usually considered a prerequisite for the initiation of hot corrosion. Some molten impurities may accelerate the hot corrosion. For example, all stainless steel and other common engineering alloys are severely attacked by molten fuel ash that contains sulfates and small amounts of vanadium (Natesan, 1976). The exact mechanism of hot corrosion has not been fully elucidated (Goebel et al., 1973; Reising and Krause, 1974; Lashko and Glezer, 1972). Many corrosion scientists now believe that the hot corrosion induced by alkali sulfates is the result of either basic or acidic fluxing of the protective oxide scales. The basic flux mechanism involves the formation of the metal sulfide, using the sulfur derived from the sulfate ions, and the subsequent rapid oxidation of the metal sulfide by one of at least two mechanisms. Nickel and chromium, presumably, corrode by the above mechanism, and the protective effect of the chromium in Ni-Cr alloys may be attributed to the fact that the chromium is hot-corroded sacrificially. The acidic flux mechanism usually entails reactions by components which reduce the oxygen ion activity in the melt to the extent that the protective oxide layers of the structural alloys become unstable and dissociate. The interactions of the oxides of molybedenum, tungsten, vanadium, and aluminum with the alkali sulfates to form molybdate, tungstate, vanadate, and aluminate ions are good examples of the acidic-fluxing form of attack.

The hot corrosion of nickel-base alloys has been studied extensively because these alloys represent the most hot-corrosion-resistant materials available for high-temperature, high-stress use (Bergman et al., 1966; Mrowec, 1971; Goebel et al., 1973; Reising and Krause, 1974; Reising, 1975, 1977; Peters et al., 1976; Vasantasree and Hocking, 1976; El-Dahshan and Stringer, 1977). At very high temperatures (>1000°C) or in the presence of significant amounts of vanadium, the rate of hot corrosion of even the nickel-base supperalloys may become quite high (Vasantasree and Hocking, 1976).

Sulfidation of structural metals is a major problem under conditions of incomplete combustion. In general, H_2S is more aggressive than SO_2 at elevated temperatures. The high rate of H_2S-induced corrosion may be attributed to two main features: (*a*) the H_2S attack generally results in the formation of low-melting metal sulfides, which may actually fuse on the metal surface during service; and (*b*) the metal sulfides are generally more highly defective than the oxides and thus tend to form less continuous, less tenacious, and less protective films. It is significant that nickel-base alloys tend to behave poorly in H_2SO-enriched atmospheres because the nickel-sulfur system has its eutectic at only 645°C.

Wet vapors of H_2S at 100°C corrode Munoz metal, naval brass, and admiralty brass at a rate of 0.012 to 0.019 mg/dm^2·day and copper and red brass at 0.30 to 0.38 mg/dm^2·day (see Leidheiser, 1971). Between 300 and 500°C, the sulfidation of copper and its alloys tends to follow linear reaction kinetics, with the rate constant strongly dependent on the partial pressure of H_2S (Takeda et al., 1968; Chattopadhyay and Sadigh-Esfandiary, 1973). The principal corrosion product is Cu_2S, although CuS may be formed at low P_{H_2S}. With Cu-Ni alloys the corrosion product also includes Ni_2S and NiS (Tideswell, 1972; Chattopadhyay and Sadigh-Esfandiary, 1973). For the sulfidation of pure copper in H_2S-argon mixtures, dissociation of the adsorbed H_2S at the gas-scale interface has been suggested as the rate-controlling step in the linear kinetics (Chattopadhyay and Sadigh-Esfandiary, 1973). The mechanism of H_2S-induced corrosion apparently is different from that in molten sulfur or sulfur vapor (see Haycock, 1959; Dravnieks and Samans, 1958).

Several studies have dealt with the high-temperature sulfidation behavior of iron-base alloys (Farber and Ehrenberg, 1952; De Santis, 1973; Narita and Nishida, 1973; Zalenko and Simkovich, 1974; Natesan, 1976).

The results show that at a H_2S concentration of about 100 ppm in the gas phase the corrosion rate for austenitic stainless steel is more than 1.0 mpy (mil per year). At any given H_2S concentration the corrosion rates for low-chromium steels are usually about 20 times higher than those for austenitic stainless steel. With low-chromium iron alloys the corrosion product is mostly FeS with some Cr_2S_3 in solid solution with the FeS. At higher chromium concentrations in the alloys, however, a multiphase scale consisting of FeS and either Cr_2S_3 or iron-chromium sulfide spinel is formed. The beneficial effects of aluminum and chromium during the sulfidation of nickel-base alloys may be attributed to the formation of a protective layer of Al_2S_3 or a mixture of Al_2S_3 and chromium sulfides. The current knowledge of the mechanisms of alloy sulfidation is rudimentary at best. Mrowec (1971) has made a valiant attempt, however, to generalize the various sulfidation behaviors of alloys and has suggested mechanisms for sulfide scale growth on metal alloys.

REFERENCES

Addleson, L. (1972). *Materials for Building,* Vol. 3. Butterworth, London, pp. 27–35.

Air Quality Criteria for Sulfur Oxides (1969). Department of Health, Education, and Welfare, National Air Pollution Control Administration, Rept. AP-50, Arlington, Va.

Ajax, R. L., Conlee, C. J., and Upham, J. B. (1967). *J. Air Pollut. Control Assoc.* **17,** 220–224.

Allen, L. A. (1948). *J. Soc. Chem. Ind.* **67,** 70–77.

Anderson, E. A. (1955). *Atmospheric Corrosion of Non-ferrous Metals.* ASTM Spec Tech. Publ. 175, p. 126.

Aoyama, Y. (1961). Werkst. Korros. **12,** 148–150.

Atteraas, L. and Hagerup, O. A. (1975). *Proceedings of the 7th Scandinavian Corrosion Congress.* Royal Norwegian Council on Scientific and Industrial Research, Oslo, Norway, pp. 57–70.

Aziz, P. M. and Godard, H. P. (1959). *Corrosion* **15,** 529t–533t.

Baines, F. (1934). *Examination into the Effects of Air Pollution on Building Stones.* Report, National Smoke Abatement Society, Manchester, England.

Barton, J. R. (1973). *Protection against Atmospheric Corrosion.* John Wiley and Sons, New York, 194 pp.

Barton, K. and Bartonova, Z. (1969). Werkst. Korros. **20,** 216–221.

Bates, J. F. and Popplewell, J. M. (1975). *Corrosion* **31,** 269–275.

Beaver, H. (1954). *Committee on Air Pollution Report.* Her Majesty's Stationary Office, London.

Bem, R. S. and Campbell, H. S. (1961). *Proceedings of the 1st International Congress on Metallic Corrosion*. Butterworths, London, pp. 630–634.

Bergman, P. A., Sims, C. T., and Beltran, A. N. (1966). ASTM Spec. Tech. Publ. 421, pp. 38–53.

Biczok, I. (1964). *Concrete Corrosion and Concrete Protection*. Akademiai Kaido, Budapest, 543 pp.

Black, H. L. and Lherbier, L. W. (1968). In *Metal Corrosion in the Atmosphere*. ASTM Spec Tech. Publ. 435, pp. 3–32.

Boon, A. G. and Lister, A. R. (1975). *Prog. Water Technol.* **I,** 289–300.

Booth, G. H. and Tiller, A. K. (1960). *Trans. Faraday Soc.* **56,** 1689–1696.

Booth, G. H. and Tiller, A. K. (1962). *Trans. Faraday Soc.* **58,** 110–115, 2510–2516.

Bothwell, M. R. (1967). In H. P. Godard, W. B. Jepson, M. R. Bothwell, and R. L. Lane, Eds., *The Corrosion of Light Metals*. John Wiley and Sons, New York, pp. 257–311.

Brandt, S. M. and Adam, L. H. (1968). In *Metal Corrosion in the Atmosphere*. ASTM Spec Tech. Publ. 435, pp. 95–128.

Braun, R. C. and Wilson, M. J. G. (1970). *Atmos. Environ.* **4,** 371–378.

Brysson, R. J., Trask, B. J., and Cooper, A. S. (1968). Am. Dyest. Rept. **57,** 15–19.

Butlin, K. R. and Postgate, J. R. (1954). In *Proceedings of the 4th Symposium Society of the General Microbiology*. Cambridge University Press, pp. 271–305.

Calami, P. (1977). *Hamilton* (Ontario) *Spectator,* Saturday, July 23, p. 23.

Cennini, C. (1398). *Il Libro dell'Arte,* XIV Sec., Neri Possa Editore, Vincenza. Cited in A. Paleni and S. Curri (1972).

Chandler, K. A. and Kilcullen, M. B. (1968). *Br. Corros. J.* **3,** 80–84.

Chattopadhyay, B. and Sadigh-Esfandiary, S. (1973). *Corros. Sci.* **13,** 747–757.

Collins, W. F. (1930). *J. Inst. Met.* **44,** 402–405.

Corey, R. C., Grabowski, H. A., and Cross, B. J. (1949). *Trans. Am. Soc. Mech. Eng.* **71,** 951–963.

De Santis, R. (1973). *Corros. Prev. Control* **20,** 7–14.

Dravnieks, A. and Samans, C. H. (1958). *J. Electrochem. Soc.* **106,** 183–191.

Duncan, J. R. and Spedding, D. J. (1973a). *Corros. Sci.* **13,** 881–889.

Duncan, J. R. and Spedding, D. J. (1973b). *Corros. Sci.* **14,** 993–1002.

Duncan, J. R. and Spedding, D. J. (1974). *Corros. Sci.* **14,** 241–249.

Dvoracek, L. M. (1976). *Corrosion* **32,** 64–68.

Ede, J. R. (1967). "Sulfur Dioxide and Vapor Phase Deacidification (abstr.), paper presented at the London Conference on Museum Climatology.

Ehlert, I. (1967). Mater. Org. **2,** 297–318.

El-Dahshan, M. E. and Stringer, J. (1977). *Corros. Sci.* **17,** 329–340.

Evans, U. R. (1960). *The Corrosion and Oxidation of Metals*. Arnold Press, London, 1094 pp.

Evans, U. R. and Taylor, C. A. J. (1972). *Corros. Sci.* **12,** 227–246.

Farber, M. and Ehrenberg, D. M. (1952). *J. Electrochem. Soc.* **99,** 427–434.

Fink, F. W., Butner, F. H., and Boyd, W. K. (1971). *Technical Economic Evaluation of Air Pollution Control Costs on Metals in the U.S.A.* Battelle Memorial Institute, Rept. PB-198-453, Columbus, Ohio.

Fleetwood, M. J. (1975). *Corros. Prev.* **22,** 16–19.

Fontana, M. G. and Greene, N. D. (1967). *Corrosion Engineering*. McGraw-Hill Book Co., New York, 391 pp.

Friend, J. P. (1973). In S. I. Rasol, Ed., *Chemistry of the Lower Atmosphere*. Plenum, New York, pp. 177–201.

Fye, C., Flaskerud, K., and Saville, D. (1969). Am. Dyest. Rept. **58,** 16.

Fyfe, D., Shanahan, C. E. A., and Shreir, L. L. (1972). *Proceedings of the 4th International Congress on Metallic Corrosion*. National Association of Corrosion Engineers, Houston, Tex., pp. 399–407.

Gilbert, P. T. (1947). *J. Inst. Met.* **74,** 139–174.

Gillette, D. G. (1975). *J. Air Pollut. Control Assoc.* **25,** 1238–1243.

Godard, H. P. (1967). In H. P. Goddard, W. B. Jepson, M. R. Bothwell, and R. L. Lane, Eds., *The Corrosion of Light Metals*. John Wiley and Sons, New York, pp. 1–218.

Goebel, J. A., Pettit, F. S., and Goward, G. W. (1973). *Metall. Trans.* **4,** 261–278.

Gregory, C. J. and Manganelli, R. M. (1970). *J. Air Pollut. Control Assoc.* **20,** 471.

Grun, R. (1937). *Der Beton*. Springer, Berlin.

Guttman, H. (1968). In *Metal Corrosion in the Atmosphere*. ASTM Spec. Tech. Publ. 435, pp. 223–239.

Guttman, H. and Sereda, P. J. (1968). In *Metal Corrosion in the Atmosphere*. ASTM Spec Tech. Publ. 435, pp. 326–359.

Hamner, N. E. (1974). *Corrosion Data Survey*. National Association of Corrosion Engineers, Houston, Tex., 283 pp.

Hart, A. B. and Cutler, A. J. B. Eds. (1973). *Deposition and Corrosion in Gas Turbines*. John Wiley and Sons, New York.

Haycock, E. W. (1959). *J. Electrochem. Soc.* **106,** 764–771.

Haynie, F. H. and Upham, J. B. (1974). In *Corrosion in Natural Environments*. ASTM Spec Tech. Publ. 558, pp. 33–51.

Heimler, B. and Vannerberg, N. G. (1972). *Corros. Sci.* **12,** 579–582.

Hendrik, T. W. (1964). *Corrosion and Its Prevention*. Manual published by the Organization for Economic Cooperation and Development, April 1964.

Henkel, F. (1954). *Dtsch. Ausschuss Stahlbeton* **118,** 39–57.

Henley, K. J. (1967). *Proceedings of the Clean Air Conference*. Blackpool, pp. 55–60.

Herman, R. S. and Castillo, A. P. (1974). In *Corrosion in Natural Environments*. ASTM Spec Tech. Publ. 558, pp. 82–96.

Hoar, T. P. and Tucker, A. J. P. (1952). *J. Inst. Met.* **81,** 665–679.

Hoey, G. R. and Dingley, W. (1971). *Can. Mining Metall. Bull.* **64,** 62–69.

Hoffman, E. (1964). *J. Oil Color Chem. Assoc.* **47,** 581–583.

Holbrow, G. L. (1962). *J. Oil Color Chem. Assoc.* **45,** 701–718.

Hudson, F. L., Grant, R. L., and Hockey, J. A. (1967). *J. Appl. Chem.* **14,** 441–447.

Iofa, Z. A. and Tomashova, G. N. (1960). *Russ. J. Phys. Chem.* **34,** 492–496.

Iofa, Z. A., Batrakov, V. V., and Cho-Ngok-Ba (1964). *Electrochim. Acta* **9,** 1645–1653.

Iverson, W. P. (1968). In A. H. Walters and J. J. Elphick, Eds., *Biodeterioration of Materials*. Elsevier, Amsterdam, pp. 28–43.

Jawed, I., Goto, S., and Kondo, R. (1976). *Cement Concr. Res.* **6,** 441–454.

Jones, J. P. and Snow, J. P. (1965). *Phytopathology* **55,** 499–500.

Kaesche, H. (1966). *Die Korrosion de Metalle*. Springer-Verlag, Berlin, 374 pp.

Karraker, D. G. (1963). *J. Phys. Chem.* **67,** 871–874.

Kawashima, A., Hashimoto, K., and Shimodaira, S. (1976). *Corrosion* **32,** 321–331.

Kempner, E. S. (1966). *J. Bacteriol.* **92,** 1843–1843.

Kimberley, A. E. (1932). *J. Res. Natl. Bur. Stand. (U.S.A.)* **8,** 159–171.

Knotkova-Cermakova, D., Bosek, B., and Vlckova, J. (1974). *Corrosion in Natural Environments*. ASTM Spec Tech. Publ. 558, pp. 52–74.

Kucera, V. (1976). *AMBIO* **5,** 243–248.

Lashko, N. F. and Glezer, G. M. (1972). *Prot. Met.* **9,** 592–596.

Legault, R. A. and Preban, A. G. (1975). *Corrosion* **31,** 117–122.

Leidheiser, H., Jr. (1971). *The Corrosion of Copper, Tin and Their Alloys*. John Wiley and Sons, New York, pp. 1–257.

Lichter, B. D. and Wagner, C. (1960). *J. Electrochem. Soc.* **107,** 168–180.

Little, A. H. (1964). *J. Soc. Dyers Color.* **80,** 527–534.

Little, A. H. and Parsons, H. L. (1967). *J. Text. Inst.* **58,** 449–462.

Llopsis, J., Gamboa, J. M., and Arizmendi, L. (1959). *Electrochim. Acta* **1,** 39–57.

Lukas, W. (1976). *Cement Concr. Res.* **6,** 225–234.

Makrides, A. C. and Hackerman, N. (1955). *Ind. Eng. Chem.* **47,** 1773–1778.

Manweiler, G. B. (1968). In *Metal Corrosion in the Natural Environment*. ASTM Spec Tech. Publ. 435, pp. 211–222.

Mattson, E. and Holm, R. (1968). In *Metal Corrosion in the Atmosphere*. ASTM Spec Tech. Publ. 435, pp. 187–210.

Mattson, E. and Kucera, V. (1971). In *Proceedings of the 2nd International Symposium on Modelling the Effects of Climate on Electrical and Mechanical Equipment*. State Research Institute of the Protection of Materials, Prague, Czechoslovakia, pp. 90–97.

Mehta, P. K. (1976). *Cement Concr. Res.* **6,** 169–182.

Miller, P. D., Krause, H. H., Vaughan, D. A., and Boyd, W. K. (1972). *Corrosion* **28,** 274–281.

Mor, E. D. and Beccaria, A. M. (1974). *Corrosion* **30,** 354–356.

Mrowec, S. (1971). *Arch. Hutn.* **16,** 27–63.

Narita, T. and Nishida, K. (1973). *Oxid. Met.* **6,** 157–180.

Natesan, K. (1976). *Corrosion* **32,** 364–370.

Olson, E. and Szybalski, W. (1960). *Corrosion* **6,** 405–414.

Ona, K., Sugano, T. and Hirai, Y. (1965). *Corros. Eng.* **14,** 16–19.

Oppenheimer, C. H. (1958). *World Oil* **147,** 144–147.

Osborn D. H. (1963). Cited in Leidheiser (1971), p. 18.

Paleni, A. and Curri, S. (1972). In A. H. Walters and E. H. H. der Plas, Eds., *Biodeterioration of Materials,* Vol. 2. Halsted Press Div., John Wiley and Sons, New York, pp. 392–400.

Parker, A. (1955). *The Destructive Effects of Air Pollution on Materials*. National Smoke Abatement Society, London.

Parker, C. D. (1955). In *Corrosion: Proceedings of a Symposium*. Melbourne, Australia, pp. 186–203.

Peters, K. R., Whittle, D. P., and Stringer, J. (1976). *Corros. Sci.* **16,** 791–804.

Petrie, T. C. (1948). *Smokeless Air* **18,** 62–64.

Postgate, J. R. (1959). *Ann. Rev. Microbiol.* **3,** 505–520.

Prowse, R. L. and Wayman, M. L. (1974). *Corrosion* **30,** 280–284.

Radcliff, A. S. and Stinger, J. (1974). *Corros. Sci.* **14,** 483–490.

Rahmel, A. (1969). In *Proceedings of the 4th International Congress on Metal Corrosion*. National Association of Corrosion Engineers, Houston, Tex., pp. 228–234.

Rahmel, A. (1973). *Corros. Sci.* **13,** 125–136.

Reising, R. F. (1975). *Corrosion* **31,** 159–163.

Reising, R. F. (1977). *Corrosion* **33,** 84–88.

Reising, R. F. and Krause, D. P. (1974). *Corrosion* **30,** 131–138.

Riederer, J. (1974). In J.-O. Willums, Ed., *New Concepts in Air Pollution Research*. John Wiley, New York, pp. 73–85.

Robertson, W. D., Nole, V. F., Davenport, W. H., and Talboom, F. P., Jr. (1958). *J. Electrochem. Soc.* **195,** 569–573.

Rosenfeld, I. L. (1972). *Atmospheric Corrosion of Metals*. National Association of Corrosion Engineers, Houston, Tex., 238 pp.

Rowlands, J. C. (1965). *J. Appl. Chem.* **15,** 57–63.

Satake, J. and Moroishi, T. (1974). *Proceedings of the 5th International Congress on Metallic Corrosion*. National Association of Corrosion Engineers, Houston, Tex., pp. 744–749.

Schaffer, R. J. (1932). *The Weathering of Natural Building Stones*. Spec Rept. 18, Her Majesty's Stationery Office, London, 149 pp.

Schikorr, G. (1941). *Korros. Metallschutz,* pp. 305–313.

Schikorr, G. (1965). *Atmospheric Corrosion Resistance of Zinc*. American Zinc Institute, New York.

Schikorr, G. and Schikorr, I. (1943). *Z. Metallkd.* **35,** 175–181.

Sculley, J. C. (1966). *Fundamentals of Corrosion*. Pergamon, Oxford, 190 pp.

Senez, J. C. (1969). In *Advances in Fermentation*. Academic Press, New York, pp. 843–857.

Sengupta, M. and De Gast, A. A. (1972). *Can. Mining Metall. Bull.* **65,** 54–58.

Sereda, P. J. (1960). *Ind. Eng. Chem.* **52,** 157–170.

Sereda, P. J. (1974). In *Corrosion in Natural Environments*. ASTM Spec Tech. Publ. 558, pp. 7–22.

Sereda, P. J. (1977). In *Sulfur in the Canadian Environment*. National Research Council, Spec Tech. Publ., Ottawa, pp. 347–413.

Sharpley, J. M. (1973). In C. C. Nathan, Ed., *Corrosion Inhibitors*. National Association of Corrosion Engineers, Houston, Tex., pp. 228–235.

Singhania, G. K., Sanyal, B., and Nanda, J. N. (1974). *Proceedings of the 5th International Congress on Metallic Corrosion*. National Association of Corrosion Engineers, Houston, Tex., pp. 769–773.

Slavin, V. S. (1963). *J. Air Pollut. Control Assoc.* **13,** 416–422.

Spedding, D. J. (1969a). *Atmos. Environ.* **3,** 341–346.

Spedding, D. J. (1969b). *Atmos. Environ.* **3,** 683.

Spedding, D. J. (1974). *Air Pollution*. Clarendon Press, Oxford, pp. 32–46.

Spedding, D. J. (1977). *Corros. Sci.* **17,** 173–178.

Speller, F. N. (1951). *Corrosion: Causes and Prevention*. McGraw-Hill Book Co., New York, 686 pp.

Spence, J. W. and Haynie, F. H. (1972). *Paint Technology and Air Pollution: A Survey and Economic Assessment*. U.S. Environmental Protection Agency, Rept. AP-103, 44 pp.

Stanners, J. F. (1972). *Proceedings of the 4th International Congress on Metallic Corrosion*. National Association of Corrosion Engineers, Houston, Tex., pp. 419–424.

Stanners, J. F. (1974). *Corrosion in Natural Environments*. ASTM Spec Tech. Publ. 558, pp. 23–32.

Starkey, R. L. (1956). *Ind. Eng. Chem.* **48,** 1429–1437.

Stern, A. C., Wohlers, H. C., Boubel, R. W., and Lowry, W. P. (1973). *Fundamentals of Air Pollution.* Academic Press, New York, pp. 99–110.

Stumer, R. (1923). *C.R.* **176,** 1316–1317.

Subrahmanyam, D. V. and Hoey, G. R. (1975). *Corrosion* **31,** 202–207.

Sury, P. (1976). *Corros. Sci.* **16,** 879–901.

Swenson, E. G. (1971). Can. Bldg. Digest 136. National Research Council, Ottawa, 4 pp.

Sydberger, T. and Vannerberg, N. G. (1972). *Corros. Sci.* **12,** 775–784.

Takeda, M., Fueki, K., and Mukaibo, T. (1968). *J. Electrochem. Soc. Jap.* **36,** 95–100.

Tanabe, Z. (1964). *Jap. Inst. Met. J.* **28,** 874–883.

Thistlethwayte, D. K. B. (1972). *The Control of Sulfides in Sewerage Systems.* Butterworths, London.

Thomas, H. E. and Alderson, H. N. (1968). In *Metal Corrosion in the Atmosphere.* ASTM Spec Tech. Publ. 435, pp. 83–94.

Thompson, D. H. (1968). In *Metal Corrosion in the Atmosphere.* ASTM Spec Tech. Publ. 435, pp. 129–140.

Tice, A. E. (1962). *J. Air Pollut. Control Assoc.* **12,** 1–7.

Tideswell, N. W. (1972). *Corrosion* **28,** 23–25.

Tracy, A. W. (1951). *Corrosion* **7,** 373–375.

Travnicek, Z. (1966). In *Proceedings of the International Clean Air Congress.* National Society for Clean Air, London, England, pp. 224–226.

Uhlig, H. H. (1967). *Corrosion and Corrosion Control,* 2nd ed. John Wiley and Sons, New York, 419 pp.

Upham, J. B. (1967). *J. Air Pollut. Control Assoc.* **17,** 398–402.

Van Aardt, J. H. P. and Visser, S. (1975). *Cement Concr. Res.* **5,** 225–232.

Vasantasree, V. and Hocking, M. G. (1976). *Corros. Sci.* **16,** 261–295.

Vernon, W. H. J. (1929). *J. Inst. Met.* **42,** 181–195.

Vernon, W. H. J. (1931). *Trans. Faraday Soc.* **27,** 255–277.

Vernon, W. H. J. (1935). *Trans. Faraday Soc.* **31,** 1668–1700.

Vernon, W. H. J. and Whitby, L. (1930). *J. Inst. Met.* **44,** 389–396.

Von Wolzogen Kuh, C. A. H. and Van der Vlugt, L. S. (1934). *Water* (Holland) **18,** 147–165.

Walton, C. J. and King, W. (1956). In *Atmospheric Corrosion of Non-Ferrous Metals.* ASTM Spec Tech. Publ. 175, pp. 21–39.

Ward, G. B. (1956). *Official Digest Federation on Paints and Varnish Production,* pp. 1089–1100.

Winkler, E. M. (1966). *Eng. Geol.* **1,** 381–400.

Wohlers, H. C. and Felstein, M. (1066). *J. Air Pollut. Control Assoc.* **16,** 19–21.

Wolynec, S. and Gabe, D. R. (1972). *Corros. Sci.* **12,** 437–450.

Yocom, J. E. (1959). *Corrosion* **15,** 541t–545t.

Yocom, J. E. and McCaldin, R. D. (1968). In A. C. Stern, Ed., Vol. I. *Air Pollution,* Academic Press, New York, pp. 617–654.

Yoshimura, K., Huang, T. C., Ebihara, H., and Shibata, N. (1964). *Radiochim. Acta* **3,** 185–191.

Zalenko, P. D. and Simkovich, G. (1974). *Oxid. Met.* **8,** 343–360.

Zeronian, S. H., Alger, K. W., and Omaye, B. T. (1973). In H. M. Englund and W. T. Berry, Eds., *Proceedings of the 2nd International Clean Air Congress.* Academic Press, New York, pp. 468–476.

Zubakin, A. A. and Salei, S. M. (1976). *Prot. Met.* **12,** 235–330.

2

EFFECTS OF SULFUR OXIDES ON ANIMALS

Mary O. Amdur

Department of Physiology,
Harvard School of Public Health,
Boston, Massachusetts, and
Department of Nutrition and Food Science,
Massachusetts Institute of Technology,
Cambridge, Massachusetts

1. TOXICOLOGY OF SULFUR DIOXIDE

1.1. Studies of Mortality and Lung Pathology

Early toxicological studies of sulfur dioxide that utilized death as the criterion of response employed concentrations far too high to have relevance to environmental exposures. Major signs of respiratory distress included coughing, labored breathing, and rhinitis. Eye irritation and conjunctivitis were also produced. Pulmonary pathology included areas of hemorrhage, consolidation, and some edema.

1.2. Effects on Mucus Secretion and Particle Clearance

Chronic exposure of animals to sulfur dixoide is capable of producing increased secretion of mucus and alterations that resemble the pathology observed in chronic bronchitis. Dalhamn (1956) found a thickening of the mucous layer in the trachea of rats exposed to 10 ppm SO_2 for 18 to 67 days. The cilia beat at their normal frequency, but were unable to move the thickened layer of mucus. The decreased rate of transport of mucus persisted for a month after cessation of exposure. This would mean that particles that are normally cleared by moving upward on the layer of mucus were being cleared more slowly from the airways. Electron microscopy demonstrated an increase in the number of goblet cells in these animals.

Reid (1963) found an increase in the number of mucus-secreting cells in the main bronchi (where they are normally present) and in the peripheral airways (from which they are normally absent) in rats exposed for 5 hr a day, 5 days a week, for 6 weeks to 300 to 400 ppm SO_2. Similar exposures for 3 months to 40 ppm, however, did not produce such changes. Cessation of exposure halted the increase in mucus-secreting cells, but an excess persisted for 3 months.

Mawdesley-Thomas et al. (1970) employed special staining techniques for the mucopolysaccharides indicative of goblet cells. Rats were given ten 6-hr exposures to 50, 100, 200, and 300 ppm SO_2. The animals were sacrificed 72 hr after the end of exposure as a pilot study had shown that the maximum number of goblet cells occurred in the bronchioles at this time. The number of goblet cells in the bronchioles increased with increasing concentration of SO_2. In the trachea the number increased with concentration up to 200 ppm and then dropped to almost a complete absence of goblet cells at 300 ppm. The highest level produced considerable epithelial damage in the trachea. The hypersecretion of mucus may result in increased amounts of mucus reaching the alveolar spaces, where it is removed by alveolar macrophages. Increase in the level of acid phosphatase activity of the alveolar macrophages was used as an indicator of increased metabolic activity. Sulfur dioxide did not alter the number of alveolar macrophages but did increase the acid phosphatase activity. This increase was observed with ten 6-hr exposures to concentrations of 25, 50, 75, 150, and 200 ppm. The increase was dose related.

Another indication of damage caused by SO_2 is an increased level of mitotic activity in the epithelium of the trachea, main bronchi, and proximal airways (Lamb and Reid, 1968). A major wave of increased mitotic activity during the first week of exposure reflects cell replacement after the epithelial cells are shed completely. Increases in mitotic activity were observed after 3 and 6 weeks of exposure. These changes reflect a general increase in cell turnover.

Lamb and Reid (1968) studied histochemical changes in the mucus of rats exposed to 400 ppm SO_2 for 3 hr a day for periods of 2 days to 6 weeks. In the normal rat, goblet cells with different types of intracellular mucus have a characteristic distribution from the trachea to the distal parts of the proximal airways. Tracheal goblet cells containing acid glycoprotein are resistant to the action of sialidase. The goblet cells of the bronchi and upper proximal airways are partially susceptible to sialidase, whereas those of the distal parts of proximal airways are completely sensitive to the enzyme. Normally there are no goblet cells in the distal airways. After only 4 days of exposure there is a marked increase in the proportion of goblet cells producing acid glycoprotein and a marked change in the susceptibility to sialidase. A resistance to sialidase develops in regions such as the proximal airways, where the cells are normally sensitive. As goblet cells form in the distal airways, those containing acid glycoprotein are sensitive to sialidase. After 3 weeks of exposure, however, these goblet cells in the distal airways become resistant. Radioautographs indicated an increased ability of the cells to take up radioactive sulfate, suggesting increased activity of the individual cells. These alterations were evident after only 2 days of exposure and progressed during the 6 weeks of exposure, by which time the uptake by tracheal cells was 10 times that of unexposed controls.

In the hope of developing an animal model larger than the rat for the study of alterations typical of chronic bronchitis, beagle dogs were exposed to 500 to 600 ppm SO_2 for 2 hr twice a week for 4 to 5 months (Chakrin and Saunders, 1974; Spicer et al., 1974). Epithelial changes included squamous metaplasia, which was most prominent and diffuse in the upper trachea and focal in the more distal portions of the respiratory tract. The squamous metaplasia may be regarded as a nonspecific response to injury which may serve a protective function. This metaplastic epithelium differed from the normal type in that it contained a sulfated mucosubstance in the apex of superficial cells. Mucous goblet cells showed marked hypertrophy and hyperplasia in the epithelium of the bronchi and bronchioles. The marked glandular hypertrophy, the dilated gland lumina, and an abundance of secretory material in the airways suggest that SO_2 caused glandular hypersecretion.

The concentrations used in all these studies were indeed high, but they demonstrate clearly that the irritant action can produce changes resembling those caused by chronic bronchitis. It would be of interest to know whether exposures to lower concentrations over longer periods of years would produce alterations of the same type. For a 2-year exposure of monkeys to 1 ppm SO_2 (Alarie et al., 1972) no pulmonary histopathology was reported. Regrettably, however, none of the specialized techniques discussed above was applied. To date, toxicologists have not used these methods, which were developed by pathologists interested in producing animal models of chronic bronchitis rather than in examining the long-term effects of lower concentrations of SO_2.

Evidence is available indicating that SO_2 can retard the rate of clearance of particles from the tracheobronchial areas. In dogs exposed to 1 ppm SO_2 for 1.5 hr twice daily 5 days a week for a 12-month period, an impairment of tracheal mucous velocity was observed. This was not accompanied by alterations in pulmonary function tests. The concentration used in these studies approximated possible short-term peak concentrations observed in urban areas (Hirsch et al., 1975).

1.3. Effects on Pulmonary Function

Inhalation of sulfur dioxide produces as an immediate response bronchial constriction and narrowing of the airways, which result in an increase in pulmonary flow resistance. This response has been observed in guinea pigs, dogs, cats, and human subjects. The increase in flow resistance is dose related. The response produced in guinea pigs by a 1-hr exposure has been used extensively as a bioassay method for comparing the irritant potencies of SO_2 and related sulfur pollutants (Amdur, 1966). Such an exposure to 0.16 ppm of SO_2 produces about a 10% increase in flow resistance. A change of this order of magnitude constitutes a measurable physiological response, not a major alteration of respiratory function. The increase in flow resistance is accompanied by a decrease in compliance (a measure of the elastic recoil of the lungs). At concentrations above 25 ppm, the respiratory frequency decreases and tidal volume increases. The increase in tidal volume is not enough, however, to compensate for the decreased frequency, so that the overall result is a decrease in minute volume. The changes occur during the first 10 min of exposure and persist for the duration of the hour exposure. At concentrations up to 100 ppm, the changes are completely reversible by the end of 1 hr after exposure.

Dogs also show an increase in flow resistance in response to short exposures to SO_2 (Balchum et al. 1960; Frank and Speizer, 1965) by nose, by tracheal cannula, and by exposure of an isolated segment of trachea while the lungs were ventilated with air. The response was greatest when the SO_2 was breathed directly via the tracheal cannula and least when only the tracheal segment was exposed. The fact that a response in the lungs was observed only to the tracheal exposure suggests a referred reflex constriction of the bronchi. The nasal resistance was also measured and found to increase in a manner roughly proportional to the concentration over a range of 7 to 61 ppm for 15 min. Above 23 ppm the nasal resistance increased progressively during the exposure period, probably as a result of swelling of the mucosa and increased secretion of mucus.

Cats have been used to study the physiological mechanism by which SO_2 produces bronchoconstriction (Nadel et al., 1965). Anesthetized cats were ventilated with a pump via a tracheal cannula. Sulfur dioxide could be administered to either the lungs or the upper airways. Exposure of the lungs and lower airways produced an increase in total pulmonary resistance during the first breath. On cessation of exposure the value returned to normal within 1 min. Exposing only the upper airway also produced an increase in flow resistance. These effects were abolished by cooling the cervical vagosympathetic nerve and reestablished by rewarming the nerve. An intravenous injection of atropine also abolished the response, which thus depends upon intact parasympathetic pathways. The rapidity of the response and its reversal point to changes in smooth muscle tone as the cause of the bronchoconstriction.

It is of interest that about 10% of both human subjects and experimental animals exposed to SO_2 seem to react with special sensitivity (Amdur, 1974). These individuals, either man or beast, are not distinguished by high initial control flow resistance values or by any specific observable pathology. It is not known, of course, what the percentage of sensitive individuals actually is in the general population. A value as high as 10% would suggest that the response of the average individual is not the proper basis for standard setting.

Alteration in flow resistance, which is typically a sensitive criterion of response to acute exposure, has not been consistently observed in response to chronic exposures. No increase in flow resistance was observed in guinea pigs exposed for a period of a year to 0.13, 1, and 6 ppm SO_2 (Alarie et al., 1970), even though these concentrations are capable of producing an acute response in a 1-hr exposure. Monkeys exposed for 2 years to 1 ppm SO_2 showed no alteration in pulmonary mechanics (Alarie et al., 1972). Dogs exposed to 5 ppm SO_2 for 21 hr a day for 225 days demonstrated a 50% increase in flow resistance and a 16% decrease in compliance (Lewis et al., 1969). As indicated above, the mechanism of the acute response is constriction of smooth muscle. Chronic increases in resistance result from other mechanisms, such as swelling of the mucosa or excess secretions.

1.4. Studies on Absorption and Distribution

Its water solubility would suggest that sulfur dioxide would be removed in large part during its passage through the upper respiratory tract. At concentrations of 15 to 20 ppm and higher this removal is indeed very efficient, and, in dogs and in rabbits it has been shown that 90 to 98% SO_2 is removed by the upper respiratory tract. Strandberg (1964) used $^{35}SO_2$ to examine the percent absorption by the upper respiratory tracts of rabbits over a concentration range of 0.05 to 700 ppm. His data indicate that at concentrations of 1 ppm and below only 2 to 10% of inhaled SO_2 was absorbed by the upper respiratory tract. These data on rabbits agree with the observation made on guinea pigs that a tracheal cannula which by passed the upper respiratory tract led to a greater increase in flow resistance (as compared to normal breathing) when the concentration was 25 ppm or greater, but did not alter the degree of response at all at a 0.4-ppm concentration (Amdur, 1966).

Studies using $^{35}SO_2$ indicate that SO_2 is readily distributed throughout the body. Uptake also occurred when only an isolated segment of trachea was exposed to SO_2. The lungs may possibly release some of the gas absorbed in this manner, as radioactivity could be detected in exhaled air samples collected at the carina or below when the lungs had

had no SO_2 exposure. Such gas must have reached the lungs via the pulmonary capillaries (Frank et al., 1967).

Some of the SO_2 inhaled reacts with disulfide bonds in the plasma to form S-sulfonate compounds in the rabbit and in human subjects. Although the reaction is reversible, plasma protein S-sulfonates are relatively stable with a biological half-life of several days (Gunnison and Benton, 1971; Gunnison and Palmes, 1973, 1974). Currently these materials are being found in lung and in aorta of rabbits.

1.5. Effects of Aerosols on Response to Sulfur Dioxide

Atmospheres polluted with sulfur dioxide almost invariably also contain particles that may or may not be capable of altering the response to SO_2. This is true of other irritant gases as well. The area of gas-aerosol interactions is a very complex one that requires a multidisciplinary approach if its complexities are ever to be understood.

The increase in flow resistance of guinea pigs has been used as a toxicological tool to study the effect of various inert particles on the degree of response produced by SO_2. Work in this area started 20 years ago (Amdur, 1957). The aerosols used in all these studies have been "inert," so defined on the basis that alone they produce no detectable increase in flow resistance. The first aerosol used was sodium chloride at a concentration of 10 mg/m^3. When the particle size was 2.5 μm, no potentiation of response was observed. When the particle size was reduced to 0.2 μm mass median diameter (MMD) (0.04 CMD) by the use of a Dautrebande D_{30} aerosol generator, potentiation of the response was observed. In all subsequent gas-aerosol work submicrometer particles were used.

Not only was the response greater when the sodium chloride aerosol was present, but also the return to control values was delayed (Amdur, 1961). A delayed return to control values is typical of the response to an irritant aerosol. It was also found that decreasing the concentration to 4 or 5 mg/m^3 decreased the potentiation observed during exposure. Postexposure resistance values were still above control levels but were less than those produced by the higher aerosol concentration. Further evidence that this residual effect was related to the presence of an irritant aerosol was obtained by reducing the aerosol exposure time to 0.5 hr at 10 mg/m^3 to yield the same total dose of aerosol (Amdur and Underhill, 1968). The postexposure resistance values were the same as those observed following a 1-hr exposure to 5 mg/m^3 sodium chloride and SO_2.

In these experiments the chamber relative humidity was 50%. The potentiation was slow to develop, that is, the response at 10 to 15 min was no different from that produced by SO_2 alone, although the difference was marked during the latter part of the hour exposure period. At a concentration of 2 ppm SO_2 and 4 mg/m^3 sodium chloride there was no potentiation of response. McJilton et al. (1973) used 1 ppm SO_2 and 1 mg/m^3 sodium chloride at 40 and at 80% relative humidity. Under conditions of high humidity, a marked potentiation became evident by the first 15 min of exposure. This indicates the critical importance of relative humidity on gas-aerosol interactions.

Additional experiments have indicated that solubility of SO_2 in a droplet plays a role in the potentiation (Amdur and Underhill, 1968). The solubility of SO_2 is greater in solutions of potassium chloride than in sodium chloride and still greater in ammonium thiocyanate. The potentiation of SO_2 by these salts is related in a rational manner to this relative solubility. Materials such as carbon, iron oxide fume, triphenylphosphate, open-hearth dust, or manganese dioxide did not potentiate the response to SO_2, even at concentrations above 10 mg/m^3. These materials are soild which would not dissolve SO_2.

The response is potentiated three to fourfold by soluble salts of metals such as manganese, ferrous iron, vanadium, and copper at concentrations of 1 mg/m^3. The potentiation is rapid in onset as opposed to the action of sodium chloride at a relative humidity of 50%. These aerosols form droplets which can dissolve SO_2. The metal ions promote the conversion of SO_2 to sulfuric acid, and measurements were made of the amount of sulfuric acid formed. At low concentrations (around 0.2 ppm) about 10% of the SO_2 was converted to sulfuric acid. When this calculated amount of sulfuric acid was combined with SO_2, the response observed to the metal aerosol-SO_2 exposures could be reproduced (Amdur, 1974).

The acute irritant effects of SO_2 were not potentiated by carbon or by dry manganese dioxide. Rylander et al. (1971), however, have shown a synergism between coal dust or manganese dioxide and SO_2 which affects clearance mechanisms in the lung. This resulted from chronic exposure of guinea pigs for 6 hr/day for 4 weeks. Neither SO_2 nor manganese dioxide alone affected the number of bacteria remaining in the lung. The combination, however, reduced the clearance. Coal dust alone was effective in reducing the clearance, and this effect was enhanced by SO_2. This emphasizes the need to employ a variety of toxicological end points to examine the effect of particles on sulfur dioxide, which, regrettably, remains a major gap in our needed information.

1.6. Possible Role in Production of Cancer

It has long been speculated that irritants may play a role in the development of lung cancer even though they themselves are not carcinogens, and available experimental evidence would appear to support this view. Peacock and Spence (1967) found that chronic exposure of mice to 500 ppm sulfur dioxide increased the incidence of tumors in a spontaneous-tumor-susceptible strain. Laskin et al. (1970) found that, when SO_2 at levels of 10 or 3.5 ppm was given to rats as a daily exposure, inhalation of benzo-a-pyrene produced squamous cell carcinoma. Inhalation of the carcinogen by rats exposed only to room air, however, did not result in the production of squamous cell carcinoma. The induction time for the appearance of the carcinomas was shorter in the rats exposed to the higher SO_2 concentration, suggesting a dose-response relationship. More work will be needed to elucidate the mechanisms of this interaction.

2. TOXICOLOGY OF SULFURIC ACID

2.1. Studies of Mortality and Lung Pathology

Studies done in the 1950s demonstrated that there is a considerable difference in sensitivity to sulfuric acid among small laboratory animals. The guinea pig is quite sensitive; rats, rabbits, and mice are much less sensitive (Treon et al., 1950). The mechanism of death in guinea pigs is severe bronchoconstriction and laryngeal spasm. Animals that survived a longer exposure showed gross and microscopic lung pathology: hemorrhage, capillary engorgement, and some edema. Damage in surviving animals was only slowly repaired (Amdur et al., 1952; Pattle et al., 1956).

It was also reported in the 1950s that the presence of an excess of ammonia greatly reduced the toxicity of H_2SO_4 (Pattle et al., 1956). This indicated that ammonium sulfate is less irritant than H_2SO_4. It also served as a warning to the workers (sometimes ignored) that in exposures to H_2SO_4 cleanliness of the chambers is of great importance to prevent the neutralization of the acid by ammonia from the urine of the animals.

2.2. Effects on Pulmonary Function

Sulfuric acid, like sulfur dioxide, produces an increase in pulmonary flow resistance (Amdur, 1958; Amdur et al., 1977a). As with any inhaled particulate material, particle size is of the utmost importance; the irritant effect increases as the particle size decreases. In guinea pigs the particle size studied have been 0.3, 0.8, 1, 2.5, and 7 μm (expressed as mass median diameter). Since the 7-μm particles would have been trapped entirely in the upper respiratory tract, their effect was to produce only a slight increase in flow resistance even at a level of 30 mg/m^3. The 2.5-μm particles produced a response that was slow in onset and appeared to be related to closure of major areas of the lung. The submicrometer particles penetrated to the lung and produced a response swift in onset, resembling that resulting from exposure to irritant gases. The most effective were the 0.3-μm particles. This, of course, is the so-called accumulation mode representing a stable size range in urban aerosols. A 1-hr exposure to 100 $\mu g/m^3$ will produce a 41% increase in flow resistance and a 26% decrease in compliance. A similar concentration of 1-μm particles produces a 14% increase in resistance and no significant change in compliance. No data are currently available on the effects of extremely small particles of H_2SO_4 such as are emitted by catalytic converters.

The alterations produced by H_2SO_4 are not rapidly reversible, as is the case with alterations resulting from irritant gases. Residual effects are still observed for at least an hour after exposure terminates.

On the basis of an equal concentration of sulfur, H_2SO_4 is a more potent irritant than SO_2. This fact has obvious implications for air pollution considerations, as SO_2 is convered to H_2SO_4 to varying degrees in the atmosphere. The greater irritant potency has been demonstrated on an acute basis by studies of the comparative responses of guinea pigs (Amdur, 1974). It has also been demonstrated in 2-yr studies of monkeys (Alarie et al., 1973, 1975). No detectable alterations in either pulmonary function or pathology resulted from 1 ppm SO_2, which would be equivalent to 1.3 mg S/m^3. Exposure to 0.7 μm H_2SO_4 at a concentration of 4.8 mg/m^3 (1.6 mg S/m^3) produced alterations in the distribution of ventilation and in arterial oxygen, as well as moderate to severe histopathology. Pulmonary flow resistance increased in some groups but not in others exposed to H_2SO_4; hence data were difficult to interpret. When H_2SO_4 was administered with either fly ash or SO_2, any effects observed were attributed to the H_2SO_4.

2.3. Effects on Clearance of Particles

Sulfuric acid inhalation can cause alterations in the deposition and clearance of particles in the respiratory tract. Such responses may well occur at concentrations below those required to produce measurable increases in pulmonary flow resistance. Fairchild et al. (1975a) exposed guinea pigs for 1 hr to H_2SO_4, followed by a 4-min, exposure to radiolabeled streptococcal aerosol. A concentration of 3020 $\mu g/m^3$ (1.8 μm MMD) induced an increase in the deposition of bacteria and a shift in regional deposition toward a greater deposition in the nasaopharyngeal region and/or laryngeal region. A concentration of 30 $\mu g/m^3$ (0.25 μm MMD) increased the deposition of particles in the trachea. No alterations of respiration were observed at these exposure levels. Exposure to 15 mg/m^3 (3.2 μm MMD) for 4 hr decreased the clearance of bacteria from the tracheobronchial area (Fairchild et al., 1975b).

In donkeys a significant transient slowing of tracheobronchial clearance of radioactively tagged ferric oxide aerosol resulted from 1-hr exposure to less than 200 $\mu g/m^3$ H_2SO_4 (0.3 μm MMD). Weekly 1-hr exposures to concentrations of 75 to 1400 $\mu g/m^3$ caused a progressive change in normal clearance patterns in two out of four animals (Schlesinger et al., 1977). In the donkey these levels do not produce measurable changes in regional deposition or pulmonary flow resistance.

3. TOXICOLOGY OF SULFATES

The sulfate ion in and of itself does not appear to be irritant. Thus it follows that the various sulfate salts vary widely in irritant potency. None of the sulfate salts is as irritant as sulfuric acid. As was indicated for H_2SO_4, the particle size of the sulfates has a profound effect upon their toxicity. Amdur and Corn (1963) studied zinc ammonium sulfate in four particle sizes ranging from 0.29 to 1.4 μm, all within the so-called respirable size range. Concentrations were from 0.25 to 3.4 mg/m^3. All concentrations tested produced an increase in flow resistance in guinea pigs exposed for 1 hr. As was the case with H_2SO_4, the values remained elevated during a postexposure period of 1 hr. The irritant potency increased as the particle size decreased at a given concentration. A family of dose-response curves was obtained. As the particle size decreased, the slope of the curve increased; thus a smaller increment in concentration produced a greater increment in response at the smaller sizes. Zinc sulfate and ammonium sulfate were studied at the smallest particle size. Both were irritant, but less so than the double salt.

Nadel et al. (1966) found that in cats zinc ammonium sulfate produced a response similar to that produced by an aerosol of histamine. A 3-min inhalation of 40 to 50 mg/m^3 increased pulmonary resistance, decreased pulmonary compliance, and increased end-expiratory transpulmonary pressure. The fact that isoproteronol prevented the changes suggested that they were due to smooth muscle contraction. Quick freezing of the lungs provided histological data to correlate with the physiological responses. Constriction was observed in the alveolar ducts and terminal bronchioles, whereas bronchi and bronchioles larger than 400 μm were not constricted. This paper, which is indeed a classic piece of work, combines physiological and anatomical data to demonstrate the locus of action of the submicrometer particles.

Amdur et al. (1977b) have since studied additional sulfate salts. The bioassay method has been uniform in measuring the increase in flow resistance produced by a 1-hr exposure in guinea pigs. It is thus possible to standardize the data by calculating the increase in flow resistance produced per microgram of sulfate. If a value of 100 is assigned to H_2SO_4, the most irritant, a rough ranking of these sulfates for irritant potency can be obtained. These data are presented in this format in Table 1. It is obvious that this ranking uses a single parameter for evaluation of irritant potency and hence is by no means definitive. It also does not address the potential effects of long-term exposures.

The effect of various salts on histamine release by rat lung and the effect of the cation on sulfate absorption by the lung have been examined by Charles and Menzel (1975a, 1975b). Isolated guinea pig lung fragments released histamine when incubated with ammonium ion. When the anion was sulfate, the histamine release was greater than with nitrate, acetate, or chloride. Sodium sulfate did not cause the release of histamine. Ammonium sulfate produced bronchoconstriction in isolated, perfused lung—a reaction not observed with sodium sulfate. The cation present could vary the uptake of sulfate by the lung. The most active

Table 1. Relative Irritant Potencies of Sulfates

Sulfuric acid	100
Zinc ammonium sulfate	33
Ferric sulfate	26
Zinc sulfate	19
Ammonium sulfate	10
Ammonium bisulfate	3
Cupric sulfate	2
Ferrous sulfate	0.7
Sodium sulfate	0.7
Manganous sulfate[a]	−0.9

[a] Resistance decreased; change N.S.

cations were ferric iron, zinc, and nickel. The absorption of sulfate was enhanced by either low or high pH. This group at Duke University Medical Center is actively working in this area. It is evident that a better understanding of the relative importance of individual sulfates will result from the combination of such pharmacological studies and studies of pulmonary function.

The fact that individual sulfates vary so widely in the responses they evoke indicates clearly that it is not sufficient to determine an unknown "soluble sulfate" (to use a term that is toxicologically meaningless) and hope that "health effects" can be meaningfully related thereto.

REFERENCES

Alarie, Y., Ulrich, C. E., Busey, W. M., Swann, H. E., and MacFarland, H. N. (1970). *Arch. Environ. Health* **21,** 769–777.

Alarie, Y., Ulrich, C. E., Busey, W. M., Krumm, A. A., and MacFarland, H. N. (1972). *Arch. Environ. Health* **24,** 115–128.

Alarie, Y., Busey, W. M., Krumm, A. A., and Ulrich, C. E. (1973). *Arch. Environ. Health* **27,** 16–24.

Alarie, Y., Krumm, A. A., Busey, W. M., Ulrich, C. E., and Kantz, R. J. (1975). *Arch. Environ. Health* **30,** 254–262.

Amdur, M. O. (1957). *Am. Ind. Hyg. Assoc. J.* **18,** 149–155.

Amdur, M. O. (1958). *Arch. Ind. Health* **18,** 407–414.

Amdur, M. O. (1961). In C. N. Davies, Ed., *Inhaled Particles and Vapors.* Pergamon Press, Oxford, pp. 281–292.

Amdur, M. O. (1966). *Arch. Environ. Health* **12,** 729–732.

Amdur, M. O. (1974). *Am. Ind. Hyg. Assoc. J.* **35,** 589–597.

Amdur, M. O. and Corn, M. (1963). *Am. Ind. Hyg. Assoc. J.* **24,** 326–333.

Amdur, M. O. and Underhill, D. W. (1968). *Arch. Environ. Health* **16,** 460–468.

Amdur, M. O., Schulz, R. Z., and Drinker, P. (1952). *Arch. Ind. Hyg. Occup. Med.* **5,** 318–329.

Amdur, M. O., Dubriel, M., and Creasia, D. A. (1977a). *Environ. Res.* (in press).

Amdur, M. O., Bayles, J., Ugro, V., and Underhill, D. W. (1977b). *Environ. Res.* (in press).

Balchum, O. J., Dybicki, J., and Meneely, G. R. (1960). *Arch. Ind. Health* **21,** 564–569.

Chakrin, L. W. and Saunders, L. Z. (1974). *Lab. Invest.* **30,** 145–154.

Charles, J. M. and Menzel, D. B. (1975a). *Arch. Environ. Health* **31,** 314–316.

Charles, J. M. and Menzel, D. B. (1975b). *Res. Commun. Chem. Pathol. Pharmacol.* **12,** 389–396.

Dalhamn, T. (1956). *Acta Physiol. Scand.* **36,** Suppl. 123, 1–161.

Fairchild, G. A., Stultz, S., and Coffin, D. L. (1975a). *Am. Ind. Hyg. Assoc. J.* **37,** 584–594.

Fairchild, G. A., Kane, P., Adams, B., and Coffin, D. L. (1975b). *Arch. Environ. Health* **30,** 538–545.

Frank, N. R. and Speizer, F. E. (1965). *Arch. Environ. Health* **11,** 624–634.

Frank, N. R., Yoder, R. E., Yokoyama, E., and Speizer, F. E. (1967). *Health Phys.* **13,** 31–38.

Gunnison, A. F. and Benton, A. W. (1971). *Arch. Environ. Health* **22,** 381–388.

Gunnison, A. F. and Palmes, E. D. (1973). *Toxicol. Appl. Pharmacol.* **24,** 266–278.

Gunnison, A. F. and Palmes, E. D. (1974). *Am. Ind. Hyg. Assoc. J.* **35,** 288–291.

Hirsch, J. A., Swenson, E. W., and Wanner, A. (1975). *Arch. Environ. Health* **30,** 249–253.

Lamb, D. and Reid, L. (1968). *J. Pathol. Bacteriol.* **96,** 97–111.

Laskin, S., Kuschner, M., and Drew, R. T. (1970). In *Inhalation Carcinogenesis.* U.S. Atomic Energy Commission Symposium Series **18,** pp. 321–350.

Lewis, T. R., Campbell, K. I., and Vaughan, T. R., Jr. (1969). *Arch. Environ. Health* **18,** 596–601.

Mawdesley-Thomas, L. E., Healey, P., and Barry, D. (1970). In W. H. Walton, Ed., *Inhaled Particles and Vapors,* Vol. III. Unwin Bros., Surrey, pp. 509–525.

McJilton, C., Frank, N. R., and Charlson, R. (1973). *Science* **182,** 503–504.

Nadel, J. A., Corn, M., Zwi, S., Flesch, J., and Graf, P. (1966). In C. N. Davies, Ed., *Inhaled Particles and Vapors,* Vol. II. Pergamon Press, Oxford, pp. 55–66.

Nadel, J. A., Salem, H., Tamplin, B., and Tokiwa, Y. (1965). *J. Appl. Physiol.* **20,** 164–167.

Pattle, R. E., Burgess, F., and Cullumbine, H. (1956). *J. Pathol. Bacteriol.* **72,** 219–232.

Peacock, P. R. and Spence, J. B. (1967). *Br. J. Cancer* **21,** 606–618.

Reid, L. (1963). *Br. J. Exp. Pathol.* **44,** 437–445.

Rylander, R., Ohrstrom, M., Hellstrom, P. A., and Bergstrom, R. (1970). In W. H. Walton, Ed., *Inhaled Particles and Vapors,* Vol. III. Unwin Bros., Surrey, pp. 535–541.

Schlesinger, R. B., Lippmann, M., and Albert, R. E. (1977). *Abstracts.* American Industrial Hygiene Association Conference, New Orleans, La., May, pp. 79–80.

Spicer, S. S., Chakrin, L. W., and Wardell, J. R. (1974). *Am. Rev. Respir. Dis.* **110,** 13–24

Strandberg, L. G. (1964). *Arch. Environ. Health* **9,** 160–166.

Treon, J. F., Dutra, F. R., Cappel, J., Sigmon, H., and Younker, W. (1950). *Arch. Ind. Hyg. Occup. Med.* **2,** 716-734.

3

HEALTH CONSEQUENCES OF HUMAN EXPOSURE

Carl M. Shy

Institute for Environmental Studies and School of Public Health, University of North Carolina, Chapel Hill, North Carolina

The sulfur cycle in nature is just as essential to the existence of life on this planet as are the carbon and nitrogen cycles. Disulfide bonds contribute significantly to the conformation of proteins, and thiol groups are vital for the catalytic functions of many enzymes (Greenberg, 1975). The purpose of this chapter is to review the health effects of human exposure to sulfur compounds in the environment, whether through occupational contact or general community encounters. The biochemistry of sulfur compounds will not be treated here, since emphasis is being given to sulfur in the environment as it affects health, rather than to endogenous metabolism of sulfur compounds.

Exposure to environmental sulfur compounds is ubiquitous in our industrial society. Fossil fuel energy sources contain significant amounts of sulfur, and in the process of combustion large amounts of oxides of sulfur are formed and may be released into the atmosphere. Many basic ores and petrochemicals contain sulfur which must somehow be disposed of in the course of smelting and refinement. Three forms of environmental sulfur are common and particularly important from a human health aspect: hydrogen sulfide, carbon disulfide, and various oxides of sulfur. The human health consequences of these three classes of sulfur compounds will be discussed in this chapter.

1. HYDROGEN SULFIDE

Hydrogen sulfide (H_2S) is a powerful asphyxiant that is much more toxic than is commonly realized. The toxicity of H_2S approaches that of hydrogen cyanide in that death may be equally rapid from poisoning with either of these gases.

1.1. Sources of Hydrogen Sulfide

Wherever sulfur is deposited, pockets of H_2S may be encountered, as in pools of water in coal, lead, gypsum, and sulfur mines. Hydrogen sulfide is also present in some natural gas fields. In extracting geothermal energy H_2S is the most troublesome pollutant, amounting to 500 ppm in the recovered steam (Science and Public Policy Program, 1975). Production and refining of high-sulfur petroleum from some of the oil fields of Mexico and Texas present opportunities for exposure to dangerous levels of H_2S; the gas begins to vaporize as soon as it reaches the surface, and this process is greatly accelerated by the heat of refining. The decay of organic matter in sewers and in wastewaters from industrial plants where animal products are handled gives rise to H_2S; thus accidental poisoning from H_2S has occurred in tanneries, glue factories, fur-dressing and felt-making plants, and beet-sugar factories. In addition, H_2S is formed in certain industrial processes in the production of sulfur dyes (brown, black, blue), carbon disulfide, and rubber containing certain sulfur compounds. Dangerous exposures are not uncommon among chemists, students, and research technicians working in ill-equipped chemistry laboratories.

1.2. Acute Toxicity of Hydrogen Sulfide

Like cyanide, H_2S produces its effect through inhibition of cytochrome oxidase. Death results from respiratory failure, usually within a few seconds after exposure to lethal doses. In concentrations of 250 to 600 ppm, pulmonary edema and bronchial pneumonia may result (Patty, 1963). A colorless gas, H_2S has the characteristic odor of rotten eggs. The sense of smell is readily fatigued by continuous exposure, however, and this loss of smell is particularly serious in industrial exposures because of the danger of rapid asphyxiation from high doses of H_2S.

A tragic air pollution episode was caused by H_2S gas in the community air of Poza Rica, Mexico, in November 1950 (McCabe and Clayton, 1952). At that time a new plant for recovery of sulfur in natural gas was opened. Hydrogen sulfide was removed from the natural gas and burned. However, because of a breakdown in the combustion system, H_2S was inadvertently released into the town during the early morning hours. Many residents were promptly afflicted with respiratory and central nervous system symptoms; 320 were hospitalized and 22 died. Most of the deaths occurred in persons who experienced central nervous system symptoms such as unconsciousness and vertigo. Persons of all ages were affected.

1.3. Other Health Effects of Hydrogen Sulfide

The most common acute effect of H_2S exposure is sensory irritation. Hydrogen sulfide can be detected at concentrations of 0.20 ppm (Gafafer, 1964), although the lower limit of odor detection may be as low as 0.072 ppm (Ryazanov, 1952). The acute responses that might be expected at given concentrations are indicated in Table 1 (Gafafer, 1964). At lower concentrations (50 to 500 ppm) H_2S is primarily a sensory and respiratory irritant, whereas exposures above 500 ppm can produce pulmonary edema, respiratory, and central nervous system paralysis.

The mildest form of poisoning is characterized by irritation of the conjunctiva, sometimes with keratitis. The appearance of a colored ring around light and increased sensitivity to all forms of light serve as early warning signs of excessive exposure. Such reactions have been reported among tunnel workers exposed to H_2S (Hamilton and Hardy, 1974; Aves, 1929). Barthelemy (1939) describes the injury to the eyes from H_2S as follows: "intense photophobia, spasm of the lids, excessive tearing, intense congestion, pain, blurred vision, the pupils contracted and reacting sluggishly, the cornea hazy, sometimes with blisters on the surface." These reactions are usually reversible upon removal from polluted air, but a severe case may result in lasting damage if corneal ulcers with scarring occur (Hamilton and Hardy, 1974).

Although recovery from acute poisoning is usually complete, exceptional cases have manifested irreversible cell damage apparently due to prolonged anoxia (Hamilton and Hardy, 1949); other cases are reported as showing permanent psychic and nervous disturbances (McCabe and Clayton, 1952; Poda, 1966). Chronic poisoning of industrial workers by H_2S is characterized by conjunctivitis, headache, attacks of dizziness, digestive disturbances, diarrhea, and loss of weight. The evidence for chronic poisoning from H_2S gas is, however, very questionable (Milby, 1962). Since H_2S is rapidly detoxified in the body through the formation

Table 1. Acute Responses to Hydrogen Sulfide

Concentration (ppm)	Response
0.2	Detectable odor
150.0	Olfactory nerve paralysis
250.0	Prolonged exposure may cause pulmonary edema
500.0	Systemic symptoms may occur in 0.5 to 1.0 hr
1000.0	Rapid collapse—respiratory paralysis imminent
5000.0	Immediate death

of sulfate, which is quickly excreted, the compound is considered to be a noncumulative poison.

Because H_2S is rapidly detoxified in the body, the primary treatment of exposed persons consists of rapid removal from the contaminated environment. Oxygen therapy should be used as a supplement and may prevent pulmonary edema from high doses. Since H_2S reacts rapidly with methemoglobin, treatment with methemoglobin-forming compounds such as sodium nitrite has been suggested (Smith and Gosselin, 1966), although successful use of sodium nitrite as an antidote for human poisoning has not yet been reported.

The threshold limit value for daily 8-hr exposure to H_2S in the United States is 10 ppm (American Conference of Governmental Industrial Hygienists, 1976). This exposure limit is largely based on repeated observations of eye effects at 20 ppm or below (American Conference of Governmental Industrial Hygienists, 1971).

2. CARBON DISULFIDE

Carbon disulfide (CS_2), a colorless liquid, was recognized by the French more than 100 years ago as an industrial poison among rubber workers (Hamilton and Hardy, 1974). In these workers CS_2 poisoning was characterized by loss of appetite, dyspepsia, visual and sensory disturbances, and sexual impotence. Prolonged exposures resulted in deep depression, memory loss, and sometimes acute mania or paranoia.

2.1. Sources of Carbon Disulfide Exposure

Carbon disulfide is used as an insecticide, as a solvent for waxes, resins, gums, and rubber, in the production of viscose rayon, and in the manufacture of optical glass. It is also encountered in the destructive distillation of coal.

For many years, CS_2 played an important role as a solvent in the rubber industry of many continental European countries, less so in England, and still less in the United States. Carbon disulfide is a solvent for sulfur chloride, which reacts with crude latex to form an elastic, heat-resistant solid useful as a rubber tire or other rubber product. In the United States crude latex is vulcanized by heat curing, a process involving the addition of flowers of sulfur to the latex mass, which is then subjected to heat and pressure. This heat cure was employed by

American manufacturers at a time when Europeans preferred the acid cure with CS_2 (Hamilton and Hardy, 1974).

In the production of viscose rayon, cellulose is derived from alkali-treated wood pulp or cotton linters and changed, by contact with CS_2 in tumbling "churns" or barrels, to a rubbery mass known as cellulose xanthate. The most severe exposure to CS_2 takes place in the churn room from leaking pipes and barrels, from discharging and scraping out the barrels, and from conveying the cellulose xanthate to the viscose mixers. A lesser exposure occurs in the spinning process (Hamilton and Hardy, 1974).

2.2. Acute Toxicity of Carbon Disulfide

Table 2 lists six representative levels of effect upon human beings, with corresponding ranges of concentration of inhaled CS_2 (Patty, 1963). The primary effect of high concentrations (1000 to 5000 ppm) is narcosis, leading to paralysis, respiratory failure, and death. Moderately high doses (500 to 1000 ppm) can produce mental symptoms, visual disturbances, dizziness, increasing sense of weariness, and slight delirium. However, acute poisoning from CS_2 is rare, and most of the reported toxicity has occurred in the viscose rayon industry from repeated exposures to low concentrations.

Carbon disulfide is absorbed mainly through the lungs into the blood stream and distributed throughout the body. Eighty-five to 90% of the CS_2 is metabolized and eliminated in the urine as inorganic sulfates or other sulfur compounds (McKee et al., 1943). Carbon disulfide has been shown to have a direct toxic action on protein metabolism and, by

Table 2. Acute Toxicity of Carbon Disulfide

Concentration (ppm)	Response
160–200	Slight or no acute effect
320–390	Slight symptoms[a] after several hours
420–510	Symptoms[a] after 0.5 hr
1150	Serious symptoms[a] after 0.5 hr
3210–3850	Dangerous to life after 0.5 hr
4815	Fatal in 0.5 hr

[a] Symptoms—slight: headache, giddiness, disturbance of vision; serious: respiratory and gastrointestinal disturbances, precordial distress, mental symptoms.

hepatic damage, to cause nervous system disease, hypercholesteremia, and early arteriosclerosis (Browning, 1965).

2.3. Chronic Effects of Carbon Disulfide

Upon prolonged exposure CS_2 has a pronounced neurotoxic action, leading to a great variety of nonspecific effects such as psychoses, polyneuritis with paralysis and atrophy, myopathy of leg muscles, and optic neuritis. The first two papers on CS_2 poisoning in the United States were published in 1892 (cited in Hamilton and Hardy, 1974) and gave histories of three rubber vulcanizers who were sent to the state hospital for the insane, suffering from acute mania. Histories of CS_2 intoxication of rubber industry workmen were being reported by 1914, with descriptions of irritability, depression, apathy, sudden outbursts of rage or fear, hallucinations, and even maniacal seizures (Vigliani, 1954). These symptoms were also observed in viscose workers (Gordy and Trumper, 1938). Under adverse working conditions during World War II, worker malnutrition and poor factory ventilation enhanced the opportunity for CS_2 poisoning. Paluch (1948) reported 148 cases in two CS_2 poisoning episodes in Polish rayon plants under conditions of enemy occupation; the author emphasized the prevalence of polyneutritis affecting especially the legs. Biopsy showed hypertrophy and atrophy of muscle fibers. Lewey (1941), on examining 120 workers employed as churn-room men and spinners in viscose manufacture, found mild forms of CS_2 poisoning in 60% of spinners and severe poisoning in 20% and 44% of churn-room men at two plants. Many parts of the central and peripheral nervous symptoms were affected, as manifested by psychic symptoms, damage to cranial nerves, pyramidal and extrapyramidal signs, and peripheral neuropathy.

Carbon disulfide vapors also affect other organ systems. The compound is irritating to the eyes, nose, and skin. Gastrointestinal symptoms similar to those of peptic ulcer are common in some cases. Heart, liver, and kidney damage is also described (Browning, 1965; Nesswetha and Nesswetha, 1967; Tiller et al., 1968). Nurminen (1976) followed a cohort of 343 men from a viscose rayon factory in Finland between 1967 and 1975 and observed a twofold excess death rate from coronary heart disease relative to a matched reference population of nonexposed paper mill workers. Trend analysis suggested decreased coronary heart disease mortality after the Finnish threshold limit value (TLV) for CS_2 in workroom air was lowered to 10 ppm.

The current TLV for workroom exposures in the United States is 20 ppm. In Finland and Czechoslovakia the TLV is 10 ppm; in the Soviet Union, 4 ppm. In view of the evidence for excess coronary heart disease mortality at concentrations above 10 ppm (Nurminen, 1976), the 20-ppm value needs to be reexamined.

Direct measurement of blood and urine for CS_2 levels is indicative of the severity of exposure (Patty, 1963). Only symptomatic treatment is available for acute and chronic CS_2 poisoning.

3. OXIDES OF SULFUR

In places where energy requirements are largely met by combustion of fossil fuels, sulfur dioxide (SO_2) and various oxidation or decay products of SO_2, such as sulfuric acid and particulate sulfates, are ubiquitous components of the atmosphere. Other important sources of atmospheric sulfur oxides include smelting of sulfur-bearing ores, refining of petroleum, manufacture of sulfuric acid, burning of refuse, paper making, and smoldering coal refuse banks (U.S. Department of Health, Education, and Welfare, 1970). Sulfur-containing coal and fuel oil used for power generation and heating account for more than 50% of the SO_2 emissions in the United States. These emissions into the ambient environment represent a very substantial component of the air pollution associated with urban and industrial activity. Sulfur dioxide in the atmosphere is converted to sulfuric acid and sulfate aerosols, and in the process is transformed from a gas to a particulate state (Tanner et al., Part I). The combination of SO_2 gas and suspended particulate matter in ambient air is often referred to as the sulfur oxide-particulate complex, or as "reducing"-type pollution, in contrast to the photochemical oxidant or "oxidizing"-type pollution characteristic of air pollution in the South Coast Air Basin of California. With the exception of the latter, air pollution in most cities of Europe and North America has been largely evaluated, from a health aspect, in terms of the sulfur oxide-particulate complex.

As opposed to community air pollution, which is best characterized as a complex mixture of sulfur oxides and particulates (with nitrogen oxides, organic aerosols, and various metallic compounds added), certain segments of the work force are exposed to relatively high concentrations of SO_2, with relatively less admixture of particulate matter or other air pollutants. Much of our knowledge of the health effects of SO_2 has been derived from these industry studies and from animal experimentation with pure SO_2 gas.

There is now considerable evidence that the toxic properties of SO_2

gas are largely relevant solely to the issue of occupational SO_2 exposures, but that community air pollution as characterized by the sulfur oxide-particulate complex presents a distinctly different qualitative and quantitative health risk. For this reason, Section 3 will be divided into a discussion of the health effects of SO_2 alone (this evidence is mainly derived from experimental animal, experimental human, and occupational health studies) and the health effects of the sulfur oxide-particulate complex, the evidence for which is largely based upon epidemiological studies.

3.1. Effects of Sulfur Dioxide Alone

Occupational Health Studies

Sulfur dioxide has a number of important industrial uses (Patty, 1963; also Chapter 1, Part I), including the manufacture of sulfuric acid and sodium sulfate, in refrigeration, bleaching, fumigating, and preserving, and as an antioxidant in melting, pouring, and heat treatment of magnesium. Some occupations that have considerable potential for contact with SO_2 include, among others (U.S. Department of Health, Education, and Welfare, 1974), blast furnace workers, diesel engine operators and repairmen, glass makers, ore smelter workers, paper makers, petroleum refining workers, sulfuric acid makers, and tannery workers. The National Institute of Occupational Safety and Health estimates that 500,000 U.S. workers could have potential exposure to SO_2 in the work place (U.S. Department of Health, Education, and Welfare, 1974).

At concentrations above 20 ppm, SO_2 has a marked irritant effect on the eyes, nose, throat, and airways, causing tearing, choking, wheezing, coughing, and sneezing (U.S. Department of Health, Education, and Welfare, 1974). Probably because of these acute irritant reactions, actual injury from high levels of industrial exposure is rare.

Chronic exposure to SO_2 is very widespread in industry. In most of these situations, exposures to a mixture of SO_2 and sulfuric acid aerosol or other gases have occurred. In a study of Norwegian pulp and paper workers, Skalpe (1964) reported excess coughing, sputum production, and shortness of breath in workers exposed to SO_2 concentrations in the range of 2 to 36 ppm. However, other investigators (Ferris et al., 1967; Anderson, 1950) failed to detect changes in respiratory disease prevalence or in lung function among paper/pulp or refining workers exposed to prolonged SO_2 levels of 10 to 20 ppm. Kehoe et al. (1932) compared

100 men having exposure to relatively pure SO_2 at concentrations averaging 20 to 30 ppm (from SO_2 used as a refrigerant) with age-matched workers in other parts of the plant and found in the exposed group a higher incidence of nasopharyngitis, alterations in senses of smell and taste, and increased sensitivity to other gases.

Lee and Fraumeni (1969) reported an eightfold excess in respiratory cancer mortality among arsenic smelter workers. Investigations revealed that persons with heavy exposure to arsenic and moderate or heavy exposure to SO_2 were most at risk. It was postulated that SO_2 or other chemicals in the work environment may have enhanced the likely carcinogenic action of arsenic. In this respect, experimental animal studies (Laskin et al., 1970; Peacock and Spence, 1967) suggest a promoting effect of SO_2 on the action of experimental carcinogens.

Bronchial asthma has been reported (Romanoff, 1939) among individuals having a predisposition to allergy and exposed to SO_2 in the refrigeration industry.

Experimental Human Exposures to Sulfur Dioxide

Several investigators (Frank et al., 1964; Nadel et al., 1964; Snell and Luchsinger, 1969; Melville, 1970; Weir and Bromberg, 1972) have demonstrated increased pulmonary flow resistance or airway resistance in healthy subjects experimentally exposed to 5 ppm SO_2 for 10 to 60 min. The effect could be completely prevented by prior administration of atropine, suggesting a reflex bronchoconstrictive effect. Airway resistance was increased more with mouth than with nose breathing, apparently because the nasal passages effectively absorb some of the inhaled SO_2 and diminish the stimulation of receptors lower in the respiratory tract. At 1 ppm SO_2 no dose-related changes in pulmonary flow resistance, subjective symptoms, or other measures of lung function could be detected in most subjects (Frank et al., 1962; Weir and Bromberg, 1972). However, other investigators (Amdur et al., 1953; Snell and Luchsinger, 1969; Bates and Hazucha, 1973; Andersen et al., 1974) have shown effects of 0.75 to 1.00 ppm SO_2 on peak expiratory flow rate, nasal flow resistance, respiratory rate, or tidal volume in some subjects. Burton et al. (1969) estimated that "hyperreactors" or individuals highly susceptible to SO_2 may constitute 10 to 20% of the healthy young adult population.

Animal Studies

The effect of SO_2 on experimental animals is qualitatively the same as on human beings, namely, irritation of mucous membranes and reflex bronchoconstriction (Amdur, 1969; also, this volume). Studies that elucidate mechanisms of action will be cited; these experiments usually employ relatively high concentrations of SO_2 (i.e., exceeding 10 ppm).

Ciliary movement in the rabbit trachea was arrested after 15-min exposure to 200 ppm SO_2 (Dalhamn and Strandberg, 1963), while rats exposed for 3 to 10 weeks to 10 ppm SO_2 showed severe morphological changes in the upper respiratory tract, with evidence of abnormal cell proliferation (Dalhamn, 1956). An excess of mucin-secreting cells developed in young rats exposed to 300 to 400 ppm SO_2 for 6 weeks (Reid, 1963), and rabbits exposed to 76 ppm SO_2 for 13 weeks showed edema of the myocardial muscle fibers, capillary enlargement and numerous perivascular hemorrhages, dystrophic changes in liver cells and in renal tubular cells, and alveolar cell proliferation in the lung (Prokhorov and Rogov, 1959).

The effect of SO_2 on susceptibility to infection has been evaluated in several studies. Guinea pigs exposed intermittently to 10 ppm SO_2 had no change in the rate of clearance of labeled bacteria from the lung (Rylander, 1969), whereas 40 ppm for 24 days appeared to promote the colonization of mycoplasma in respiratory tissue of rats (Battigelli and Fraser, 1969). Exposure of mice to 20 ppm SO_2 for 7 days increased the amount of pneumonia after infection with influenza virus (Fairchild et al., 1972); however, hamsters repeatedly exposed to 650 ppm both before and after virus infection showed unaltered susceptibility to influenza.

Continuous exposure of guinea pigs and monkeys for 1 year to 0.1, 1, and 5 ppm SO_2 revealed no detrimental changes in pulmonary function, body weight, growth, survival, or gross or microscopic organ structure (Alarie et al., 1970, 1972).

Ninety to 95% of SO_2 is removed from inhaled air in the nasopharynx (Balchum et al., 1959; Frank et al., 1969). Mouth breathing tends to increase SO_2 uptake at lower levels of the respiratory tract, especially at higher ventilatory rates (Frank et al., 1969).

Summary of Sulfur Dioxide Effects

Sulfur dioxide alone is not a potent pulmonary irritant unless concentrations exceed 10 to 20 ppm—levels that are encountered virtually never in ambient air and rarely in workroom environments. The apparent reason for this low potency is that inhaled SO_2 is nearly completely absorbed in the upper airways and does not reach the lung. A few sensitive individuals may show minor changes in lung function at concentrations of 1 to 2 ppm, but long-term animal exposures at these levels reveal no apparent detrimental effects. The 1976 threshold limit value in the United States for occupational exposure to SO_2 was 5 ppm (American

Conference of Governmental Industrial Hygienists, 1976), but a revision of the TLV to 2 ppm has been proposed (U.S. Department of Health, Education, and Welfare, 1974).

3.2. Effects of the Sulfur Oxide-Particulate Complex

Experimental Studies

Amdur (1969, 1973) and Lewis et al. (1972) have ably summarized the toxicological literature on various atmospheric transformation products of SO_2, including sulfuric acid and several sulfate salts such as zinc sulfate, ammonium sulfate, and zinc ammonium sulfate. These compounds are more potent respiratory tract irritants than SO_2, as demonstrated in terms of mortality and lung pathology in experimental animals as well as in studies of pulmonary function in experimental animals and in human subjects. The irritant potency of SO_2 transformation products is greater at submicronic particle sizes and at high relative humidities. The presence of particulate matter capable of oxidizing SO_2 to sulfuric acid causes a three- to fourfold potentiation of the irritant response in the respiratory tract. Aerosols causing this potentiation were soluble salts of ferrous iron, manganese, and vanadium. Studies of SO_2 emitted into the atmosphere have shown that this compound is transformed to sulfuric acid and particulate sulfates—a transformation that is likely to increase its irritant potency.

Epidemiological Studies

An extensive review of this subject was prepared by a panel for the National Academy of Sciences (1975). The toxicological and the epidemiological literature on sulfur oxides present an interesting paradox. On the one hand, numerous epidemiological investigations have shown consistent relationships between excess mortality and disease in populations exposed to moderately elevated levels of atmospheric SO_2 and particulates. Yet, in toxicological studies, concentrations of these pollutants are well below those known or expected to produce adverse effects on human health, of and by themselves. Battigelli (1968) examined alternative hypotheses to explain why SO_2 might be more toxic in polluted air than in experimental exposures of animals or human beings and found that none of the hypotheses could resolve the dilemma. Frank (1968) countered with the argument that "the constituents of air pollution are many, are complex, and are not completely known; they have potential for additive, cumulative, and synergistic effect; they may interact with age, smoking, climate, and ill health to cite only a partial list. By contrast, the variables in a toxicological experiment are kept few and regulated."

Numerous epidemiological studies (cited in National Academy of Sciences, 1975) conducted in England, the United States, Canada, Japan, and several continental European countries have demonstrated an association between community exposure to the sulfur oxide-particulate complex and increased risk of death and of acute and chronic disease.

Mortality was significantly increased during a series of air pollution episodes reported between 1930 and 1962 (Firket, 1932; Schrenk et al., 1949; Ministry of Health, 1954; Scott, 1963; Greenburg et al., 1963). As shown in Table 3, the number of deaths attributed to high levels of air pollution, usually characterized by measurements of SO_2 and particulate matter or smoke, was calculated by comparing the number of deaths during the episode with the number expected from similar time periods in other years or before and after the episodes. Excess deaths were attributed mainly to bronchitis, pneumonia, and cardiac diseases and were found particularly among persons aged 45 years and older. These episodes drew the attention of the medical community to the public health dangers of uncontrolled air pollution.

Other investigators examined possible relationships between variations in day-to-day sulfur oxide-particulate concentrations and daily mortality (Martin and Bradley, 1960; Martin, 1964; Brasser et al., 1967; Watanabe and Kaneko, 1971; Buechley et al., 1973; Schimmel and Greenburg, 1972; Lebowitz, 1973). These analyses tend to agree in finding higher mortality on days of greater pollution, but the association is by no means simple. Many cofactors such as temperature extremes, season, influenza epidemics, and holiday weekends have strong effects on day-to-day mortality and may enhance or diminish the effect of air

Table 3. Selected Air Pollution Episodes

Date	Place	Attributed Mortality
December 1930	Meuse Valley, Belgium	63
October 1948	Donora, Pennsylvania	20
December 1952	London, England	4000
November 1953	New York City	200
December 1962	London, England	700

pollution. Such cofactors are variously taken into account, so that investigators seldom agree on the magnitude of the air pollution effect.

More difficult is the evaluation of excess mortality and disease resulting from long-term residence in relatively polluted communities. Comparison of mortality and morbidity rates of populations in different geographical units increases the difficulty of isolating the effect of air pollution, since information on other disease factors such as smoking habits, length of residence in the community, occupation, socioeconomic status, and ethnic background is difficult to obtain; yet these factors may significantly influence reported mortality and morbidity rates. Nevertheless, an impressive number of studies conducted by different investigators in various countries and under variable conditions have shown a consistent association between residence in areas having high sulfur oxide-particulate air pollution and mortality (Table 4) or chronic respiratory disease (Table 5). Although the possibility of intervening variables explaining some of the air pollution-health effect associations in Tables 4 and 5 exists, the consistency of these associations cannot always be explained by these other factors. In particular, in the studies of chronic bronchitis prevalence cited in Table 5 information was generally obtained on smoking, socioeconomic level, age, sex, and other potential confounders, and these factors were taken into account in the analysis. The weight of evidence suggests that, although cigarette smoking exerts an overall stronger adverse effect on life expectancy and disease rates, air pollution from the sulfur oxide-particulate complex is at least additive to, and possibly enhances, the effect of cigarette smoking on the risk of early mortality or chronic respiratory disease.

Persons with preexisting chronic respiratory disease or asthma appear to be particulary susceptible to short-term exposures to air pollution concentrations above average levels. An extensive series of studies on the effects of sulfur oxides and particulates on bronchitic patients attending chest clinics in Britain was conducted by Lawther et al. (1970) between 1955 and 1970. These studies showed that exacerbations of disease were associated with high concentrations of smoke and SO_2, although disease aggravation was more clearly associated with relative rather than absolute increases in measured pollution. With decreasing levels of pollutants in British air, it has been difficult since 1969 to relate changes in the subjects' symptom status to variations in air pollution levels. Other investigators (Clifton et al., 1959; Gregory, 1970; Martin, 1964) reported associations of absence because of sickness, rates of physician consultation, and hospital emergency admissions with periods of high air pollution. Studies performed in North America have yielded variable results, with Burrows et al. (1968) finding little influence of SO_2

on bronchitic patients in Chicago once the effect of temperature was accounted for, while Bates et al. (1962, 1966) noted better lung function in Canadian armed services veterans followed for 4 years who lived in the least polluted of four Canadian cities.

Asthmatics represent another segment of the population that appears to be particularly susceptible to short-term peak concentrations of sulfur oxides and particulates. During air pollution episodes, asthmatics were more frequently affected than persons with other diseases (Schrenk et al., 1949), and increased emergency room visits for asthma have been recorded during several air pollution episodes (Glasser et al., 1967; Chiaramonte et al., 1970). Several investigators have reported a greater asthma prevalence in more polluted communities (Zeidberg et al., 1961; Sultz et al., 1970; Yoshida et al., 1976), and Cohen et al. (1972) found a temporal relationship between temperature-adjusted asthma attack rates and 24-hr sulfur oxide-particulate concentrations among a group of asthmatics living near a coal-fired power plant. In some cases environmental factors appeared to be playing a role, but specific air pollutants could not be incriminated, as in "New Orleans asthma" (Weill et al., 1964; Carroll, 1968) and the "Tokyo-Yokohama asthma" of American servicemen stationed in Japan (Phelps, 1965; Meyer, 1976). The evidence suggests a continuum in the dose-response relationship between sulfur oxide-particulate concentration and asthma attack rates, with increasing numbers of asthmatics affected at higher air pollution levels, but with other intervening factors such as temperature and season greatly modifying the response to air pollution.

The effect on children of community air pollution from sulfur oxides and particulates has been investigated extensively, because of the long-term implications of air pollution-induced impairment in childhood. An impressive number of studies, as shown in Table 6, consistently demonstrate an association between air pollution exposure and rates of acute respiratory disease and of impaired lung function in children. The frequency and the severity of acute lower respiratory disease such as acute bronchitis, pneumonia, or "chest colds" appear to be more strongly related to air pollution exposure than the case for upper respiratory tract disease. British investigators (Colley and Reid, 1970; Kiernan et al., 1976) have conducted a long-term prospective study of children born throughout England and Wales in one week of March 1946, to evaluate the effects of growing up in communities with different air pollution levels. An association between the prevalence of current cough with lower respiratory illness in childhood and current smoking habits was observed in the follow-up examinations. At age 20, and more so at age 25 years, the prevalence of cough, adjusted for social class and

Table 4. Association of Geographical Differences in Mortality with Long-Term Residence in Areas of High Air Pollution

Reported Study	Characteristics of the Study	Findings
Pemberton and Goldberg (1954)	1950–1952 bronchitis mortality rates for males 45 years and older in county boroughs of England and Wales	Sulfur oxide levels (sulfation rates) were consistently correlated with bronchitis death rates in the 35 county boroughs analyzed.
Buck and Brown (1964)	1955–1959 bronchitis mortality rates in 214 areas of Britain, evaluated with respect to SO_2, particulates, social index, and population density	Social index and SO_2 accounted for 36% of the variation in bronchitis mortality within county and noncounty boroughs and in urban districts. Within London boroughs, social index was most important factor.
Zeidberg et al. (1967)	1949–1960 deaths for each cause in Nashville, Tenn., categorized into three levels of air pollution and three levels of economic class	Within the middle social class, total respiratory disease mortality, but not bronchitis and emphysema mortality, was significantly associated with sulfation rates and social index. White infant mortality rates were significantly related to sulfation rates.

Winklestein et al. (1967, 1968)	120 census tracts in Buffalo stratified into four levels of particulate pollution and cross-stratified into five levels of economic status	Within economic level, death rates of white males aged 50–69 years, for all causes and for chronic respiratory disease, corresponded to the gradient of particulate, but not SO_2, pollution.
Lave and Seskin (1970)	Multiple regression analysis of mortality as a function of air pollution, population density, socioeconomic level, age, and race, in 114 Standard Metropolitan Statistical Areas of the United States, 1960	Air pollution (particulates and sulfates) was a statistically significant variable in the regression models for total death rates and death rates for infants less than 1 year old.
Watanabe and Kaneko (1971)	1965–1966 mortality rates in Osaka, Japan, stratified by air pollution level	There was a stepwise increase in total mortality and deaths from circulatory disease in areas of greater pollution.

Table 5. Studies of Air Pollution and Prevalence of Chronic Respiratory Symptoms

Reported Study	Characteristics of the Study	Findings
Ferris and Anderson (1962), Anderson and Ferris (1965), Anderson et al. (1965)	Questionnaire and ventilatory function survey of random sample of adult population in Berlin, N.H., and Chilliwack, British Columbia, 1961–1963	No apparent excess in symptom prevalence in association with air pollution. Diminished pulmonary function in residents of more polluted community, though differences in occupation, climate, and ethnic factors may account for these findings.
Holland et al. (1965)	Questionnaire and ventilatory function survey of outdoor telephone workers aged 40–59 in London, rural England, and East and West Coasts of United States	Increased prevalence of respiratory symptoms, adjusted for smoking and age, a larger volume of morning sputum, and a lower average ventilatory function in London workers, and in the English compared with the American workers.
Petrilli et al. (1966)	1961–1962 study of respiratory symptoms in nonsmoking women aged 65 and over who had not worked in industry and were long-time residents of their neighborhoods	Strong association between chronic respiratory symptom prevalence and area gradient for particulates and SO_2. Comparability between residents of suburban and industrialized areas not assured by exclusion of persons on social welfare assistance.

Study	Description	Findings
Bates et al. (1966)	Comparison of symptom prevalence, work absence, and ventilatory function in Canadian veterans residing in four Canadian cities	Lower prevalence of symptoms and work absences and better ventilatory function in veterans living in the least polluted city.
Lambert and Reid (1970)	Postal survey of nearly 10,000 British residents, aged 35–64, using standardized questionnaire	Among smokers, prevalence of chronic respiratory symptoms increased with increasing air pollution. No apparent symptom-pollution association in nonsmokers.
Ferris et al. (1973)	Repeat survey of residents of Berlin, N.H., 6 years after the original 1961 study	Slightly lower prevalence of chronic respiratory symptoms and improved ventilatory function tests after standardization for age, sex, and smoking habits. Results attributed to decrease in air pollution levels.
Tsunetoshi et al. (1971)	Prevalence survey (Medical Research Council Questionnaire) of nine areas of Osaka and Hyogo prefectures, Japan, residents aged 40 and over	Multiple regression analysis revealed increasing prevalence of chronic bronchitis, adjusted for age, sex, and smoking, corresponding to the area gradient of air pollution.

Table 6. Air Pollution and Acute Respiratory Disease in Children

Reported Study	Characteristics of the Study	Findings
Douglas and Waller (1966)	Evaluation of over 3000 children aged 15 years, born in March 1946 and followed regularly as part of a comprehensive health survey; social class generally equal among the air pollution categories	Lower respiratory infections consistently related to an area index of pollution; frequency and severity of illness increased with pollution index across areas. No association with upper respiratory illness.
Lunn et al. (1967, 1970)	Respiratory illness in 5-year-old children living in four districts of Sheffield, England, 1963–1965; study repeated 4 years later on same children	Distinct increase in acute morbidity of both upper and lower respiratory tracts in relation to neighborhood air pollution. Four years later, after improvements in air quality, neighborhood differences in illness had narrowed and were statistically not significant.
Colley and Reid (1970)	Respiratory disease prevalence in over 10,000 children 6–10 years old, England and Wales, 1966	Definite gradient of past bronchitis and current cough, from lowest rates in rural areas to highest rates in most heavily polluted areas. Differences clearer in children of semiskilled and unskilled workers. No effect on upper respiratory illness rates.
Colley and Reid (1970), Kiernan et al. (1976)	Reexamination at ages 20 and 25 years of the cohort of children studied by Douglas and Waller (1966) from different areas of England and Wales	Dominant effect of cigarette smoking on respiratory symptom prevalence, followed by a history of lower respiratory tract illness at less than 2 years of age. Air pollution and social class had only small and not statistically significant effects. Tendency for air pollution effect to be greatest in lower social class.

Collins et al. (1971)	Death rates in children 0–14 years old during the period 1958–1964 in relation to social and air pollution indices in 83 county boroughs of England and Wales	Partial correlation analysis suggests that indices of domestic and industrial pollution account for a greater part of the area differences in mortality from bronchopneumonia and all respiratory diseases among children 0–1 year of age.
Nelson et al. (1964)	Retrospective study of lower respiratory disease frequency in children 0–12 years old living in four communities of the Salt Lake basin, 1967–1970	Increased frequency of total lower respiratory diseases, croup, and bronchitis (adjusted for age, sex, social class, residence duration) in children from highest pollution area. No significant pollution effect in families with less than 3 years' residence in the community. Responses validated against physicians' records.
Finklea et al. (1974)	Incidence of acute respiratory illness, 1969–1970, in nursery school children aged 2–5 years, determined at 2-week intervals; study population stratified into three air pollution categories	Greater incidence of upper and lower respiratory illness among children living in the two higher air pollution areas. Rates adjusted for social level, residential mobility, season of year, and smoking habits of parents.
Love et al. (1974)	Prospective survey of acute respiratory disease in children residing in three communities in the New York City area; illness rates determined at 2-week intervals, September 1970–May 1971	Increased incidence of lower respiratory tract illness (adjusted for social class, age, sex, smoking of parents, residential mobility) in children from the two more polluted communities.
Irwig et al. (1974)	Prospective 4-year study of peak flow rates and respiratory illness in 11,000 children aged 5–11 years from 28 areas of England and Scotland, 1973–1977	During first year of study a statistically significant gradient of prevalence of lower respiratory disease corresponding to the air pollution gradient in 10 of the areas studied. Allowance made for differences in age, sex, and social class.

Table 6. (Continued)

Reported Study	Characteristics of the Study	Findings
Hammer et al. (1976)	Retrospective survey of lower respiratory illness in children aged 0–12 years, living in four New York City area communities, 1969–1972	Rates of total respiratory illness, croup, bronchitis, and other chest infections were significantly higher among black and white children residing in higher air pollution communities. Differences in family size, crowding, parental smoking, and social level could not explain the findings.
Toyama (1964), Watanabe and Kaneko (1971)	Peak flow rates in Japanese school children residing in Kawasaki and Osaka	Lower peak flow rates in children from more polluted communities. Improved peak flow rates when air pollution levels declined.
Anderson and Larsen (1966)	Peak flow rates and school absence rates in 6–7-year-old children from three towns in British Columbia	Statistically significant decrease in peak flow rates in two towns affected by Kraft pulp mill emissions. No effect on school absences. Ethnic differences not studied.
Lunn et al. (1967, 1970)	Ventilatory function in 5- and 11-year-old children living in four districts of Sheffield, England, 1963–1965, and reexamined in 4 years	Lower than predicted ventilatory function only in the most polluted community. Reexamination 4 years later failed to show this effect, but air quality had improved in the interval.

Holland et al. (1969)	Peak flow rates in over 10,000 school children in northwest London, stratified into two levels of air pollution exposure, November 1966–March 1967	Independent and additive effects of social class, family size, history of respiratory illness, and community air pollution level on peak flow rates were found. Area of residence (as index of air pollution exposure) had greatest influence on peak flow rates.
Shy et al. (1973)	Ventilatory function in elementary school children from communities selected to represent an air pollution gradient in Cincinnati, New York City, and Chattanooga, 1967–1971	Ventilatory function adjusted for age, sex, height, and social class was lower in white, but not in black, children living in more polluted areas of Cincinnati, in white children in Chattanooga, and in white children aged 9–13, but not aged 5–8 years in New York. Lack of differences in 5–8-year-old New York City children may be due to improved air quality, though conclusions are speculative. Differences were generally small and not always consistent.

Table 7. Expected Health Effects of Air Pollution on Selected Population Groups

	Value ($\mu g/m^3$) Causing Effect			
Pollutant	Excess Mortality and Hospital Admissions	Worsening of Patients with Pulmonary Disease	Respiratory Symptoms	Visibility and/or Human Annoyance Effects
SO_2[a]	500 (daily average)	500–250[b] (daily average)	100 (annual arithmetic mean)	80 (annual geometric mean)
Smoke[a]	500 (daily average)	250 (daily average)	100 (annual arithmetic mean)	80 (annual geometric mean)[c]

[a] British Standard Practice (Ministry of Technology, 1966). Values for sulfur dioxides and suspended particulates apply only in conjunction with each other. They may have to be adjusted when translated into terms of results obtained by other procedures.
[b] These values represent the differences of opinion within the committee of experts.
[c] Based on high-volume samplers.

smoking habits, was related to air pollution but was more strongly affected by current smoking and history of childhood illness.

In nearly all reported studies (only some of which are cited in Table 6), children residing in more polluted communities show diminished ventilatory function when compared with children of similar age, sex, race, and social class living in less polluted areas. The differences are seldom large enough to be clinically significant in terms of manifest symptoms or respiratory disease. When other respiratory challenges such as recurrent infections, cigarette smoking, and occupational exposures are added in adulthood, however, compromised ventilatory function may well increase the risk of respiratory disease. Burrows et al. (1977) observed close relationships between a history of respiratory disorders in childhood and prevalence of symptoms and ventilatory function impairment as adults.

The World Health Organization (1972) assembled a committee of experts to evaluate the cumulative evidence on the effects of air pollution on health and to assess the levels of sulfur oxides and particulates that are associated with adverse effects. The conclusions of the committee are summarized in Table 7, which is reproduced from its report. It was emphasized that most of the health effects "were associated with the presence of both suspended particulates and sulfur dioxide in temperate climates and at relatively low altitudes, where both pollutants occur simultaneously from the burning of fossil fuel." Although application of these conclusions to other physical and environmental conditions seemed "reasonable as a first approximation", the committee recommended further studies to evaluate effects in different environments. The committee also cautioned that, in interpreting the data of Table 7, "one must bear in mind that a numerical value associated with a given effect does not mean that all exposed individuals will thus be affected." The data simply indicate that effects were reported at the stated concentrations and that the number of those affected was large enough to be statistically different from the number in the control groups.

REFERENCES

Alarie, Y., Ulrich, C. E., Busey, W. M., Swan, H. E., Jr., and MacFarland, H. N. (1970). "Long-Term Continuous Exposure of Guinea Pigs to Sulfur Dioxide," *Arch. Environ. Health* **21,** 769–777.

Alarie, Y., Ulrich, C. E., Busey, W. B., Krumm, A. A., and MacFarland, H. N. (1972). "Long-Term Continuous Exposure to Sulfur Dioxide in Cynomolgus Monkeys," *Arch. Environ. Health* **24,** 115–128.

Amdur, M. O. (1969). "Toxicological Appraisal of Particulate Matter, Oxides of Sulfur, and Sulfuric Acid," *J. Air Pollut. Control Assoc.* **19,** 638.

Amdur, M. O. (1973). "Animal Studies." In *Proceedings of the Conference on the Health Effects of Air Pollutants.* National Academy of Sciences, U.S. Senate Committee on Public Works, Committee Print, Serial No. 93-15, U.S. Government Printing Office, Washington, D.C., pp. 175–205.

Amdur, M. O., Melvin, W. W., and Drinker, P. (1953). "Effects of Inhalation of Sulphur Dioxide by Man," *Lancet* **2,** 758–759.

American Conference of Governmental Industrial Hygienists (1971). *Documentation of the Threshold Limit Values for Substances in Workroom Air, 3rd ed.* American Conference of Governmental Industrial Hygienists, Cincinnati, Ohio.

American Conference of Governmental Industrial Hygienists (1976). *Threshold Limit Values for Chemical Substances in Workroom Air Adopted by ACGIH for 1976.* American Conference of Governmental Industrial Hygienists, Cincinnati, Ohio.

Andersen, I., Lundqvist, G. R., Jensen, P. L., and Proctor, D. F. (1974). "Human Response to Controlled Levels of Sulfur Dioxide," *Arch. Environ. Health* **28,** 31–39.

Anderson, A. (1950). "Possible Long-Term Effects of Exposure to Sulfur Dioxide," *Br. J. Ind. Med.* **7,** 82–86.

Anderson, D. O. and Ferris, B. G., Jr. (1965). "Air Pollution Levels and Chronic Respiratory Disease," *Arch. Environ. Health* **10,** 307–311.

Anderson, D. O. and Larsen, A. A. (1966). "The Incidence of Illness among Young Children in Two Communities of Different Air Quality: A Pilot Study," *Can. Med. Assoc. J.* **95,** 893–904.

Anderson, D. O., Ferris, B. G., Jr., and Zickmantel, R. (1965). "The Chilliwack Respiratory Survey, 1963. Part III: The Prevalence of Respiratory Disease in a Rural Canadian Town," *Can. Med. Assoc. J.* **92,** 1007–1016.

Aves, C. M. (1929). "Hydrogen Sulfide Poisoning in Texas," *Tex. State J. Med.* **24,** 761.

Balchum, O. J., Dybicki, J., and Meneely, G. R. (1959). "Adsorption and Distribution of $S^{35}O_2$ Inhaled through the Nose and Mouth by Dogs," *Am. J. Physiol.* **197,** 1317–1321.

Barthelemy, H. L. (1939). "Ten Years' Experience with Industrial Hygiene in Connection with Manufacture of Viscose Rayon," *J. Ind. Hyg.* **21,** 141.

Bates, D. V., and Hazucha, M. (1973). "The Short-Term Effects of Ozone on the Human Lung." In *Proceedings of the Conference on Health Effects of Air Pollutants.* Oct. 3–5, U.S. Senate Committee on Public Works, Committee Print, Serial No. 93-15, U.S. Government Printing Office, National Academy of Sciences, Washington, D.C. pp. 507–540.

Bates, D. V., Woolf, C. R., and Paul, G. I. (1962). "Chronic Bronchitis: A Report on the First Two Stages of the Coordinated Study of Chronic

Bronchitis in the Department of Veterans Affairs, Canada," *Med. Serv. J. Can.* **18,** 211.

Bates, D. V., Gordon, C. A., Paul, G. I., Place, R. E. G., Snidal, D. P., and Woolf, C. R. (with special sections contributed by M. Katz, R. G. Fraser, and B. B. Hale) (1966). "Chronic Bronchitis: Report on the Third and Fourth Stages of the Coordinated Study of Chronic Bronchitis in the Department of Veterans Affairs, Canada," *Med. Serv. J. Can.* **22,** 5.

Battigelli, M. C. (1968). "Sulfur Dioxide and Acute Effects of Air Pollution," *J. Occup. Med.* **10,** 500.

Battigelli, M. C. and Fraser, D. A. (1969). *The Long-Term Effects of Sulfur Dioxide on Ciliary Activity of the Trachea and Microflora of the Respiratory Tract of Rodents.* The American Petroleum Institute, Environmental Affairs, Med. Res. Rept. EA 7101, Washington, D.C.

Brasser, L. J., Joosting, P. E., and Von Zuilen, D. (1967). *Sulfur Dioxide—to What Level Is It Acceptable?* Research Institute for Public Health Engineering, Rept. G-300, Delft, Netherlands. (Originally published in Dutch, September 1966).

Browning, E. (1965). *Toxicology and Metabolism of Industrial Solvents.* Elsevier, Amsterdam.

Buck, S. F. and Brown, D. A. (1964). *Mortality from Lung Cancer and Bronchitis in Relation to Smoke and Sulfur Dioxide Concentration, Population Density, and Social Index.* Tobacco Research Council, Res. Paper 7, London, England.

Buechley, R. W., Riggan, W. B., Hasselblad,.V., and Van Bruggen, J. B. (1973). "SO_2 Levels and Perturbations in Mortality: A Study in the New York-New Jersey Metropolis," *Arch. Environ. Health* **27,** 134–137.

Burrows, B., Kellog, A. L., and Buskey, J. (1968). "Relationship of Symptoms of Chronic Bronchitis and Emphysema to Weather and Air Pollution," *Arch. Environ. Health* **16,** 406–413.

Burrows, B., Knudson, R. J., and Lebowitz, M. (1977). "The Relationship of Childhood Respiratory Illness to Adult Obstructive Airway Disease," *Am. Rev. Respir. Dis.* **115,** 751–760.

Burton, G. G., Corn, M., Gee, J. B. L., Vasallo, C., and Thomas, A. P. (1969). "Response of Health Men to Inhaled Low Concentrations of Gas-Aerosol Mixtures," *Arch. Environ. Health* **18,** 681–692.

Carroll, R. E. (1968). "Epidemiology of New Orleans Epidemic Asthma," *Am. J. Public Health* **58,** 1677–1683.

Chiaramonte, L. T., Bongiorno, J. R., Brown, R., and Laano, M. E. (1970). "Air Pollution and Obstructive Respiratory Disease in Children," *N.Y. State J. Med.* **70,** 394–398.

Clifton, M., Kerridge, D., Pemberton, J., Moulds, W., and Donoghue, J. K. (1959). "Morbidity and Mortality from Bronchitis in Sheffield in Four

Periods of Severe Air Pollution." In *Proceedings of the International Clean Air Conference,* National Society for Clean Air, London, pp. 189–192.

Cohen, A. A., Bromberg, S., Buechley, R. W., Heiderscheit, L. T., and Shy, C. M. (1972). "Asthma and Air Pollution from a Coal-Fueled Power Plant," *Am. J. Public Health* **62,** 1181–1188.

Colley, J. R. T. and Reid, D. D. (1970). "Urban and Social Origins of Childhood Bronchitis in England and Wales," *Br. Med. J.* **2,** 213–217.

Collins, J. J., Kasap, H. S., and Holland, W. W. (1971). "Environmental Factors in Child Mortality in England and Wales," *Am. J. Epidemiol.* **93,** 10–22.

Dalhamn, T. (1956). "Mucous Flow and Ciliary Activity in the Trachea of Healthy Rats and Rats Exposed to Respiratory Irritant Gases (SO_2, H_3N, HCHO)," *Acta Physiol. Scand.* **36,** Suppl. 123, 125-126, 142–143.

Dalhamn, T. and Strandberg, L. (1963). "Synergism between Sulphur Dioxide and Carbon Particles: Studies on Adsorption and on Ciliary Movements in the Rabbit Trachea *in vivo,*" *Int. J. Air Water Pollut.* **7,** 517–529.

Douglas, J. W. B. and Waller, R. W. (1966). "Air Pollution and Respiratory Infection in Children," *Br. J. Prev. Soc. Med.* **20,** 1-8.

Fairchild, G. A., Roan, J., and McCarroll, J. (1972). "Atmospheric Pollutants and the Pathogenesis of Viral Respiratory Infections," *Arch. Environ. Health* **25,** 174.

Ferris, B. G., Jr., and Anderson, D. O. (1962). "The prevalence of Chronic Respiratory Disease in a New Hampshire Town," *Am. Rev. Respir. Dis.* **86,** 165–177.

Ferris, B. G., Burgess, W. A., and Worcester, J. (1967). "Prevalence of Chronic Respiratory Disease in a Pulp Mill and a Paper Mill in the United States," *Br. J. Ind. Med.* **24,** 26–37.

Ferris, B. G., Jr., Higgins, I. T. T., Higgins, M. W., and Peters, J. M. (1973). "Chronic Nonspecific Respiratory Disease in Berlin, New Hampshire, 1961 to 1967: A Follow-up Study," *Am. Rev. Respir. Dis.* **107,** 110–122.

Finklea, J. F., French, J. G., Lowrimore, G. R., Goldberg, J., Shy, C. M., and Nelson, W. C. (1974). "Prospective Surveys of Acute Respiratory Disease in Volunteer Families: Chicago Nursery School Study, 1969–1970." In *Health Consequences of Sulfur Oxides: A Report from CHESS, 1970–1971.* U.S. Environmental Protection Agency, Office of Research and Development, EPA-650/1-74-004, U.S. Government Printing Office, Washington, D.C. pp. 4–37 to 4–55.

Firket, J. (1931). "The Cause of the Symptoms Found in the Meuse Valley during the Fog of December 1930," *Bull. Roy. Acad. Med.* **11,** 683–739.

Frank, N. R. (1968). "Discussion of Paper by Dr. Battigelli," *J. Occup. Med.* **10,** 512–514.

Frank, N. R., Amdur, M. O., Worcester, J., and Whittenberger, J. L. (1962). "Effects of Acute Controlled Exposure to SO_2 on Respiratory Mechanics in Healthy Male Adults," *J. Appl. Physiol.* **17,** 252–258.

Frank, N. R., Amdur, M. O., and Whittenberger, J. L. (1964). "A Comparison of the Acute Effects of SO_2 Administered Alone or in Combination with NaCl Particles on the Respiratory Mechanics of Healthy Adults," *Int. J. Air Water Pollut.* **8,** 125–133.

Frank, N. R., Yoder, R. E., Brain, J. D., and Yokoyama, E. (1969). "SO_2 (35 S Labeled) Adsorption by the Nose and Mouth under Conditions of Varying Concentration and Flow," *Arch. Environ. Health* **18,** 315–322.

Gafafer, W. M., Ed. (1964). *Occupational Diseases: a Guide to Their Recognition.* U.S. Department of Health, Education, and Welfare, Public Health Service, Publ. 1097, U.S. Government Printing Office, Washington, D.C.

Glasser, M., Greenburg, L., and Field, F. (1967). "Mortality and Morbidity during a Period of High Levels of Air Pollution: New York, Nov. 23 to 25, 1966," *Arch. Environ. Health* **15,** 684–694.

Gordy, S. T. and Trumper, M. (1938). "Carbon Disulfide Poisoning with a Report of Six Cases," *J. Am. Med. Assoc.* **110,** 1543–1549.

Greenberg, D. M., Ed. (1975). Preface to *Metabolism of Sulfur Compounds,* Vol. VII of *Metabolic Pathways,* 3rd Academic Press, New York, p. xiii.

Greenburg, L., Jacobs, M. B., Drolette, B. M., Field, F., and Braverman, M. M. (1963). "Report of an Air Pollution Incident in New York City, November 1953," *Public Health Rept.* **78,** 1061–1065.

Gregory, J. (1970). "The Influence of Climate and Atmospheric Pollution on Exacerbations of Chronic Bronchitis," *Atmos. Environ.* **4,** 453–468.

Hamilton, A. and Hardy, H. L. (1949). *Industrial Toxicology,* 2nd ed. Hoeber, New York.

Hamilton, A. and Hardy, H. L. (1974). *Industrial Toxicology,* 3rd ed. Publishing Sciences Group, Acton, Mass.

Hammer, D. I., Miller, F. J., Stead, A. G., and Hayes, C. G. (1976). "Air Pollution and Childhood Lower Respiratory Disease. I: Exposure to Sulfur Oxides and Particulate Matter in New York (1972). In Asher J. Finkel and Ward C. Duel, Eds., *Clinical Implications of Air Pollution Research.* Publishing Sciences Group, Acton, Mass., pp. 321–337.

Holland, W. W., Reid, D. D., Seltser, R., and Stone, R. W. (1965). "Respiratory Disease in England and the United States: Studies of Comparative Prevalence," *Arch. Environ. Health* **10,** 338–343.

Holland, W. W., Halil, T., Bennett, A. E., and Elliott, A. (1969). "Factors Influencing the Onset of Chronic Respiratory Diseases," *Br. Med. J.* **1,** 205–208.

Irwig, L., Altman, D. G., Gibson, R. J. W., and Florey, C. du V. (1974). "Air Pollution: Methods to Study Its Relationship to Respiratory Disease in British School Children," paper given at the International Symposium on Recent Advances in the Assessment of the Health Effects of Environmental Pollution, Paris, June.

Kehoe, R. A., Machle, W. F., Kitzmiller, K., and Le Blanc, T. J. (1932). "On the Effects of Prolonged Exposure to Sulphur Dioxide," *J. Ind. Hyg.* **14,** 159–173.

Kiernan, K. E., Colley, J. R. T., Douglas, J. W. B., and Reid, D. D. (1976). "Chronic Cough in Young Adults in Relation to Smoking Habits, Childhood Environment and Chest Illness," *Respiration* **33,** 236–244.

Lambert, P. M. and Reid, D. D. (1970). "Smoking, Air Pollution, and Bronchitis in Britain," *Lancet* **1,** 853–857.

Laskin, S., Kuschner, M., and Drew, R. T. (1970). "Studies in Pulmonary Carcinogenesis." In M. G. Hanna, Jr., P. Nettescheim, and J. R. Gilbert, Eds., *Inhalation Carcinogenesis: Proceedings of a Biology Division, Oak Ridge National Laboratory Conference Held in Gatlinburg, Tenn., Oct. 8–11, 1969.* U.S. Atomic Energy Commission, Division of Technical Information, pp. 321–351.

Lave, L. B., and Seskin, E. P. (1970). "Air Pollution and Human Health: The Quantitative Effect, with an Estimate of the Dollar Benefit of Pollution Abatement, Is Considered," *Science* **169,** 723–733.

Lawther, P. J., Waller, R. E., and Henderson, M. (1970). "Air Pollution and Exacerbations of Bronchitis," *Thorax* **25,** 525–539.

Lebowitz, M. D. (1973). "A Comparative Analysis of the Stimulus-Response Relationship between Mortality and Air Pollution-Weather," *Envir. Res.* **6,** 106–118.

Lee, A. M., and Fraumeni, J. F. (1969). "Arsenic and Respiratory Cancer in Man: An Occupational Study," *J. Natl. Cancer Inst.* **42,** 1045–1052.

Lewey, F. H. (1941). "Neurological, Medical, and Biochemical Signs and Symptoms Indicating Chronic Industrial CS_2 Absorption," *Ann. Intern. Med.* **15,** 869.

Lewis, T. R., Amdur, M. O., M. D., Fritzhand, and Campbell, K. I. (1972). *Toxicology of Atmospheric Sulfur Dioxide Decay Products.* Environmental Protection Agency, Publ. AP-111, Research Triangle Park, North Carolina.

Love, G. J., Cohen, A. A., Finklea, J. F., French, J. G., Lowrimore, G. R., Nelson, W. C., and Ramsey, P. B. (1974). "Prospective Surveys of Acute Respiratory Disease in Volunteer Families: 1970–1971 New York Studies." In *Health Consequences of Sulfur Oxides: A Report from CHESS, 1970–1971.* U. S. Environmental Protection Agency, Office of Research and Development, EPA–650/1–74–004, U. S. Government Printing Office, Washington, D.C., pp.5–49 to 5–69.

Lunn, J. E., Knoweldon, J., and Handyside, A. J. (1967). "Patterns of Respiratory Illness in Sheffield Infant School Children," *Br. J. Prev., Med.* **21,** 7–16.

Lunn, J. E., Knoweldon, J., and Roe, J. W. (1970). "Patterns of Respiratory Illness in Sheffield Junor School children: A Follow-up Study," *Brit. J. Prev. Soc. Med.* **24,** 223–228.

Maritn, A. E. (1964). "Mortality and Morbidity Statistics and Air Pollution," *Proc. Roy. Soc. Med.* **57,** 969–975.

Martin, A. E. and Bradley, W. (1960). "Mortality, Fog and Atmospheric Pollution." *Mon. Bull. Ministry Health* **19,** 56–59.

McCabe, L. C. and Clayton, G. D. (1952). "Air Pollution by Hydrogen Sulfide in Poza Rica, Mexico," *Arch. Ind. Hyg. Occup. Med.* **6,** 199–213.

McKee, R. W., Kipper, C., Fountain, J. H., Riskin, A. M., and Drinker, P. (1943). "A Solvent Vapor Carbon Disulfide: Absorption, Elimination, Metabolism and Mode of Action," *J. Am. Med. Assoc.* **122,** 217–222.

Melville, G. N. (1970). "Changes in Specific Airway Conductance in Healthy Volunteers Following Nasal and Oral Inhalation of SO_2," *West Indian Med. J.* **19,** 231–235.

Meyer, G. W. (1976). "Environmental Respiratory Disease (Tokyo–Yokohama Asthma): The Case for Allergy." In Asher J. Finkel and Ward C. Duel, Eds., *Clinical Implications of Air Pollution Research.* Publishing Sciences Group, Acton, Mass., pp. 177–182.

Milby, T. H. (1962). "Hydrogen Sulfide Intoxication: Review of the Literature and Report of Unusual Accident Resulting in Two Cases of Nonfatal Poisoning," *J. Occup. Med.* **4,** 431–437.

Ministry of Health (1954). *Mortality and Morbidity during the London Fog of December 1952.* Rep. 95 on Public and Medical Subjects, Her Majesty's Stationery Office, London.

Minstry of Technology (1966). *National Survey of Smoke and Sulphur Dioxide: Instruction Manual.* Warren Spring Laboratory Stevenage, Herts.

Nadel, J. A., Salem, H., Tamplin, B., and Tokiwa, Y. (1965). "Mechanism of Bronchoconstriction during Inhalation of Sulfur Dioxide," *J. Appl. Physiol.* **20,** 164–167.

National Academy of Sciences (1975). *Air Quality and Stationary Source Emission Control.* Report by the Commission on Natural Resources, National Academy of Sciences, National Academy of Engineering, National Resource Council, U.S. Senate Committee on Public Works, Serial 94-4, U.S. Government Printing Office, Washington, D.C.

Nelson, W. C., Finklea, J. F., House, D. E., Calafiore, D. C., Hertz, M. B., and Swanson, D. H. (1974). "Frequency of Acute Lower Respiratory Disease in Children: Retrospective Survey of Salt Lake Basin Communities, 1967–1970." In *Health Consequences of Sulfur Oxides: A Report from CHESS, 1970–1971.* U.S. Environmental Protection Agency, Office of Research and Development, EPA-650/1-74-004, U.S. Government Printing Office, Washington, D.C., pp. 2–55 to 2–73.

Nesswetha, L. and Nesswetha, W. (1967). In H. Brieger and J. Teisinger, Eds., *Toxicology of Carbon Disulfide.* Excerpta Medica Foundation, Amsterdam.

Nurminen, M. (1976). "Survival Experience of a Cohort of Carbon Disulphide Exposed Workers from an Eight-Year Prospective Follow-up Period," *Int. J. Epidemiol.* **5,** 179.

Paluch, E. (1948). "Two Outbreaks of Carbon Disulfide Poisoning in Rayon Staple Fiber Plants in Poland," *J. Ind. Hyg. Toxicol.* **30,** 37.

Patty, F. A. (1963). In D. W. Fassett and D. D. Irish, Eds., *Industrial Hygiene and Toxicology,* Vol. II; *Toxicology,* 2nd ed., rev. Wiley-Interscience, New York.

Peacock, P. R. and Spence, J. B. (1967). "Incidence of Lung Tumors in LX Mice Exposed to (1) Free Radicals; (2) SO_2," *Br. J. Cancer* **21,** 606–618.

Pemberton, J. and Goldberg, C. (1954). "Air Pollution and Bronchitis," *Br. Med. J.* **2,** 557–570.

Petrilli, F. L., Agnese, G., and Kanitz, S. (1966). "Epidemiologic Studies of Air Pollution Effects in Genoa, Italy," *Arch. Environ. Health* **12,** 733--740.

Phelps, H. W. (1965). "Follow-up Studies on Tokyo-Yokohama Respiratory Disease," *Arch. Environ. Health* **10,** 143–147.

Poda, G. A. (1966). "Hydrogen Sulfide Can Be Handled Safely," *Arch. Environ. Health* **12,** 795.

Prokhorov, Y. D. and Rogov, A. A. (1959). "Histopathological and Histochemical Changes in the Organs of Rabbits after Prolonged Exposure to Carbon Monoxide, Sulfur Dioxide and Their Combinations," *Gig. Sanit.* **24,** 22–26 (Russian). Also in *USSR Literature on Air Pollution and Related Occupational Diseases-a Review,* Vol 5 translated by B. S. Levine. U.S. Public Health Service, 1961, pp. 81–86.

Reid, L. (1963). "An Experimental Study of Hypersecretion of Mucus in the Bronchial Tree," *Brt. J. Exp. Pathol.* **44,** 437–445.

Romanoff, A. (1939). "Sulfur Dioxide Poisoning as a Cause of Asthma," *J. Allergy* **10,** 166–169.

Ryazanov, V. (1952). *Limits of Allowable Concentrations by Atmospheric Pollutants,* Book 1 (translated by B. S. Levine). U.S. Department of Commerce, Washington, D.C.

Rylander, R. (1969). "Alterations of Lung Defense Mechanisms against Airborne Bacteria," *Arch. Environ. Health* **18,** 551–555.

Schimmel, H. and Greenburg, L. (1972). "A Study of the Relation of Pollution to Mortality: New York City, 1963–1968," *J. Air. Pollut. Control Assoc.* **22,** 607–616.

Schrenk, H. H., Heimann, H., Clayton, G. D., Gafafer, W., and Wexler, H. (1949). *Air pollution in Donora, Pennsylvania: Epidemiology of the unusual smog episode of October 1948.* Public Health Bull. 306, U.S. Government Printing Office, Washington, D.C., 173 pp.

Science and Public Policy Program (1975). *Energy Alternatives: A Comparative Analysis.* University of Oklahoma, Norman, Okla. U.S. Govt. Printing Office, Washington, D.C., Stock 041-011-00025-4.

Scott, J. A. (1963). "The London Fog of December 1962," *Med. Officer* **109,** 250–252.

Shy, C. M., Hasselblad, V., Burton, R. M., Nelson, C. J., and Cohen, A. (1973). "Air Pollution Effects on Ventilatory Function of U.S. School Children: Results of Studies in Cincinnati, Chattanooga, and New York," *Arch. Environ. Health* **27,** 124–128.

Skalpe, T. O. (1964). "Long-Term Effects of Sulphur Dioxide Exposures in Pulp Mills," *Br. J. Ind. Med.* **21,** 69–73.

Smith, R. P. and Gosselin, R. E. (1966). "On the Mechanism of Sulfide Inactivation by Methemoglobin," *Toxicol. Appl. Pharmacol.* **8,** 159.

Snell, R. E. and Luchsinger, P. C. (1969). "Effects of Sulfur Dioxide on Expiratory Flow Rates and Total Respiratory Resistance in Normal Human Subjects," *Arch. Environ. Health* **18,** 693–698.

Sultz, H. A., Feldman, J. G., Schlesinger, E. R., and Mosher, W. E. (1970). "An Effect of Continued Exposure to Air Pollution on the Incidence of Chronic Childhood Allergic Disease," *Am. J. Public Health* **60,** 891–900.

Tiller, J. R., Schilling, R. S. F., and Norris, J. N. (1968). "Occupational Toxic Factor in Mortality from Coronary Heart Disease," *Br. Med. J.* **4,** 407–411.

Toyama, T. (1964). "Air Pollution and Its Health Effects in Japan," *Arch. Environ. Health* **8,** 153–173.

Tsunetoshi, Y., Shimizu, T., Takahashi, H., Tchinosawa, A., Ueda, M., Nakayama, N., and Yamagata, Y. (1971). "Epidemiological Study of Chronic Bronchitis with Special Reference to Effects of Air Pollution," *Int. Arch. Arbeitsmed.* **29,** 1–27.

U.S. Department of Health, Education, and Welfare (1970). *Air Quality Criteria for Sulfur Oxides.* Public Health Service, National Air Pollution Control, Publ. AP-50, U.S. Government Printing Office, Washington, D.C.

U.S. Department of Health, Education, and Welfare (1974). *Criteria for a Recommended Standard: Occupational Exposure to Sulfur Dioxide.* National Institute for Occupational Safety and Health.

Vigliani, E. C. (1954). "Carbon Disulfide Poisoning in Viscose Rayon Factories," *Br. J. Ind. Med.* **11,** 235–244.

Watanabe, H. (1965). "Air Pollution and Its Health Effects in Osaka," paper presented at 58th Annual Meeting of Air Pollution Control Association, Toronto, Canada, June 20–24.

Watanabe, H. and Kaneko, F. (1971). "Excess Death Study of Air Pollution." In H. M. Englund and W. T. Beery, Eds., *Proceedings of the Second International Clean Air Congress.* Academic Press, New York, pp. 199–200.

Weill, H., Ziskind, M. M., Debres, V., Lewis, R., Horton, R. J. M., and McCaldin, R. O. (1964). "Further Observations on New Orleans Asthma," *Arch. Environ. Health* **8,** 184–187.

Weir, F. W. and Bromberg, P. A. (1972). "Further Investigation of the Effects of Sulfur Dioxide on Human Subjects." In *Abstracts of the Society of Toxicology,* p. 87, March.

Winklestein, W., Kantor, S., Davis, E. W., Maneri, C. S., and Mosher, W. E. (1967). "The Relationship of Air Pollution and Economic Status to Total Mortality and Selected Respiratory System Mortality in Men. I: Suspended Particulates," *Arch. Environ.* Health **14,** 162–171.

Winklestein, W.,Kantor, S., Davis, E. W., Maneri, C. S., and Mosher, W. E.

(1968). "The Relationship of Air Pollution and Economic Status to Total Mortality and Selected Respiratory System Mortality in Men. II: Oxides of Sulfur," *Arch. Environ. Health* **16,** 401–405.

World Health Organization (1972). *Air Quality Criteria and Guides for Urban Air Pollutants.* Report of a WHO Expert Committee, World Health Organization Tech. Rept. Series 506, Geneva.

Yoshida, R., Motomiya, K., Saito, H., and Funabashi, S. (1976). "Clinical and Epidemiological Studies on Childhood Asthma and Air Polluted Areas in Japan." In Asher J. Finkel and Ward C. Duel, Eds. *Clinical Implications of Air Pollution Research,* Publishing Sciences Group, Acton, Mass., pp. 165-176.

Zeidberg, L. D., Prindle, R. A., and Landau, E. (1961). "The Nashville Air Pollution Study. I: Sulfur Dioxide and Bronchial Asthma: A Preliminary Report," *Am. Rev. Respir. Dis.* **84,** 489–503.

Zeidberg, L. D. Horton, R. J. M., and Landau, E. (1967). "The Nashville Air Pollution Study. V: Mortality from Diseases of the Respiratory System in Relation to Air Pollution," *Arch. Environ. Health* **15,** 214–224.

4

EFFECTS OF AIRBORNE SULFUR POLLUTANTS ON PLANTS

Samuel N. Linzon

Phytotoxicology Section, Air Resources Branch, Ontario Ministry of the Environment, Toronto, Ontario

1. INTRODUCTION

This chapter will discuss the effects on plant life caused by airborne sulfur pollutants: sulfur dioxide (SO_2), hydrogen sulfide (H_2S), and sulfuric acid (H_2SO_4), and by acid precipitation. The most information is available on SO_2, and the effects of this gas on forest trees, horticultural plants, and lichens will be described. General reviews of the effects of various sulfur pollutants have been written by Thomas (1951), Daines (1968), Barrett and Benedict (1970), Guderian and Van Haut (1970), Van Haut and Stratmann (1970), Linzon (1972), the U.S. Environmental Protection Agency (1973), and Heck and Brandt (1975).

This chapter on the effects of sulfur pollutants is based on personal experience and a literature review of published papers to the end of 1976.

The effects of airborne sulfur pollutants on plants may be divided into three categories: acute, chronic, and subtle (physiological and biochemical). This chapter discusses mainly the visible acute and chronic effects on plants, while another chapter (see Hällgren, this volume) addresses itself to the effects of SO_2 on physiological and biochemical processes in plants, especially on the photosynthetic mechanism.

Environmental and other factors influence the responses of plants to sulfur pollutants, and these factors and the kinds of plants that are most susceptible will be discussed. The effects of SO_2 on host-parasite relationships, with respect to increasing or decreasing virulence of the pests or changes in the susceptibility of the host plant, will be described. Also, since a particular air pollutant rarely occurs alone in the ambient atmosphere, the effects of SO_2 in combination with other pollutants will be examined.

The concentrations of SO_2 reported in this chapter are either in parts per million (ppm) or micrograms per cubic meter ($\mu g/m^3$) depending on the usage in the literature reviewed. Where conversion is required 1 ppm $= 2600\ \mu g/m^3$ or $1\ \mu g/m^3 = 3.82 \times 10^{-4}$ ppm based on 25°C and 760 mm Hg.

Sulfur is an essential element for plants. It is a constituent of the amino acids cystine, cysteine, and methionine, which are three of the building blocks of plant proteins. It is also a constituent of the plant vitamins thiamine and biotin, and of other biochemical constituents such as glutathione, coenzyme A, and cytochrome *c*. Normally, sulfur is taken up by plants from soil in the sulfate form and assimilated into various compounds usually after being chemically reduced. Sulfur dioxide absorbed from the air has been shown to rapidly undergo oxidation to sulfates inside plant tissues. De Cormis (1969) found that tomato plants soon after exposure to $^{35}SO_2$ had 98% of the ^{35}S in the form of sulfates. De Cormis also reported the subsequent evolution of $H_2{}^{35}S$, especially if the plants were illuminated. It is apparent that sulfur in its various forms is necessary in the intermediary metabolism of plants.

Background levels of sulfur dioxide in the atmosphere are 4 $\mu g/m^3$ or less (Georgii, 1970) and there is no evidence that these levels can cause harm to the environment. However, sulfur in excessive amounts can have a deleterious effect on plant life, and this chapter reviews these effects and the concentrations of sulfur pollutants over various time periods that can cause effects.

Sulfur dioxide is emitted from various industrial sources, and the adverse effects on vegetation in the vicinity of these sources have been well documented. Some of the industrial sources are zinc and lead smelters (Katz et al., 1939), nickel and copper smelters (Linzon, 1958a), iron concentrators (Gordon and Gorham, 1963), petroleum refineries (Linzon, 1965), pulp and paper mills (Linzon et al., 1973), and fossil fuel thermal generating stations (Jones et al., 1973; Linzon, 1975).

Estimates of the cost of air pollution damage to vegetation have varied considerably. Wolozin and Landau (1966) estimated an annual loss of \$500 million in the United States from air pollution injury to crop plants with approximately one third of this total loss attributable to SO_2 injury. Waddell (1970) estimated that sulfur oxide damage to plants and the environment in the United States was between \$35 million and \$75 million annually. In 1973, Barrett and Waddell estimated the costs of direct damage to vegetation from all pollutants to be \$120 million annually in the United States with sulfur oxides accounting for $13 million of the total. The total estimated cost of sulfur oxide damage, which included effects on human health, property, and materials, was over $8 billion.

Although direct vegetation losses represent only a small part of the total economic losses caused by sulfur pollution, the effects on vegetation are important for other reasons, as follows:

1. Vegetation is more sensitive to SO_2 than is human health. Thus, by establishing air quality standards for SO_2 to safeguard vegetation, protection to human health should ensue. At the same time, however, it must be recognized that human health is more susceptible to other sulfur pollutants such as sulfates and sulfuric acid aerosols and to a combination of SO_2 and particulate matter than to pure SO_2 alone.
2. The identification of visible SO_2 injury to plants assists control officials to determine the source of the phytotoxic emissions for industrial abatement purposes.
3. Both animals and human beings are dependent on vegetation for their sustenance. Thus decreases in agricultural crop yields caused by sulfur pollutants ultimately affect human welfare. Damage to forests by sulfur pollutants not only interferes with the ecosystem web but also reduces the quantity of wood and paper products available to man. Furthermore, sulfur pollutant damage to ornamentals diminishes the aesthetic quality of the landscape.

2. SULFUR DIOXIDE DOSES RELATED TO PLANT EFFECTS

The first visible evidence of SO_2 injury to plants is discernible in the foliage. The stems, bud, and reproductive parts of plants are visibly more resistant to SO_2 than are the foliar parts. It is important to stress the difference between acute, chronic, and subtle effects of SO_2 on plant life. Sulfur dioxide enters leaves mainly through the stomata and is toxic to the metabolic processes taking place in the mesophyll cells (Linzon, 1972). Acute injury is caused by a rapid accumulation of bisulfite and sulfite. When the oxidation product, sulfate, accumulates beyond a threshold value that the plant cells can tolerate, chronic injury occurs. It is estimated that sulfate is about 30 times less toxic than sulfite (Thomas, 1951).

Definitions vary for acute and chronic injury and for subtle effects caused by air pollutants, and to avoid different interpretations for these terms with respect to SO_2 the following definitions will apply in this chapter as interpreted by this author. *Acute injury* is macroscopic necrotic injury to plant tissue visible within hours or days after exposure to short-term (less than 24 hr) high concentrations of SO_2. *Chronic injury* is macroscopic chlorotic injury (sometimes changing to necrotic injury) to plant tissue usually developing over a long period of time (from over 1 day to 1 or more years) from exposure to variable concentrations of SO_2. Subtle effects are measured physiological or biochemical changes, and/or reductions in plant growth or yield in the absence of macroscopic injury.

Table 1 differentiates the three types of plant effects (acute, chronic, and subtle) caused by SO_2.

2.1. Acute Effects

Acute injury to broad leaves takes the form of lesions on both surfaces, usually occurs between veins, and is often more prominent toward the petiole. The injury is local. The metabolic processes are completely disrupted in the dead or necrotic areas with the surrounding green tissue remaining functional. The tissue immediately adjacent to the veins is extremely resistant. In some cases, injury can occur on the margins of the leaves. Young leaves rarely display necrotic markings, whereas fully expanded leaves are most sensitive to acute SO_2 injury. The oldest leaves are moderately sensitive. In monocotyledonous leaves the injury can occur at the tips and in lengthwise areas between the main veins. In conifers acute injury usually appears as a bright orange-red tip necrosis on current-year needles, often with a sharp line of demarcation between the injured tips and the normally green bases. Occasionally the injury may occur as bands at the tip, middle, or base of the needles (Linzon, 1972).

The anatomical effects of acute injury on foliage can be seen under the microscope (Katz and Shore, 1955). Initially there are water-soaked, flaccid areas of diffuse, grayish green coloration—the chlorophyll appears to have diffused from the chloroplasts into the cytoplasm. This is followed by desiccation and shrinkage of the affected cells. The green pigments are decomposed, and the affected leaf area assumes a bleached, ivory, tan, orange-red, reddish brown, or brown appearance, depending upon the plant species, the time of the year, and the weather conditions.

Acute injury to forest trees in the field has been attributed to the following SO_2 doses: 0.95 ppm for 1 hr, 0.55 ppm for 2 hr, 0.35 ppm for 4 hr, and 0.25 ppm for 8 hr (Dreisinger, 1965). For acute injury to occur,

Table 1. Differentiation between Three Types of Plant Effects Caused by Sulfur Dioxide

Acute Effects: Caused by High Concentrations of SO_2 during Short-Term Exposures (less than 24 hr)	Chronic Effects: Caused by Variable Concentrations of SO_2, Usually over a Long Period of Time	Subtle Effects: Caused by Variable Concentrations of SO_2
Necrotic lesions on foliage of the current year	Chlorotic markings on foliage, sometimes developing to necrosis; on perennial conifers usually on foliage older than current year	No foliar markings
Foliar lesions appear within hours or days after exposure	Foliar injury usually develops slowly over a long period of time	May be possible to measure effects on physiological processes—photosynthesis, respiration, and transpiration; on biochemical processes—enzyme activities and chemical compositions; on pollen tube elongation; or on pollen germination
May cause premature abscission of injured foliage	On perennial conifers usually cause premature abscission of oldest affected foliage	
May cause reduction in growth and loss in yield	May cause reduction in growth and loss in yield	May cause reduction in growth and loss in yield
Rarely result in plant mortality, unless from recurring acute fumigations	May result in plant mortality from slowly developing injuries (especially on perennial conifers and lichens)	No plant mortality

other environmental and plant factors are important. These include sunlight, moderate temperature, high relative humidity, adequate soil moisture, and plant genotype and stage of growth. If these factors are not conducive to injury, the plants will escape harm even in the presence of doses two or three times higher than those noted above. Conversely, if the predisposing factors are especially conducive to plant injury, the SO_2 doses could be reduced by about 25% to 0.70 ppm for 1 hr, 0.40 ppm for 2 hr, 0.26 ppm for 4 hr, and 0.18 ppm for 8 hr (Dreisinger and McGovern, 1970).

Katz and McCallum (1939) subjected various plant species to different SO_2 concentrations over a number of exposure periods. The threshold for injury was found to be 0.30 ppm SO_2 for 8 hr for western larch *(Larix occidentalis)* and 0.14 ppm SO_2 for 12 hr for Douglas fir *(Pseudotsuga taxifolia)*.

In a 1-hr experimental exposure Zimmerman and Crocker (1934) reported that 0.66 ppm SO_2 injured buckwheat *(Fagopyrum esculentum)*. Karnosky (1976) produced acute injury to foliage of trembling aspen *(Populus tremuloides)* in artificial fumigations with 0.35 ppm SO_2 for a period of 3 hr. Metcalfe (1941) damaged begonia varieties in fumigations of 0.25 ppm SO_2 for 1 hr under very humid conditions. Berry (1967) reported injuring foliage of eastern white pine *(Pinus strobus)* at a concentration of 0.25 ppm SO_2 in 1 hr. Later, Berry (1971, 1974) injured several pine species in artificial fumigations utilizing 0.25 ppm SO_2 in 2-hr exposure periods. Murray et al. (1975) induced moderate to severe injury on several Kentucky bluegrass cultivars *(Poa pratensis)* in artificial fumigations of 0.20 ppm SO_2 for 2 hr.

The acute effects referred to above may be considered responses to medium doses of SO_2. In the literature there are reports of responses of plants to much lower and to substantially higher SO_2 doses. For example in artificial fumigation experiments SO_2 doses as low as 0.03 ppm for 1 hr (Costonis, 1971) and 0.025 ppm for 6 hr (Houston, 1974) have been reported to injure extremely sensitive strains of eastern white pine; and conversely SO_2 doses as high as 3.0 ppm for 1 hr (O'Connor et al., 1974), 2.0 ppm for 2 hr (Hill et al., 1973) and 2.5 ppm for 3 hr (Brennan and Leone, 1972) have been found to be required in order to injure tolerant plant species as *Acacia pruinosa,* several native desert plant species, and several chrysanthemum varieties, respectively. In interpreting the results of the very low levels or high levels of SO_2 used in the short-term artificial fumigations, careful consideration must be given to the experimental conditions as outlined in the referenced papers.

Sulfur dioxide concentrations that caused acute injury to vegetation in short-term periods are listed in Table 2. These SO_2 doses were in many cases the minimum found by the investigators to cause injury. Higher concentrations of SO_2 or longer periods of time usually caused more severe effects. However, in some of the reports the concentrations and the exposure periods used were arbitrarily selected for the purposes of the experiment; and although injury occurred, the SO_2 doses reported were not necessarily threshold levels for the plant species tested.

Table 2. Dose Responses to Sulfur Dioxide—Acute Effects

SO_2 Concentration (ppm)[a]	Exposure Period (hr)	Plant Response	Reference
		A—Responses to Low SO_2 Doses	
0.03	1	Injury to sensitive eastern white pine	Costonis (1971)
0.025	6	Injury to sensitive eastern white pine	Houston (1974)
0.05–0.12	4–8	Injury to peanut	Applegate and Durrant (1969)
		B—Responses to Medium SO_2 Doses	
0.25	1	Injury to begonia	Metcalfe (1941)
0.25	1	Injury to eastern white pine	Berry (1967)
0.20	2	Injury to several Kentucky blue-grass cultivars	Murray et al. (1975)
0.25	2	Injury to eastern white, red, and jack pines	Berry (1971)
0.25	2	Injury to Virginia, shortleaf, slash, and loblolly pines	Berry (1974)
0.25	4	Injury to broccoli	Tingey et al. (1973)
0.35	3	Injury to trembling aspen	Karnosky (1976)
0.54	3	Injury to mountain ash	Spierings (1967)
0.66	1	Injury to buckwheat	Zimmerman and Crocker (1934)
0.95	1	Injury to foliage of forest trees[a]	Dreisinger (1965)
0.55	2		
0.35	4		
0.25	8		
0.70	1	Injury to forest trees under sensitive environmental conditions[b]	Dreisinger and McGovern (1970)
0.40	2		
0.26	4		
0.18	8		
0.30	8	Injury to western larch	Katz and McCallum (1939)
0.14	12	Injury to Douglas fir	
		C—Responses to High SO_2 Doses	
2.0	2	Injury to 10 of 87 native desert species	Hill et al. (1973)
2.0	2	Injury to Chinese elm	Temple (1972)
2.0	4	Injury to 10 weed species	Benedict and Breen (1955)

Table 2. (Continued)

SO_2 Concentration (ppm)[a]	Exposure Period (hr)	Plant Response	Reference
2.5	3	Injury to 7 of 16 chrysanthemum varieties	Brennan and Leone (1972)
3.0	1	Injury to *Acacia pruinosa*	O'Connor et al. (1974)

[a] 1 ppm = 2600 $\mu g/m^3$ at 25°C and 760 mm Hg.
[b] Field observations. Other references report plant responses observed in artificial fumigation experiments.

Table 2 is separated into three parts; A—Responses to Low SO_2 Doses, B—Responses to Medium SO_2 Doses, and C—Responses to High SO_2 Doses. Not all the literature on SO_2 dose response is cited in this table but pertinent references are included.

2.2. Chronic Effects

Chronic injury appears as a yellowing or chlorosis of the leaf, sometimes from lower to upper surfaces on broad leaves. Occasionally only a bronzing or silvering will occur on the undersurface of the leaves (Brisley et al., 1959). In perennial conifers, chronic injury affects older needles and appears as a yellowish green color, that changes to reddish brown, starting at the tips and developing toward the base (Linzon, 1969). The rate of metabolism is reduced in leaves displaying chronic injury.

Chronic injuries develop slowly on coniferous perennial foliage, with the greatest increase in injuries in the Sudbury sulfur-fume effects area occurring on the 1-year-old needles of eastern white pine trees (Linzon, 1971). Continued chronic injury to perennial foliage of coniferous trees results in premature needle abscission, reduced radial and volume growth, and early death of the trees.

Long-term chronic effects on trees are related to a variety of SO_2 exposures which include short-term high concentrations, short-term and long-term periods of sublethal concentrations, and even SO_2-free periods when the plant life can recuperate by translocating and assimilating accumulated sulfur. To correlate the chronic effects on forest trees with atmospheric SO_2 levels, it is preferable to use the average concentration for the total period of exposure, rather than the average of SO_2 fumigation periods only. Attacks by intermittent fumigations are unpredictable as to frequency of occurrence, concentration of SO_2, and duration of the fumigation at any particular location. Chronic effects develop slowly, and the response of the receptor is influenced by other environmental factors in addition to SO_2. Moreover, the vegetation affected is exposed throughout the growing season to the vagaries of the environment, (sun, rain, wind, and drought) all of which have an influence on plant response to SO_2.

The effects of SO_2 on vegetation at Sudbury, Canada (Linzon, 1958a, 1971; Dreisinger, 1965), may be compared to the effects observed at Biersdorf, Germany (Guderian and Stratmann, 1962). Table 3 compares the two areas.

Table 4 compares the SO_2 data for Sudbury and Biersdorf for the years 1959 and 1960. Starting at a distance from the sources within the severe injury zones, Skead (Sudbury) and Station III (Biersdorf), and continuing out to the zones of light injury, Kukagami Lake and Station IV, and then with further distance to the zone of very light injury, Grassy Lake and Station V, it is apparent that the frequency of occurrence of short-term high concentrations of SO_2 was much greater at Biersdorf. For example, close to the source at Biersdorf (Station I, 325 m SE) a maximum 0.5-hr concentration of 6.6 ppm SO_2 was recorded in 1960, whereas at Sudbury the maximum 0.5-hr SO_2 concentration recorded during a 10 year period, 1954 to 1963, was 3.64 ppm which occurred at Skead (16 miles NE); the maximum concentration recorded at Garson, a station closer to Sudbury (5 miles NE) was 3.06 ppm (Dreisinger, 1965). It should be pointed out also that instruments with a low precision were used to measure atmospheric SO_2 at Biersdorf, with concentrations below 0.1 ppm not being taken into account (Stratmann, 1963). At Sudbury the SO_2 instruments had a detectability level down to 0.01 ppm.

The U.S. Environmental Protection Agency (1973), in reviewing the Biersdorf investigation reports, concluded that "many short-term high concentration episodes were responsible for the injury and growth reductions that occurred." From Table 4 this conclusion would appear to be correct. However, it should be pointed out that the Biersdorf studies were conducted in the vicinity of an iron ore concentrator that emitted only about 10 tons of SO_2 daily (Table 3). The source was located in a valley extending in a northwest-southeast direction, and from the bottom of the valley to the top of the ridges on the sides was a distance of 200 m. The tops of the chimney stacks were only 50 to 70 m above the bottom of the valley. Although vegetation injury was severe in

Table 3 Comparison between Vegetation Studies at Sudbury (Canada) and at Biersdorf (Germany), 1960

Item	Sudbury	Biersdorf
Source	3 nickel-copper smelters	1 iron concentrator
Location of source	Not in a valley—undulant topography	In a valley with ridges 200 m high
Stack height	Ca. 160 m above ground level	Ca. 60 m above bottom of valley
Tons of ore treated annually	15 million	0.5 million
Sulfur content in ore (%)	32–33	0.2–2.0
Emission (tons SO_2/day)	6000	10
Maximum 0.5-hr SO_2 concentration (ppm)	3.64 (1954–1963)	6.6 (1959–1960)
Type of vegetation effects study	Natural forest vegetation	Experimental plantings in wooden buckets
Degree of damage	Severe injury to a distance of 25 miles northeast of Sudbury (located central to three smelters); sharp improvement beyond 25 miles to mostly no injury beyond 40 miles	Very heavy to complete injury close to source (325 m), gradually improving with distance to mostly no injury beyond 2 km
Size of area affected	About 700 square miles—severe to moderate injury; an additional 1600 square miles—sporadic injury	About 2 square miles total

the immediate vicinity of the Biersdorf plant, SO_2 damage extended for only a little more than 2 km from the source (Guderian and Stratmann, 1962). The air-monitoring and experimental vegetation stations were established in five locations at increasing distances in a southeast direction, with the fifth and last station in the SO_2-affected area located only 1900 m from the source. Control stations were located at distances of 4500 and 6000 m from the iron concentrator. Thus high concentrations of SO_2 occurred at the vegetation stations in the affected area, but their occurrences were infrequent as seen by the low frequency of SO_2 values

Table 4. Comparison between Sudbury[a] and Biersdorf[b] for SO_2 Recorded during the Growing Seasons of 1959 and 1960

Air Sampling Station	Maximum 0.5-hr SO_2 Concentration (ppm)	Average Concentration for SO_2 Fumigation Periods Only (ppm)	Frequency of SO_2 Readings (Average for 2 Years) (%)	Average SO_2 Concentration[c] for Total Measurement Period (Including Zero Readings) (ppm)	Vegetation Effects
Sudbury					
Skead (16 miles NE)	2.5	0.20	21.9	0.044	Acute and chronic injury
Kukugami Lake (27 miles NE)	0.96	0.11	18.7	0.020	Mostly chronic and little acute injury
Grassy Lake (40 miles NE)	0.52	0.09	12.0	0.011	Very little chronic injury
Biersdorf					
III (725 m SE)	2.4	0.24	20.5	0.049	Heavy damage
IV (1350 m SE)	1.9	0.18	11.5	0.021	Light damage and partly no damage
V (1900 m SE)	1.7	0.15	8.0	0.012	Very light damage, mostly no damage

[a] Data provided by P. C. McGovern.
[b] Guderian and Stratmann (1962).
[c] Calculated.

monitored: less than 10% of the time at a distance of 1900 m from the source.

The situation in the Sudbury area can be compared to that at Biersdorf, but the events were completely different. At Sudbury three large nickel and copper smelters discharged about 6000 tons of SO_2 daily into the atmosphere (Table 3). Damage to white pine forests was severe, with both acute and chronic effects occurring to a distance of about 25 miles northeast of Sudbury (Linzon, 1958a). Beyond this distance the condition of white pine improved rapidly where the concentrations of SO_2 were consistently lower than 0.25 ppm.

The SO_2 data and the occurrence of potentially injurious fumigations (PIF) for the stations northeast of Sudbury for a 10-year period (1954 to 1963) are shown in Table 5. From the table it can be seen that the average frequency of fumigations with SO_2 over 0.25 ppm was only 1.1% of the time at Kukagami Lake for the 10-year period, with the average SO_2 concentration for the total measurement period being 0.017 ppm at this station. At Skead on the other hand, the average frequency of SO_2 readings over 0.25 ppm was 8.56% for the 10-year period, with the average SO_2 concentration for the total measurement period being 0.045 ppm. More potentially injurious fumigations occurred at Skead than at Kukagami Lake during the 10-year period (Dreisinger, 1965). There were 224 days with PIF at Skead, or an average of about 22 days during each growing season, whereas at Kukagami Lake the total was 23 days, or an average of about 2 days during each growing season, when acute injury to vegetation could occur. Actually, tree observations revealed for the 10-year period that acute injury occurred during 11 of the 23 PIF at Kukagami Lake, or only on an average of about once per year.

Forest injury in the vicinity of Kukagami Lake showed, based on long-term studies conducted at Portage Bay 25 miles northeast of Sudbury (Linzon, 1971) infrequent acute injury to the current year's foliage on white pine trees, but a gradual increase in chronic injury to the 1-year-old foliage as a result of continual exposure to concentrations of SO_2 below 0.25 ppm. At Kukagami Lake SO_2 concentrations below 0.25 ppm occurred during 15% of the total of 16% of the time that SO_2 was measured, showing the importance of all concentrations of SO_2 in the development of chronic injury.

Table 6 shows the results of forest studies conducted over a 10-year period (1953 to 1963) in the sulfur-fume effects area near Sudbury (Linzon, 1971). Very little acute injury to current year's needles occurred in the Sudbury forests northeast of Sudbury. The greatest increase in foliar injuries during the growing season was found for the 1-year-old needles. The expression of chronic injury to the older foliage

Table 5. Sulfur Dioxide Measurements in the Sudbury Area for a 10-Year Period (1954–1963)[a]

Air Sampling Station (Distance and Direction from Sudbury)	Average Number of Days per Year with PIFs[b]	Maximum 0.5-hr SO_2 Concentration (ppm)	Average Frequency of SO_2 Readings (%)			Average SO_2 Concentration for Total Measurement Period (ppm)
			Over 0.25 ppm	Trace to 0.25 ppm	Zero	
Skead (16 miles NE)	22.4	3.64	8.56	12.24	79.20	0.045
Kukugami Lake (27 miles NE)	2.3	1.24	1.10	15.10	83.80	0.017
Grassy Lake (40 miles NE)	0.4	0.63	0.34	9.96	89.70	0.008

[a] Data from Dreisinger (1965).

[b] PIF = potentially injurious fumigation (SO_2 concentrations sufficiently high to cause acute injury to vegetation).

Table 6. Degree of Forest Damage at Various Distances from Sudbury Smelters[a]

Forest Sampling Station[a] (Distance and Direction from Sudbury)	Trees with Current Year's Foliage Injured in August 1963 (%)	Trees with 1-Year-Old (1962) Foliage Injured		Trees with 2-Year-Old Foliage		Net Annual Average Gain or Loss in Total Volume, 1953–1963 (%)	Annual Average Mortality, 1953–1963 (%)	Degree of SO_2 Damage	Average SO_2 Concentration for Total Measurement Period 1954–1963[b] (ppm)
		June 1963 (%)	August 1963 (%)	Injured in June 1963 (%)	Lacking in August 1963 (%)				
West Bay (19 miles NE)	2.0	38.0	77.9	96.0	20.6	−1.3	2.6	Acute and chronic injury	0.045
Portage Bay (25 miles NE)	1.1	21.5	55.6	77.0	15.2	−0.5	2.5	Mostly chronic and little acute injury	0.017
Grassy to Emerald Lake (40–43 miles NE)	0.4	2.5	16.7	37.5	9.1	+1.8	1.4	Very little chronic injury	0.008
Lake Matinenda (93 miles W)	0.6	0.3	2.1	10.1	3.9	+2.1	0.5	Control: no SO_2 injury	0.001[c] (Sturgeon Falls)
Correlation coefficient (r)	0.96*	0.96*	0.93**	0.90**	0.94**	0.90**	0.81		

[a] Linzon (1971).
[b] Dreisinger (1965).
[c] Data for 5-month growing season—1971.
* $p < 0.05$.
** $p < 0.10$.

was different from the appearance of acute injury to the current year's foliage. The long-term chronic effects resulted in chlorosis of the older foliage which developed slowing into necrosis, starting at the tips and progressing basepitally. The continuous development of injuries on the older needles was reflected in early abscission of the oldest foliage, reduced radial and volume growth, and premature death of the trees. Also shown in the table are the correlation coefficients *(r)* between average annual SO_2 concentrations and different injurious effects observed on white pine trees. All the effects tabulated were significant at the 10% level or less except for tree mortality, which may be explained by the high incidence of trees dying in both the West Bay and Portage Bay areas that were the forest sampling stations located closest to the SO_2 sources in line with the prevailing wind.

It is apparent that chronic effects on forest growth were prominent (Linzon, 1971) where SO_2 air concentrations averaged 0.017 ppm (44 $\mu g/m^3$), the arithmetic mean for the total 10-year measurement period, and chronic effects were slight where SO_2 annual concentrations averaged 0.008 ppm (21 $\mu g/m^3$). The long-term chronic effects on coniferous trees were related to a variety of SO_2 exposures, which included short-term high concentrations, short-term and long-term periods of sublethal concentrations, and even SO_2-free periods in which the trees could recuperate by translocating and assimilating accumulated sulfur.

Chemical analyses of vegetation (Table 7) in the Sudbury area showed the accumulation and buildup of sulfur in foliage in vegetation growing close to the SO_2 sources. Also shown in the table are the correlation coefficients *(r)* between average atmospheric SO_2 concentrations and the average contents of total sulfur analyzed in the foliage of jack pine and white pine birch trees. The accumulation of sulfur in the foliage was significant at the 5% level or less. Both high concentration and low concentration fumigations of SO_2 over varying periods of time contributed to the accumulation of sulfur in coniferous foliage which lead to the development of chronic injury in older needles (Linzon, 1973a).

In 1973, 10 years after the first 10-year period (1953 to 1963) of forest studies reported by Linzon (1971), McGovern and Balsillie (1974a) conducted a field survey of forest injury in the area northeast of Sudbury and found chronic effects similar to those described by Linzon.

In Czechoslovakia Materna et al. (1969) reported the occurrence of moderate chronic injury to foliage of spruce trees at Celna, under the influence of an average concentration of SO_2 of 0.019 ppm (50 $\mu g/m^3$) for the years 1966 and 1967. In this report the authors stressed the importance of both long-term concentrations of SO_2 and other environmental factors in producing injury to plants.

Table 7. Results of Chemical Analyses for Total Sulfur (%) in Foliage at Various Distances from Sudbury Smelters—1971[a]

Location	Collection Month, 1971	Total Sulfur Content in Foliage (%)			Average Atmospheric SO_2 Concentrations for Total Measurement Period, June, July, and August 1971 (ppm)
		Jack Pine			
		Current Foliage	1-Year-Old Foliage	White Birch Foliage	
Skead (16 miles NE)	June	—	0.19	0.33	
	July	—	0.20	0.61	
	August	0.23	0.28	0.68	
	Average[b]	0.23	0.22	0.54	0.045
Kukagami Lake (27 miles NE)	June	—	0.15	0.21	
	July	—	0.22	0.26	
	August	0.15	0.14	0.40	
	Average[b]	0.15	0.17	0.29	0.020
Grassy Lake (40 miles NE)	June	—	0.13	0.26	
	July	—	0.11	0.22	
	August	0.14	0.15	0.25	
	Average[b]	0.14	0.13	0.24	0.006
Blind River (100 miles W) control area	June	—	0.10	0.13	
	July	—	0.07	0.12	
	August	0.12	0.12	0.16	
	Average[b]	0.12	0.10	0.14	0.001 (Sturgeon Falls)
Correlation coefficient	r	0.98*	0.98*	0.98*	

[a] Linzon (1973a).
[b] Average sulfur levels in foliage from McGovern and Balsillie (1973).
* $p < 0.05$.

Epiphytic lichens, which are perennial and evergreen are extremely sensitive to SO_2 because they are continually exposed to the gases in a polluted environment. In studies conducted on the occurrence of lichens at Sudbury (Leblanc et al., 1972) the number of epiphytes found growing on balsam poplar (*Populus balsamifera*) trees was drastically reduced in zones where the growing season mean levels of SO_2 were over 0.02 ppm and slightly reduced in zones where the mean levels of SO_2 were over 0.01 ppm. Similarly, in Sweden (Skye, 1964) it was found that the survival of lichens was less in areas with an annual SO_2 concentration of approximately 0.015 ppm, and in the Tyne Valley, England (Gilbert, 1969), a low species diversity of lichens was reached when the annual average of SO_2 was above 0.016 ppm.

Table 8 lists the chronic effects found by investigators when plants were subjected to an average low concentration of SO_2 for periods of time greater than 1 day.

2.3. Subtle Effects

Early investigators studying the effects of SO_2 on vegetation in Europe concluded that invisible injury, or physiological disturbance or effects on the growth or yields of plants could occur in the absence of visible markings (Wislicenus, 1901; Stoklasa, 1923). This theory was refuted by several investigations carried out in North America in the 1930s (Hill and Thomas, 1933, Swain and Johnson, 1936; Setterstrom et al., 1938; Katz et al., 1939). These investigators found that nonmarking concentrations of SO_2 emitted under controlled conditions for long periods of time did not affect carbon dioxide assimilation, stomatal behavior, chemical composition, or the rate of growth of the exposed plants. Similarly, in more recent years, Brisley et al. (1959) reported no loss in yield from cotton plants exposed to SO_2 in which visible injury was absent, and Davis, in 1972, found no yield loss from soybeans unless there were visible manifestations resulting from SO_2 fumigations.

Other studies conducted in England reported losses in yields of S23 ryegrass *(Lolium perenne)* in the absence of visible lesions after exposure to SO_2. Bleasdale (1973), in summarizing the results of work done in 1951, reported that ryegrass plants exposed to ambient air in one glasshouse weighed between 16 and 57% less than plants grown in similar air in another glasshouse that had 98 to 100% of the SO_2 removed by water scrubbing. There is a possibility that other contaminants remaining in the water-scrubbed air may have contributed to the observed yield effects. However, the work of Bell and Clough (1973) has

Table 8. Dose Response to SO_2—Chronic Effects

SO_2 Concentration (ppm)[a]	Exposure Period	Plant Response	Reference
0.07	3 days	Injury to sensitive eastern white pine	Banfield (1972)
0.076	14 days	Injury to barley	Mandl et al. (1975)
0.05	24 days	Injury to beech	Keller and Bucher (1976)
0.067	26 weeks	Depressed yields of ryegrass by 52%	Bell and Clough (1973)
0.035	Annual average (5-month growing period)	Injury to eastern white pine in field[b]	Roberts (1976)
0.017	10-year average (5-month growing season)	Prominent injury to old foliage of eastern white pine trees in forest[b]	Linzon (1971)
0.008	10-year average (5-month growing season)	Slight injury to old foliage of eastern white pine trees in forest[b]	
0.019	Monthly average over 2 years	Medium foliar injury to spruce trees in forest[b]	Materna et al. (1969)
0.015	Annual average	Reductions in lichens and bryophytes in field[b]	Skye (1964)
0.016	Annual average	Very low diversity in lichen and bryophyte species in field[b]	Gilbert (1969)
0.02	Growing season average	Epiphyte species drastically reduced in forest[b]	Leblanc et al. (1972)

[a] 1 ppm = 2600 $\mu g/m^3$ at 25°C and 760 mm Hg.
[b] Field observations. Other references report plant responses observed in artificial fumigation experiments.

placed additional emphasis on the role of low concentrations of SO_2 in reducing growth and yields of S23 ryegrass. Under controlled conditions, ryegrass plants were exposed to air that was purified by both activated charcoal and an absolute filter. In one chamber SO_2 was injected at a controlled rate so that plants were exposed over a 26-week period to a mean concentration of 191 $\mu g/m^3$ (0.067 ppm) SO_2, while in the control chamber the SO_2 levels were 14 $\mu g/m^3$ (0.005 ppm). At the end of 26 weeks the S23 plants in the polluted chamber were smaller and chlorotic (but had no lesions) as compared to the healthy controls. In addition, a reduction in dry weight of about 52% occurred in the polluted chamber in comparison to the control.

Subtle effects, without macroscopic visible injury to plant tissues, have been reported in other investigations in recent years. Bennett and Hill (1974) found that subnecrotic pollutant exposures could repress photosynthetic rates. They reported a 2% depression in CO_2 uptake by alfalfa at 0.25 ppm SO_2 for 1 hr and a 21% depression at 0.50 ppm SO_2 for 1 hr.

Seedlings of red pine *(Pinus resinosa)* in the cotyledon stage of development were inhibited by SO_2 fumigations without the occurrence of visible injury (Constantinidou et al., 1976). The chlorophyll content in cotyledons and the dry weight of primary needles were reduced in a 2-hr exposure to 0.50 ppm SO_2. The chlorophyll content in cotyledons was affected also by 1.0 ppm SO_2 for 30 min.

Keller (1976) introduced the term "latent" for subtle effects and reported the results of long-term exposures to a low concentration of 0.05 ppm SO_2. The effects included decreased photosynthesis of Scots pine *(Pinus sylvestris),* decreased pollen germination in white fir *(Abies alba),* and increased peroxidase activity of needle homogenates of spruce *(Picea excelsa).*

Godzik and Linskens (1974) found an increase in the total amount of free amino acids and a concurrent reduction in protein synthesis in primary leaves of Widusa bean *(Phaseolus vulgaris)* when artificially fumigated with 0.70 ppm SO_2. Some of the effects were observed after only 1 hr of fumigation and before the occurrence of visible injury.

Subtle effects of SO_2 have been reported on pollen tube elongation in *in vitro* studies. Karnosky and Stairs (1974) found highly significant decreases in pollen tube length of *Populus deltoides* in fumigations of 0.30 ppm SO_2 for 4 hr. At greater SO_2 doses frequent bursting of pollen tubes occurred. Masaru et al. (1976) reported a 55% inhibition of pollen tube lengths in *Lilium longiflorum* in fumigations of 0.71 ppm SO_2 for 1 hr.

From the preceding account it is apparent that the question of the occurrence of subtle effects in the absence of visible injury caused by SO_2 is not completely resolved. However, investigations conducted in the early 1970s generally indicate that subtle effects can occur without the presence of visible injury. These subtle effects include changes in physiological and biochemical processes and effects on pollen which occur during short- or long-term exposures to SO_2. Some of the subtle effects are transient with the plants recovering after removal of the gas. It may be assumed that subtle effects are occurring in many places, especially in urban areas, as a result of long-term exposure to low levels of atmospheric SO_2. Although these subtle effects are difficult to document or quantify, it is believed that vegetation in urban areas would generally be more vigorous and would display better growth if it was not exposed to SO_2.

Table 9 lists the results of experiments conducted to determine the occurrence of subtle effects on plants in the presence of controlled SO_2 doses.

3. ENVIRONMENTAL AND OTHER FACTORS AFFECTING RESPONSES OF PLANTS TO SULFUR DIOXIDE

The environmental factors that are conducive to optimum plant growth are usually the same factors that abet SO_2 injury. These factors include sunlight, moderate temperature, high relative humidity, wind, and adequate soil moisture. In addition, time of day and season, and plant factors such as genotype, nutrition, stage of growth, and tissue maturation determine the sensitivity of a particular species to SO_2 injury.

Vegetation is most susceptible to SO_2 during the active growth months of June, July, and August. If the environmental factors and the growth stages of the plants are not conducive to injury, the plants will escape injury even in the presence of potentially injurious concentrations of SO_2.

Most investigators have shown a direct relationship between open stomata and the absorption of SO_2 and subsequent leaf injury. When the stomata are closed, either at night because of darkness or during the day because of other factors, plants are more resistant to SO_2. Majernik and Mansfield (1970) reported that, when the relative humidity was greater than 40% at 18°C, SO_2 concentrations from 0.25 to 1.0 ppm caused a stimulation in stomatal openings which increased the absorption of SO_2.

In 1923 Swain stated that relative humidity was one of the most important factors in affecting the response of plants to atmospheric SO_2. From 70 to 100% relative humidity there was not much difference in sensitivity, but resistance increased below 70% and became very pronounced below 50%. Zimmerman and Crocker (1934) found that plants in a turgid condition were sensitive to SO_2, whereas wilted plants were extremely resistant. Katz and Ledingham (1939) stated that a high relative humidity was related to increased leaf turgor, which favored the

Table 9. Dose Response to SO_2—Subtle Effects

SO_2 Concentration (ppm)[a]	Exposure Period	Plant Response	Reference
0.25	1 hr	2% Depression in CO_2 uptake by alfalfa	Bennett and Hill (1974)
0.50	1 hr	21% Depression in CO_2 uptake by alfalfa	
0.70	1 hr	Increased amino acids and decreased protein synthesis in bean	Godzik and Linskens (1974)
0.71	1 hr	55% Inhibition of pollen tube length of lily	Masaru et al. (1976)
1.0	30 min	Reduced chlorophyll in cotyledons of red pine	Constantinidou et al. (1976)
0.5	2 hr	Reduced chlorophyll in cotyledons and decreased dry weight of primary needles of red pine	
0.30	4 hr	Effects on pollen tube elongation of poplar	Karnosky and Stairs (1974)
Ambient air measurements for last 100 days of experiment showed SO_2 exceeded 0.09 ppm on only 1 day and was below 0.07 ppm for 88 days	187 days	Ryegrass grown in water-scrubbed air with 98–100% of SO_2 removed had 48% more dry weight than plants grown in polluted air	Bleasdale (1973)
0.067[b]	26 weeks	Depressed yields of ryegrass by 52% (chlorosis of leaves but no lesions)	Bell and Clough (1973)
0.05	31–60 days	Decreased photosynthesis in Scots pine	Keller (1976)

Table 9. (Continued)

SO_2 Concentration (ppm)[a]	Exposure Period	Plant Response	Reference
0.05	141 days	Decreased pollen germination in white fir	
0.05	9 months	Increased peroxidase activity in spruce needle homegenates	

[a] 1 ppm = 2600 $\mu g/m^3$ at 25°C and 760 mm Hg.
[b] The results from Bell and Clough (1973) are reported also in Table 8 (Chronic Effects) since chlorosis (but no lesions) developed on the exposed foliage.

opening of the stomata; this, in turn, affected the rate of absorption of SO_2.

Several investigators have found that middle-aged, fully grown, and highly functional leaves were the most sensitive to SO_2. The youngest, newly developing leaves on alfalfa were the most resistant, followed by the oldest, maturing leaves (Katz and Ledingham, 1939). This is apparently true for most broadleaf plants. Barley, however, was much more susceptible to SO_2 when the leaves were developing and tillering out rapidly than when they were more mature and at the flowering stage. In coniferous trees this writer has observed that current-year needles were most susceptible to SO_2 in August after the needles had attained their full development.

Soil moisture is important to prevent plants from becoming wilted. The stomata are almost closed on wilted plants, and thus the plants are more resistant to SO_2. With adequate soil moisture, variations within wide limits do not appreciably affect the susceptibility of plants (Katz and Ledingham, 1939).

Moderate temperatures are conducive to plant injury by SO_2. Wells (1917) considered that temperatures over 40°F were necessary for plant injury to occur. Provided that other factors were not limiting, temperatures between 65 and 105°F had little effect on plant response to SO_2.

Daylight is important for stomatal opening. Light in excess of 3000 footcandles did not have any marked influence on the susceptibility of alfalfa, but below this value decreases in light intensity resulted in less absorption of SO_2, thus increasing the resistance to SO_2 injury (Katz and Ledingham, 1939).

The nutritional condition of plants also has a bearing on susceptibility to SO_2. Leone and Brennan (1971) found that tobacco and tomato plants grown at optimal levels of supplied nitrogen or sulfur were more susceptible to injury by SO_2 than plants deficient in these nutrients. An overabundance of nitrogen increased resistance, whereas an oversupply of sulfur increased susceptibility, as compared to optimal levels.

4. SUSCEPTIBILITY AND RESISTANCE OF PLANTS TO SULFUR DIOXIDE

Different plant species and varieties, and even individuals of the same species, may vary considerably in their sensitivity or tolerance to SO_2. Susceptibility lists have been made by several investigators. These lists should be used only as guides, however, since variations can occur because of differences in geographical location, climate, and plant stage of growth and maturation.

O'Gara (1922) listed the sensitivity of about 100 plants to SO_2 as determined by fumigation experiments that were validated to some extent by Thomas and Hendricks (1956). Alfalfa, barley, endive, and cotton were most sensitive, with privet being 15 times more tolerant to SO_2.

Katz et al. (1939) published lists, based on field observations and fumigation experiments, of the relative susceptibilities of plants to SO_2 in British Columbia. Larch, birch, ninebark, alfalfa, and lettuce were most sensitive; red cedar, silver maple, spiraea, field corn, and asparagus, most tolerant.

Weeds are as sensitive to air pollutants as are commercial plants, and knowledge of their sensitivities to SO_2 is useful in field studies. Benedict and Breen (1955) selected 10 weeds that occur commonly throughout the United States and determined their sensitivities to SO_2 in fumigation experiments. Chickweed was most sensitive; mustard, annual bluegrass, sunflower, Kentucky bluegrass, pigweed, and cheeseweed were intermediate in sensitivity; and lamb's-quarters, dandelion, and nettleleaf goosefoot were tolerant.

Zimmerman and Hitchcock (1956) determined the comparative susceptibilities of a number of plants to SO_2 in fumigation experiments. Chicory, alfalfa, geranium, buttonbush, and eggplant were most sensitive, whereas Jerusalem cherry, tulip, milo maize, and corn (field and sweet) were most tolerant.

Zahn (1961) classified crops into three resistance groups according to their tolerance limits. Clover-type fodder plants were most sensitive to SO_2; wheat, leafy vegetables (excluding cabbage), beans, strawberries, and roses were moderately sensitive; and roots and cabbage were least sensitive.

At Winnipeg, Manitoba, Linzon (1965) found Manitoba maple and trembling aspen to be most susceptible, white elm and choke cherry intermediate, and balsam poplar and green ash resistant to SO_2.

At Sudbury, Ontario, Dreisinger (1965) reported on the susceptibility of cultivated plants and native forest trees to SO_2, based on field observations. Buckwheat, red clover, trembling aspen, jack pine, eastern white pine, white birch, and bracken fern were most sensitive, whereas cabbage, corn, cedar, spruce, and maple were most tolerant.

Barrett and Benedict (1970) listed plants that are relatively sensitive to SO_2. They stated that most trees, with the possible exception of larch, are more resistant to SO_2 than the sensitive weeds and garden and crop plants. They considered pumpkin and squash to be the most sensitive garden plants.

In controlled fumigation experiments, Temple (1972) found resistance to SO_2 increased among four species of urban trees in the following order: Chinese elm, Norway maple, ginkgo, and pin oak.

In cities, the following trees have been observed by this writer to be resistant to SO_2 air pollution: ailanthus, pin oak, gingko, Carolina poplar, London plane, Norway maple, and little-leaf linden. Table 10 lists trees according to their sensitivities and tolerances to atmospheric SO_2 (Linzon, 1972).

A recent compilation of the susceptibility of woody plants to sulfur

Table 10. Sensitivities of Trees to Sulfur Dioxide[a,b]

Sensitive	Intermediate	Tolerant
Black willow	Austrian pine	Balsam poplar
Chinese elm	Balsam fir	Carolina poplar
Douglas fir	Basswood	Grand fir
Eastern white pine	Catalpa	Little-leaf linden
Jack pine	Choke cherry	Lodgepole pine
Largetooth aspen	Eastern cottonwood	London plane
Manitoba maple	Englemann spruce	Red oak
Trembling aspen	Mountain maple	Silver maple
Western larch	Red pine	Sugar maple
Western yellow pine	Western hemlock	Western red cedar
White ash	Western white pine	White cedar
White birch	White elm	White spruce

[a] Listed in alphabetical order in each column.
[b] Linzon (1972).

dioxide was prepared by Davis and Wilhour (1976). These authors conducted a detailed review of European and North American literature and presented many original tables from the papers examined.

5. EFFECTS OF SULFUR DIOXIDE ON FOREST ECOSYSTEMS

5.1. Introduction

Atmospheric SO_2 may affect a forest ecosystem in various ways. Forest communities have evolved and become established through the selective pressures of the environment. The addition of a new pressure, such as increased atmospheric sulfur, can alter the delicate balance of the ecological system. This alteration may be beneficial or deleterious, depending on the magnitude of the addition and the state of the receptor.

Smith (1974) distinguished three major categories of forest ecosystem effects arising from exposures to various concentration of atmospheric pollutants. In Class I, under conditions of low dosage, the vegetation and soils of forest ecosystems may act as a sink for contaminants with no detectable effects occurring except an increase in nutrient levels, accompanied possibly by a stimulatory (fertilizing) effect. In Class II, under conditions of intermediate dosage, individual trees may be adversely or subtly affected by nutrient stress, reduced photosynthesis rate, and predisposition to entomological or microbial organisms. In Class III, under conditions of high dosage, acute morbidity or mortality of specific trees occurs and may seriously alter the structure and function of the forest ecosystem.

Woodwell (1970), in discussing the effects of chronic irradiation on forest ecosystems, considered the impact of other stresses such as fire, herbicides, and oxides of sulfur to be similar and predictable. Structural changes in the ecosystem included first a reduction in the diversity of the forest by the elimination of sensitive species, then the elimination of the tree canopy, and finally the survival of resistant shrubs and herbs widely recognized as serial or successional species.

5.2. Forest Ecosystem Effects in the Vicinity of Nickel and Copper Smelters

In the Sudbury area of Ontario, three large nickel and copper smelters discharged approximately 6000 tons SO_2/day into the surrounding atmosphere. Forest effects in the area were excessive with severe injury to trees occurring up to 25 miles northeast of Sudbury (Linzon, 1958a).

On the basis of studies of over 6000 white pine trees on 42 sample plots during a 10-year period, the Sudbury area of Ontario was segregated into three fume zones: inner, intermediate, and outer (Linzon, 1966). In the inner fume zone, an area of about 720 square miles, white pine trees displayed acute and chronic foliar injuries which resulted in reduced radial and volume growth and in excessive tree mortality. In the intermediate fume zone, an area of about 1600 square miles, some chronic SO_2 injury was present, while in the outer fume zone atmospheric contamination was too dilute to cause visible injuries. These three zones may be compared to the three categories of forest ecosystem effects distinguished by Smith (1974): Class III (high dosage), Class II (intermediate dosage), and Class I (low dosage).

Chronic injuries develop slowly on coniferous perennial foliage with the greatest increase in injuries in the Sudbury sulfur-fume effects area having occurred on the 1-year-old needles of eastern white pine trees (Linzon, 1971). An artificial defoliation experiment indicated that of the three ages of needles on white pine, the 1-year-old was the most important for the tree's welfare (Linzon, 1958b). Current-year needles on white pine trees were infrequently injured by acute SO_2 injury in the Sudbury area because of the few high concentration potentially injurious fumigations that occurred in any one location at a distance from the Sudbury sources.

Chemical analysis of vegetation in the Sudbury area showed the accumulation and buildup of sulfur in foliage in vegetation growing close to the SO_2 sources. Both high concentration and low concentration fumigations of SO_2 over varying periods of time contributed to the accumulation of sulfur in foliage and lead to the development of chronic injury in older coniferous needles (Linzon, 1973a).

Over three times as many eastern white pine trees died annually in the inner fume zone as in a control area remote from the sources of SO_2 fumes (Linzon, 1971). The white pine forests in the inner fume zone were found to exhibit a net average annual negative growth in volume, since the trees that died during the investigation period exceeded the volume added by the surviving trees. Radial increment growth for the period 1940 to 1960 as measured on cores extracted in 1963 from 20 dominant living white pine trees on each of 42 sample plots showed a gradual decline in the inner fume zone, whereas a constant pattern was maintained in the other areas (Linzon, 1973b). An estimate was made of the loss in income to the producers of wood in the Sudbury area. For white pine alone which represents only 7.6% of the total productive forest area in parts of the Sudbury district, a loss of $117,000 was estimated to occur annually in the inner fume zone (Linzon, 1971).

It has been observed in the Sudbury area that other biological effects can also occur as a result of atmospheric contamination. The disease blister rust *(Cronartium ribicola)* of white pine was found to be practically nonexistent in areas close to the source of the sulfur fumes and the incidence of weevil insect injuries was less than in control areas (Linzon, 1958a, 1971). In addition it has been observed that felled trees persist on the soil surface for lengthy periods with little subsequent decay. Preliminary results from a study currently in progress show that the number of colonies of bacteria, actinomycetes, and fungi in soil decrease with proximity to the sulfur fume sources (D. Balsillie, and S. Bisessar, unpublished study).

Studies conducted on the occurrence of lichens in the Sudbury area (Leblanc et al., 1972) showed that in zones close to the SO_2 sources, the number of epiphytes found growing on *Populus balsamifera* trees was drastically reduced.

A study conducted in a northeasterly direction from the Falconbridge smelter in the Sudbury area (Gorham and Gordon, 1960) revealed strong sulfate accumulation in the surface soils up to 1 mile, and in pond and lake water up to 2 miles, from the source. Floristic studies showed a reduction in the number of terrestrial plant species close to the smelter. *Polygonum cilinode* and *Sambucus pubens* were reported to be the most tolerant herbaceous plant species, and *Acer rubrum* and *Quercus rubra* were the most tolerant tree species, occurring within 1 mile of the pollution source. The most sensitive herbaceous and tree species were *Vaccinum myrtilloides* and *Pinus strobus* respectively, which were not encountered for a distance of 16 miles northeast of the smelter.

A floristic survey of aquatic vegetation in the Sudbury area (Gorham and Gordon, 1963) found that sulfate concentrations in lake and pond waters were highest within 4 to 5 miles of the three smelters in the area. In general, there were fewer aquatic plant species in lakes where the sulfate level was about 1 meq/L and the pH was below 5.0. However, anomalies were encountered in which the pH was close to neutral but the sulfate level was high and the flora impoverished. *Leptodictyum riparium* was most tolerant of smelter pollution of lakes, whereas *Potomotegon epihydrus* var. *nuttalii* was most sensitive, not being found for a distance of 19 miles southwest of Copper Cliff.

Studies conducted on the degree of sulfur and heavy-metal contamination of soil and vegetation in the Sudbury area (McGovern and Balsillie, 1973) revealed that within 5 miles of the three smelters there was an intense zone of contamination by sulfur, copper, iron, and nickel.

Between 5 and 10 miles some reduction in the level of contamination was found, and from 10 to 20 miles there was a major decrease in the levels of these elements. In addition to soil and vegetation, a snow sampling program has been conducted annually in the Sudbury area. The results of the snow sampling during the winter of 1972 showed that sulfur levels in snow were elevated over an area of about 1500 square miles, whereas levels of copper, iron, and nickel were high in an area of about 400 square miles. The majority of the pH readings of snow were consistently between 4.0 and 5.0 regardless of distance from the three smelters, indicating that other factors such as long-range transport of pollutants also were contributory.

5.3. Forest Ecosystem Effects in the Vicinity of an Iron Sintering Plant

In the Wawa area of Ontario, an iron sintering plant emitted annually about 200,000 tons of SO_2 into the atmosphere. Since the prevailing winds are southwesterly during the vegetation growing season, the forests toward the northeast have suffered.

The sinter plant was constructed in 1939, and operations were greatly expanded in 1949. High-sulfur ore (15% S) was burnt and the waste gases were emitted from a 150-ft-high stack. This resulted in emissions of high concentrations of atmospheric SO_2 which killed most of the higher forms of vegetation for about 10 miles to the northeast. Beyond this "total kill" area, vegetation suffering partial kill and heavy damage gradually merged with vegetation displaying less injury, until normal conditions prevailed about 30 miles to the northeast. In 1958 two new stacks, each 250 ft in height, were erected; and, starting in 1962, the plant utilized ore with a low sulfur content (less than 4%) during the growing season. These recent developments have resulted in reduced SO_2 concentrations in the air and concomitant reductions in foliar injury to vegetation.

Each year, usually in August, the Ontario Ministry of Natural Resources has mapped the area of forest damage from the air. Three categories of damage were mapped, based on effects on white birch trees, and these categories were designated "total kill," "heavy kill," and "light damage." In 1970 the areas in square miles for the three categories were 41.9 (total kill), 73.8 (heavy kill), and 227.4 (light damage), for a total of 343.1 square miles of affected forest. The 1970 mapping is shown in Figure 1.

Much of the damage mapped from the air is permanent residual forest damage. In order to assess the annual occurrence of SO_2 injuries to

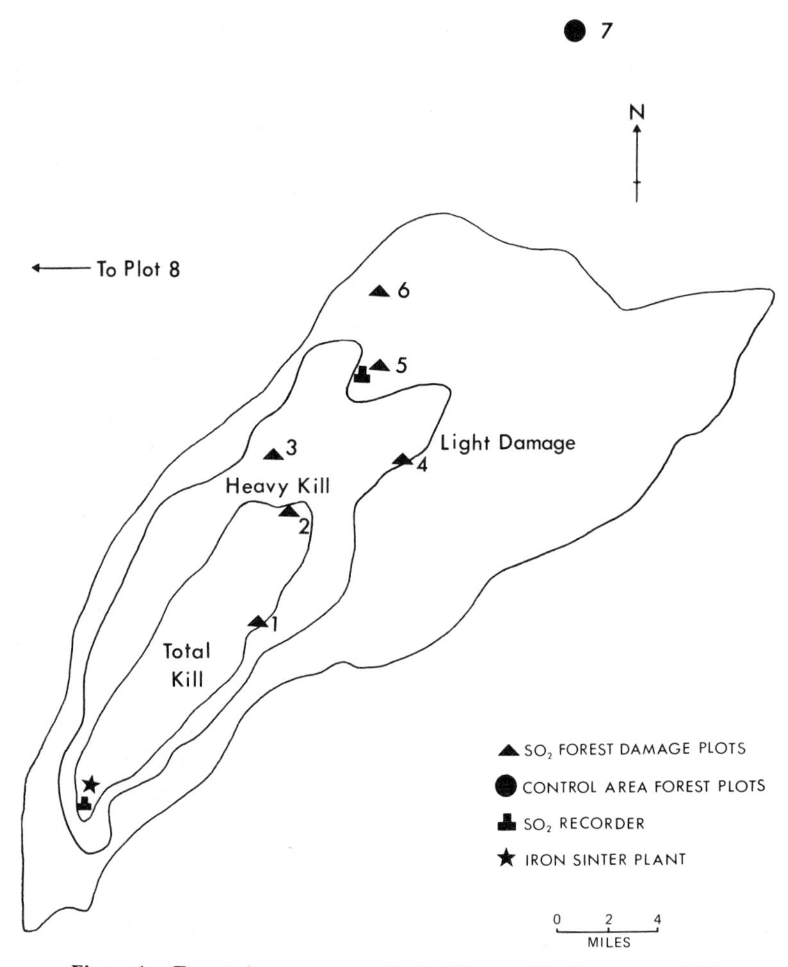

Figure 1. Forest damage survey in the Wawa sulfur fume area, 1970.

vegetation and the sulfur levels in air and vegetation in the Wawa area, a ground survey project was started in 1969 by the Phytotoxicology Section of the Ontario Ministry of the Environment. Six plots, each one fortieth of an acre in size, were established in a transect line northeast of the sintering plant. Two plots were located in the periphery of the "total kill" area at distances of 10 and 12 miles; two plots were established within the "heavy kill" area at distances of 16 and 19 miles; and two plots were located inside the "light damage" area at distances of 22 and 24 miles northeast of Wawa. In addition, two check plots were located at distances beyond the influence of sulfur fumes originating from the sinter plant (Figure 1).

In successive years varying degrees of injury on vegetation were observed at the close-in plots. The levels of sulfur in the air detected by lead peroxide candles were remarkably consistent from year to year at the established forest plots. In addition, a close relationship was found between sulfur levels in the air, as detected by candles, and the sulfur levels in vegetation, as determined by chemical analyses, with both decreasing with distance from the source (McGovern and Balsillie, 1974b). These sulfur levels in the air and vegetation for the years 1970 to 1972 are shown in Table 11. Statistics applied to these data showed that the correlation coefficients (*r*) between average sulfation rate in the air and average concentration of total sulfur in foliage were significant at the five % level or less.

An ecological survey was conducted in the Wawa Area in 1960 (Gordon and Gorham, 1963). Within about 10 miles northeast from the sinter plant studies conducted on ridgetops showed that beyond this distance ground flora declined markedly from about 20 to 40 species per 40-m^2 quadrat to 0 to 1 species within 2 miles of the pollution source. At

Table 11. Air Candle Sulfation Rates and Total Sulfur Contents in Vegetation Samples—Wawa, 1970, 1971, and 1972[a]

Plot No.	Distance and Direction from Sinter Plant	Average Sulfation Rate (mg SO_3/100 cm^2·day)			Average Concentration of Total Sulfur (%) in Vegetation		
		1970	1971	1972	1970	1971	1972
1	10 miles NE	1.03	1.48	1.55	0.24	0.44	0.51
2	12 miles NE	0.61	0.88	0.87	0.25	0.36	0.44
3	16 miles NE	0.57	0.36	0.42	0.22	0.27	0.36
4	19 miles NE	0.25	0.39	0.38	0.19	0.30	0.28
5	22 miles NE	0.35	0.26	0.35	0.22	0.30	0.34
6	24 miles NE	0.25	0.20	0.19	0.17	0.25	0.26
7 (Control)	38 miles NE	0.08	0.12	0.11	0.17	0.21	0.23
8 (Control)	35 miles NW	0.04	0.04	0.02	0.18	0.26	0.21
Correlation coefficient (*r*)		0.82*	0.78*	0.94**			

[a] McGovern and Balsillie (1974b).
* $p < 0.05$.
** $p < 0.01$.

the same time sulfate in lake and pond waters increased greatly from normal levels of about 0.2 to 0.3 meq/L to more than 0.5 meq/L within 11 miles northeast and up to 2.0 meq/L within 2 miles of the source. Waters within about 5 miles northeast were strongly acidic (pH 3.2 to 3.8) but were low in calcium. The most sensitive tree species was *Pinus strobus,* not observed within 30 miles of the sinter plant, while *Picea glauca,* P. *mariana,* and *Populus tremuloides* were not observed within 15 miles. As was the case at Sudbury, *Polygonum cilinode* and *Sambucus pubens* were reported to be the most tolerant herbaceous plant species.

During the summers of 1965 and 1966 a study of epiphytic flora was conducted in the Wawa area (Rao and Leblanc, 1967). No epiphytes were recorded within 16.1 km northeast of the sinter plant. Beyond the epiphyte "desert," corticolous lichens gradually appeared. Nearest to the sinter plant, *Bacidia chlorococca* and *Cladonia coniocraea* were the only species, and were present only on the bases of trees. At a greater distance they gradually appeared on the trunks of trees. There appeared to be a sequence of increasing sensitivity from crustose to foliose to fruticose lichens. The maximum number of species found on one tree occurred at 43.5 km northeast of the sinter plant.

The forest ecosystem effects in the Wawa area typify the classical changes that may be expected from the addition of an adverse stress. The type of industry at Wawa, the use of high-sulfur ore initially, the remote location of the industry in a northern boreal forest, a wide plain bordered by hills running northeast from Wawa, and a prevailing southwest wind during the growing season all contributed to this classic situation. The three classes of low, intermediate, and high dosage with accompanying forest effects are readily apparent in the Wawa forest damage area. The successive deteriorating effects on the tree, shrub, herbaceous, and microflora layers of the plant community are evident to the northeast and laterally from this main axis. These severe effects caused by atmospheric sulfur and followed by fires (in the "total kill" area) and erosion have led to large barren areas. The soils on the hillsides in the "total kill" area have eroded toward the central plain. The tops of these hills are almost barren and support little vegetation, whereas in the pockets of soil below more diverse forms of vegetation occur, especially hummocks of grass. As was found in the Sudbury area, wood decay of fallen trees in the "total kill" area is slow, and trees that have been dead for many years are still in good condition.

6. EFFECTS OF SULFUR DIOXIDE ON LICHENS

Sulfur dioxide pollution has adversely affected lichen populations in many areas of Europe and North America. Over 200 papers have been published on the effects of air pollution on lichens since 1958 (Hawksworth, 1973).

Lichens are composed of two plants, an alga and a fungus, in a symbiotic relationship. They are perennial, evergreen, slow growing, and capable of living for millennia. However, in general, lichens are probably the plant group most sensitive to sulfur pollution, because they can be continually subjected to gases in the atmosphere over long periods of time. Lichens lack cuticles and stomata and absorb substances in solution over their entire surface area. In addition, unlike many other plants, lichens do not shed toxin-laden parts each year.

Sulfur pollution not only has caused the depletion of lichen vegetation in certain areas but also has resulted in changes in the distribution of different species (Hawksworth et al., 1973). Lichens form an integral part of an ecosystem, and their depletion could lead to hardships to reindeer and caribou dependent on them for winter fodder (Schofield and Hamilton, 1970).

There are three major types of lichens: fruticose (pendulous or stalked), foliose (leaflike), and crustose (adhering closely to substratum) with sensitivities to SO_2 generally declining as the morphology of the lichen approaches the prostrate form. However, exceptions occur, with some crustose species being more sensitive than foliose or fruticose species (Hawksworth, 1973). Furthermore, the epiphytic lichens (growing on other plants, particularly trees) are more sensitive to SO_2 than terrestrial lichens.

In western Europe, *Lecanora conizaeoides* has been found to be extremely resistant to SO_2, and in polluted areas with mean levels between 80 and 130 $\mu g/m^3$ of SO_2 in the air this crustose lichen is the predominant species found on tree bark surfaces (Hawksworth, 1973). At lower mean levels of SO_2, below 60 $\mu g/m^3$ more fruticose and foliose lichens are present on trees but certain sensitive species are lacking so that the cover of the total lichen vegetation may amount to only 60 to 80%. Gilbert (1969) found the sensitive *Evernia prunasti* lichen species to be absent in the city center of Newcastle and to distances of 12 km from the city center, where the annual mean SO_2 levels ranged from 65 to 200 $\mu g/m^3$. Beyond 12 km from the city center, where the mean SO_2 level was below 60 $\mu g/m^3$, *Evernia prunasti* populations on ash trees increased steadily with distance from the city.

The effects of SO_2 on lichens are attributable for the most part to long-term absorption of sulfur from dilute solutions on the thallus surface with a resultant buildup of toxic concentrations of sulfates within the tissues. Gilbert (1969) found 3290 ppm sulfur in *Parmelia saxatilis* collected 6 km from the center of a town, compared to 225 ppm sulfur in the same lichen species collected 34 km from the center.

Both shelter and a high pH of the medium in which lichens are present can alleviate the effects of sulfur pollution (Gilbert, 1969). Certain sheltered situations reduce the levels of SO_2 to which the lichens are exposed and permit their survival. High pH values also reduce the effects of sulfur pollution, with Laundon (1967) finding in London, England, that a large number of lichens could survive on limestome substrates. The explanation for the pH effect is that the degree of ionization of sulfurous acid varies with pH (Vass and Ingram, 1949). In solution, SO_2 is distributed in three forms: above pH 5 the sulfite ion predominates, between pH 2 and 5, the bisulfite ion predominates, and below pH 2 sulfurous acid increasingly develops. At a pH of 3.2 in which only 5% of unionized sulfurous acid was present Gilbert (1968) experimentally found the most toxic effects to bryophytes (mosses).

Adverse effects of long-term exposure of lichens to atmospheric SO_2 were found also by Leblanc (1969), who studied the epiphytic population of *Populus balsamifera* in the Sudbury area. He examined lichen populations in four pollution zones surrounding the Sudbury smelters, based on over 10 years of air sampling for SO_2 by Dreisinger (1965). In zone I (heavy pollution), where the average concentration of SO_2 was over 0.03 ppm, only two species of lichens were found. In zone II (medium pollution), with an average concentration between 0.02 and 0.03 ppm, three lichen species were collected. In zone III (light pollution), with an average concentration between 0.01 and 0.02 ppm, the number of lichen species increased to 12, and in zone IV (very light pollution), with an average concentration between 0.005 and 0.01 ppm, 28 species of lichens were found.

Similarly, in Sweden, Skye (1964) found that the number of species of lichens was less in areas with annual SO_2 concentrations of approximately 0.015 ppm, and in the Tyne Valley, England, Gilbert (1969) reported a low diversity of lichen species when the annual average of SO_2 was above 0.016 ppm.

In 1972 Leblanc et al. elaborated on the epiphytic study conducted in the Sudbury area by classifying the zones of pollution and frequency of occurrence of epiphytes utilizing an index of atmospheric purity (IAP). The IAP method was first developed by De Sloover and Leblanc (1968) and later modified by Leblanc and De Sloover (1970) by using the following formula: $IAP = \Sigma_n (Q \times F)/10$, where n is the number of species at a site, F is the frequency coverage of each species at each site, and Q is the average number of epiphytes concurrently present with a species at all sites. Five IAP zones were classified surrounding the Sudbury nickel smelters, and the lichen species found in the zones were related to their sensitivity to SO_2. *Bacidia chlorococca, Lecanora saligna,* and *Parmelia sulcata* were found in all pollution zones, but none of these was recorded within a radius of 7 km from Sudbury. *Xanthoria fallax, X. polycarpa, Physica orbicularis,* and *Parmelia rudecta* were most sensitive and were first encountered beyond 14 km from Sudbury.

Gilbert (1968) found the most sensitive indicators of sulfur pollution to be certain bryophytes which disappeared from the tops of sandstone walls in the Newcastle area when the winter average (September to April) of SO_2 exceeded 50 $\mu g/m^3$. Some common lichens from the same habitat (e.g., *Parmelia saxatilis*) were slightly more resistant being limited by a winter average of about 56 $\mu g/m^3$ of SO_2. Some of the first species of moss to disappear were *Grimmia pulvinata, Campothecium sericium,* and *Lophocolea bidentata,* none of which appears in the built-up area of towns in the Tyne Valley. Slightly more resistant was *Hypnum cupressiforme.* As the center of pollution was approached the most resistant mosses, *Leptobryum pyriforme* and *Funarea hygrometrica,* were found. In comparing the harmful effects of SO_2 on mosses, Gilbert concluded that the winter average (September to April) of SO_2 was more important than the annual average or the highest monthly average. The winter average could be twice the summer average because of increased domestic fuel consumption. In transplant experiments he found that sensitive species of mosses could survive up to 2 months of heavy pollution so that recovery was possible from the highest month, whereas few species were visible after a winter's pollution.

It is apparent from the preceding account that lichens and bryophytes adequately demonstrate the long-term injurious chronic effects of sulfur pollution. Epiphytic lichens, in being continually exposed to atmospheric pollution and having no stomata and no control over shedding plant portions, are subject to long-term accumulation of sulfates which ultimately may cause their demise.

7. EFFECTS OF SULFUR DIOXIDE ON HOST-PARASITE RELATIONSHIPS

The effects of SO_2 on the relationship between plant hosts and various organisms have been reported by a number of investigators. Heagle (1973) summarized the interactions between several air pollutants, including sulfur dioxide, and plant-parasite relationships.

There are a number of possible host-parasite interaction effects that may occur when a pollutant is introduced into a particular habitat. The severity of the parasitism may be increased or decreased as a result of the action of the pollutant on the virulency of the parasite or on the susceptibility of the host. In addition, the effects of the pollutant on the host may be intensified or reduced by the presence of the parasite. Considering the large number of host-parasite-pollutant combinations that can occur, the research required to determine the effects of these interactions is practically limitless. However, there are a number of reports of experiments that have been conducted and of observations that have been made during field studies. From the reports to date there appears to be no consistent effect from such interactions.

Generally, excess SO_2 in the atmosphere tends to decrease the severity or incidence of host-parasite relationships. Rust diseases appear to be particularly sensitive to SO_2. Johansson (1954) reported less wheat stem rust caused by *Puccinia graminis* in an industrialized area of Sweden than in nonindustrialized areas. Scheffer and Hedgcock (1955) found decreased effects by a number of rust diseases *(Cronartium, Coleosporium, Melampsora,* and *Peridermium)* where trees were injured by SO_2. Linzon (1958a) found the disease blister rust of white pine, caused by *Cronartium ribicola,* to be practically absent in forests to distances of 25 miles northeast of the Sudbury smelters in line with the prevailing wind. With increasing distances from the Sudbury SO_2 sources, blister rust increased in incidence. In controlled-environment experiments, Weinstein et al. (1975) found that SO_2 reduced the incidence and severity of bean rust disease *(Uromyces phaseoli)* on pinto bean.

Fungi infecting plant foliage have been reported to be reduced in incidence by atmospheric SO_2 in the vicinity of industries in several countries. These leaf fungi included *Microsphaera alni* (Kock, 1935); *Hypodermella laricis, Lophodermium pinastri,* and *Hypodermella* spp. (Scheffer and Hedgcock, 1955); *Diplocarpon rosae* (Saunders, 1966); *Venturia inaequalis* (Pyzybylski, 1967); *Hysterium pulicare* (Skye, 1968); *Lophodermium juniperinum* and *Rhytisma acerinum* Barkman et al., 1969); and *Microsphaera alni* (Hibben and Taylor, 1975).

Conversely, some researchers have found that SO_2 can increase the severity of foliar diseases. Tanaka et al. (1975) reported that sooty leaf mold disease, *Rhizosphaera kalkhoffii,* was more severe on *Pinus densiflora* trees that were inoculated and transplanted to areas with high concentrations of atmospheric SO_2 than on trees in areas with "no air pollution."

In the Sudbury area of Ontario, it has been observed that fallen trees persist on the soil surface for lengthy periods with little subsequent decay. Studies conducted by the Ontario Ministry of the Environment in the Sudbury area showed that the numbers and kinds of fungi on fallen trees, and the colonies of bacteria, actinomyces, and fungi in the soil, decreased with proximity to the SO_2 sources (D. Balsillie and S. Bisessar, unpublished study). Heagle (1973) cited conflicting reports on the numbers of fungi in soil near European SO_2 sources, with Sobotka (1964) finding more soil fungi and Mrkva and Grunda (1969) finding fewer.

Whether or not the reduction in the severity of host-parasite relationships may be considered as a favorable consequence, it must be regarded as an expression of an unbalanced natural condition caused by sulfur pollution.

8. EFFECTS OF SULFUR DIOXIDE IN COMBINATION WITH OTHER AIR POLLUTANTS

Rarely is plant life in nature exposed to the influence of only one air pollutant. The bulk of an emission from a particular industrial source may consist of only one pollutant, but other pollutants may be conjointly present. Thus it is important to know the concentrations of various phytotoxic pollutants in the atmosphere and the effects on plant life from gas mixtures. Some controlled-environment research has been conducted in which plants have been subjected to combinations of SO_2 and ozone (O_3), SO_2 and nitrogen dioxide (NO_2), and SO_2 and hydrogen fluoride (HF). The various results obtained have been classified as additive (equal to the sum of the effects of the individual pollutants), synergistic (greater than the additive effects), or antagonistic (less than the additive effects).

Before 1966 some observations and experiments had indicated that pollutant mixtures might influence plants differently from the action of a single pollutant. Then, in 1966, Menser and Heggestad published the results of exposing tobacco plants to mixtures of SO_2 and O_3, which stimulated intensified studies of pollutant combinations on plant life by several investigators in the next decade.

8.1. Sulfur Dioxide and Ozone

Menser and Heggestad (1966) found that three varieties of tobacco plants suffered from 25 to 38% leaf damage upon exposure to a combination of 0.24 ppm SO_2 and 0.027 ppm O_3 for 2 hr, whereas either pollutant alone at approximately the same concentration and for the same time period caused no injury to any of the three tobacco varieties. The authors suggested that the injury to the tobacco plants was caused by a synergistic effect of SO_2 and O_3. The leaf injury caused by the combination of the two gases resembled typical O_3 injury rather than a combination of SO_2 and O_3 symptoms. Heck (1968) reported that a combination of 0.10 ppm SO_2 and 0.03 ppm O_3 in a 4-hr exposure acted synergistically to cause moderate to severe injury on the foliage of Bel W-3 tobacco plants.

The response of plants to gas mixtures is dependent not only on the plant species but also on the ratio of the concentrations of the gases in the mixture. For example, Tingey et al. (1973) found that foliar injury on broccoli showed an additive response to a mixture of 0.25 ppm SO_2 and 0.10 ppm O_3 for 4 hr, whereas Bel W-3 tobacco showed a synergistic response under the same regime. However, if the ratio of the contaminants was changed to 0.10 ppm SO_2 and 0.10 ppm O_3, the reverse occurred, broccoli showing a synergistic response and tobacco an additive response in 4-hr exposure periods.

There have been a number of reports of the effects of combined SO_2 and O_3 mixtures on eastern white pine. Table 12 compares some of the reported results of controlled artificial fumigations on sensitive eastern white pine trees. Of the four references listed in Table 12, only one reported an antagonistic response to the mixture of the two gases when compared to the responses of the individual pollutants, whereas the other three references found synergistic responses. It is of interest to note that SO_2 alone at very low levels and for short-time periods was found by some of these investigators to be able to injure the new needles of eastern white pine. However, it must be recognized that the injuries were induced on extremely sensitive strains of eastern white pine and under the experimental conditions outlined in the papers cited.

Low concentrations of SO_2 and/or O_3 were reported by Applegate and Durrant (1969) to injure the foliage of peanut plants in controlled experiments, the response to a mixture of the two gases being synergistic.

Karnosky (1976) studied the effects of SO_2 and O_3 on trembling aspen and found the response to a mixture of the two gases to be synergistic. Whereas it required either 0.35 ppm SO_2 alone for 3 hr or 0.05 ppm O_3 alone for 3 hr to injure the foliage of trembling aspen, a combination of 0.20 ppm SO_2 and 0.05 ppm O_3 for 3 hr caused more injury than the sum of the individual pollutants. In fact, 0.20 ppm SO_2 alone for 3 hr caused no injury to trembling aspen.

The effects of mixtures of SO_2 and O_3 described above refer to foliar injury on host plant species. Tingey and Reinert (1975) reported results of studies in which plant growth was measured also. Although no synergistic responses were found, tobacco plants suffered growth reduc-

Table 12. Response of Eastern White Pine to Sulfur Dioxide and Ozone Mixtures

$SO_2 + O_3$ (ppm)	Exposure Period	Plant Response	Overall Response	Reference
0.05 + 0.05	2 hr	Less injury than occurred from exposure to 0.05 ppm SO_2 alone for 2 hr (O_3 not toxic at 0.05 ppm for 2 hr)	Antagonistic	Costonis (1973)
0.025 + 0.05	6 hr	Greater injury than occurred from exposure to 0.025 ppm SO_2 alone for 6 hr (O_3 not toxic at 0.05 ppm for 6 hr)	Synergistic	Houston (1974)
0.05 + 0.10	1–12 days	100% of trees injured (SO_2 alone was toxic at 0.07 ppm for 3 days but not toxic at 0.05 ppm for 4 days, and O_3 not toxic at 0.10 ppm for 4 days)	Synergistic	Banfield (1972)
0.10 + 0.10	4–8 hr/day, 5 days/week, for 4–8 weeks	Greater injury than occurred from exposure to 0.10 ppm SO_2 alone or 0.10 ppm O_3 alone for the same time period	Synergistic	Dochinger et al. (1970)

tions from the mixture of the gases equal to the additive effects of the individual gases.

8.2. Sulfur Dioxide and Nitrogen Dioxide

The potential harm that may be caused by mixtures of low levels of SO_2 and NO_2 was strikingly demonstrated by the results of experiments conducted by Tingey et al. (1971). A gaseous mixture of 0.10 ppm SO_2 and 0.10 ppm NO_2 caused synergistic effects with greater than 5% leaf injury being induced on five of six plant species treated in 4-hr exposure periods. Bel W-3 tobacco plants displayed 9% leaf injury after a 4-hr exposure to a mixture of 0.05 ppm SO_2 and 0.10 ppm NO_2. The leaf damage caused by the SO_2–NO_2 mixture consisted of upper surface flecks typical of O_3 injury, and some undersurface silvering on some species that resembled SO_2 injury reported by Brisley et al. (1959). The similarities of the symptoms caused by gas mixtures in controlled experiments to those due to various pollutants in the field make differentiation of observed field injuries difficult.

White et al. (1974) reported a synergistic response of alfalfa to a mixture of 0.15 ppm SO_2 and 0.15 ppm NO_2 in a 2-hr exposure which depressed the apparent photosynthetic rate by 7%, whereas 0.25 ppm of SO_2 alone or 0.40 ppm NO_2 alone was required to cause a 2% depression.

8.3. Sulfur Dioxide and Hydrogen Fluoride

Few reports have been published on plant responses to mixtures of SO_2 and HF. A recent paper by Mandl et al. (1975) described studies in which barley, sweet corn, and bean plants were subjected to SO_2, HF, or mixtures of the two gases for different periods of time. A synergistic response was found to a mixture of low concentrations of the two gases. For example, barley leaves displayed greater injury in a mixed gas experiment than the sum of the injuries caused by the individual pollutants in a 27-day exposure period. The concentrations of the pollutants utilized were approximately 0.08 ppm SO_2 and 0.6 $\mu g/m^3$ HF.

8.4. Discussion

From the preceding account it is apparent that very low concentrations of SO_2 for periods of a few hours caused foliar injury to sensitive plants

Table 13. Pollutant Combinations that Caused Synergistic Responses in Vegetation

Pollutant Combination	Exposure Period (hr)	Plant Injured	Reference
0.02–0.03 ppm SO_2 + 0.008–0.01 ppm O_3	4–5	Peanut	Applegate and Durrant (1969)
0.025 ppm SO_2 + 0.05 ppm O_3	6	Eastern white pine	Houston (1974)
0.10 ppm SO_2 + 0.03 ppm O_3	4	Tobacco	Heck (1968)
0.10 ppm SO_2 + 0.10 ppm O_3	4	Horticultural crops	Tingey et al. (1973)
0.10 ppm SO_2 + 0.10 ppm NO_2	4	Horticultural crops	Tingey et al. (1971)
0.20 ppm SO_2 + 0.05 ppm O_3	3	Trembling aspen	Karnosky (1976)
0.24 ppm SO_2 + 0.027 ppm O_3	2	Tobacco	Menser and Heggestad (1966)

when combined with another air pollutant, whereas the same concentrations of SO_2 alone either were not toxic or caused minimal injury. Table 13 lists some of the low concentration pollutant combinations that caused synergistic effects in short time periods on exposed plants. These experiments indicate that a low concentration of 0.10 ppm SO_2 in combination with either O_3 or NO_2 for periods of 4 hr can injure a wide variety of plants. The pollutant levels and time periods utilized in these experiments occur quite frequently in the ambient air, and thus it is possible that plant effects observed in the field which have been attributed to individual pollutants may have been caused by gaseous mixtures.

9. EFFECTS OF HYDROGEN SULFIDE ON PLANTS

The concentrations of hydrogen sulfide (H_2S) required to cause injury to vegetation are far in excess of the concentrations that are readily obnoxious or injurious to human beings. In addition, the concentrations of H_2S that plants can tolerate are much higher than phytotoxic concentrations of SO_2.

The most significant work to date on the effects of H_2S on plant life was reported by McCallan et al. in 1936. Of 29 plant species tested, seven were injured at concentrations of 20 to 40 ppm (30,000 to 60,000 μg/m³) over a 5-hr exposure period. These plant species were buckwheat *(Fagopyrum esculentum),* Turkish tobacco *(Nicotiana glauca),* aster *(Aster macrophyllus),* cucumber *(Cucumis sativus),* poppy *(Papaver somniferum),* tomato *(Lycopersicon esculentum),* and radish *(Raphanus sativus).* Another seven plant species were quite resistant, not being injured at concentrations ranging from 60 to 400 ppm H_2S for 5 hrs. These plant species were carnation *(Dianthus caryophyllus),* purslane *(Portulaca oleracea),* Boston fern *(Nephrolepis exaltata),* apple *(Malus pumila),* cherry *(Prunus serotina),* strawberry (*Fragaria* spp.), and peach *(Prunus persica).*

The symptoms of injury caused by H_2S are generally confined to the youngest foliage, with the leaf margins being scorched. With less intense injury interveinal portions of the leaf show effects.

Temperature and soil moisture were found by McCallan et al. (1936) to be important contributory factors to the sensitivity and resistance of plants to H_2S. Wilted plants were resistant, and higher temperatures increased the sensitivity of plants to injury by H_2S.

10. EFFECTS OF SULFURIC ACID ON PLANTS

Combustion of sulfur-containing fuels yields SO_2 in quantities 40 to 80 times that of SO_3 (U.S. Department of Health, Education, and Welfare, 1969). Sulfur dioxide is oxidized in the atmosphere catalytically or photochemically. The photooxidation rate of SO_2 in air and sunlight is between 0.1 and 0.2%/hr. Between 3100 and 4100 Å, the photooxidation rate is more rapid when hydrocarbons and nitrogen oxides are present. The SO_3 molecules combine with water vapor in the air to form sulfuric acid (H_2SO_4) aerosols.

Few ambient air measurements have been made of H_2SO_4 mist. A high concentration of 50 μg/m³ and an average of 25 μg/m³ were found in Los Angeles (U.S. Department of Health, Education, and Welfare, 1969). Acid smog injury to vegetation observed in Los Angeles was described by Thomas et al. (1952). Spots were formed on the upper surface of the leaves which later could extend through the leaf. The fog droplets settled on the leaf, the moist leaf surfaces having a pH of 3 or less. The fog-smog type of injury was most typically seen on Swiss chard and beets.

Experiments utilizing 30 to 65 ppm (108 to 234 mg/m³) of H_2SO_4 mist for 4 hr failed to cause injury to alfalfa and sugar beets (Thomas et al., 1952). These aerosols, composed of droplets of about 1 μ in diameter, were so dense that visibility was reduced to 3 ft. When coarser aerosols

(5 to 15 μ) were used and the leaf surfaces wetted with a fine spray of water, the H_2SO_4 droplets caused a spotted type of injury. Without wetting the concentrated H_2SO_4 droplets rested as spheres on the leaves without producing injury. The addition of water was necessary to reduce the surface tension of the strong acid and make it act on the leaf tissues.

11. EFFECTS OF ACID PRECIPITATION ON PLANTS

Beginning with the publication by Odén in 1968 of the extent of the acidity problem in Scandinavia, studies and research on this phenomenon have steadily increased. International conferences were convened in 1975 at Columbus, Ohio, in the United States (Dochinger and Seliga, 1976) and in 1976 at Telemark, Norway (Overrein, 1976), to bring up to date the present status of acid precipitation on a global basis.

Acid rain has the potential to cause serious widespread effects on terrestrial ecosystems in certain areas of the world. However, the experience gained from studies conducted in the Scandinavian countries in documenting the effects of acid rain on air, water, and soil quality has stimulated action in North America to plan comprehensive monitoring programs and research studies on ecological effects. The results of these studies will identify the extent of the problem and will lay the framework for air quality standards to be established and for control action to be taken.

In the Scandinavian countries, the northeastern United States, and in a localized area near Sudbury, Ontario, acid rain has been found to have had serious implications with respect to increasing the acidity of lakes and reducing fish populations. It has been postulated that decreased productivity of forests in susceptible regions in Sweden is attributable to natural rainfall acidified by industrial emissions. However, not all scientists agree that acid rain can be directly related to decreased forest productivity in southern Scandinavia (Tamm, 1976). The reasons for the uncertainty include factors such as the variability in tree growth due to site, competition, tree age and genotype, and the fact that the nitrogenous component of acid rain can act as a fertilizer to improve tree growth.

In controlled experiments using simulated acid rain, a number of effects on plants and host-parasite interactions have been demonstrated. Shriner (1976) reported that simulated acid rain of pH 3.2 resulted in (1) an 86% inhibition in telia production of fusiform rust *(Cronartium fusiforme)* on willow oak *(Quercus phellos)*; (2) a 66% inhibition of root knot nematode *(Meloidogyne hapla)* on field-grown red kidney beans *(Phaseolus vulgaris);* (3) a 10% decrease in the severity of bean rust

(Uromyces phaseoli) on kidney bean; (4) inhibition of nitrogen-fixing bacteria *(Rhizobium)* nodulation of kidney bean and of soybean *(Glycine max)* by an average of 73%; (5) marked cuticular erosion of leaf surfaces of willow oak and kidney bean; (6) stimulation of development of halo blight *(Pseudomonas phaseolicola)* of kideny bean if applied to the host before inoculation; and (7) inhibition of halo blight of kidney bean if applied to the host after infection.

Wood and Bormann (1974, 1975, 1976) reported (1) foliar tissue damage of yellow birch *(Betula alleghaniensis)* seedlings, using artificial acid mists of pH 3.0, and significant growth decreases at pH 2.3 (1974); (2) increased foliar leaching of potasium, magnesium, and calcium from sugar maple *(Acer saccharum)* and pinto bean seedlings using artificial acid mists of pH 4.0, and tissue damage at pH 3.0 (1975); and (3) a 20% increase in growth of eastern white pine seedlings subjected to artificial acid rain at a pH of 2.3 compared to rains with higher pHs. The growth increase was attributed to the nitrogenous content of the artificial acid rain. However, soil acidity increased, and magnesium, calcium and potassium cations were leached from the soil at the low pHs (1976).

Shen-Miller et al. (1976) reported increased uptake of cadmium by soybean plants treated with simulated acid rain of pH 2.1. It was postulated that the acid rain altered cell wall and membrane structures, making the plants more susceptible to other pollutants.

Expectations are that, unless relevant standards for sulfates are established to prevent increasing acidity of precipitation, forested areas in susceptible regions of the world will be liable to degradation in the future. Research utilizing simulated acid rain has shown that a number of individual effects can occur on subjected plants, such as leaf cuticular erosion, foliar lesions, leaching of foliar nutrients, and alterations in host-parasite or host-pollutant interactions. In agriculture, these effects are not expected to pose as great a problem as in forestry because of the continual addition of fertilizers and the practice of more intensive crop management. However, these effects together with the leaching of calcium and other metallic cations from forest soils, and the release of previously bound metals (e.g., aluminum) to build up to potentially toxic concentrations present a potential threat to forest production in susceptible areas in the next quarter century.

12. AIR QUALITY STANDARDS FOR SULFUR DIOXIDE

In formulating air quality standards for SO_2 for a particular jurisdiction, many factors must be taken into consideration. Foremost are the scientific criteria which describe the effects of SO_2 for given concentrations and time periods on exposed receptors. These criteria are generally assembled from information published in scientific and technical reports. The published data may be based on effects in the field caused by ambient atmospheric concentrations of SO_2 or on experiments conducted under controlled environmental conditions.

Generally, the effects of SO_2 observed in the field, under natural conditions, provide the best basis for interpretation. Difficulties are encountered in attempting to extrapolate the results from experimental artificial fumigations to natural ecosystems. The reason for the difficulties is that the varying environmental conditions that occur out-of-doors are almost impossible to duplicate in experimental fumigations. The temperature, relative humidity, light quantity and quality, wind, soil moisture, plant responsiveness, SO_2 concentrations, and presence of other air pollutants are continuously fluctuating in the field at the site of the plant-environment interface.

Despite these difficulties, published data on artificial fumigation work indicated that effects on plant life were caused by some extremely low SO_2 doses. Conversely, the literature showed that some plants were notably resistant and could tolerate very high concentrations of SO_2. Examination of Table 2 showed injury occurring on extremely sensitive strains of eastern white pine at a dose as low as 0.03 ppm SO_2 for 1 hr, whereas 3.0 ppm for 1 hr was required to cause minimal injury to *Acacia pruinosa*. Many jurisdictions have an air quality standard for SO_2 close to 0.30 ppm for 1 hr which is 10 times higher than one of the lowest dose responses reported and 10 times lower than one of the highest dose-responses reported.

In evaluating published dose-response data, the reviewer must use judgment in determining the quality and significance of the reported work. A number of questions require answers, such as the following: can the results of the experiment be repeated, were the plants established for a sufficient period of time to overcome the shock of transplanting or grafting, and was the SO_2 monitoring instrument reliable and calibrated?

Other factors to be considered by administrators in establishing air quality standards for SO_2 are somewhat beyond the scientific criteria stage. Regional, economic, and social factors may be important considerations. What degree of effects is a jurisdiction willing to accept? If it will accept some adverse effects, the air quality standards can be set above threshold levels. If, on the other hand, a jurisdiction desires to safeguard the environment and other receptors against any effects, the air quality standards will have to be set below threshold levels.

Table 14. Air Quality Standards for Sulfur Dioxide

Location	SO_2 Air Quality Standards, Criteria, or Objectives
Canada	0.34 ppm for 1 hr (maximum acceptable)
	0.02 ppm annual average (maximum acceptable)
Alberta	0.20 ppm for 30 min
	0.17 ppm for 1 hr
	0.01 ppm annual average
British Columbia	0.30 ppm for 3 hr (for mining industry)
	0.03 ppm annual average
	0.17 ppm for 1 hr (other industries)
	0.01 ppm annual average
Ontario	0.25 ppm for 1 hr
	0.02 ppm annual average
Quebec	0.50 ppm for 1 hr
	0.02 ppm annual average
United States	0.50 ppm for 3 hr (secondary)
	0.03 ppm annual average (primary)
California	0.50 ppm for 1 hr
Florida	0.10 ppm for 1 hr
	0.003 ppm annual average
Missouri	0.25 ppm for 1 hr (not more than once in any 4 days)
Montana	0.25 ppm for 1 hr (not more than once in any 4 consecutive days)
	0.02 ppm annual average
New Mexico	0.02 ppm annual average
New York	0.25 ppm for 1 hr (not to be exceeded 1% of time in any 12 consecutive months)
North Dakota	0.28 ppm for 1 hr
	0.02 ppm annual average
Vermont	0.10 ppm for 1 hr
	0.02 ppm annual average
Washington	0.25 ppm for 1 hr (not to be exceeded more than 2 times in 7 consecutive days)
	0.02 ppm annual average
AIHA	0.30 ppm for 1 hr
(Recommended 1970)	0.03 ppm annual average
Argentina	0.03 ppm for 30 days
Columbia	0.03 ppm annual average
Belgium	0.06 ppm annual average
Denmark	0.30 ppm for 30 min (not to be exceeded 1% of the time)

Table 14. (Continued)

Location	SO_2 Air Quality Standards, Criteria, or Objectives
Finland	0.28 ppm for 30 min 0.03 ppm annual average
Italy	0.30 ppm for 30 min
Sweden	0.30 ppm for 30 min (not to be exceeded more than 1% of time during a month)
Switzerland	0.30 ppm for 30 min (summer guideline)
West Germany	0.10 ppm for 30 min not exceeding 2.5% of time
Proposed by VDI (Knabe, 1976) (to protect most sensitive plants)	0.02 ppm growing season average
WHO	0.28 ppm for 30 min 0.03 ppm annual average
Israel	0.30 ppm for 30 min
Japan	0.12 ppm for 1 hr
Bulgaria	0.20 ppm for 20 min
USSR	0.20 ppm for 20 min
Czechoslovakia	0.20 ppm for 30 min
East Germany	0.20 ppm for 30 min
Hungary	0.20 ppm for 30 min
Romania	0.30 ppm for 20 min
Poland	0.35 ppm for 20 min

Table 14 lists air quality standards for SO_2 presently in force or proposed or recommended in several jurisdictions throughout the world. From this table it can be seen that many jurisdictions have air quality standards for SO_2 close to 0.30 ppm for 1 hr. Several investigators, however, have reported acute injury to vegetation from SO_2 doses standards for SO_2 close to 0.30 ppm for 1 hr. Several investigators have reported acute injury to vegetation from SO_2 doses approximately in this range or at somewhat higher levels (Table 2, Part B).

In considering other factors, administrators may reduce current SO_2 standards. These other factors include (1) synergism, in which the concentration of SO_2 required to cause injury to vegetation can be lowered significantly when combined with other air pollutants such as O_3 or NO_2; and (2) acidification, in which the oxidation of SO_2 to SO_3 and hydrolysis to H_2SO_4 contributes to acid precipitation, which has the potential to cause effects on soils and plant life. However, the present state of knowledge on the precise role that ambient levels of atmospheric SO_2 play in these two phenomena is incomplete.

Although data have been published in the literature to support a standard of 0.25 ppm SO_2 for 1 hr in order to protect vegetation from acute effects with an adequate margin of safety, the most important precedent for this standard was established by the decisions laid down by the Trail Smelter Arbitral Tribunal in the years 1938 to 1941 (Dean and Swain, 1944). Limitations in SO_2 concentrations were stipulated by the tribunal to prevent injury to vegetation in the vicinity of the Trail smelter in British Columbia, Canada and in the state of Washington, in the United States. Although investigations in the Trail area showed that short-term exposures of 1.0 ppm of SO_2 for 1 hr, or longer exposures of 0.30 ppm for several hours were required to injure vegetation, the following stipulation was made on March 11, 1941: "If the Columbia Gardens recorder (about 6 miles southeast of Trail) indicates 0.30 ppm or more of SO_2 for two consecutive twenty minute periods during the growing season, and the wind direction is not favorable, emissions shall be reduced by 4 tons of sulfur per hour or shut down completely when the turbulence is bad, until the recorder shows 0.20 ppm or less of SO_2 for three consecutive twenty minute periods."

Actually, the tribunal's stipulations became stricter with time. In 1938 the tribunal recommended cutbacks to achieve 0.50 ppm when the recorder showed 1.0 ppm of SO_2 for 1 hr, and 1939 it required reductions to less than 0.30 ppm SO_2 for 1 hr when the recorder indicated a concentration of SO_2 of 0.50 ppm for 40 mins. Apparently, the rationale in 1941 was that 0.30 ppm or more of SO_2 for 40 min during the growing season under unfavorable weather conditions could result in persistent potentially injurious concentrations of SO_2. By cutting back production the SO_2 concentrations were reduced to tolerable levels.

With regard to a long-term standard for SO_2, usually established for a period of 1 year, some controversy exists as to the importance of chronic injury on vegetation and whether chronic injury is the ultimate result of short-term SO_2 fumigations only, or the result of all SO_2 exposures. Section 2.2 in this chapter reviews the data on this subject and presents sufficient support for the occurrence of chronic injury and that this injury may develop over a long period of time under the influence of various environmental factors and the total period of exposure to all SO_2 concentrations including SO_2-free periods.

Establishing a satisfactory annual average of SO_2 is important for the protection of all receptors, including man, animals, vegetation, water and fish. An adequate annual average air quality standard for SO_2

combined with an appropriate short-term standard will help to prevent the following:

1. Chronic forest damage, which includes accumulation of excess sulfur in perennial foliage on trees, progressively increasing injuries to older foliage, early abscission of older foliage, reduced radial and volume of growth of trees, and premature tree mortality.
2. Killing of lichens and changes in the species composition of lichen populations in sulfur-polluted areas. Lichens provide fodder for reindeer and caribou.
3. Adverse effects on lake quality and fish populations caused by long range transport of SO_2 and its oxidation production SO_3 which contribute to acid precipitation.

It may be concluded from the evidence to date that acute injury to native vegetation does not occur below 0.70 ppm (1820 $\mu g/m^3$) of SO_2 for 1 hr or 0.18 ppm (468 $\mu g/m^3$) of SO_2 for exposures of 8 hr. However, acute injury may occur if SO_2 persists for several hours at concentrations over 0.25 ppm (650 $\mu g/m^3$). Furthermore, prominent chronic injury or slight chronic injury to natural forests may occur from average concentrations of SO_2 as low as 0.017 ppm (44 $\mu g/m^3$) or 0.008 ppm (21 $\mu g/m^3$), respectively, over entire growing seasons, in which the SO_2 fumigations are of variable intensities.

REFERENCES

Applegate, H. G. and Durrant, L. C. (1969). *Environ. Sci. Technol.* **3,** 759–760.

Banfield, W. M. (1972). *Phytopathology* **62,** 493.

Barkman, J. J., Rose, F., and Westhoff, V. (1969). In *Air Pollution: Proceedings of the First European Congress on the Influence of Air Pollution on Plants and Animals.* Wageningen, pp. 237–241.

Barrett, L. B. and Waddell, T. E. (1973). U.S. Environmental Protection Agency, Publ. AP-85, Washington, D.C.

Barrett, T. W. and Benedict, H. M. (1970). In J. S. Jacobson and A. C. Hill, Eds., *Recognition of Air Pollution Injury to Vegetation; A Pictorial Atlas.* Air Pollution Control Association, Pittsburgh, Pa., Sect. C, 17 pp.

Bell, J. N. B. and Clough, W. S. (1973). *Nature* **241,** 47–49.

Benedict, H. M. and Breen, W. H. (1955). In *Proceedings of the Third National Air Pollution Symposium.* Pasadena, Calif., pp. 177–190.

Bennett, J. H. and Hill, A. C. (1974). In M. Dugger, Ed., *Air Pollution Effects on Plant Growth.* American Chemical Society, Washington, D.C., Chap. 10, pp. 115–127.

Berry, C. R. (1967). *Phytopathology* **57,** 804.

Berry, C. R. (1971). *Phytopathology* **61,** 231–232.

Berry, C. R. (1974). *Phytopathology* **64,** 207–209.

Bleasdale, J. K. A. (1973). *Environ. Pollut.* **5,** 275–285.

Brennan, E. and Leone, I. A. (1972). *Plant Dis. Rept.* **56,** 85–87.

Brisley, H. R., Davis, C. R., and Booth, J. A. (1959). *Agron. J.* **51,** 77–80.

Constantinidou, H., Kozlowski, T. T., and Jensen, K. (1976). *J. Environ. Qual.* **5,** 141–144.

Costonis, A. C. (1971). *Phytopathology* **61,** 717–720.

Costonis, A. C. (1973). *Eur. J. Forest Pathol.* **3,** 50–55.

Daines, R. H. (1968). In *Proceedings of the Symposium on Air Quality Criteria.* New York, pp. 84–92. (Reprinted from *J. Occup. Med*).

Davis, C. R. (1972). *J. Air Pollut. Control Assoc.* **22,** 964–966.

Davis, D. D. and Wilhour, R. G. (1976). U.S. Environmental Protection Agency, Bull. EPA-600/3-76-102, Corvallis, Ore., 71 pp.

Dean, R. S. and Swain, R. E. (1944). U.S. Department of the Interior, Bull. 453, 304 pp.

De Cormis, L. (1969). In *Air Pollution: Proceedings of the First European Congress on the Influence of Air Pollution on Plants and Animals.* Wageningen, pp. 75–78.

De Sloover, J. and Leblanc, F. (1968). In *Proceedings of the Symposium on Recent Advances in Tropical Ecology.* Varanasi, India, pp. 42–56.

Dochinger, L. S. and Seliga, T. A., Eds. (1976). *Proceedings of the First International Symposium on Acid Precipitation and the Forest Ecosystem.* U.S. Department of Agriculture, Forest Service, Gen. Tech. Rept. NE-23, 1074 pp.

Dochinger, L. S., Bender, F. W., Fox, F. L., and Heck, W. W. (1970). *Nature* **225,** 476.

Dreisinger, B. R. (1965). Paper 65–121, presented at the 58th Annual Meeting of the Air Pollution Control Association, Toronto, 21 pp.

Dreisinger, B. R. and McGovern, P. C. (1970). In S. N. Linzon, Ed., *Proceedings: Impact of Air Pollution on Vegetation, Specialty Conference, Toronto, Ontario,* Air Pollution Control Association, Pittsburgh, Pa., pp. 11–28.

Georgi, H. W. (1970). *J. Geophys. Res.* **75,** 2365–2371.

Gilbert, O. L. (1968). *New Phytol* **67,** 15–30.

Gilbert, O. L. (1969). In *Air Pollution: Proceedings of the First European Congress on the Influence of Air Pollution on Plants and Animals.* Wageningen, pp. 223–235.

Godzik, S. and Linskens, H. F. (1974). *Environ. Pollut.* **7,** 25–38.

Gordon, A. G. and Gorham, E. (1963). *Can. J. Bot.* **41,** 1063–1078.

Gorham, E. and Gordon, A. G. (1960). *Can. J. Bot.* 38, 307–312.

Gorham, E. and Gordon, A. G. (1963). *Can. J. Bot.* **41,** 371–378.

Guderian, R. and Stratmann, H. (1962). *Forschungber. Landes Nordrhein-Westfalen* **1118,** 7–102.

Guderian, R. and Van Haut, H. (1970). *Staub-Reinhalt* **30,** 22–35.

Hawskworth, D. L. (1973). *Air Pollution and Lichens.* University of Toronto Press, Chap. 3, pp. 38–76.

Hawksworth, D. L., Rose, F., and Coppins, B. J. (1973). *Air Pollution and Lichens.* University of Toronto Press, Chap. 16, pp. 330–367.

Heagle, A. S. (1973). *Ann. Rev. Phytopathol.* **11,** 365–388.

Heck, W. W. (1968). In *Proceedings of the Symposium on Air Quality Criteria.* New York, pp. 64–67. (Reprinted from *J. Occup. Med.)*

Heck, W. W. and Brandt, C. S. (1975). In A. C. Stern, Ed., *Air Pollution, Vol. 1, 3rd Ed., Academic Press, New York, Chap. 14.*

Hibben, C. R. and Taylor, M. P. (1975). Environ. Pollut. **9,** 107–114.

Hill, A. C., Barrett, T. W., Hill, S., and Lamb, C. (1973). Paper 73–109, presented at the 66th Annual Meeting of the Air Pollution Control Association, Chicago, Ill., 21 pp.

Hill, G. R., Jr. and Thomas, M. D. (1933). *Plant Physiol.* **8,** 223–245.

Houston, D. B. (1974). *Can. J. Forest Res.,* **4,** 65–68.

Johansson, O. (1954). Lic. dvh. vid K. Lantbrukshögskolan, Uppsala.

Jones, H. C., Cunningham, J. R., McLaughlin, S. B., Lee, N. T., and Ray, S. (1973). Paper 73–110, presented at the 66th Annual Meeting of the Air Pollution Control Association, Chicago, Ill, 24 pp.

Karnosky, D. F. (1976). *Can. J. Forest Res.* **6,** 166–169.

Karnosky, D. F. and Stairs, G. R. (1974). *J. Environ. Qual.* **3,** 406–409.

Katz, M. and Ledingham, G. A. (1939). In *Effect of Sulphur Dioxide on Vegetation.* National Research Council, Bull. 815, Ottawa, Canada, Chap. XI, pp. 262–287.

Katz, M. and McCallum, A. W. (1939). In *Effect of Sulphur Dioxide on Vegetation.* National Research Council, Bull. 815, Ottawa, Canada, Chap. X, pp. 244–261.

Katz, M. and Shore, V. C. (1955). *J. Air Pollut. Control Assoc.* **5,** 144–150.

Katz, M., et al. (1939). National Research Council, Bull. 815, Ottawa, Canada, pp. 1–447.

Keller, T. (1976). Preprint, presented at XVI IUFRO World Congress, Oslo, Norway, 12 pp.

Keller, T. and Bucher, J. (1976). *Schweiz. Z. Forstwese* **127,** 476–484.

Knabe, W. (1976). *Ambio* **5,** 213–218.

Kock, G. (1935). *Z. Pflanzenk.* **15,** 44–45.

Laundon, J. R. (1967). *Lichenologist* **3,** 227.

Leblanc, F. (1969). In *Air Pollution: Proceedings of the First European*

Congress on the Influence of Air Pollution on Plants and Animals. Wageningen, pp. 211–221.

Leblanc, F. and De Sloover, J. (1970). *Can. J. Bot.* **48,** 1485–1496.

Leblanc, F., Rao, D. N., and Comeau, G. (1972). *Can. J. Bot.* **3,** 519–528.

Leone, I. A. and Brennan, E. (1971). Paper 71–64, presented at the 64th Annual Meeting of the Air Pollution Control Association, Atlantic City, N.J., 21 pp.

Linzon, S. N. (1958a). Joint Publ., Ontario Department of Lands and Forests and Ontario Department of Mines, Toronto, 45 pp.

Linzon, S. N. (1958b). *For. Chron.* **34,** 50–56.

Linzon, S. N. (1965). *For. Chron.* **41,** 245–250.

Linzon, S. N. (1966). *J. Air Pollut. Control Assoc.* **16,** 140–144.

Linzon, S. N. (1969). In N. J. Lacasse and W. J. Moroz, Eds., *Handbook of Effects Assessment, Vegetation Damage*. C.A.E.S., Pennsylvania State University, University Park, Pa., Sec. VIII, 13 pp.

Linzon, S. N. (1971). *J. Air Pollut. Control Assoc.* **21,** 81–86.

Linzon, S. N. (1972). *For. Chron.* **48,** 182–186.

Linzon, S. N. (1973a). Paper 73–107, presented at the 66th Annual Meeting of the Air Pollution Control Association, Chicago, Ill., 19 pp.

Linzon, S. N. (1973b). Preprint, Fourth Joint Chemical Engineering Conference, Vancouver, B. C., pp. 1–17.

Linzon, S. N. (1975). *HortScience* **10,** 494–495.

Linzon, S. N., McIlveen, W. D., and Temple, P. J. (1973). *Water, Air, Soil Pollut. J.* **2,** 129–134.

Majernik, O. and Mansfield, T. A. (1970). *Nature* **227,** 377–378.

Mandl, R. H., Weinstein, L. H., and Keveny, M. (1975). *Environ. Pollut.* **9,** 133–143.

Masaru, N., Syozo, F., and Saburo, K. (1976). *Environ. Pollut.* **11,** 181–187.

Materna, J., Jirgle, J., and Kucera, J. (1969). *Ochr. Ovzdusi* **6,** 84–93.

McCallan, S. E. A., Hartzell, A., and Wilcoxon, F. (1936). *Contrib. Boyce Thompson Inst.* **8,** 189–197.

McGovern, P. C. and Balsillie, D. (1973). *Effects in the Sudbury area (1972)*. Ontario Ministry of the Environment, Sudbury, 50 pp.

McGovern, P. C. and Balsillie, D. (1974a). *Effects in the Sudbury area (1973)*. Ontario Ministry of the Environment, Sudbury, 47 pp.

McGovern, P. C. and Balsillie, D. (1974b). *Effects in the Wawa area (1973)*. Ontario Ministry of the Environment, Sudbury, 30 p.

Menser, H. A. and Heggestad, H. E. (1966). *Science* **153,** 424–425.

Metcalfe, C. R. (1941). *Ann. Appl. Biol.* **28,** 301–315.

Mrkva, R. and Grunda, B. (1969). *Acta Univ. Agric. Brno* **38,** 247–270.

Murray, J. J., Howell, R. K., and Wilton, A. C. (1975). *Plant Dis. Rept.* **59,** 852–854.

O'Connor, J. A., Parbery, D. G., and Strauss, W. (1974). *Environ. Pollut.* **7,** 7–23.

Odén, S. (1968). Swedish National Science Research Council, Ecology Committee, Bull. 1, 68 pp.

O'Gara, P. J. (1922). *Abstr. Ind. Eng. Chem.* **14,** 744.

Overrein, L. N. (1976). "Report from the International Conference on the Effects of Acid Precipitation in Telemark, Norway, June 14–19, 1976," *AMBIO* **5,** 200–252.

Przybylski, Z. (1967). *Postepy Nauk Roln.* **2,** 111–118.

Rao, D. N. and Leblanc, F. (1967). *Bryologist* **70,** 141–157.

Roberts, B. R. (1976). *Envir. Pollut.* **11,** 175–180.

Saunders, P. J. W. (1966). *Ann. Appl. Biol.* **58,** 103–114.

Scheffer, T. C. and Hedgcock, G. G. (1955). U.S. Department of Agriculture, Forest Service, Tech. Bull. 1117, Washington, D. C., 49 pp.

Schofield, E. and Hamilton, W. L. (1970). *Biol. Conserv.* **2,** 278–280.

Setterstrom, C., Zimmerman, P. W., and Crocker, W. (1938). *Contrib. Boyce Thompson Inst.* **9,** 179–198.

Shen-Miller, J., Hunter, M. B., and Miller, J. (1976). *Plant Physiol.,* Annual Meeting Supplement, **57,** 50.

Shriner, D. S. (1976). U.S. Department of Agriculture, Forest Service, Gen. Tech. Rept. NE-23, pp. 919–925.

Skye, E. (1964). In *Svensk Naturvetenskap.* Statens Naturvetenskapliga Forskningsrad, Stockholm, pp. 327–332.

Skye, E. (1968). *Acta Phytogeorgr. Suec.* **52,** 1–123.

Smith, W. H. (1974). *Environ. Pollut.* **6,** 111–129.

Sobotka, A. (1964). *Lesn. Cas., Praha* **10,** 987–1002.

Spierings, F. (1967). *Atmos. Environ.* **1,** 205–210.

Stoklasa, J. (1923). Urban and Schwartzenberg, Berlin.

Stratmann, H. (1963). *Forschungber. Landes Nordrhein-Westfalen,* No. 1184, 69 pp.

Swain, R. E. (1923). *Ind. Eng. Chem.* **15,** 296–301.

Swain, R. E. and Johnson, A. B. (1936). *Ind. Eng. Chem.* **28,** 42–47.

Tamm, C. O. (1976). *AMBIO* **5,** 235–238.

Tanaka, K., Okada, T., and Hanami, K. (1975). Japan Forestry Society, pp. 290–292.

Temple, P. J. (1972). *J. Air Pollut. Control Assoc.* **22,** 271–274.

Thomas, M. D. (1951). *Ann. Rev. Plant Physiol.,* pp. 293–322.

Thomas, M. D. and Hendricks, R. H. (1956). In P. L. Magill, F. R. Holden, and C. Ackley, Eds., *Air Pollution Handbook.* McGraw-Hill Book Co., New York, Sec. 9, 44 pp.

Thomas, M. D., Hendricks, R. H., and Hill, G. R. (1952). In L. McCabe, Ed., *Air Pollution,* McGraw-Hill Book Co., New York, pp. 41–47.

Tingey, D. T. and Reinert, R. A. (1975). *Environ. Pollut.* **9,** 117–125.

Tingey, D. T., Reinert, R. A., Dunning, J. A., and Heck, W. W. (1971). *Phytopathology* **61,** 1506–1511.

Tingey, D. T., Reinert, R. A., Dunning, J. A., and Heck, W. W. (1973). *Atmos. Environ.* **7,** 201–208.

U.S. Department of Health, Education, and Welfare (1969). *Air Quality Criteria for Sulfur Oxides.* Publ. AP-50, Washington, D.C., 178 pp.

U.S. Environmental Protection Agency (1973). *Revised Chapter 5, Air Quality Criteria for Sulfur Oxides.* Document EPA-R3-73-030, Research Triangle Park, North Carolina, 43 pp.

Van Haut, H. and Stratmann, H. (1970). *Landes Nordhrein-Westfalen.* Verlag W. Girardet, Essen.

Vass, K. and Ingram, M. (1949). *Fd. Mf.* **24,** 414.

Waddell, T. E. (1970). U.S. National Air Pollution Control Administration, unpublished report, pp. 1–60.

Weinstein, L. H., McCune, D. C., Aluisio, A. L., and Van Leuken, P. (1975). *Environ. Pollut.* **9,** 145–155.

Wells, A. E. (1917). *Ind. Eng. Chem.* **9,** 640–646.

White, K. L., Hill A. C., and Bennett, J. H. (1974). *Environ. Sci. Technol.* **8,** 574–576.

Wislicenus, H. (1901). *Z. Angew. Chem.* **28,** 689–712.

Wolozin, H. and Landau E. (1966). *J. Farm Econ.* **48,** 394–405.

Wood, T. and Bormann, F. H. (1974). *Envir. Pollut.* **7,** 259–268.

Wood, T. and Bormann, F. H. (1975). *AMBIO* **4,** 169–171.

Wood, T. and Bormann, F. H. (1976). U.S. Department of Agriculture, Forest Service, Gen. Tech. Rept. NE-23, pp. 815–825.

Woodwell, G. M. (1970). *Science* **168,** 429–433.

Zahn, R. (1961). *Staub.* **23,** 343–352.

Zimmerman, P. W. and Crocker, W. (1934). *Contrib. Boyce Thompson Inst.* **6,** 455–470.

Zimmerman, P. W., and Hitchcock, A. E. (1956). *Contrib. Boyce Thompson Inst.* **18,** 263–279.

5

PHYSIOLOGICAL AND BIOCHEMICAL EFFECTS OF SULFUR DIOXIDE ON PLANTS

Jan-Erik Hällgren

Department of Plant Physiology, University of Umeå, Umeå, Sweden

1. INTRODUCTION

Studies of the effects of sulfur dioxide (SO_2) on plants were first initiated in the late nineteenth century. The existing literature is therefore voluminous, and it has frequently been reviewed (1–12). Most studies have tended to concentrate on the visible symptoms produced by SO_2, and we may now conclude that these, although not always easily distinguishable from symptoms of other harmful substances or of diseases, are fairly well known (13–15).

Such descriptive studies do not of themselves provide a basis for a mechanistic understanding of the physiological reactions to SO_2 as an air pollutant. However, the rapidly increasing interest and activity in environmental research have led to important advances in our knowledge of the physiological effects of SO_2, and several reviews of plant responses to SO_2 on metabolic levels have been published (16–21). Lichens and bryophytes, because of their well-documented sensitivity to air pollutants, have been the subject of considerable interest, and the effects of SO_2 on these organisms have also been reviewed (22–26).

In this chapter only primary effects of pollutant SO_2 on the photosynthesis of higher plants will be discussed. The intention is not to reexamine all details covered in recent reviews. Much available information will probably be excluded, and a few aspects may perhaps be overemphasized. Since our understanding of the effects of SO_2 on higher plants is still incomplete, part of the discussion may appear fragmentary and some parts even speculative.

2. GENERAL ASPECTS

Sulfur dioxide has a special position among air pollutants since sulfur is one of the essential plant nutrients (27). As a consequence, both harmful and beneficial effects are conceivable (e.g., 28–33). However, the manner in which SO_2 affects the metabolism of the plant is not well understood, partly because of the limited understanding of the normal metabolic processes of plants, the way in which they are integrated and controlled, and also because of the metabolic effects of SO_2 itself. The biochemical effects of SO_2 involve many processes that occur at different times and on different levels of organization, and therefore any investigation of them must use different types of experimental systems (34).

At the level of the plant, or the plant leaf, the metabolic process of primary interest is photosynthetic carbon dioxide fixation. Since 90 to 95% of the dry weight of plants is derived from this process (35), the impairment of photosynthesis has been a popular candidate for explaining plant growth reduction caused by SO_2 (17). However, plant growth is the end result of a complex interaction of many processes, any of which may limit the rate and all of which are influenced, either directly or indirectly, by environmental factors. Sulfur dioxide may then be regarded as one of these environmental factors (36). It should also be noted that photosynthesis is seldom measured by those concerned with SO_2 and the growth of plants, and the growth or growth rate of the plant is seldom mentioned in photosynthetic experiments with SO_2. When permanent injury to plant foliage occurs, it is generally agreed that a comparable reduction in photosynthetic rate will take place (1,3). It is also known that apparent photosynthesis can be inhibited before visible injury occurs, and that it can recover from inhibition caused by unfavorable SO_2 concentrations (37–42). Such inhibition of photosynthesis may temporarily cause metabolic alterations, but these need not be followed by visible injuries (13). Conversely, it could be argued that a temporary inhibition of photosynthesis need not imply a shortage of aviable nutrients or assimilates at the growing region and hence should not express itself as reduced dry weight. One might also argue that SO_2-induced changes in metabolism might become evident on a later occasion. Impaired CO_2 assimilation by leaves could, for example, impair the formation of reproductive organs or fruit development (43–47).

Interpretations of the results of SO_2 experiments are difficult because the plant responses and photosynthetic responses depend not only on the concentration or duration of exposure but also on several environmental factors such as light, temperature, humidity, nutritional and water status, and previous history (37). The evaluation is further complicated by the diversified reactions of different species (46–53) or different individuals within the same species (54–60), and the disparate responses of leaves and other organs in different stages of development or age (7,61–63). It must be stressed that the physiological status of the plant under study is very important. The status of the plant will determine not only photosynthesis but also the uptake of SO_2 from the bulk air, and the developmental stage will influence the source-sink relationship and the fixation and translocation of carbon and sulfur compounds within the plant (64,65). High amounts of sulfur may accumulate in rapidly growing parts (e.g., in the shoot apex) (65). Sulfur anions or radicals, which are known to affect cell division, may be formed and may inhibit growth (66).

The vegetative and reproductive growth of the plant is further regulated by several plant growth hormones. So far, extremely little is known about SO_2 effects on the formation, translocation, or action of auxins, gibberellins, cytokinins, abscisic acid, or ethylene, although a limited number of experiments *in vitro* and *in vivo* may suggest that SO_2 interacts with some of these substances (67–70). Several observed effects *in vivo*, for example, leaf abscission (71,72), earlier senescence (73), and impaired flowering (13,45) and seed germination (45,47), may well be explained by SO_2 effects on the complicated interaction of the natural plant hormones. More than 20 years ago it was stated that more work on the effects of air pollution on plant regulators was needed (74). That statement is still valid.

The influence of environmental stress factors on plant growth and photosynthesis is evident (see Linzon, this volume). However, only a few experimental investigations have been concerned with combined environmental effects, such as water stress, mineral deficiency, climatic stress, and several air pollutants in combination, although the latter is the prevailing situation in urban and industrial environments. To evaluate SO_2 effects in the field is an unavoidable but extremely difficult task. It is known that sensitive plants such as *Pinus silvestris* are negatively affected at mean concentrations of SO_2 over 80 $\mu g\ m^3$ during the vegetation season in the Ruhr area (12). Certain lichen species are even more sensitive, and the most sensitive species may be affected at 13–26 $\mu g/m^3$ (24,75–77). It has also been concluded by Knabe that measurable economic effects on agriculture and horticulture can be expected at mean concentrations somewhere between 80 and 125 $\mu g\ SO_2/m^3$ in Germany (12), and experiments have shown that acute injury to the foliage of sensitive species develops after fumigation with low SO_2

concentrations (78,79). One of the most sensitive clones of *Pinus strobus* shows visible effects on 4- to 5-week-old needles from 130 $\mu g\ SO_2/m^3$ for 1 hr (80).

Other selected sensitive species have also been shown to react to low SO_2 concentrations (81). It is beyond the scope of this chapter to examine field and laboratory experiments measuring growth and visible damages. It may be mentioned, however, that Knabe has recently described and suggested a form of standardized analyses that might be recommended for use in field investigations (82). In addition, it is evident that much experimental work is needed on different levels of plant organization (83).

On the ecophysiological side of plant research, identification of the factors that determine the rate of absorption of SO_2 and the metabolic effects of absorbed components is of ultimate importance (64). This chapter will begin with a brief discussion of the uptake of SO_2 from air and then examine some of the subsequent effects on the photosynthetic mechanism.

3. SULFUR DIOXIDE EXCHANGE BETWEEN PLANT AND ATMOSPHERE

Different plant species have different capacities to remove SO_2 from the surrounding air (11), and their sorbtive properties are basically prescribed by their genetic constitutions and expressed through the morphological structure and physiological processes. The uptake of SO_2 by the plant leaves can be expressed and analyzed in an analogous manner to the uptake of CO_2 or any other gaseous compound (84–86). To analyze exchanges of CO_2 and water vapor between the atmosphere and the plant, experimental methods have been developed that permit the analyses to be performed in terms of potentials; fluxes, and resistances (86). Unsworth et al. (87) give an excellent and comprehensive description of an analytical approach by which the responses of plants to their physical environment can be separated from physiological changes induced by pollutants such as SO_2. Bennett and Hill (88) and others (e.g., 89–100) have described the interactions between air and vegetation by using resistance analogues to model the uptake of air pollutants in the field. These methods are very useful for studying the photosynthesis and respiration of plants, for detailed examination of the activity of various plant organs, and also for pinpointing the influence of environment and pollutants on plant growth. Unfortunately, this approach has seldom been used in SO_2 fumigation experiments, thus contributing to the circumstance that our knowledge about this important pollutant is limited. Indeed, the lack of well-defined experimental conditions makes it difficult to perform meaningful comparisons of photosynthetic responses to SO_2 between species.

In the typical ambient concentration range, uptake of SO_2 increases linearly with increasing concentration (100); however, this does not always seem to be the case at higher concentrations (e.g., 54,62,101). The rate of uptake of SO_2 per unit leaf area and time, divided by the concentration of SO_2 in the surrounding air, has the dimensions of velocity. The term is therefore called the velocity of deposition (V_g). The reciprocal of V_g is the total resistance to mass transfer (r). The total resistance is the sum of the external and internal resistances within the leaf, since it is assumed that these act sequentially. The principal resistances to CO_2 flux are located in the laminary boundary layer, the stomata, the cuticle, the intercellular spaces, and the liquid phase in the cell wall and cytoplasm of the mesophyll cells (86). Some important differences exist between CO_2 and SO_2, because of the different chemical characteristics and metabolic roles of the two compounds (88). Furthermore, although the principal resistances to intake of gases by the leaves are known, we do not know exactly the relative importance of each of these resistances in determining the total resistance (102). It has been assumed that the resistances to the intake of SO_2 may be the same as the resistance to the loss of water (102). In that case the total resistance is mainly the sum of the boundary layer and stomata resistances, adjusted for SO_2 diffusion. Numerous factors will affect the gaseous exchange, and the steady-state absorption rate will also depend upon the plant's ability to metabolize, translocate, or otherwise remove the active pollutant from the absorbing solution (100).

3.1. Laminary Boundary Layer Resistance

The boundary layer constitutes the main aerodynamic resistance to the transfer of SO_2 or any other gas. The thickness of the boundary layer, and thus its resistance, are mainly a function of wind speed. The resistance is also much influenced by the dimensions, anatomy, and orientation of the leaf. In the immediate vicinity of the leaf, there is a transition from turbulence to molecular diffusion of the gas. The resistance of transfer in the boundary layer is therefore proportional to $D^{-2/3}$ (103), where D represents the molecular diffusion coefficient of the gas.

The very variable boundary layer resistance is connected in series with other resistances that are under physiological control. Quite obviously, it is of utmost importance that the boundary layer resistance be small in relation to the leaf resistances when photosynthetic responses to SO_2 are to be observed sensitively (87).

3.2. Stomatal Resistance

The stomata constitute a variable diffusion resistance between the atmosphere and the interior of the leaf. The stomatal resistance depends on several morphological and physiological factors and is also influenced by numerous environmental factors, of which light and water are the most important. The transfer of gases through the stomata is generally considered to proceed by molecular diffusion, and consequently the resistance is inversely proportional to the molecular diffusion coefficient. However, the possibility of mass transfer should not be totally excluded (104). The physiological mechanisms regulating stomatal resistance and the action of different substances, including SO_2, on stomata have been reviewed (105–108).

As early as 1921, Loftfield (109) could show that lucerne was susceptible to SO_2 only if the stomata were open. The vast majority of investigations since then confirm this result, and it is generally accepted that the primary factor controlling SO_2 uptake by plant leaves is the degree of opening of the stomata. Nevertheless, one cannot always expect to find a close correlation between the stomatal apertures, or the number of stomata (110), and the susceptibility to SO_2. The reason is, as pointed out by Mansfield (107), that there exist several other resistances to SO_2 uptake, connected in series as well as in parallel to the stomatal resistance, and that during some circumstances these might be the factors limiting SO_2 uptake.

The Action of Sulfur Dioxide on the Stomata

It was earlier believed that SO_2 had no direct effect on the stomata. Later, in a series of experiments, Majernic and Mansfield (111–114) determined more accurately the stomatal responses to SO_2 in *Vicia faba*. They demonstrated that with increasing SO_2 concentration (650 to 2600 $\mu g/m^3$) there was, at high relative humidity, a decrease in the viscous flow resistance of stomata. At low relative humidity and the same range of SO_2 concentrations, the stomata closed. Prolonged exposure to SO_2 could cause irreversible effects, and the stomata could lose their ability to close. Similar stimulatory effects were observed on the stomata of *Zea mays* and *Hordeum vulgare* (112).

Using concentrations of SO_2 between 72 and 1430 $\mu g/m^3$, Biscoe et al.

(92,115) could also show that the stomata of SO_2-treated *Vicia faba* and *Zea mays* plants had lower diffusion resistances than those of untreated plants. Exposure to 290 μg SO_2/m^3 throughout the day suppressed the normal diurnal variation in stomatal aperture. The minimum concentration to which the stomata responded was 72 $\mu g/m^3$, but from other experiments in which the SO_2 level was gradually increased from zero, it could be suggested that the stomata responded to 58 and 28 $\mu g/m^3$ in *Vicia* and *Zea*, respectively. These authors' results with water-stressed plants are of great importance, since they showed SO_2 to stimulate stomatal opening under these conditions. If the plants were transferred to clean air the following day, normal stomatal behavior was recovered. The stimulation of stomatal opening by SO_2 clearly will have undesirable consequences for the plant. First, the increase in transpiration can lead to early and more severe water deficit during dry conditions. Second, the pollutant will have easier access to the mesophyll tissue, and hence the photosynthesis and metabolic processes may be more drastically affected.

The results reported in the literature are somewhat contradictory. High concentrations of SO_2 have been reported to result in a decline or oscillation of the transpiration (1,13,40,70,116–118). A closure reaction induced by SO_2 during conditions that would have favored stomatal opening in *Vicia faba* has been observed in *Nicotiana* and *Pelargonium* (70,116,118) plants. Bonté et al. (118) pointed out that the different reactions observed in different species could also be noticed between individuals of the same clone. Generally, drought is considered to be protective of the plant since the stomata close during this condition. Schramel (119) fumigated plants with SO_2 under drought conditions and observed an increase in stomatal aperture, although the diffusive resistance of the leaves remained approximately the same as in the control plants. He believed that the protective effect of drought might be associated with higher diffusive resistances of leaves, resulting in a lower adsorption of the toxic gas, since the effect could not be explained by the influence of SO_2 on stomata. Markowski et al. (120) concluded that the size of stomatal apertures, as well as the diffusive resistances of the leaves, was correlated to plant resistance to SO_2 as manifested by visible injuries.

Indirect evidence for an action of SO_2 on stomata comes from the observed effects on transpiration. In several investigations, an initial increase in transpiration has been noticed during SO_2 exposure (121–124). From experiments with *Pinus silvestris* it has been reported that this initial increase in transpiration can exceed 100% (124). Coniferous plants seem to be highly sensitive: prolonged exposure of such plants to

SO_2 may cause permanent impairment of stomatal regulation, leading to the loss of their ability to close the stomata (125). Halbwachs (117,126) has also concluded that SO_2 concentrations usually considered harmless may cause a permanent increase in water loss in coniferous plants. However, in some investigations with *P. silvestris* stimulation of the transpiration has not been detected (127), or has been evident only at higher SO_2 concentrations (128).

A question of central interest is the long-term effects of low SO_2 concentrations on the stomata under natural conditions. Bell and Mudd (60) investigated SO_2 effects on two varieties of *Lolium perenne* with different sensitivities to SO_2 but did not find any significant differences in stomatal resistance between fumigated and control plants after short or after long fumigation times. Moreover, Cowling and Lockyer (129) found no effect of 50 $\mu g\ SO_2/m^3$ on the transpiration coefficient for the culitvar S23 of *L. perenne*.

The Mechanism of Sulfur Dioxide Action on the Stomata

Clearly, the action of SO_2 on stomata cannot be understood from the existing information, and thus remains to be explained (107). The fact that the guard cells accumulate sulfur after fumigation with SO_2, as shown by microautoradiography (122), does not provide any explanation as to the effects found. Biscoe et al. (92) have suggested that SO_2 may have a direct effect on the turgor balance between the guard cells and the subsiduary cells, thus causing an opening reaction. Explanations in terms of CO_2 balance (106) may offer an alternative clue to the action of SO_2 if the CO_2 concentration in the vicinity of the stomata is decreased by stimulated photosynthesis. However, this has not generally been observed in photosynthetic measurements with intact plants, although it could be suggested from studies on chloroplasts (130).

It has also been demonstrated that HSO_3^- disturbs the fluxes of K^+, Cl^-, and H^+ (131,132). The influence of SO_2 on these ion fluxes associated with stomatal movements has, however, not been investigated. On the whole, the mechanism of SO_2 influence on stomatal opening and closure is totally unknown, and there is a need for well-designed experiments to obtain more information (107).

3.3. Cuticular Resistance

The cuticle covering the epidermis of the leaf is not fully impermeable to gases, and SO_2, as well as many other substances, may enter and leave the leaf by diffusion through this barrier. The rate is, however, expected to be slow since the cuticle normally offers considerable resistance to diffusion. Consequently, cuticular diffusion is usually ignored in comparison with the parallel diffusion through stomata. Spedding (89) has calculated the leaf cuticle resistance to SO_2 and considers it to be in the same range as that for CO_2. Fowler, cited in Unsworth et al. (87), has pointed out, however, that the total cuticular resistance to SO_2 transfer sometimes is much less than the corresponding resistance to CO_2 or water vapor transfer.

Another unsolved problem is the sorbtion of SO_2 in water films on the leaf surface. Some investigations indicate that the surface wetness is of importance when the dry deposition rate is calculated (89,95), though calculations of the fixation of SO_2 by oxidation show that this process is relatively unimportant (133). When the cuticle is wet for long periods, as during nighttime (89,95) or during the winter, the deposition rate of SO_2 may be considerably enhanced in spite of the fact that the stomata are "closed." Quantitative uptake of SO_2 in conifers has been measured during the winter period (134), and specific so-called winter injuries have also been described in conifers (135). These results merely confirm that, even if the rate of SO_2 entry is expected slow, stress damage may occur after exposure during periods when the SO_2 concentration is high. In addition to the SO_2 level, the pH of water in which SO_2 can be dissolved will also influence the values of stomatal and cuticular resistance (89). If the permeability of the cuticle to SO_2 is low, any SO_2 adsorbed on the leaf surfaces is more likely to be removed by rainfall than to be absorbed through the cuticle and thus contribute to the wet deposition. Emission of gaseous sulfur compounds (136–138) or leakage of sulfur from leaves by rain (139) also contributes to the removal of deposited sulphur.

The Action of Sulfur Dioxide on Leaf Surfaces

Sulfur dioxide may also react with chemical compounds making up the cuticle and thus change the cuticular resistance. It has been suggested that SO_2 stimulates the production of waxes, thereby building up the cuticular structure, and the turbidity of hot water extracts has been used as an indicator method (140–143). From scanning electron micrographs it can be seen that SO_2 alters the surface of conifer needles (144). The stomatal pores of many conifers are located at the bottom of a chamber shaped like a cup or an inverted cone. This antechamber is filled with wax tubes. An increase in wax production here may increase the stomatal resistance since it has been estimated that about one third of this resistance is accounted for by the wax tubes in the antechamber (145). The effects of SO_2 deposition on plant leaf surfaces were discussed by Saunders (146) in 1971, and the pH effects have more recently been investigated (e.g., 147–150). It can be noted that the acidic effects of SO_2 are associated with an increased loss of nutrient cations, especially K^+, from the leaves (148,149). The effects of SO_2 and particulate matter on plant leaf surfaces are more thoroughly discussed by Linzon (this volume).

3.4. Internal Resistance

The removal capacity of the sink is a function of SO_2 solubility in the water layer surrounding the cells of the mesophyll. The solubility of SO_2 in water is governed by Henry's law. Since SO_2 is very soluble in water, and since the substomatal cells and cavities usually are saturated with water, it is likely that the internal concentration drives SO_2 into solution in the mesophyll water. During normal conditions the resistance inside the leaves to SO_2 is considered very small. However, Spedding estimated the internal resistance to SO_2 and surprisingly found it to be of the same magnitude as that observed for CO_2 (89). No further measurements of the mesophyll resistance have been found in the literature, but since the physiological activities are of importance this resistance can be expected to be highly variable and to differ between plant species and physiological states of plants (87). Gases dissolved in water diffuse more slowly than in air, and the length of the path in the liquid phase has to be considered as a diffusion barrier, which may also affect the relative importance of the stomata. However, the mesophyll resistance is not a purely diffusive resistance if we consider that SO_2 from the intercellular spaces must cross the cell wall, the cell membrane, and a thin layer of cytoplasm before it reaches the chloroplast envelope. While diffusing inward into the cell and the chloroplast to exert a primary effect on photosynthesis, SO_2 may disturb the integrity of the membranes and may also disrupt the ionic environment of the cell and thereby exert secondary effects.

Also, the chemical fate of the SO_2 molecule in the liquid phase must then be considered. The membranes will certainly be more permeable to SO_2 ($SO_2 \cdot H_2O$) than to the charged HSO_3^- and SO_3^{2-} molecules that rapidly form when SO_2 dissolves in water (151). The undissociated acid (H_2SO_3) cannot be distinguished from SO_2 + water by its Raman spectrum (152), and there is some doubt that it will exist in solution (19). Depending on the pH of the water phase, a number of ionic states in equilibrium can be assumed (151,153). Puckett et al. (153,154) have pointed out that the characteristics of the different forms of dissolved SO_2 can be understood by examining their standard electrode potentials. These show that SO_2

may act as a reducing or an oxidizing agent, depending on the pH of the medium (153). It is evident from the data that all forms of dissolved SO_2 become increasingly better oxidizing agents as the pH of the medium is lowered (154). The pH of the substomatal water phase is probably slightly acidic; the pH values inside cells and chloroplasts are variable and very difficult to measure, but optimum pH values for the enzymes of the reductive pentose phosphate cycle are considered to be around 7 to 8 for most plants. In this pH range HSO_3^- and SO_3^{2-} will be present, but SO_3^{2-} will be predominant at the higher pH (151). However, the effective redox potentials are also dependent on the concentration and the nature of other solutes, and thermodynamic predictions based on reduction potentials are limited if the kinetics of a reaction are unfavorable.

Early estimates (155) of the concentration of the dissolved SO_2 in the aqueous solution surrounding the cells are disputable (156). Since the equilibrium is nonlinear and strongly temperature dependent, one cannot, as often has been done, use a single approximation and apply the estimation to a wider range of concentrations. On the basis of experimental data, Niboer et al. (156) calculated that at 18°C 0.2 ppm SO_2 v/v in air could correspond to 5 ppm (w/v) in solution. During the same conditions 1 ppm SO_2 in air could give rise to 10 ppm in solution. What are the actual concentrations inside the mesophyll cells and in the chloroplasts, after fumigation with SO_2, at different dosages? From fumigation experiments (4 hr/day for 30 days at 23°C) at an ambient concentration of 500 $\mu g/m^3$, Steinhart et al. (98) calculated the SO_2 uptake sufficient to produce visible damage, which they called the damage uptake horizon (DUH). They estimated that the DUH for alfalfa was approximately 48×10^{-8} $\mu g/cm^3$ of mesophyll water. Their experiments indicated that saturation concentration in the mesophyll was reached within about 1 hr at the conditions used. However, SO_2 exchange outside the mesophyll tissue or incorporation into the metabolic pathway was not included in their model. One might argue that the concentration of SO_2 within the plant is an undeterminable quantity; however, there are several good reasons that warrant an answer to this question.

4. BUFFERING CAPACITY

The acidic effects of SO_2 can, to some extent, be buffered in the plant leaf, presumably by plant leaf proteins (157). In most cells the major buffering substances operative in the pH range of 6 to 8 are phosphate compounds (with a pK_a of 6.8 to 7.0) and certain amino acids like -SH of cystein, both free and in proteins (158). Some of the effects of SO_2 have been attributed to acidifying effects, either as sulfurous acid or after oxidation to sulfuric acid (159). It is evident that the plant buffering system may be momentarily overloaded when high concentrations of SO_2 are used in experiments, even if the fumigation period is short. The plant has a better capacity, and is able to buffer more acid, in fumigations utilizing low SO_2 concentrations (157). Hence, if two experiments with quite different SO_2 concentrations are compared, there is a risk that one may find quite contrasting results and draw completely different conclusions. It has been suggested that the depression of CO_2 assimilation is due to sulfurous acid accumulation resulting from the oxidation of SO_2 on the surface of cells, and the accompanying lowering of pH. If this occurs, the recovery of photosynthesis will be associated with the neutralization of the sulfurous acid by the plant and a return to the normal pH of the cell surface (37). The pH change is usually considered to be slow (157). However, it has been reported that homogenates of leaves exposed to SO_2 have a lower buffer capacity than do unexposed ones (142,161).

Field and laboratory studies with lichens show that the sensitivity of these organisms is closely related to the buffering capacity of the lichen species and the buffering capacity of the substrate (162–164). The most sensitive species are generally those with the lowest buffering capacity for acid substances (162,164). It must be noted, however, that the overall sensitivity of a lichen species depends on a multiplicity of factors, and one cannot simply regard the buffering capacity as the sole determinant of field sensitivity (24,163,165). There appears to be no clear correlation between the SO_3^{2-} uptake of lichens and their field or laboratory sensitivities to SO_2 (166). It has also been proved in numerous experiments that lowering the pH per se does not cause effects comparable to those found when SO_2 and derived anions are present (e.g., 24,167–169).

5. SULFITE OXIDATION

Undoubtedly SO_3^{2-} is oxidized to SO_4^{2-} in the plant leaf, and this ability has been correlated with resistance to toxicity (170). Aerobic oxidation of SO_3^{2-} is known to be enhanced by ultraviolet radiation, by catalysts such as metal ions, and by several enzymes, including several oxidases as well as cytochrome oxidase, peroxidase, ferredoxin, and ferredoxin-NADP reductase (171). It may be assumed that part of the SO_3^{2-} formed by SO_2 uptake probably oxidizes directly according to the redox conditions prevailing in the cells (18).

Studies indicate that the predominant site for SO_3^{2-} oxidation is the chloroplast (130,171–173). In isolated chloroplasts the SO_3^{2-} oxidation is shown to be dependent on the electron transport in photosynthesis, since the oxidation is greatly inhibited by 3-(3,4-dichlorophenyl)-1,1-climethylurea (DCMU) (130,171,172). This is probably initiated through the univalent reduction of oxygen, and it has been convincingly demonstrated that the photooxidation of SO_3^{2-} is powerfully inhibited by superoxide dismutase (SOD) (171–173). The chloroplast enzyme SOD catalyzes the dismutation of $O_2^{\cdot-}$ into O_2 and H_2O_2, and therefore inhibits the reaction in which $O_2^{\cdot-}$ participates. An extensive review of O_2 radicals, hydrogen peroxide, and O_2 toxicity is given by Fridovich (174), and free radicals in photosynthesis are also discussed by Loach and Hales (175). It is known that the strong reductant in photosystem I (PS I) can also mediate the reduction of O_2 (e.g., 176,177). The electron transport from water to oxygen is probably a physiological reaction somehow involved in the regulation of photosynthesis (176–178) or, alternatively, an electron leakage caused by inefficient NADP reduction, thus forcing electrons to oxygen (179). Asada and Kiso (171) studied the effects of SO_3^{2-} on the oxygen uptake by spinach chloroplasts. From their results they could propose that a superoxide anion formed by the univalent reduction of oxygen on the reducing side of the electron transport system was the initiator of the aerobic oxidation of SO_3^{2-}.

Oxidation of SO_3^{2-} can be initiated by any system that induces univalent reduction of oxygen or univalent oxidation of SO_3^{2-}. In illuminated chloroplasts the donation of one electron from SO_3^{2-} to PS I or PS II may initiate the reaction. Libera et al. (130) showed that 0·25 to 5/mM SO_3^{2-} could stimulate the photoreduction of ferricyanide and NADP. The inhibition of oxygen evolution and the reduction of ferricyanide by pretreatment of the chloroplasts with high concentrations of Tris buffer could also be compensated for by the addition of SO_3^{2-}. This was not seen, however, when NADP reduction was measured. From this study these authors suggested that low SO_3^{2-} concentrations could stimulate electron transport by donating electrons to PS II. They also observed an enhancement of the CO_2 fixation at SO_3^{2-} concentrations below 1/mM. However, the increased CO_2 fixation was not accompanied by a stimulated ATP production. In a later study Ziegler and Libera (173) demonstrated that the enhanced CO_2 fixation also could be completely abolished by adding SOD. They have suggested that SO_3^{2-} is acting in a manner similar to that of ascorbate, and that it is oxidized by the superoxide radical to form HSO_3 radical, which in turn acts as an electron acceptor for PS I, being reduced to SO_3^{2-} again.

It might be suggested that if superoxide or hydroxyl radicals are intermediates in water oxidation in PS II, both radicals may initiate the oxidation. The SO_3^{2-} oxidation may also be initiated by hydrogen peroxide stimulated by peroxidase (180) and/or cytochrome oxidase (180,181). Hydrogen peroxide formed by the Mehler reaction (182) in chloroplasts was, however, not expected by Asada and Kiso to be the initiator of the SO_3^{2-} oxidation, since the photooxidation was not affected by cyanide (171). They also noticed that catalase did not affect the reaction, but ferredoxin plus NADP and cytochrome *c*, as well as SOD, prevented the oxidation.

It has also been reported that chlorophyll initiates the oxidation of SO_3^{2-} in response to illumination (183); however, Asada and Kiso considered that the rate of this reaction was much lower than the photooxidation rate induced by PS II.

Experiments *in vitro* show that SO_3^{2-} can be oxidized to SO_4^{2-} by a free radical chain reaction during the nonenzymatic oxidation of methionine (69). The free radical oxidation of SO_3^{2-} is accelerated in the presence of IAA (69,184) or tryptophan and Mn^{2+} and oxygen (69,184,185). What are the consequences for the plant *in vivo* of the fact that the indole compounds also are oxidized in this reaction? The SO_3^{2-} can be oxidized enzymatically by sulfite oxidase, which is found in plant leaf mitochondria (186,187). This enzyme is Mg^{2+} and cytochrome *c* dependent. It has also been suggested that the SO_2-induced reduction in mitochondrial respiration was related to the oxidation of SO_3^{2-} (188).

Evidently, there exist several systems in plant leaves that can mediate the SO_3^{2-} oxidation, although photooxidation has been suggested as the most important one (171). It is hoped that more information will become available to explain the free radical reactions connected with SO_3^{2-} oxidation and photosynthetic electron transport. Of what relevance here is the reported SO_2-induced stimulation of peroxidases (143,190,191) or inhibition of catalase (191,192)? The participation of sulfite oxidase is unclear, and the SO_4^{2-} accumulation in relatively high amounts after SO_2 fumigation may be due to a sulfite oxidase in the cytoplasm (18).

5.1. Reduction of Sulfite and Sulfate

The SO_4^{2-} formed by oxidation of SO_3^{2-} will be reduced to the S^{2-} level via the assimilatory pathway of sulfate reduction, reviewed by Shiff and

Hodson (193). The chloroplast is the demonstrated site also for sulfate reduction, and this process appears to be connected with the photosynthetic electron transport that delivers ATP for SO_4^{2-} activation. Trebst and Schmidt have published a reduction rate of 3 μM SO_4/mg chlorophyll·hr for spinach chloroplasts (194,195). The intermediates in the pathway presented are normally bound to carriers, and it is then unlikely that free SO_3^{2-} would exist in the chloroplast (193). However, it is known that SO_3^{2-} can undergo a number of exchange reactions (196), and one might assume that a dose-effect relationship exists between bound and free forms of SO_3^{2-} and S^{2-} (193). Since the presence of soluble sulfite reductase(s) in chloroplasts has been established (cf. 197), it has been attractive to explain SO_3^{2-} reduction by this system (198). The action of this protein, active for free SO_3^{2-}, seems to be connected, via ferredoxin, to the photosynthetic electron transport chain (197). Whether this enzyme also can catalyze the $6e^-$ transfer in the normal sulfate pathway (193,197,199) is not yet known. Hence it is also possible that the soluble sulfite reductase that has been found functions in some parallel pathway. Several reductases more or less specific for SO_3^{2-} have been described, and evidently some of them do not function in the normal sulfate reduction pathway (193).

Chloroplast preparations of plant leaves can reduce SO_3^{2-} to S^{2-} or H_2S (194,195,198–200). It has also been reported that plants exposed to SO_2 can release measurable quantities of H_2S into the surrounding atmosphere in the light, but not in the dark (136). The photoreduction of SO_2 to S^{2-}, protonated in the plasma to H_2S, is inhibited by DCMU, which shows the requirement for PS II activity as an electron donor. The photoreduction of H_2S from sulfate or thiosulfate by the blue-green alga *Synechococuss lividus* has been investigated by Sherdian (201). The results indicate that, in addition to water, oxidative metabolism may also act as a source of electrons to reduce the photooxidant under certain conditions.

If the SO_2 is reduced, the sulfide can be used directly for formation of cystein (202) in a reaction which is the quantitatively most important step for sulfur incorporation into organic compounds (203). Another important reaction in sulfur metabolism is the formation of 6-sulfoquinovose (193,204), found in the sulfolipids of plants, particularly in the chloroplast membrane. In this case, SO_3^{2-} could be used without prior reduction or oxidation.

Experiments with *Lemna minor* indicate that, although these plants can reduce SO_3^{2-} directly to S^{2-} (H_2S) *in vitro,* this reaction plays no significant role *in vivo* (206). It is suggested from the experiments with *Lemna* that the effects of SO_2 cannot be attributed to a setting up of toxic concentrations of SO_3^{2-} (207). Fankhauser et al. (207b) concluded that SO_2-sulfur enters the sulfoquinovose of the sulfolipids and the sulfur amino acids after oxidation to SO_4^{2-} and subsequent reduction.

It is evident that important parts of SO_3^{2-} and SO_4^{2-} metabolism take place within the chloroplasts, interconnected with photosynthetic electron transport and phosphorylation. As SO_2 readily affects photosynthesis, it will certainly also have indirect consequences for the sulfur metabolism of the leaf and the plant. Unfortunately, in most experiments where the intention has been to study the fate of incorporated SO_2 (e.g., $^{35}SO_2$) very high concentrations of SO_2 have been used. In some cases these SO_2 concentrations probably totally inhibited the photosynthetic process. Most of the available information concerning the uptake and fate of SO_2 has been derived from experiments using simple fractionation techniques and analyses of tissue (208–210). The radiosulfur investigations demonstrate that $^{35}SO_2$ absorbed at sublethal uptake rates accumulates as SO_4^{2-} (cf. 18) and is metabolized in essentially the same way as SO_4^{2-} supplied via the roots (211). The $^{35}SO_2$ has been shown to be incorporated into many sulfur-containing metabolites and is found in nearly all places where protein accumulates (18). The transport of sulfur within the plant has been described by Läuchli (212) and characterized by Smith (213).

An increase in plant productivity has been attributed to this improvement of the sulfur supply to metabolism, especially if SO_2 is fed to plants growing in sulfur-deficient soils (32,33,129,211). It can be noted that sulfur deficiency also induces changes in the photosynthetic rates and in the CO_2 exchange in leaves of *Beta vulgaris* (214). Most of this decrease in the rate of CO_2 uptake could be attributed to an increase in the mesophyll resistance, and initially sulfur deficiency had only small effects on stomatal resistance (214). A supply of sulfur from SO_2 might then also stimulate photosynthesis by decreasing the mesophyll resistance.

Although work with higher plants has indicated that SO_2 can function as a source of sulfur, this finding has been contradicted by studies with the green alga *Chlorella pyrenoidosa* (215,216). Under conditions of SO_4^{2-} deficiency the effects of HSO_3^- on this alga are uniformly negative (216).

The relation between the sulfur content in the plant leaves and the first visible signs of injury (e.g., 13,14), as well as estimates of threshold concentrations of SO_2, has been reported from numerous experiments (e.g., 217–219). Threshold values are very difficult to assess, and the variables that can affect the development of symptoms and the determinations are so numerous that further refinements of these estimates can hardly be regarded as meaningful. On the cell level the view of the threshold for biological response to a toxic agent advanced by Dinman (220) has been considered when the potential toxicity of SO_2 is discussed (221).

6. EFFECTS ON PHOTOSYNTHESIS

Although the sulfur accumulation in leaves sometimes exhibits a certain relationship to photosynthesis, it does not always seem to coincide with the degree of leaf or plant sensitivity established by visible injuries (62). Leaves are reported to be more sensitive during morning hours than later in the day when net photosynthesis and SO_4^{2-} accumulation reach their maxima (1,62). In this connection it may be of interest that potato foliage, whose stomata remain open at night, has been reported to be more susceptible to SO_2 toxicity during the dark hours (222). Setterström and Zimmerman (223) concluded that both prefumigation in the dark and growing plants under heavy shade increased the susceptibility to SO_2. The reports that plants are much more sensitive during the morning hours have led to the assumption that the sensitivity of the leaf is associated with its sugar content (3). However, there is only sparse and indirect evidence to support this hypothesis. Other more or less specified conditions that maximize or minimize the effect of SO_2 on the photosynthesis of higher plants and lichens have already been described (1–10,13,37–42,49,70,223,224).

The inhibition of photosynthesis is often regarded as the first sign of SO_2 action on plants (e.g., 225–228). In some species, however, physiological processes such as nitrogen fixation (229) or other processes mediated by sensitive enzymes (230) may be equally or even more rapidly inhibited by SO_2 (cf. 21). An initial stimulation of photosynthesis at low SO_2 concentration has been observed in studies with chloroplasts (130,173), algae (132,229), lichens (226), and higher plants (42), even by acid (230a). Generally the effect is short lived and is followed by a substantial inhibition at prolonged fumigation time or at elevated SO_2 concentration (42). The magnitude of this inhibition generally differs between species (38,49) and also between leaves of different age (38). Not only the inhibition but also the recovery rates differ between plants and plant leaves of different ages (38,40). Sulfur dioxide sensitivity as a function of leaf age and the photosynthetic responses of *Phaseolus vulgaris*, a reportedly SO_2-sensitive species, and *Zea mays*, which is regarded as SO_2-tolerant, studied by Sij and Swanson (40). At 2600 $\mu g/m^3$ for 1 hr the kinetics of inhibition of photosynthesis for the two species were similar; however, at 7800 $\mu g/m^3$ apparent photosynthesis was inhibited less in *Z. mays* than in *P. vulgaris*. It could also be noted that the recovery level was higher in the species more tolerant to SO_2. There is experimental evidence that photosynthesis decreases at much lower SO_2 concentrations in combination with other pollutants (42,231,-231a). Bull and Mansfield (42) showed that the net photosynthesis in detached leaves of *Pisum sativum* was reversibly inhibited by 130 to 260 $\mu g\ SO_2/m^3$. Lichens, which are the most sensitive plants on the basis of field studies, do not seem to differ markedly from higher plants in photosynthetic studies under laboratory conditions. In short-time fumigations the most sensitive lichen species yet studied is *Lobaria pulmonaria,* whose photosynthesis was inhibited after exposure to 1300 $\mu g\ SO_2/m^3$ for 14 hr (41). The studies (132,156,226) of lichens submerged in aqueous solutions are more difficult to evaluate (232) but may indicate that lower SO_2 concentrations affect photosynthesis.

It has generally been difficult to explain the observed results on photosynthesis and the underlying mechanism of action. Most studies of the SO_2 action on photosynthesis have been related to the photosynthetic performance of the plant, for example, the net CO_2 exchange. From these studies, however, information about the underlying mechanisms can hardly be obtained. To clarify the action of SO_2 in detached leaves, chloroplasts or isolated membrane or enzyme systems have been used. Although only a selection of a plant's numerous physiological or biochemical processes may be monitored in any experiment, such studies are needed to gain an understanding of the potential for SO_2 reactivity based on a reasonable chemical model. A view of the potential reactivity of SO_2 with biological systems has been advanced by Petering and Shih (221). Although numerous experiments with plants have been ambivalent concerning SO_2 toxicity at low concentrations, and it is evident that there are several plausible explanations of the SO_2 action on photosynthesis, several investigators have stressed the acid effect and the one on chlorophyll and membranes. Others consider that the central point of attack may be represented by the competitive inhibition between CO_2 and SO_2 in relation to the CO_2-fixing enzyme. Other plausible explanations, such as interference with redox reactions and electron flow and phosphorylation in photosynthesis, or attack of proteins, enzymes, or blocking of metal-binding sites, have also been offered. Other well-known reactions of aqueous SO_2 have also been shown to be involved in its phytotoxicity. The elucidation of the complex effects of SO_2 is still in its infancy, but some interpretations may be discussed.

7. EFFECT ON CHLOROPHYLL AND OTHER PIGMENTS

About 20% of the total lipid content of chloroplasts consists of chlorophyll, which is an integrated functional brick in the chloroplast membrane. In the living system chlorophyll exists in a highly organized state. Chlorophylls may undergo several (photo)chemical reactions such as oxidation, reduction, phaeophytinization, and reversible bleaching (cf. 232a).

Since the studies of Willstätter and Stoll (233), the formation of phaeophytin, caused by acid substances, has been known. Chlorophylls are converted to phaeophytins by acid substances, whereby Mg^{2+} is split off and replaced by atoms of hydrogen. This transformation has been observed in studies both of lichens (24) and of higher plants exposed to high SO_2 concentrations (e.g., 227–229). For a review of lichen studies, see Le Blanc and Rao (24). The phaeophytin formation in plants is known to be catalyzed enzymatically (235). It is also known that light accelerates the phaeophytinization (235a), and this has been suggested as one explanation of why the SO_2 effects are more pronounced in the light than in the dark (24). Chlorophyll *a* and *b* have received most attention in pigment studies; chl *a* tends to be destroyed at a faster rate than chl *b*, as can be seen from *in vivo* and *in vitro* studies (e.g., 54,234). This difference in stability can easily be demonstrated by use of acidic agents such as SO_2, which converts chl *a* to phaeophytin *a*. However, rapid phaeophytin formation has been observed only in studies with high SO_2 concentrations and with low pH. These situations rarely exist in the field, and phaeophytin formation may have very little relevance to the decrease of photosynthesis found in plants exposed to SO_2 (236).

Recently it was demonstrated that the effect of SO_2 on pigment breakdown and photosynthesis was a specific effect and was not a function of increased acidity. From experiments with *Pinus contorta*, Malhotra (236a) reported that concentrations below 100 ppm SO_2 in solution had no effect on chl *a* or phaeophytin *a*. However, at lower concentrations of SO_2 (10 to 50 ppm) a significant increase in chlorophyllase activity was detected, and chl *b* was converted to the corresponding chlorophyllid *b*. The enzyme converts chlorophylls to chlorophyllids by removal of the phytol group. Chlorophyllase, which was discovered by Willstätter and Stoll (237), is intimately associated with chlorophyll, and its activation is also influenced by light (237a). However, it is possible that this enzyme is associated with a lipoprotein chlorophyll complex and thus is inactive *in vivo*. The maximum activity is obtained when the enzyme is solubilized (237a). The influence of SO_2 on this hydrolytic reaction mediated by chlorophyllase in plants remains to be fully investigated.

The *in vivo* chlorophyll destruction by SO_2 is complex to evaluate, and the vast majority of work merely confirms that there is a net pigment destruction influenced by pH, light, and other factors. Hence the mechanism of chlorophyll destruction is not explained, although the most likely reaction is an oxidation of the pigment molecule. In a series of investigations by Puckett and co-workers, it has been shown that SO_2 induces Chlorophyll bleaching associated with a spectral absorption shift toward shorter wavelengths. Light is required for bleaching to occur, and the effects resemble those found with other oxidants, such as permanganate (153,156). A nonenzymatic oxidation of chlorophylls results initially in the formation of allomerized derivates (238), and these exhibit a spectral absorption maximum of 10 to 20-nm shorter wavelengths as compared to chl *a*. However, the exact mechanism of oxidation and attack by SO_2 on chlorophylls *in vivo* is not known, although it may be due to an effect on the redox potentials of the pigment-carrier complexes. One possible explanation of the SO_2-induced irreversible photooxidation of chlorophyll *in vivo* is that SO_2 or formed radicals inhibit the electron transport chain, thus inhibiting reversible reduction of reaction centra. This might ultimately lead to an oxidation of the light-harvesting antennae of chlorophyll.

From *in vitro* studies at pH 4 Peiser and Yang (238a) have suggested that free radicals produced by aerobic oxidation of bisulfite are involved in the destruction of chlorophylls. The light requirement of this destruction could be partially substituted for by Mn^{2+}, and the destruction was further stimulated by linoleic acid. When linoleic acid hyperoxide (LOOH) was present, chlorophyll was rapidly destroyed by 0.5 m*M* HSO_3^-. Addition of α-tochopherol prevented the chlorophyll destruction. According to Peiser and Yang, free radicals produced from LOOH decomposition by HSO_3^- were responsible for the chlorophyll destruction (238a).

The chlorophyll content of chloroplast membranes fluctuates during the season. Reports as to the extent to which these natural variations can be changed or delayed or enhanced by SO_2 are scarce (239). The natural changes are due to adaptations in the structure and function of the chloroplast membrane to climatic stress. One can assume that during periods of adaptation (e.g., during spring or autumn) the chloroplast membrane may be more sensitive to attacks by radicals or oxidizing or reducing agents. Although most studies have dealt with chlorophyll degradation, it has been noticed that chlorophyll formation can also be inhibited by high SO_2 concentrations (239a). Since several factors determine the actual chlorophyll contents of leaves, the effects of SO_2 are not always evident in the field situation.

Jensen (240) measured chlorophyll content in SO_2-fumigated *Populus* cuttings, but found no reduction in chlorophyll even though visible injury was observed.

The observed changes in pigment differ widely from one species to another, and even between varieties of the same species (e.g., 60). There are also different responses within the same algal species in various lichens (156). The different effects on chlorophyll synthesis or degradation may be important factors determining the resistance of the plant to the pollutant (60). However, several investigations have shown that the rate of photosynthesis in short-term experiments is affected before any effect on the chlorophyll can be detected (132,154,225,226,229,42).

In addition to chlorophyll, other important photosynthetic pigments can also be affected by SO_2 (241,242). Degradation of β-carotene has been shown to reflect the action of SO_2 (242,243), in contrast to earlier findings (244). Measurements of chlorophyll content are normally included with the measurements of photosynthesis. As pointed out, measurements of the chlorophyll contents of leaves can hardly be a means to explain the action of SO_2 on photosynthesis. A more promising method might be the use of chlorophyll fluorescence measurements, which are discussed below.

8. EFFECTS ON PHOTOSYNTHETIC ELECTRON TRANSPORT AND PHOSPHORYLATION

When the studies of SO_2 action on photosynthesis in isolated chloroplasts are considered, some of the reported results may seem contradictory. Part of the discrepancy may be due to the use of different methods and experimental conditions. Different kinds of chloroplast populations (245) or chloroplasts isolated from plants in different stages of development have probably been used. Malhotra has shown that the degree of inhibition of oxygen evolution by SO_2 in pine needle chloroplasts is dependent on the developmental stage of the needles used for chloroplast preparation (246). The state of the chloroplast preparations may be one of the most important parameters when the effects of SO_2 are examined.

It has been shown convincingly that the oxygen evolution of intact spinach chloroplasts is inhibited by SO_2, HSO_3^-, SO_3^{2-}, and SO_4^{2-} (246–248). Low SO_4^{2-} concentrations inhibit oxygen evolution when coupled to CO_2 or phosphoglycerate reduction (248). Conversely, SO_4^{2-} at the same concentration does not inhibit oxygen evolution in envelope-free chloroplasts (248). However, rapid inhibition is seen in intact chloroplasts, where the Calvin cycle is functioning as the terminal electron and hydrogen acceptor, whereas this is not the case in "broken" chloroplasts, where added ferricyanide or NADP serves as "artificial" electron acceptor (248). The inhibition of oxygen evolution in entire spinach chloroplasts by SO_4^{2-} or other sulfur anions has been shown to be prevented by the addition of orthophosphate (246) or pyrophosphate (247). There are also a number of investigations showing that the sulfur anions have no direct inhibitory effect on the rate of photosynthetic electron transport (249–257). Provided that artificial electron acceptors are used, some investigations have even shown that the reduction of ferricyanide or NADP can be accelerated by the addition of SO_3^{2-} or SO_4^{2-}, in sonicated or envelope-free chloroplasts, in the presence of ADP and Mg^{2+} (257).

The marked reversal of the inhibition after addition of phosphate (247) and the finding that SO_3^{2-} and SO_4^{2-} can stimulate electron transport by the uncoupling of photophosphorylation (257) offer one explanation for the SO_2 inhibition of oxygen evolution in intact chloroplasts. Inhibition of photophosphorylation will ultimately lead to inhibition of the Calvin cycle. This will result in an accumulation of reduced NADP, and a decrease in electron transport and oxygen evolution.

Although the inhibitory action of SO_2 and the sulfur anions on photophosphorylation has been well documented, the mechanism of inhibition is not fully understood. Clearly, SO_3^{2-} and SO_4^{2-} compete with phosphate in photophosphorylation, since this has been demonstrated in all phosphorylating systems tested: cyclic and noncyclic phosphorylation, postillumination ATP formation, and acid base-driven phosphorylation (252). These anions have also been shown to inhibit ATP formation in mitochondria and bacterial chromatophores (253,254). Ballantyne (253) suggested that SO_2 may change the ratio of oxidized to reduced sulfydryl compounds in mitochondria, and inferred that the *in vivo* effect was not directly related to SO_3^{2-} effects on ATP synthesis. Pick and Avron (252) suggested that SO_4^{2-} could act as an energy-transfer inhibitor during photophosphorylation. ATP synthesis takes place on the so-called coupling factor 1 (CF 1), which is the probable target for an energy-transfer inhibitor. Jagendorf has reviewed the mechanism of photophosphorylation and proposed an interesting hypothesis for this SO_4^{2-} inhibition (255).

Roy and Moudrianakis (256) have further reported that the light-dependent formation of coupling factor ADP complex from AMP and P_i

was inhibited by SO_4^{2-}. It has also been suggested by Datta et al. (257) that, when ADP and ATP are present, a state of CF 1 is achieved which can be inhibited by SO_4^{2-} and permanganate. In the presence of ADP and Mg^{2+}, SO_4^{2-} has been shown to cause an irreversible inactivation of CF 1, with an end point at about 50% (258). Work by Grebarnier and Jagendorf (259) also indicates that SO_4^{2-} causes an uncoupling of photophosphorylation. These authors suggest that SO_4^{2-} causes an alteration in the interaction between CF 1 and the thylacoid membranes which leads to an increase in the permeability of these membranes to protons.

Although the mechanism of SO_4^{2-} inhibition is not fully understood, a number of investigations may serve as good starting points for further experimentation (255). It must also be stated that the vast majority of investigations undertaken to explain the mechanism of photophosphorylation by using SO_4^{2-} as an analogue of PO_4^{2-} have not been concerned primarily with SO_3^{2-} or other sulfur anions. Also, the concentrations of SO_4^{2-} used have sometimes been high, and it is difficult to make reliable comparisons with the SO_4^{2-} concentrations found in plant leaves after prolonged exposure to SO_2. Hence the extent to which one may explain the decrease of photosynthesis by inhibition of photophosphorylation is disputable. Sij and Swanson (40) tested C3 and C4 species with 1 ppm SO_2 for short times, and suggested that the observed effects on apparent photosynthesis could result from this uncoupling effect of the sulfur anions. Silvius et al. (247) showed that HSO_3^- was much more inhibitory to oxygen evolution than were both SO_3^{2-} and SO_4^{2-}. However, each of these ions inhibited cyclic and noncyclic photophosphorylation to a similar extent. Silvius et al. therefore doubted that the extent of the inhibition of photophosphorylation could entirely explain the inhibition of photosynthetic oxygen evolution by SO_2. In searching for mechanisms involving ATP, the activation of SO_4^{2-} in the sulfate reduction pathway must also be borne in mind.

The results of Ziegler et al. (130) regarding sulfite oxidation, which were discussed previously, must also be considered. They demonstrated that the CO_2 fixation and ATP concentration (turnover rate) in chloroplasts were decreased by SO_3^{2-} in concentrations above 1 mM and that the inhibition of CO_2 fixation was not altered after the addition of SOD. Ziegler (260) has demonstrated that RUBP-carboxylase is inhibited by SO_3^{2-}, and it was shown earlier that SO_4^{2-} can also affect this enzyme (261). The interference with CO_2 fixation also helps to explain the SO_2 effects on oxygen evolution in entire chloroplasts, since a decrease in CO_2 fixation will indirectly affect oxygen evolution by PS II.

Several other possible sites and targets for SO_2 inhibition in the photosynthetic process, however, have been suggested by work in other laboratories. It was stated earlier that dissolved SO_2 can act either as an oxidizing or a reducing agent, depending on the pH (e.g., 153). From a thermodynamic view there is a possibility that the reduction of these compounds interferes with reactions in the photosynthetic electron transport chain (262). As a nucleophilic agent, the sulfite sulfur, even in low concentrations, interacts directly with the pyridinium moiety of, for example, NAD and NADP (263,264). Nicotinamide adenine dinucleotide rapidly forms a reversible adduct with HSO_3^- in a manner analogous to its reaction with other nucleophiles such as H^- and CN^- (264). It must be noted that the equilibrium and rate constants may vary markedly, depending on the environment of the reaction. The formation constants for $NAD \cdot SO_3$ and $FAD \cdot SO_3$ seem to increase substantially when these coenzymes are protein bound (cf. 221).

When SO_3^{2-} combines with NADP, it gives a hydroxypyridine-4-sulfonic acid that is capable of forming an undissociable complex with dehydrogenases and hence can inhibit the hydrogen transfer (263). It can be suggested that SO_3^{2-} may interfere with redox reactions in the photosynthetic process mediated by these nucleotides, since SO_3^{2-} and HSO_3^- are reactive, and essentially active as nucleophiles toward electrophilic centers.

In contrast to SO_4^{2-}, SO_3^{2-} is a strong ligand and binds to iron/heme-containing enzyme centers (265). The formation of metal complexes may, at least in part, help to explain SO_2 blockage of metalloenzyme-mediated processes in photosynthesis (156). Moreover, earlier investigations (e.g., 266) reported an increase in water-soluble "chloroplast iron" which was proportional to the duration of fumigation. The chloroplast iron is now known to represent ferredoxin and cytochromes in the photosynthetic electron transport chain. Also, SO_3^{2-} is known to bind to vitamin B_{12} efficiently (267), and this too has been used analytically (226). At a moderate rate HSO_3^- *in vitro* is known to react also with menadiol, a derivative of vitamin K (cf. 221). The influence of SO_2 on the vitamin contents of plants has recently been investigated (268). Ferredoxin, cytochromes, and vitamins occupy key positions in plant metabolism and potosynthesis, and the effects of SO_3^{2-} and HSO_3^- on these and similar metabolites are therefore of special intrest, when searching for explanations of SO_2 inhibition on photosynthesis. Direct extrapolation of the *in vitro* experiments is not possible, however, and the relative importance of these reactions *in vitro* needs scrutiny.

It is also tempting to speculate that free radical reactions with HSO_3^- and SO_3^{2-}, mediated by photochemical and enzymatic processes or metals, could lead to destructive processes inhibiting electron transport and oxygen evolution in photosynthesis.

9. MEMBRANE EFFECTS

Evidence has lately accumulated indicating that the anions formed from SO_2 affect the membranes of chloroplasts and other organelles. The membrane effects of bisulfite compounds were summarized by Brinkman et al. (269). Lüttge et al. (131) futher investigated the effects of bisulfite compounds and HSO_3^- on photosynthesis and ion uptake in leaf slices. Transient changes in H^+ flux were shown to be inhibited by 5×10^{-4} *M* glyoxal bisulfite (GBS) or HSO_3^-. Also, the Cl^- transport across cell membranes was strongly inhibited at this concentration. It was suggested that bisulfite compounds caused a nonspecific effect on the membranes at concentrations lower than those required for inhibition of enzyme activities *in vivo*. The uncoupling effect on phosphorylation caused by these inhibitors may be one example of the nonspecific alteration of the membrane integrity (131).

Murray and Bradbeer (270) also concluded that sulfonate compounds attack primarily the chloroplast membrane. α-Hydroxysulfonates are formed through the reactions of SO_3^{2-} with, for example, aldehydes. The α-hydroxysulfonates have received considerable interest in plant physiological studies (e.g., 271). In C3 plants the sulfonates either have no effect on CO_2 fixation or, under some conditions, stimulate it (272). In C4 species, on the other hand, these compounds generally depress photosynthesis (273). Sulfonates are, however, potent enzyme inhibitors, and glycolate oxidase, a key enzyme in photorespiration, has been the subject of special interest, as will be discussed below (cf. 274). It has been pointed out by Ziegler (18) that the different actions of SO_3^{2-} and sulfonates on RuBP and PEP carboxylase indicate that sulfonates cannot serve as a model for SO_2 action. Glyoxal bisulfite and α-hydroxysulfonates can be detected in leaves after SO_2 fumigation (275,276). However, the concentrations used in these experiments were very high, and the relevance of the reactions to the HSO_3^- or SO_3^{2-} concentrations more likely to be found in cells has not been clarified. However, the reactivity of the SO_3^{2-} is such that one might expect reactions with phenolic compounds and quinones, as well as with other products such as aldehydes, especially α,β-saturated aldehydes and ketones (151). The reaction with aldehydes and ketones is applicable to C5 and C6 sugars formed in the dark reaction of photosynthesis. Sulfur dioxide reactions have been known for many years, and sulfite has often been used by food chemists to prevent enzymatic browning in stored tissues, for example, because of its inhibitory effect on phenol oxidase (277). Another well-known reaction is the thiamine cleavage by SO_3^{2-} (cf. 227). Fumigation experiments with SO_2 have also shown that the thiamine content of plant leaves is decreased (70,278), and it has been suggested that all thiamine-dependent enzymes could be generally inhibited by high SO_2 concentrations (279).

One of the most important aspects of protein and membrane disruption is the cleavage of disulfide linkages by SO_3^{2-}. This reaction is well characterized in simple systems (e.g., 280). Cecil and Wake (280) have pointed out that certain disulfide bonds in proteins are readily broken (e.g., cystine), whereas others are highly resistant. Since the structures and functions of several proteins are highly dependent on the integrity of the disulfide bonds, breakage of these bonds should gradually deactivate several enzymes and alter membrane proteins. Grill and Esterbauer (281) investigated the effect of SO_2 on pine and spruce trees and found that the S-H content was two to four times higher in damaged trees than in those suffering no injury. However, it can be argued that this is not a direct demonstration of effects on the disulfide linkages (19).

The HSO_3^- reacts chemically with the amino acids tryptophan and methionine (282), and this attack may be important in addition to any breakage of disulfide bonds.

Conformational changes and splitting of proteins not involving disulfide linkages have been demonstrated (132). Although somewhat conflicting results have been presented, Pahlich (159) concluded that only the membrane-associated form of glutamatic-oxaloacetic transaminase (GOT) was affected by SO_2, the cytoplasmatic form of the enzyme being unaffected. He also suggested that the results in respect to glutamate dehydrogenase (GDH) could be explained by the sulfur anion altering the mitochondrial membrane. As mentioned above, SO_2 is known to disturb the fluxes of ions through membranes and to cause K^+ efflux from plant membranes (132, 148, 156).

The observed leakage of K^+ and primary photosynthetic products (132,156,283) is also consistent with the destruction of protein structure in the chloroplast membranes. It is evident from studies with lichens that the semipermeability of the membranes is destroyed by high concentrations of SO_2. Naturally, the observed curtailment of photosynthesis, ATP formation, and so forth could be explained by this effect. However, the leakage of K^+ and other ions is obviously not a sufficiently sensitive parameter whereby to assess the influence of SO_2 during short-term exposures (156). Lichen studies have shown that K^+ was released into the incubation medium only at high concentrations of SO_2. With an increase in the incubation time, however, release of this ion was evident at much lower concentrations of SO_2 (156). The efflux of K^+ seems to be a general SO_2 response that, it has been suggested, can be used as a simple method of evaluating the SO_2 sensitivities of different lichen species (156). Niboer et al. (156) have shown that the K^+ efflux can be detected before any chlorophyll degradation occurs. However, it must be noted that lichens, during rewetting after a dry period, show a leakage of ions, photosynthetic products, and nitrogen products (284).

From ultrastructural investigations of SO_2-fumigated plants, it is evident that effects on chloroplast membranes can be observed before visible injury of the leaves occurs (285–288). The first symptoms of membrane damage are swelling of the grana thylakoids and the fret channels between the grana and granulation of the stroma. It is interesting that Wellburn et al. (285) found this swelling to be reversible after short-term fumigation with SO_2. In a study of the biochemical activity and ultrastructural organization of pine needle chloroplasts, Malhotra (246) confirmed that the decrease of oxygen evolution produced by 500 ppm SO_2 in solution was accompanied by ultrastructural changes in the chloroplasts. Also, in lichens the decrease in photosynthesis has been correlated with changes in chloroplast organization (289). Occasionally, degradation of starch within the chloroplasts seems to follow the SO_2 action.

10. EFFECTS ON CHLOROPHYLL FLUORESCENCE

To obtain more information about SO_2 effects *in vivo,* alternative methods such as measurements of chlorophyll fluorescence may be potentially useful (290). Several fluorescence parameters are measurable, although the fluorescence spectra and the relative fluorescence yield are the most readily studied (cf. 291). In living cells the intensity of chl *a* fluorescence changes as a function of time. This characteristic variable fluorescence, the so-called Kautsky effect (291), consists of a fast change and a slower fluctuation. The fast change is sensitive to the rate of electron transport through PS II. The slow change is thought to be related to temporary alterations in the thylacoid membrane ultrastructure and pigment systems associated with ionic transport and photophosphorylation; see the review by Papageorgio (291). The characteristic variable fluorescence has been used as a technique to study the action of metal contaminants and SO_2 in chloroplasts and algae (290). The advantage of this technique is that it is sensitive, nondestructive, rapid, and potentially informative.

To have an effect on the fluorescence properties of chl *a in vivo,* the pollutant must affect the photosynthetic electron transport and/or the pigment composition. The effect of the pollutant will be reflected in the kinetics and yield of the variable fluorescence, as well as in the shape of the excitation and emission spectra. The variable fluorescence will be affected if SO_2 causes blockage on either side of the oxidizing or reducing of the PS II photoreaction. The effects of SO_2 on the variable fluorescence and the fluorescence spectra have been studied in our laboratory. At pH 8.0 and 1.0 m*M* SO_3^{2-} there were increases in the fluorescence yield of spinach chloroplasts, but the opposite effect was observed at pH 6.2, where HSO_3^- dominates (292). Arndt (290) has noticed both a slight SO_3^{2-} stimulation of fluorescence at low concentrations and a decrease at higher concentrations (>1 m*M* 10^{-3} *M*), indicating two different modes of action of this compound on the electron transport chain in photosynthesis. Kessler and Zunft (293) further noted that SO_4^{2-} did not show any effect on fluorescence, while 5×10^{-4} *M* SO_3^{2-} produced a slight decrease of the steady-state fluorescence.

It may be mentioned that delayed light emission (luminiscence, delayed fluorescence), which also originates from PS II and reflects the primary events in photosynthesis, seems to be unexplored in studies connected with SO_2 action.

11. EFFECTS ON ENZYMES INVOLVED IN CARBON DIOXIDE ASSIMILATION

Most of the available information about the modes of SO_2 action on CO_2-fixing enzymes comes from the extensive work by Ziegler and her collaborators. From the earlier work (260) it is known that RuBP-carboxylase is inhibited by SO_3^{2-}. The pattern of inhibition is competitive with respect to SO_3^{2-} and HCO_3^- but noncompetitive with respect to RuBP and MG^{2+}. From the experiments *in vitro* the K_i values have been determined to be 14 and 9.5 m*M*, respectively, for the noncompetitive type of inhibition but only 3.0 m*M* for the competitive one (260). This important finding represents a key mechanism by which SO_2 and solution products of SO_2 directly compete with CO_2 in the entire chloroplast. Ziegler has also pointed out that the competitive type of inhibition with SO_2 seems to be a general phenomenon (18). One may ask whether the isolated form of RuBP-carboxylase (K_m 4.0 m*M*) is also the active form *in vivo*, and different opinions presently exist (cf. 294–296). In any case it is probable that the same type of inhibition as is found *in vitro* should exist *in vivo*. Inhibition caused by the theoretically assumed reaction with the keto group of RuBP has been stated to be of minor importance (260). It has also been assumed that the competitive form of inhibition will dominate in the chloroplast under conditions with low internal CO_2 concentrations. The question as to the internal concentration of CO_2 (and SO_2) in the chloroplast is unresolved (e.g., 297, 298).

Low SO_3^{2-} concentrations stimulate the CO_2 fixation in chloroplasts (130). According to Ziegler, this can be explained by the effects of SO_3^{2-} on electron transport (discussed earlier). In a field with light saturation or nearly light saturation one should not expect electron transport to be the rate-limiting factor for photosynthesis, and hence the suggested effect may then play a minor role. Furthermore, under these conditions there is usually a close correlation between RuBP-carboxylase activity and net photosynthetic rate, and the inhibitory action of SO_2 on RuDP-carboxylase should be more significant. In this context it should also be mentioned that Wellburn et al. (230) found a decrease in RuBP-carboxylase content in seedlings of *Pisum* which was caused by exposure to SO_2.

Phosphoenolpyruvate (PEP) carboxylase (299) and malic enzyme (300,301), considered to act as a CO_2 pump increasing the CO_2 concentration in bundle sheath cells of C4 plants, are inhibited by SO_3^{2-}. The k_i values of SO_3^{2-} for PP-carboxylase of *Zea mays* are high: 84.5 m*M* with respect to Mg^{2+} and 27 m*M* with respect to HSO_3^-, and no inhibition at all with respect to PEP at concentrations up to 10 m*M* SO_3^{2-}. Hence from these studies PEP-carboxylase can be regarded as far less sensitive to SO_3^{2-} than RuBP-carboxylase.

The action of SO_3^{2-} on NAD- and NADP-dependent malate dehydrogenase takes several forms (301). First, a strong inhibition takes place, and the results indicate that both the NAD and the activated NADP-dependent enzyme are much more sensitive than both RuBP- and PEP-carboxylase. A residual activity of 10 to 15% is, however, maintained even at high SO_3^{2-} concentrations; this may be enough to make possible the metabolism of the ocaloacetate occurring through the action of PEP-carboxylase (301). Second, SO_3^{2-} causes a splitting of the enzyme, probably by the decomposition of disulfide proteins through the cleavage of S-S bonds in polypeptides.

Inhibition studies on PEP-carboxylase and NAD-malate dehydrogenase in extracts of *Zea mays* have shown that malate formation is not impaired by low SO_3^{2-} concentrations, since PEP-carboxylase is relatively insensitive toward SO_3^{2-}, but is evidently the rate-limiting step of the reaction sequence. This cannot, however, be interpreted to mean that C4 plants are less sensitive to SO_2, since the Calvin cycle is the ultimate CO_2-fixing pathway in all green plants (302).

In addition to the inhibition of the CO_2-fixating enzymes, further changes in the metabolism of assimilated carbon have been reported. The regulatory aspects of photosynthetic carbon metabolism have been extensively reviewed (302), and it has also recently been suggested (303) that SO_2 could interfere with the regulatory process in the Calvin cycle. Very often the effects of SO_2 on photosynthesis have been interpreted in terms of an inhibition rather than a loss of control. Anderson and Avron (304) investigated the light modulation of enzyme activity in chloroplasts from *Pisum sativum*. Modulation of the activity of several dehydrogenases and ribulose-5-P-kinase apparently involves a component preceding ferredoxin in the photosynthetic electron transport chain. The capacity for light regulation of glucose-6-P-dehydrogenase and NADP-linked malic dehydrogenase was strongly reduced after treatment of the particulate fraction from broken chloroplasts with 10 μM SO_3^{2-} in the light or in the dark. The results indicate that light-generated vicinal-dithiols seem to be involved in the modulation of the enzyme activity. It must be noted that neither enzyme was inhibited directly; also, the low concentration seems to eliminate the possibility that the effect could be attributed to inhibition of the electron transport system. From these results it seems possible that SO_3^{2-} concentrations generated *in vivo* from SO_2 can disrupt metabolic regulation in higher plants (304). Ziegler et al. (303) have studied the direct action of SO_3^{2-} on the substrate kinetics of NADP-linked glyceraldehyde-3-phosphate dehydrogenase in spinach chloroplasts and noted that SO_3^{2-} reduced the maximum activities and suppressed the enzyme form with the higher V_{max}. It was mentioned previously that the reduction of 3-PGA was also inhibited by SO_4^{2-} in entire chloroplasts. In this context it is tempting to speculate that the inhibition of phosphoglycerate kinase by low SO_4^{2-} concentrations observed in yeast (305) also occurs in chloroplasts, and that this would help to explain the inhibition of PGA reduction.

It has been reported that the relative amounts of PGA and sugar phosphates decrease in chloroplast treated with SO_3^{2-}, while the labeling of aspartate and malate increases (306). It is possible that this could be explained by an activation of fructose-1,6-diphosphatase (13). However, the complex relationship between the regulatory mechanisms involved in the reductive pentose phosphate cycle may offer other plausible explanations (cf., e.g., 302,307).

Reports on general changes in the main constituents of plant metabolism more or less directly connected with the photosynthetic process are very difficult to evaluate. In many cases a decrease in the amount of product formed simply reflects impaired photosynthesis, or altered stomatal resistance or transpiration, but a change in the contents of starch (207b,308), sucrose (239,309–312), amino acids (209,313,315), and so forth can also be attributed to qualitative changes in the general metabolic activity or growth rate of the plant. Inhibition of an enzyme need not be complete, and hence the remaining activity may be sufficient to metabolize the substrate delivered. In many cases, however, inhibition or inactivation of an enzyme will decrease the rate of a particular metabolic reaction. The reduction in the rate of this reaction may modify the pathway in which it occurs and result in a depletion of the final product. This, in turn, will cause the accumulation of precursors or intermediates. The result may, however, also involve an activation or stimulation of alternative pathways. It has been demonstrated that the enzyme glutamate dehydrogenase (GDH) was stimulated by SO_2 toward reductive amination, while the oxidative deamination by this enzyme was inhibited (316). The action of SO_2 may also involve a direct disruption of the enzyme structure, or a direct effect on a catalytic site or an indirect effect on cofactors, as described above. Evidently, some enzymes are rapidly stimulated by SO_2, for example, peroxidase (21,-189,190,243) and polyphenol oxidase (191,317), whereas other enzymes are mostly inhibited (cf. 230). An enhanced uptake of SO_2 from air will also result in an expenditure of energy for detoxification and activation, since both SO_3^{2-} and SO_4^{2-} can be metabolized by plant tissue. The interference of SO_3^{2-} with regulatory processes in plant metabolism, governed by environmental changes, may certainly contribute to the various actions of SO_2.

12. EFFECTS ON PHOTORESPIRATION

Photorespiration is considered to be the oxidative and irreversible biosynthesis and metabolism of glycolate; see Tolbert and Ryan (318) for a review. Specific sulfonates induce accumulation of glycolate in plant leaves and leaf slices treated with these compounds (274), and it is generally believed that the inhibition of glycolate oxidase is responsible for this effect (274). Sulfonates act as bisulfite addition compounds and presumably exchange their aldehyde moiety with glyoxylate generated in the glycolate oxidase reaction to form glyoxylate bisulfite (319). This sulfonate is the probable inhibitor of the enzymatic reaction, regardless of which sulfonate is supplied (274). *In vivo,* 50% or greater inhibition of glycolate oxidase activity occurs at a concentration of 10^{-4} *M*, and HSO_3^- has been found to be nearly as effective (320). It has also been shown that fumigation with high concentrations of SO_2 results in an accumulation of glycolic acid in leaves of *Hordeum vulgare* (239). The blocking of the oxidation of glycolate and photorespiration leads to an increase in net photosynthesis (318). Since the function of and the necessity for photorespiration are obscure, it has been claimed that this process is unessential and wasteful and that the action of SO_2 should then be to the advantage of the plant (239). However, plants incapable of photorespiration have not been found, and some experiments suggest that photorespiration in air is a protective metabolic process (318). It has been considered to be unavoidable because of the RuBP-oxygenase reaction (321,322). Sij and Swanson (40) considered a specific effect of SO_2 in photorespiration unlikely since the photosynthetic responses of the C3 and C4 species tested were similar. During photosynthesis glycolate is formed in the chloroplast and is converted to glycine in the peroxisomes, and the light-stimulated CO_2 evolution has been attributed to the conversion of glycine to serine in the peroxisomes (323). In labeling experiments with leaf slices or attached leaves treated with HSO_3^- or SO_2, serine and glycine have been reported to decrease, increase, and remain unchanged (131,239,306). Since there exists an alternative pathway of serine and glycine formation, a decrease in these compounds cannot be regarded simply as a consequence of glycolate oxidase inhibition (318).

Spinach chloroplasts have very little glycolate oxidase activity, and it was shown previously that 0.7 m*M* sulfonate did not change the proportion of $^{14}CO_2$ fixed into glycolate, although CO_2 fixation was inhibited by 50% (270). It is also known that C4 species show a greater sensitivity to the sulfonates than do C3 species, although the reason is not understood (274). Recently, it was demonstrated that incorporation into glycolate increased not only in leaf slices but also in chloroplasts for both C4 and C3 species treated with SO_3^{2-} (306). No good explanation of the presumed enhancement of glycolate synthesis by SO_3^{2-} in chloroplasts has been suggested so far.

We can conclude that some experiments indicate that SO_2 may be a potent inhibitor of glycolate oxidase and that it also may affect glycolate synthesis. The consequences for the plant are totally unknown, however, and the action of SO_2 on photorespiration in intact plants has not yet been investigated.

13. EFFECTS ON RESPIRATION

The effect of SO_2 on respiration has usually been regarded as a secondary one caused by impaired photosynthesis. But it has also been suggested that an increase in respiration is a direct consequence of SO_2 injury (7,324). De Koning and Jegier (325) reported that SO_2 (5 ppm) increased the rate of respiration in *Euglena* cells, whereas photosynthesis was inhibited. Keller and Müller (124) showed that the rate of respiration increased initially in pine and spruce plants exposed to SO_2. The same effect has been noted also in lichens and in bryophytes, although the species differences are equally striking (41,165). The effects on enzymes noted above indicate that this pollutant may well have the ability to alter respiratory processes. The documentation of SO_2 effects on respiration is scarce, and more thorough investigations are clearly desirable.

14. CONCLUDING REMARKS

Most articles conclude with a brilliant summary of the facts that have been presented. In this case the conclusion is very simple: the knowledge of SO_2 action on photosynthesis in plants deserves further development. The mechanism of SO_2 action on photosynthesis has been studied in chloroplast preparations from plant leaves. To make it possible to extrapolate some of the results to *in vivo* conditions, intact chloroplasts or at least well-characterized chloroplast populations must be recommended. One of the main reasons is that the photosynthetic process *in vivo* occurs within the chloroplast, and all enzymes of the CO_2-reducing cycle either are localized in the stroma or are loosely affixed to the thylacoid membranes. Therefore the use of heterogeneous populations of chloroplasts, which partly or wholly have lost their envelopes and stroma substance, may lead to confusing results in studies with added ions such as SO_3^{2-} and HSO_3^-. Even with intact chloroplasts, extrapolation to *in vivo* conditions is difficult. However, most studies of the action of SO_2 on photosynthesis *in vivo* have been related only to the photosynthetic performance of the plant. To clarify the action of SO_2 on photosynthesis, the fundamental processes of photosynthetic electron transport and carbon reduction should be investigated in parallel with studies of SO_2 flux, the absorbed quantum flux, and the accompanying processes of CO_2 flux between air and leaf, determined by the photosynthetic and (light) respiratory processes. The photosynthetic CO_2 flux at limiting light intensities as a function of the absorbed quantum flux gives the photochemical efficiency, or the quantum yield. Such studies offer possibilities for relating the basic photochemical research on chloroplasts to the photosynthetic characteristics of the intact plant. In this way realistic comparisons of the photosynthetic characteristics of plants of different physiological conditions, growth patterns, structures, and shapes could be made (see 326).

ACKNOWLEDGMENTS

This work was supported by a grant from the National Swedish Environment Protection Board. I wish to thank the following persons for valuable comments on the manuscript: Dr. A. Dunberg, Prof. L. Eliasson, Dr. K. Gezelius, Dr. K-R. Sundström, Dr. B. Martin, and Dr. G. Öquist. Appreciation is extended also to A. Ehman and I. Nilsson for technical assistance.

REFERENCES

1. Katz, M. (1949). *Ind. Eng. Chem.* **41,** 2450–2465.
2. Thomas, M. D. (1951). *Ann. Rev. Plant Physiol.* **2,** 293–322.
3. Thomas, M. D. (1961). World Health Organization, Monograph Series 46, Geneva, pp. 233–278.
4. Webster, C. C. (1967). *The Effects of Air Pollution on Plants and Soil.* Agriculture Research Council of London.
5. Garber, K. (1967). *Luftverunreinigung und ihre Wirkungen.* Gebrüder Borntraeger, Berlin-Nikolassee, pp. 1–279.
6. Brandt, C. S. and Heck, W. W. (1968). In A. C. Stern, Ed., *Air Pollution,* Vol. 1. Academic Press, New York, Chap. 14, pp. 401–443.
7. Berge, H. (1970. *Handbuch der Pflanzenkrankheiten,* Vol. 1(4). P. Parey, Berlin and Hamburg, pp. 1–169.
8. Linzon, S. N., Ed. (1970). *Impact of Air Pollution on Vegetation.* A.P.C.A., Pittsburgh, Pa.
9. Treshow, M., Ed. (1970). *Environment and Plant Response.* McGraw-Hill Book Co., New York.
10. Tamm, C. O. and Aronsson, A. (1972). Royal College of Forestry, Res. Note 12, Stockholm.
11. Jensen, K. F., Dochinger, L. S., Roberts, B. R., and Townsend, A. M. (1976). In J. P. Miksche, Ed., *Proceedings in Life Sciences: Modern Methods in Forest Genetics.* Springer-Verlag, Berlin, pp. 189–216.
12. Knabe, W. (1976). *AMBIO* 5–6, pp. 213–218.
13. van Haut, H. and Stratman, H. (1970). *Farbtafelatlas über Schwefeldioxid-Wirkungen an Pflanzen.* Verlag W. Girardet, Essen.
14. Jacobson, J. S. and Hill, A. C. (1970). *Recognition of Air Pollution Injury to Vegetation. A Pictorial Atlas. Environment Quality and Safety.* A.P.C.A., Pittsburgh, Pa.
15. Hindawi, I. J. (1970). *Air Pollution Injury to Vegetation.* U.S. Department of Health, Education, and Welfare, AP-71.

16. McCune, D. C. (1973). *Adv. Chem.*, Series 122, pp. 48–62.
17. Ziegler, I. (1973). In F. Coulston, Ed., *Env. Qual. Safety.* Vol. II. Stuttgart, pp. 182–208.
18. Ziegler, I. (1975). *Res. Rev.*, pp. 79–104.
19. Mudd, J. B. (1975). In J. B. Mudd and T. T. Kozlowski, Eds., *Responses of Plants to Air Pollution.* Academic Press, New York.
20. Malhotra, S. S. and Hocking, D. (1976). *New Phytol.* **76,** 227–237.
21. Horsman, D. C. and Wellburn, A. R. (1976). In T. A. Mansfield, Ed., *Effects of Air Pollutants on Plants.* Cambridge University Press, pp. 192–196.
22. Ferry, B. W., Baddeley, M. S., and Hawksworth, D. L. (1973). *Air Pollution and Lichens.* Athlone Press, London.
23. Gilbert, O. L. (1973). In V. Ahmadjian and M. S. Hale, Eds., *The Lichens.* Academic Press, New York, Chap. 13, pp. 443–469.
24. Le Blanc, F. and Rao, D. N. (1975). In J. B. Mudd and T. T. Kozlowski, Eds., *Responses of Plants to Air Pollution.* Academic Press, New York, pp. 237–272.
25. Hawksworth, D. L. (1976). *Natural History Book Reviews,* pp. 8–15.
26. Gilbert, O. L. (1976). In L. Kärenlampi, Ed., *Proceedings of the Kuopio Meeting on Plant Damages Caused by Air Pollution.*
27. Lougham, B. C. (1964). *Agrochimica* **8**(3), 189–91.
28. Bleasdale, J. K. A. (1952). *Nature* **169,** 376–377.
29. Reinert, R. A., Tingey, D. T., Heck, W. W., and Wycliff, C. (1969). *Agron. Abstr.* **61,** 34.
30. Tingey, O. T., Heck, W. W., and Reinert, R. A. (1971). *J. Am. Soc. Hort. Sci.* **96,** 36–71.
31. Thomas, M. D., Hendricks, R. H., Collier, T. R., and Hill, G. R. (1943). *Plant Physiol.* **18,** 345–371.
32. Faller, N., Herwig, K., and Kühn, H. (1970). *Plant Soil* **33,** 171–91.
33. Cowling, D. W., Jones, L. H. P., and Lockyer, D. R. (1973). *Nature* **243,** 479–80.
34. Weinstein, L. H., McCune, D. C. (1971). *J. Air Pollut. Control Assoc.* **21,** 410.
35. Zelitch, I. (1975). In R. Marcelle, Ed., *Environmental and Biological Control of Photosynthesis.* W. Junk, The Hague, pp. 251–262.
36. Taylor, O. C. (1975). *Hort. Sci.* **10**(5), 501–504.
37. Daines, R. H. (1968). *J. Occup. Med.* **10**(9), 516–523.
38. Vogl, M., Börtitz, S., and Polster, H. (1964). *Arch. Forstw.* **13**(10), 1031–1043.
39. Jensen, K. F. and Kozlowski, T. T. (1974). *Proceedings of the Third North American Forest Biology Workshop.* Fort Collins, Colo., p. 359.

40. Sij, J. W. and Swanson, C. A. (1974). *J. Environ. Qual.* **3**(2), 103–107.
41. Türk, R., Wirth, V., and Lange, O. L. (1974). *Oecologia* **15,** 33–64.
42. Bull, J. N. and Mansfield, T. A. (1974). *Nature* **250,** 443–44.
43. Hedgcock, G. G. (1912). *Torreya* **12,** 25–30.
44. Brisley, H. R., Davis, C. R., and Booth, J. A. (1959). *Agron. J.* **51,** 77–80.
45. Krutikow, A. (1969). *Las Polski* **13/14,** 23–24.
46. Fujiwara, T. (1970. *J. Jap. Soc. Hort. Sci.* **3,** 13–17.
47. Mamajev, S. A. and Shkarlet, O. D. (1972). *Mitt. Forstl. Bundesversuchanst. Wien* **97,** 443–50.
48. Scurfield, G. (1960). *For. Abstr.* **21**(3), 339–347, **21**(4), 517–28.
49. Vogl, M. and Börtitz, S. (1965). *Flora* **155,** 347–352.
50. Wentzel, K. F. (1969). In *Air Pollution: Proceedings of the 1st European Congress on the Influence of Air Pollution on Plants and Animals.* Wagenigen, pp. 357–370.
51. Ranft, H. and Dässler, H. G. (1970). *Flora* **159**(6), 573–588.
52. Brennan, E. and Halisky, P. M. (1970). *Phytopathology* **60**(11), 1544–1546.
53. Dochinger, L. S. (1971). *Mitt. Forstl. BundesVersuchsanst. Wien* **92,** 7–32.
54. Börtitz, S. (1964). *Biol. Zentralbl.* **83,** 501–513.
55. Schönbach, H., Dässler, H. G., Enderlein, H., Bellman, E., and Kästner, W. (1964). *Züchter* **34**(8), 312–316.
56. Dochinger, L. S., Bender, F. W., Fox, F. O., and Heck, W. W. (1970). *Nature* **225,** 476.
57. Hodges, G. H., Menser, H. A., Jr., and Ogden, W. B. (1971). *Agron. J.* **63,** 107–111.
58. Menser, H. A., Hodges, H. A., and McKee, G. G. (1973). *J. Environ. Qual.* **2,** 253–258.
59. Taylor, G. E., Jr. and Murdy, W. A. (1975). *Bot. Gaz.* **136**(2), 212–215.
60. Bell, J. N. B. and Mudd, C. H. (1976). In T. A. Mansfield, Ed., *Effects of Air Pollutants on Plants.* Cambridge University Press, pp. 87–103.
61. Van Haut, H. (1961). *Staub-Reinhalt* **21**(2), 52–56.
62. Guderian, R. (1970). *Z. Pflanzenkr.* **77**(6, 200–220, 289–308, 387–399).
63. Cracker, L. E. and Starbruck, Y. S. (1973). *Environ. Res.* **6,** 91.
64. Garsed, S. G. and Read, D. J. (1977). *New Phytol.* **78,** 111–119.
65. Garsed, S. G. and Read, D. J. (1974). *New Phytol.* **73,** 299.
66. Bleasdale, J. K. A. (1973). *Environ. Pollut.* **5,** 275–85.
67. Meudt, W. J. (1971). *Phytochemistry* **10,** 2103.
68. Yeh, R., Hemphill, D., and Sell, H. (1971). *Phytochemistry* **49,** 162.
69. Yang, S. F. and Saleh, M. A. (1973). *Phytochemistry* **12,** 1463–1466.
70. Fischer, K. (1971). *Mitt. Forstl. Bundesversuchanst. Wien* **92,** 209–23.
71. Pelz, E. and Materna, J. (1964). *Arch. Forstwes.* **13**(2), 177–216.

72. Hällgren, J. E. and Nyman, B. (1977). *Studia Forestalia Suecica,* 137 (in press).
73. Stoklasa, J. (1923). *Die Beschadigungen der Vegetation durch Rauchgase und Fabriksexhalationen.* Urban und Schwarzenberg, Berlin.
74. Hull, H. M., Went, F. W., and Yamada, N. (1954). *Plant Physiol.* **29,** 182–187.
75. Hawksworth, D. L. and Rose, F. (1970). *Nature* **227,** 145–148.
76. LeBlanc, F., Rao, D. N., and Comeau, G. (1972). *Can. J. Bot.* **50,** 519–528.
77. Taoda, H. (1972). *Jap. J. Ecol.* **22,** 125–133.
78. Tingey, D. T., Reinert, R. A., Dunnings, J. A., and Heck, W. W. (1973). *Atmos. Environ.,* pp. 201–208.
79. Houston, D. B. (1974). *Can. J. For. Res.* **4,** 65–68.
80. Costonis, A. C. (1970). *Phytopathology* **60**(6), 994–99.
81. Tingey, D. T. and Reinert, R. A. (1975). *Environ. Pollut.* **9,** 117–25.
82. Knabe, W. (1976). *Proceedings of the XVI International Union of Forest Research Organizations World Congress, Oslo, Norway.* Norwegian Forest Research Institute Ås-NLH, Norway, Division II 564–78.
83. Heck, W. W., Taylor, O. C., and Heggestad, H. E. (1973). *J. Air Pollut. Control Assoc.* **23,** 257–66.
84. Chamberlain, A. C. (1966). *Proc. Roy. Soc.* **A290,** 236.
85. Monteith, J. L., Ed. (1976). *Vegetation and the Atmosphere,* Vols. 1 and 2. Academic Press, New York.
86. Sesták, Z., Catsky, J., and Jarvis, P. G., Eds. (1971). *Plant Photosynthetic Production Manual of Methods.* W. Junk, The Hague.
87. Unsworth, M. H., Biscoe, P. V., and Black, V. (1976). In T. A. Mansfield, Ed., *Effects of Air Pollutants on Plants.* Cambridge University Press, pp. 5–16.
88. Bennett, J. H. and Hill, A. C. (1975). In J. B. Mudd and T. T. Kozlowski, Eds., *Responses of Plants to Air Pollution.* Academic Press, New York, pp. 273–306.
89. Spedding, J. D. (1969). *Nature* **224,** 1229–1230.
90. Waggoner, P. E. (1971). *Bioscience* **21,** 455–459.
91. Bennett, J. H., Hill, A. C., and Gates, D. M. (1973). *J. Air Pollut. Control Assoc.* **23,** 957–962.
92. Biscoe, P. V., Unsworth, M. H., and Pinckney, H. R. (1973). *New Phytol.* **72,** 1299–1306.
93. Garland, J. A., Clough, W. S., and Fowler, D. (1973) *Nature* **242,** 256–257.
94. Garland, J. A., Atkins, D. H. F., Reading, C. J., and Caughey, S. J. (1974). *Atmos. Environ.* **8,** 75–79.
95. Fowler, D. and Unsworth, M. H. (1974). *Nature* **249,** 389–90.
96. Murphy, B. D. (1976). In *Proceedings of the American Meteorological*

Society Third Symposium on Atmospheric Turbulence, Diffusion and Air Qualty, October 1976, Raleigh, N.C.

97. Murphy, C. E., Sinclair, T. R., and Knoerr, K. R. (1976). In *Proceedings of the Conference on Metropolitan Physical Environments.* U.S. Department of Agriculture, Forest Service, Northeastern Forest Experiment Station, Upper Darby, Pa.
98. Steinhart, I., Fox, D. G., and Marlatt, W. E. (1976). In: *Proceedings of the Fourth National Conference on Fire and Forest Meteorology, St. Louis, Mo. November 76.* U.S. Department of Agriculture, Forest Service, Gen. Tech. Rept. RM-32, pp. 209–213.
99. Garland, J. A. and Branson, J. R. (1977). *Tellus* (in press).
100. Hill, A. C. (1971). *J. Air Pollut. Control Assoc.* **21**(6), 341–346.
101. Jensen, K. F. and Kozlowski, T. T. (1975). *J. Environ. Qual.* **4**(3), 379–382.
102. Waggoner, P. E. (1976). In J. L. Monteith, Ed., *Vegetation and the Atmosphere,* Vol. 1. Academic Press, New York, p. 224.
103. Thom, A. S. (1968). *Quart. J. Roy. Meteorol. Soc.* **94,** 44–55.
104. Jarvis, P. G. and Slayter, R. O. (1966). C.S.I.R.O. Division of Land Research, Tech. Paper 29, pp. 1–16.
105. Meidner, H. and Mansfield, T. A. (1968). *Physiology of Stomata.* McGraw-Hill Book Co., London.
106. Raschke, K. (1975). *Ann. Rev. Plant Physiol.* **26,** 309–340.
107. Mansfield, T. A. (1976). In: H. Smith, Ed., *Commentaries in Plant Science,* Pergamon Press, New York, pp. 13–22.
108. Meidner, H. and Willmer, C. (1976). In: H. Smith, Ed., *Commentaries on Plant Science,* Pergamon Press, New York, pp. 137–151.
109. Loftfield, J. V. G. (1921). Carnegie Institution, Publ. 314, p. 1.
110. Zimmerman, P. W. and Hitchcook, A. E. (1956). *Contrib. Boyce Thompson Inst. Plant Res.* **18,** 263–279.
111. Mansfield, T. A. and Majernic, O. (1970). *Environ. Pollut.* **1,** 149–154.
112. Majernic, O. and Mansfield, T. A. (1970). *Nature* **227,** 377–378.
113. Majernic, O. and Mansfield, T. A. (1971). *Phytopathol. Z.* **71,** 123–128.
114. Majernic, O. and Mansfield, T. A. (1972). *Environ. Pollut.* **3,** 1–7.
115. Unsworth, M. H., Biscoe, P. V., and Pickney, H. R. (1972). *Nature* **239,** 458–459.
116. Menser, H. A. and Heggestad, H. E. (1966). *Science* **153**(3734), 424–425.
117. Halbwachs, G. (1971). *Ber. Dtsch. Bot. Ges.* **84,** 507.
118. Bonté, J., De Cormis, L., and Louguet, P. (1975). *C. R. Acad. Sci. Paris,* Series D, pp. 2377–2380.
119. Schramel, M. (1975). *Bull. Acad. Polon. Sci.* **23**(1), 57–63.
120. Markowski, A., Gresiak, S., and Schramel, M. (1975). *Bull Acad. Polon. Sci.* **23**(9), 637–45.

121. Heiling, A. (1933). *Phytopath. Z.* **5,** 435–000.
122. Weigl, J. and Ziegler, H. (1962). *Planta* **58,** 435.
123. Godzik, S. and Piskornik, Z. (1966). *Bull. Acad. Polon. Sci.* **14**(3), 181–184.
124. Keller, H. and Müller, J. (1958). *Forstwiss. Forsch.* **10,** 5–63.
125. Vogl, M. (1964). *Biol. Zentralbl.* **83,** 587.
126. Halbwachs, G. (1967). *Allg. Forstztg.* **78**(9), 196–197.
127. Klemm, W. (1966). *Biol. Zentralbl.* **85,** 781–783.
128. Börtitz, S. and Vogl, M. (1967). *Arch. Forstwes.* **16,** 663–666.
129. Cowling, D. W. and Lockyer, D. R. (1976). *J. Exp. Bot.* **27**(98), 411–417.
130. Libera, W., Ziegler, H., and Ziegler, I. (1973). *Planta* **9,** 269.
131. Lüttge, U., Osmond, C. B., Ball, E., Brinckmann, E., and Kinze, G. (1972). *Plant Cell Physiol.* **13,** 505.
132. Puckett, K. J., Richardson, D. H. S., Flora, W. P., and Nieboer, E. (1974). *New Phytol.* **73,** 1183–1192.
133. Granath, L. (1977). IMI/MISU AC report (to be published).
134. Materna, J. and Kohout, R. (1969. *Commun. Inst. For. Chechoslov.* **6,** 38–47.
135. Huttunen, S. (1975). *Acta Univ. Ouluensis* (Finland), Series A, No. 33, *Biol.,* No. 2.
136. De Cormis, L. (1969). In *Air Pollution: Proceedings of the 1st European Congress on the Influence of Air Pollution on Plants and Animals.* Wageningen, pp. 75–78.
137. Materna, J. (1966). *Arch. Forstwes.* **66**(15), 691–92.
138. Krouse, H. R. (1977). *Nature* **265,** 45–46.
139. Raybould, C. C., Unsworth, M. H., and Gregory, P. J. (1977). *Nature* **267,** 146–147.
140. Härtel, O. (1953). *Zentralbl. Gesamte Forstwes.* **72,** 1, 12–21.
141. Härtel, O. (1972). *Oecologia* **9,** 103–11.
142. Grill, D., Esterbauer, H., and Beck, G. (1975). *Phytopathol. Z.* **82,** 182–184.
143. Keller, Th., Schwager, H. and Yeemeiler, D. (1976). *Eur. J. For. Pathol.* **6**(4), 244–249.
144. Grill, D. (1973). *Phytopathol. Z.* **78,** 75–80.
145. Jeffree, C. E., Johnson, R. P. C., and Jarvis, P. G. (1971). *Planta* **98,** 1–10.
146. Saunders, P. J. W. (1971). In T. F. Preece and C. H. Dickinson, *Ecology of Leaf Surface Microorganisms.* Academic Press, New York, pp. 81–89.
147. Squire, G. R. and Mansfield, T. A. (1972). *New Phytol.* **71,** 1033–43.
148. Wood, T., and Borman, F. H. (1975). AMBIO **4,** 169–71.
149. Fairfax, J. A. W. and Lepp, N. W. (1975). *Nature* **255**(5506), 324–325.
150. Fairfax, J. A. W. and Lepp, N. W. (1976). In L. Kärenlampi, Ed.,

Proceedings of the Kuopio Meeting in Plant Damage Caused by Air Pollution, pp. 26–36.

151. Schroeter, L. C. (1966). *Sulfur Dioxide: Applications in Foods, Beverages and Pharmaceuticals.* Pergamon Press, Oxford.
152. Falk, M. and Giguère, P. A. (1958). *Can. J. Chem.* **36,** 1141.
153. Puckett, K. J., Nieboer, E., Flora, W. P., and Richardson, D. H. S. (1973). *New Phytol.* **72,** 141–154.
154. Richardson, D. H. S. and Puckett, K. J. (1973). In B. W. Ferry et al., Eds., *Air Pollution and Lichens.* Athlone Press, Chap. 13, pp. 283–298.
155. Saunders, P. J. W. (1966). *Ann. Appl. Biol.* **58,** 103–114.
156. Nieboer, E., Richardson, D. H. S., Puckett, K. J., and Tomasini, F. D. (1976). In T. A. Mansfield, Ed., *Effects of Air Pollutants on Plants.* Cambridge University Press, pp. 61–85.
157. Thomas, M. D., Henricks, R. H., and Hill, G. R. (1944). *Plant Physiol.* **19,** 212.
158. Raven, J. A. and Smith, F. A. (1976). *Current Advances in Plant Science,* Vol. 8, No. 5, pp. 649–656.
159. Pahlich, E. (1975). *Atmos. Environ.* **9,** 261–63.
160. Nielsen, J. P. (1938). Cited from Ref. 37.
161. Grill, D. and Härtel, O. (1972). *Mitt. Forstl. Bundesversuchsanst. Wien* **97,** 367–386.
162. Skye, E. (1968). *Acta Phytogeogr. Suec.* **52,** 1–123.
163. Wirth, V., and Türk, R. (1975). *Flora* **164,** 133–143.
164. Barkman, J. J. (1968). *Phytosociology and Ecology of Cryptogamic Epiphytes.* Assen, Netherlands.
165. Ferry, B. W. and Baddeley, M. S. (1973). In B. W. Ferry et al., Eds. *Air Pollution and Lichens.* Athlone Press.
166. Ferry, B. W. and Baddeley, M. S. (1976). In D. H. Brown, D. L. Hawksworth, and R. H. Bailey, Eds., *Lichenology: Progress and Problems.* Academic Press, New York, Chap. 16).
167. Gilbert, O. L. (1968). *New Phytol.* **67,** 15.
168. Hill, D. J. (1971). *New Phytol.* **70,** 831.
169. Taoda, H. (1973). *Hikobia* **6,** 224–250.
170. MacLeod, R. M., Farkas, W., Fridovitch, I., and Handler, P. J. (1961). *J. Biol. Chem.* **236,** 1841.
171. Asada, K. and Kiso, K. (1973). *Eur. J. Biochem.* **33,** 253–257.
172. McCord, J. M. and Fridovitch, I. (1969). *J. Biol. Chem.* **244,** 6056.
173. Ziegler, I. and Libera, W. (1975). *Z. Naturforsch.* **30,** 634–37.
174. Fridovitch, I. (1976). In W. A. Pryor, Ed., *Free Radicals in Biology,* Vol. 1, Chap. 6, Academic Press, New York.

175. Loach, P. A. and Hales, B. J. (1976). In W. A. Pryor, Ed., *Free Radicals in Biology,* Vol. 1, Chap. 5, Academic Press, New York.
175. Loach, P. A. and Hales, B. J. (1976). In W. A. Pryor, Ed., *Free Radicals in Biology,* Vol. 1, Chap. 5.
176. Egneus, H., Heber, U., Matthiesen, U., and Kirk, M. (1975). *Biochim. Biophys. Acta* **408**(3), 252–68.
177. Radmer, R. J. and Kok, B. (1976). *Plant Physiol.* **58**(3), 336–40.
178. Heber, U. (1973). *Biochim. Biophys. Acta* **305,** 140–152.
179. Patterson, C. O. P. and Myers, J. (1973). *Plant Physiol.* **51,** 104–109.
180. Fridovich, I. and Handler, P. (1961). *J. Biol. Chem.* **236,** 1836–1840.
181. Yang, S. F. (1967). *Arch. Biochem. Biophys.* **122,** 481–487.
182. Mehler, A. H. (1951). *Arch. Biochem. Biophys.* **33,** 65–77.
183. Fridovich, I. and Handler, P. (1960). *J. Biol. Chem.* **235,** 1835–1838.
184. Yang, S. F. (1970). *Biochemistry* **9,** 5008–5014.
185. Yang, S. F. (1973). *Environ. Res.* **6,** 395–402.
186. Tager, J. M. and Rautanen, N. (1956). *Physiol. Plant* **9,** 665.
187. Arrigoni, O. (1959). *Ital. J. Biochem.* **8,** 181.
188. van Auken, O. W., Rovlands, J. R., and Hendersson, A. O. (1976). *Plant Phys.,* Suppl. 5.
189. Keller, Th. (1974). *Eur. J. For. Pathol.* **4,** 11–19.
190. Keller, Th. (1976). *Schweiz. Z. Forstwes.* **127,** 237–251.
191. Nikolaevskiy, V. S. (1966). *Okrana Prior. Urale* **5,** 19–23.
192. Thomas, M. D., Henricks, R. H., and Hill, G. R. (1950). *Ind. Eng. Chem.* **42,** 2231–2235.
193. Schiff, J. A. and Hodson, R. C. (1973). *Ann. Rev. Plant Phys.* **24,** 381–414.
194. Trebst, A. and Schmidt, A. (1969). In H. Metzner, Ed., *Progress in Photosynthesis Research,* Vol. 3, p. 1510, Tübingen. International Union of Biological Sciences.
195. Schmidt, A. and Trebst, A. (1969). *Biochim. Biophys. Acta* **180,** 529.
196. Schmidt, A. (1968). *Arch. Mikrobiol.* **84,** 77–86.
197. Hennies, H. (1975). *Z. Naturforsch.* **30,** 359–362.
198. Roy, A. B. and Trudinger, P. A. (1970). *The Biochemistry of Inorganic Compounds of Sulphur.* University Press, Oxford.
199. Asahi, T. (1964). *Biochim. Biophys. Acta.* **82,** 58–66.
200. Silvius, J. E., Baer, C. H., Dodrill, S., and Patrick, H. (1976). *Plant Physiol.* **57,** 799–801.
201. Sherdian, R. P. (1973). *J. Phycol.* **9,** 437–445.
202. Smith, L. K. and Thompson, J. F. (1969). *Biochem. Biophys. Res. Commun.* **35,** 939.
203. Thompson, J. F. (1967). *Ann. Rev. Plant Physiol.* **18,** 59.

204. Lehmann, J. and Benson, A. A. (1964). *J. Am. Chem. Soc.* **86,** 4469–72.

205. Brunold, Chr. and Erismann, K. H. (1973). *Ber. Schweiz. Bot. Ges.* **83,** 213–222.

206. Brunold, Chr. and Erismann, K. H. (1975). *Experientia,* p. 3213.

207. Fankhauser, H. (1975). Lizentiatsarbeit, Universität Bern. Cited from Ref. 207b.

207b. Fankhauser, H., Brunold, Chr., and Erismann, K. H. (1976). *Oecologia* **23,** 201–209.

208. Wentzel, K. F. (1975). *Angew. Bot.* **49,** 223–28.

209. Jäger, H-J. (1976). *Eur. J. For. Pathol.* **6,** 25.

210. Roberts, E. R. (1976). *Environ. Pollut.* **11**(3), 175–80.

211. Faller, N. (1972). In *Isotopes and Radiation in Soil-Plant Relationships, Including Forestry.* Proceedings of the IAEA/FAO, Symposium, Vienna, p. 51.

212. Läuchli, A. (1972). *Ann. Rev. Plant Physiol.* **23,** 197.

213. Smith, I. K. (1976). *Plant Physiol.* **58**(3), 358–362.

214. Terry, N. (1976). *Plant Physiol.* **57,** 477–479.

215. Das, G. and Runeckles, V. C. (1974). *Environ. Res.* **7,** 353–362.

216. Das G., and Runeckles, V. C. (1975). *J. Exp. Bot.* **26**(94), 705–712.

217. O'Gara, P. J. (1922). *Ind. Eng. Chem.* **14,** 744.

218. Stratmann, H. (1961). *Staub-Reinhalt* **21**(2), 61–64.

219. Zahn, R. (1963). *Z. Pflanzenkr.* **70**(2), 81–95.

220. Dinman, B. D. (1972). *Science* **1975,** 495–497.

221. Petering, D. H. and Shih, N. T. (1975). *Environ. Res.* **9,** 55–65.

222. Nielsen, J. Parkinson (1938). Cited from ref. 37.

223. Setterström, C. and Zimmerman, P. W. (1939). *Contrib. Boyce Thompson Inst.* **10,** 151–181.

224. Bennett, J. and Hill, A. C. (1973). *J. Environ. Qual.* **2**(4), 526–530.

225. Showman, R. E. (1972). *Bryologist* **75,** 335–341.

226. Hill, D. J. (1974). *New Phytol.* **73,** 1193–1205.

227. Inglis, F. and Hill, D. J. (1974). *New Phytol.* **73,** 1207.

228. Roberts, B. R., Townsend, A. M., and Dochinger, L. S. (1971). *Plant Physiol.* **47,** 30 (Suppl.).

229. Hällgren, J-E. and Huss, K. (1975). *Physiol. Plant.* **34,** 171–176.

230. Wellburn, A. R., Capron, T. M., Chan, H-S., and Horsman, D. C. (1976). In T. A. Mansfield, Ed., *Effects of Air Pollutants on Plants,* Cambridge University Press, pp. 105–114.

230a. Ferenbaugh, R. W. and Bilderback, D. E. (1974). *Am. J. Bot.* **61** (Suppl.).

231. Reinert, R. A., Heagle, A. S., and Heck, W. W. (1975). In J. B. Mudd and

T. T. Kozlowski, Eds., *Responses of Plants to Air Pollution*. Academic Press, New York, pp. 159–178.

231a. White, K. L., Hill, A. C., and Bennett, J. H. (1974). *Environ. Sci. Technol.* **8,** 574–576.

232. Kershaw, K. A. (1972). *Can. J. Bot.* **50,** 543.

232a. Vernon, L. P. and Seely, G. R., Eds. (1966). *The Chlorophylls*. Academic Press, New York.

233. Willstätter, R. and Stoll, A. (1913). In *Untersuchungen über Chlorophyll*. Julius Springer, Berlin.

233a. Dörries, W. (1932). *Ber. Dtsch. Bot. Ges.* **50,** 47.

234. Katz, M. and Shore, V. C. (1955). *J. Air Pollut. Control Assoc.* **5**(8), 144–150, 182.

234a. Dässler, H. G. (1970). *Forstl. Rauchschaden Forsch. Tharandt. Mitt.*, p. 102.

235. Chichester, C. C. and Nakayama, T. O. M. (1965) in T. W. Goodwin, Ed., *Chemistry and Biochemistry of Plant Pigments*. Academic Press, London, pp. 439–457.

235a. Krasnowskij, A. A. (1969). *Prog. Photosyn. Res.* **2,** 709.

236. Arndt, U. (1971). *Environ. Pollut.* **2,** 37–48.

236a. Malhotra, S. S. (1977). *New Phytol.* **78,** 101–109.

237. Willstätter, R. and Stoll, A. (1910). *Liebigs Ann.* **378,** 18.

237a. Holden, M. (1961). *Biochem. J.* **78,** 359.

238. Aronoff, S. (1953). *Adv. Food Res.* **4,** 133.

238a. Peiser, G. and Yang, S. F. (1975). *Plant Physiol.*, Suppl. A, **56**(2), 13; Suppl. B, **57**(5), 47.

239. Bethge, H. (1958). In *Schriftenreihe des Vereins für Wasser-, Boden- und Luft-hygiene,* 13; pp. 3–10.

239a. Spedding, D. J. and Thomas, W. J. (1973). *Aust. J. Biol. Sci.* **26,** 281–286.

240. Jensen, K. F. (1975). U.S. Department of Agriculture, Forest Service, Res. Note NE-209, Northeast Forest Experiment Station, Upper Darby, Pa.

241. Dässler, H. (1972). In *Wirkungen von Luftverunreinigungen auf Waldbaume*. Forst. Bundesversuchanst, Vienna, pp. 353–366.

242. Ricks, G. R. and Williams, R. J. H. (1975). *Environ. Pollut.* **8,** 97–106.

243. Horsman, D. C. and Wellburn, A. R. (1975). *Environ. Pollut.* **8,** 123–133.

244. Kändler, U. and Ullrich, H. (1964). *Naturwiss.* **51,** 518.

245. Hall, D. O. (1972). *Nature New Biol.* **235,** 125–126.

246. Malhotra, S. S. (1976). *New Phytol.* **76,** 239.

247. Silvius, J. E., Ingle, M., and Baer, C. H. (1975). *Plant Physiol.* **56,** 434–437.

248. Baldry, C. W., Cockburn, W., and Walker, D. A. (1968). *Biochim. Biophys. Acta* **153,** 476–483.

249. Asada, K., Kitoh, S., Deura, R., and Kazai, Z. (1965). *Plant Cell Physiol.* **6,** 615.
250. Asada, K., Deura, R., and Kasai, Z. (1968). *Plant Cell Physiol.* **9,** 143–146.
251. Hall, D. O. and Telfer, A. (1969). *Prog. Photosynth. Res.* **3,** 1281–1287.
252. Pick, U. and Avron, M. (1973). *Biochim. Biophys. Acta* **325,** 297–303.
253. Ballantyne, D. J. (1973). *Phytochemistry* **12,** 1207–1209.
254. Gromet-Elhanan, Z. and Leiser, M. (1973). *Arch. Biochem. Biophys.* **159,** 583–589.
255. Jagendorf, A. T. (1975). In Govindjee, Ed., *Bioenergetics of Photosynthesis.* Academic Press, New York, pp. 414–492.
256. Roy, H. and Moudrianakis, E. N. (1971). *Proc. Natl. Acad. Sci. U.S.* **68,** 2720–2724.
257. Datta, D. B., Ryrie, I. J., and Jagendorf, A. T. (1974). *J. Biol. Chem.* **249,** 4404.
258. Ryrie, I. and Jagendorf, A. T. (1971). *J. Biol. Chem.* **246,** 582–588.
259. Grebanier, A. E. and Jagendorf, A. T. (1977). *Biochim. Biophys. Acta* **459,** 1–9.
260. Ziegler, I. (1972). *Planta* **103,** 155–163.
261. Paulsen, J. M. and Lane, M. D. (1966). *Biochemistry* **5,** 2350.
262. Osmond, C. G. and Avadhani, P. N. (1970). *Plant Physiol.* **45,** 228.
263. Pfleiderer, G., Jeckel, D., and Wieland, T. (1956). *Biochem. Z.* **328,** 187.
264. Shih, N. T. and Petering, D. H. (1973). *Biochem. Biophys. Res. Commun.* **55,** 1319.
265. Williams, R. J. P. (1973). *Biochem. Soc. Trans.* **1,** 1–26.
266. Noack, K., Wehner, O., and Griessmeyer, H. (1929). *Z. Angew. Chem.* **42,** 123.
267. Haywood, G. C. Hill, H. A. O., Pratt, J. M., Vanston, N. J., and Williams, R. J. P. (1965). *J. Chem. Soc.* **6,** 485–493.
268. Unziker, H. J., Jäger, H. J., and Steubing, H. J. (1975). *Angew. Bot.* **49**(3–4), 131–140.
269. Brinkmann, E., Lüttge, U., and Fischer, K. (1971). *Ber. Dtsch. Bot. Ges.* **84**(9), 523–524.
270. Murray, D. R. and Bradbeer, J. W. (1971). *Phytochemistry* **1,** 1999–2003.
271. Zelitch, I. (1971). *Photosynthesis, Photorespiration and Plant Productivity,* Academic Press, New York.
272. Zelitch, I. (1969). *Plant Physiol.* **41,** 1623–1631.
273. Zelitch, I. (1973). *Plant Physiol.* **51,** 299–305.
274. Zelitch, I. (1976). In H. Smith, Ed., *Commentaries on Plant Science.* Pergamon Press, pp. 51–61.
275. Tanaka, H., Takanashi, T., and Yatazawa, M. (1972). *Water, Air Soil Pollut.* **1,** 205–211.

276. Jiráček, V. Macháčková, I., and Koštír, J. (1972). *Experientia* **28**(9), 1007–1008.
277. Loomis, W. D. (1974). In S. P. Colowick and N. O. Kaplan, Eds., *Methods of Enzymology,* Vol. XXXI; *Biomembranes.* Part A, Sec. IV, Academic Press, New York, pp. 528–544.
278. Jäger, H. J. and Pahlich, E. (1976). *Int. J. Environ. Anal. Chem.* **4**(4), 257–262.
279. Haisman, D. R. (1974). *J. Sci. Food Agric.* **25,** 803–810.
280. Cecil, R. and Wake, R. G. (1962). *Biochem. J.* **82,** 401.
281. Grill, D. and Esterbauer, H. (1973). *Eur. J. For. Pathol.* **3,** 65–71.
282. Inous, M. and Hauatsu, H. (1971). *Chem. Pharm. Bull.* **19,** 1286.
283. Phillips, M. L. and Puckett, K. J. (1976). In L. Kärenlampi, Ed., *Proceedings of the Koupio Meeting on Plant Damage Caused by Air Pollution.* Kuopio, p. 143.
284. Farrar, J. F. and Smith, D. C. (1976). *New Phytol.* **77,** 115–125.
285. Wellburn, A. R., Majernik, O., and Wellburn, A. M. (1972). *Environ. Pollut.* **3,** 37–49.
286. Fischer, K., Kramer, D., and Ziegler, H. (1973). *Protoplasma* **76,** 83–96.
287. Godzik, S. and Sassen, M. M. A. (1974). *Phytopathol. Z.* **79,** 155–159.
288. Thomson, W. W. (1975). In J. B. Mudd and T. T. Kozlowski, Eds., *Responses of Plants to Air Pollution.* Academic Press, New York, pp. 179–194.
289. Ikonen, S. and Kärenlampi, L. (1976). In L. Kärenlampi, *Proceedings of the Kuopio Meeting on Plant Damage Caused by Air Pollution.* Kuopio, pp. 37–45.
290. Arndt, U. (1974). *Environ. Pollut.* **6,** 193.
291. Papageorgiou, G. (1975). In Govindjee, Ed., *Bioenergetics of Photosynthesis.* Academic Press, New York, pp. 319–371.
292. Hällgren, J-E. (1977). Manuscript in preparation.
293. Kessler, E. and Zumft. W. G. (1973). *Planta* **111,** 41–46.
294. Bahr, J. T. and Jensen, R. G. (1974). *Plant Physiol.* **53,** 39–44.
295. Andrews, T. J., Badger, M. R., and Lroimer, G. H. (1975). *Arch. Biochem. Biophys.* **171,** 93–103.
296. Lilley, R. McCl. and Walker, D. A. (1975). *Plant Physiol.* **55,** 1087–1092.
297. Evans, G. C. (1972). In *The Quantitative Analysis of Plant Growth Studies in Ecology,* Vol. 1. Blackwell, Oxford, Chap. 29, pp. 499–513.
298. Prioul, J. L., Reyss, A., and Chartier, P. (1975). In R. Marcelle, Ed., *Environmental and Biological Control of Photosynthesis.* W. Junk, pp. 17–27.
299. Ziegler, I. (1973). *Phytochemistry* **12,** 1027.
300. Ziegler, I. (1974). *Biochim. Biophys. Acta* **364,** 28.

301. Ziegler, I. (1974). *Phytochemistry* **13,** 2403.

302. Kelly, G. J., Latzko, E., and Gibbs, M. (1976). *Ann. Rev. Plant Physiol.* **27,** 181–205.

303. Ziegler, I., Marewa, A., and Schoepe, E. (1976). *Phytochemistry* **15,** 1627–1632.

304. Anderson, L. E. and Avron, M. (1976). *Plant Physiol.* **57,** 209–213.

305. Meyer, M. C. and Westhead, E. W. (1976). *FEBS Lett.* **71**(1), 25–28.

306. Libera, W., Ziegler, I., and Ziegler, H. (1975). *Z. Pflanzenphysiol.* **74,** 420–433.

307. Walker, D. A., Slabas, A. R., and Fitzgerald, M. P. (1976). *Biochim. Biophys. Acta* **440,** 147–162.

308. Börtitz, S. (1968). *Biol. Zentralbl.* **87,** 63–70.

309. Vogl, M., Tzschacksch, O., Enderlein, H., Börtits, S., and Haedicke, E. (1972). *Biol. Zentralbl.* **91**(5), 601–612.

310. Börtitz, S. (1969). *Arch. Forstwes.* **18,** 123–131.

311. Koštír, J., Macháčková, I., Jiráček, V., and Bučhar, E. (1970). *Experientia* **26,** 604–605.

312. Nikolaevsky, V. S. (1968). *Fiziol. Rast.* **15,** 110–115.

313. Arndt, U. (1970). *Staub-Reinhalt* **30,** 256–259.

314. Godzik, S. and Linskens, H. F. (1974). *Environ. Pollut.* **7,** 25–38.

315. Jäger, H. J., Pahlich, E., and Stenbing, L. (1972). *Angew. Bot.* **40,** 199–211.

316. Pahlich, E., Jäger, H. J., and Steubing, L. (1972). *Angew. Bot.* **46,** 183–197.

317. Rabe, R. and Kreeb, K. (1976). *Angew. Bot.* **50**(1–2), 70–78.

318. Tolbert, N. E. and Ryan, F. J. (1976). In R. A. Burris and C. C. Black, Eds., *CO_2 Metabolism and Plant Productivity.* University Park Press, Baltimore, pp. 141–159.

319. Corbett, J. R. and Wright, B. J. (1971). *Phytochemistry* **10,** 2015–2024.

320. Zelitch, I. (1959). *J. Biol. Chem.* **234,** 3077–3081.

321. Lorimer, G. H. and Andrews, T. J. (1973). *Nature* **243,** 359.

322. Bowes, G. and Ogren, W. L. (1972). *J. Biol. Chem.* **247,** 2172–2176.

323. Canvin, D. T., Lloyd, N. D. H., Fock, H., and Przybylla, K. (1976). In R. H. Burris and C. C. Black, Eds., *CO_2 Metabolism and Plant Productivity.* University Park Press, Baltimore, pp. 161–176.

324. Taniyama, T. and Arikado, H. (1969). *Crop. Sci. Soc. Jap. Proc.* **38,** 597.

325. De Koning, H. W. and Jegier, Z. (1968). *Atmos. Environ.* **2**(4), 321–326.

326. Björkman, O. (1974). In A. C. Giese, Ed., *Photophysiology,* Vol. VIII. Academic Press, New York, pp. 1–63.

6

CHEMISTRY OF POLLUTANT SULFUR IN NATURAL WATERS

Jerome O. Nriagu
Canada Centre for Inland Waters, Burlington, Ontario, Canada

John D. Hem

U.S. Geological Survey, Menlo Park, California

1. INTRODUCTION

Sulfur is one of the four chalcogens (or so-called ore-forming elements) in Group VIA of the periodic system and is characterized by the $[Ne]3S^23p^4$ electronic configuration. It preferentially exerts the maximum valence of 6 toward strongly electronegative elements such as oxygen and fluorine but also has a strong capacity to function as negatively bivalent toward strongly electropositive elements such as hydrogen and the heavy metals. Valence states between 2− and 6+ are also common with the result that sulfur and its derivatives are stable over a wide range of chemical conditions encountered at or near the earth's surface. Over 2000 sulfur-bearing minerals are known (e.g., see Palache et al., 1951); the most important ones are shown in Table 1.

Sulfur may enter into several types of bonds in order to complete its S^2p^4 subshell. Two electrons may be acquired; two electron-pair bonds may be formed; one electron-pair bond may be coupled to the addition of one electron as in HS^-; a large variety of bonds may be formed by the hybridization of one or more of the orbital electrons. Its coordination number is thus highly variable, ranging from unity in thiocyanates to 17 in high-chalcocite, β-Cu_2S (see Wuensch, 1972). Similarly, the coordination polyhedra range from highly regular in some common configurations to the extremely distorted forms found in complex sulfides (Wuensch, 1972).

In general, the most oxidized (sulfate) and the most reduced (sulfide) states are the most stable chemically. The effective ionic radius of sulfur in the 2− oxidation state is 0.29 Å and in the 6+ state is 1.84 Å (Pauling, 1940). Thus the S^{6+} ion must have a very high charge intensiy. The S^{2-} ion can form covalent bonds with adjacent S^{2-} ions, and various polymerized structures can result in which the effective oxidation number of sulfur is intermediate between 2− and 0. The sulfur incorporated into protein and other forms of biological material is generally in a reduced state, but it is sometimes impossible to assign exact oxidation states to the carbon, sulfur, and nitrogen ions in such structures when they participate in covalent bonding.

The sulfate ion, which consists of four tetrahedrally coordinated oxygen ions grouped around a central S^{6+}, forms compounds of low solubility with alkaline earth metals. Some of these compounds are important factors in sulfur mobility and thus in the geochemical cycle of

Table 1. Important Sulfur-Bearing Minerals[a]

Mineral Name	Composition	Sulfur Content (%)	System
	Sulfides		
Argentite	Ag_2S	12.9	Isomeric
Arsenopyrite	FeAsS	19.7	Orthorhombic
Bismuthinite	Bi_2S_3	18.8	Orthorhombic
Bornite	Cu_5FeS_4	25.5	Isomeric
Bravoite	$(Ni,Fe)S_2$	~36	Isomeric
Chalcocite	Cu_2S	20.2	Orthorhombic
Chalcopyrite	$CuFeS_2$	35.0	Tetragonal
Cinnibar	HgS	13.8	Trigenal
Cobaltite	CoAsS	19.3	Isomeric
Covellite	CuS	33.6	Hexagonal
Cubanite	$CuFe_2S_3$	35.4	
Digenite	$Cu_{2-X}S$	<20	Orthorhombic
Dimorphite	As_4S_3	24.3	
Galena	PbS	13.4	Isomeric
Grecnockite	CdS	22.2	Hexagonal
Greigite	Fe_3S_4	43.4	Isomeric
Mackinawite	FeS	34.1	Tetragonal
Marcasite	FeS_2	53.4	Orthorhombic
Molybdenite	MoS_2	40.1	Hexgonal
Oldhamite	CaS	44.4	Isomeric
Orphiment	As_2S_8	39.0	Monoclinic
Pentlandite	$(Fe,Ni)_9S_8$	~36	Isomeric
Pyrite	FeS_2	53.4	Isomeric
Pyrrhotite	$Fe_{1-x}S$	~36	Hexagonal
Realgar	AsS	29.9	Monoclinic
Sphalerite	ZnS	33.0	Isomeric
Stibnite	Sb_2S_3	28.6	Orthorhombic
Stannite	Cu_2FeSnS_4	29.8	Tetragonal
Troilite	FeS	36.5	
Tungstenite	WS_2	25.9	Hexagonal
Violarite	Ni_2FeS_4		Isomeric
	Sulfosalts		
Andorite	$PbAgSb_3S_6$	22.0	—
Argyrodite	Ag_8GeS_6	17.1	Isomeric
Boulangerite	$Pb_5Sb_4S_{11}$	18.9	Monoclinic
Bournonite	$PbCuSbS_3$	19.7	Orthorhombic
Chalcostibite	$CuSbS_2$	25.7	Orthorhombic
Cosalite	$Pb_2Bi_2S_5$	16.2	—
Enargite	Cu_3AsS_4	32.6	Orthorhombic

Table 1. (Continued)

Mineral Name	Composition	Sulfur Content (%)	System
Livingstonite	$HgSb_4S_7$	24.6	—
Plagionite	$Pb_5Sb_8S_{17}$	21.3	Monoclinic
Polybasite	$(Ag,Cu)_{16}Sb$	~17.9	Monoclinic
Proustite	Ag_3AsS_3	19.4	Trigonal
Pyrargyrite	Ag_3SbS_3	17.8	Trigonal
Sulvanite	Cu_3VS_4	34.7	Isomeric
Tennantite	$(CuFe,Zn,Ag)_{12}As_4S_{13}$	~28	Isomeric
Tetrahedrite	$(Cu,Fe,Zn,Ag)_{12}Sb_4S_{13}$	~25	Isomeric
Native sulfur	S	100	Orthorhombic
	Sulfates		
Alum	$(K,Na,NH_4)Al(SO_4)_2\cdot 12H_2O$	~39.9	Isomeric
Alunite	$KAl_3(SO_4)_2(OH)_6$	15.4	Trigonal
Anglesite	$PbSO_4$	10.6	Orthorhombic
Anhydrite	$CaSO_4$	23.5	Orthorhombic
Barite	$BaSO_4$	13.7	Orthorhombic
Bloedite	$Na_2Mg(SO_4)_2\cdot 4H_2O$	19.2	Monoclinic
Brochantite	$Cu_4(SO_4)(OH)_6$	7.1	Monoclinic
Celestite	$SrSO_4$	17.4	Orthorhombic
Copiapite	$CuSO_4\cdot 5H_2O$	12.8	Triclinic
Epsomite	$MgSO_4\cdot 7H_2O$	13.0	Orthorhombic
Gypsum	$CaSO_4\cdot 2H_2O$	18.6	Monoclinic
Halotrichite	$Fe(II)Al_2(SO_4)_4\cdot 22H_2O$	14.4	Monoclinic
Hexahydrite	$MgSO_4\cdot 6H_2O$	14.0	Monoclinic
Jarosite	$KFe_3(SO_4)_2(OH)_6$	12.8	Trigonal
Kainite	$KM_g(SO_4)Cl\cdot 3H_2O$	12.9	Monoclinic
Kieserite	$MgSO_4\cdot H_2O$	23.2	Monoclinic
Langbeinite	$K_2Mg_2(SO_4)_3$	23.2	Isomeric
Mascagnite	$(NH_4)_2SO_4$	24.3	Orthorhombic
Melanterite	$FeSO_4\cdot 7H_2O$	11.5	Monoclinic
Mendozite	$NaAl(SO_4)_2\cdot 11H_2O$	14.6	Monoclinic
Mirabilite	$Na_2SO_4\cdot 10H_2O$	10.0	—
Polyhalite	$K_2Ca_2Mg(SO_4)_4\cdot 2H_2O$	21.2	Triclinic
Picromerite	$K_2Mg(SO_4)_2\cdot 6H_2O$	15.9	Monoclinic
Thenardite	Na_2SO_4	22.6	Orthorhombic

[a] Based mostly on the compilation by Staples (1972).

the element. Sulfuric acid owes some of its strong acid properties, according to the views of some theoretical chemists, to the strong tendency for repulsion of protons from the inner hydration sheath of the S^{6+} ion, with its small size and high charge intensity.

In reduced systems the stable species in water is H_2S. This substance is a gas (boiling point −60.7°C) at ordinary temperatures. It is approximately 100 times as soluble in water at 25°C as is oxygen at 1 atm of pressure. The dissolved gas has weakly acidic properties and dissociates in two steps, first to HS^- (K_d near $10^{-7.00}$) and then, at high pH, to S^{2-} (near $10^{-13.00}$).

Many metals, particularly some of those in the transition series, form sulfides that have very low aqueous solubilities. This fact is fundamentally responsible for the accumulation of many of the economically important deposits of metallic ores. Heterogeneous sulfide systems may also exert a controlling influence on the concentration and distribution of heavy metals in natural waters (Hutchinson, 1957).

2. SOURCES OF SULFUR

The sulfur in natural waters comes from a wide variety of natural and anthropogenic sources. Sulfur in various forms is a common constituent of most rocks, the weathering of which acts to supply the sulfur in natural waters. Table 2 shows the sulfate contents of groundwaters draining the various rock types. Berner (1971a) estimates that rock weathering currently accounts for about 35% of the sulfate in the world-average river; the weathering ratio for sulfur in sediments and igneous-metamorphic rocks is about 46:1 (Holser and Kaplan, 1966), implying that the weathering of sedimentary rocks is the primary source of sulfur in natural waters. Higher weathering inputs may be expected in surface waters where evaporites, sulfide deposits, and coals/lignites are exposed. Although attempts have been made to relate the sulfur contents of surface waters to the lithology of the drainage areas (e.g., see Talling and Talling, 1965; Fleetwood, 1969; Berner, 1971a; Hitchon and Krouse, 1972), there has been no systematic endeavor to model the influence of the various rock types on the concentration of sulfur in natural waters. This must be considered an important oversight since information on the background sulfur levels is needed before the impacts of the anthropogenic sulfur inputs can be fully assessed. It is significant, however, that the geological controls on the sulfur contents of the surface waters of the Mackenzie River drainage basin (Canada) and of the surface waters of

Table 2. Sulfate Contents of Groundwaters[a]

Source	Rock	Mean Sulfate Concentration (ppm)	Range
Igneous	Granitic	6.0	0.1–468
	Diorite, andesite, and syenite	(?)	0.1–115
	Basic and ultrabasic	5	0–61
	Ephemeral springs, Sierra Nevada	1.0[b]	—
	Perennial springs, Sierra Nevada	2.4[b]	—
Sedimentary	Sandstone, arkose, and graywacke	22	1.4–362
	Siltstone, clay, and shale	54	1.5–2420
	Limestone and dolomite	20	0.2–707
	Unconsolidated sand and gravel	25	1.0–1910
	Miscellaneous	—	6.1–1570
Metamorphic	Quartzite and marble	13	2.0–130
	Slate, schist, gneiss, etc.	31	0–132

[a] Data from White et al. (1963) unless otherwise specified.
[b] From Garrels and Mackenzie (1967).

the USSR can be assessed using isotope ratio measurements (Hitchon and Krouse, 1972; Chukrov et al., 1975).

The atmosphere is also an important contributor of sulfur to natural waters. Sulfur cycles through the atmosphere as gases and in particulate form and may be carried into the natural waters in rainfall and dry fallout and in subsequent surface runoff. The principal natural sources of the atmospheric sulfur supply include sea-salt sprays, biogenic emissions from soils, and coastal marine waters and volcanogenic effluvia (see Moss, Part I). Sea sprays and biogenic H_2S account for 25 to 40%, whereas volcanic emissions generally are responsible for less than 10% of the sulfur in present-day lakes and rivers. Sulfur associated with sea-salt sprays can be identified by (*a*) the $SO_4^{2-}:Cl^-$ ratio, which approximates that of seawater (about 0.15) in coastal regions but decreases with distance inland, and (*b*) high enrichment with respect to ^{34}S ($\delta^{34}S$ for seawater being about 20‰). Sulfur of biogenic origin is generally characterized by low $\delta^{34}S$ values (Grey and Jensen, 1972; Nriagu and Coker, 1977). Thermal springs and hydrothermal solutions tend to contain elevated levels of sulfur (see below) and may lead to a significant enrichment of surface waters with sulfur on a local and even regional scale. Similarly waters in contact with natural gases and/or petroleum reservoirs may be enriched in sulfur.

In addition to the source influence, the background concentrations of sulfur in natural waters may be affected by evaporation and the endogenic cycling of the sulfur. Extended evaporative water losses in endorheic regions such as in western Australia and the north-central United States may result in saline lakes with sulfur concentrations that generally exceed 1000 ppm (see Hutchinson, 1957; White et al., 1963). The effects of endogenic processes on sulfur levels in natural waters are discussed below. It suffices to mention here that large-scale variations have occurred in both the sulfate concentrations and the $\delta^{34}S$ of ocean water during the past geologic epochs, variations caused largely by the transfer of sulfur between the sulfide and sulfate reservoirs in the oceans (Holser and Kaplan, 1966; Holland, 1973).

Sulfur pollution of natural waters comes from many sources: discharges of domestic sewage, industrial wastes, and mine waters; urban and road drainage; leaching of sulfur-containing fertilizers; enhanced weathering of exposed coal and sulfide ore minerals after mining operations; and inadvertent spillage of sulfur products. To these must be added the enhanced atmospheric contributions resulting from the burning of fossil fuels and from refuse incineration, as well as the reduction of pollutant sulfur in sewage and industrial wastes in near-shore marine muds. Anthropogenic sources now account for 20 to over 90% of the sulfur in surface waters; the anthropogenic contribution of sulfur to the average river water has been estimated to be 40% of the total concentration (Friend, 1973). The forms in which the sulfur is introduced into natural waters are diverse and include sulfuric acid and its salts, sulfite and other sulfur oxides, sulfur halides, and a wide range of organosulfur compounds.

There is ample evidence that the sulfur contents of natural waters are increasing because of pollution. The available historical records show that between 1850 and 1967 the sulfate levels in Lake Erie increased at a rate of about 2 mg/l each decade (Beeton, 1965; Nriagu, 1975); the recent increases in the sulfate concentrations in the other Great Lakes are shown in Figure 1. Ackermann et al. (1970) showed that between 1906 and 1968 the sulfate level in the Illinois River rose from 46.6 to about 120 ppm; the concentration in the Ohio River also increased from 45 to 61 ppm between 1951 and 1966. Berner (1971a) and Friend (1973) believe that the sulfur content of the average river has increased by 40 to 60% because of pollution. Some spectacular side effects have occurred in Scandinavia and parts of the northeastern United States, where atmospheric sulfur pollution is widely implicated in the general reduction of the pH of surface waters, to below 4.0 in many instances (Dochinger and Seliga, 1976; Haar and Coffey, 1975). Although the impact of sulfur

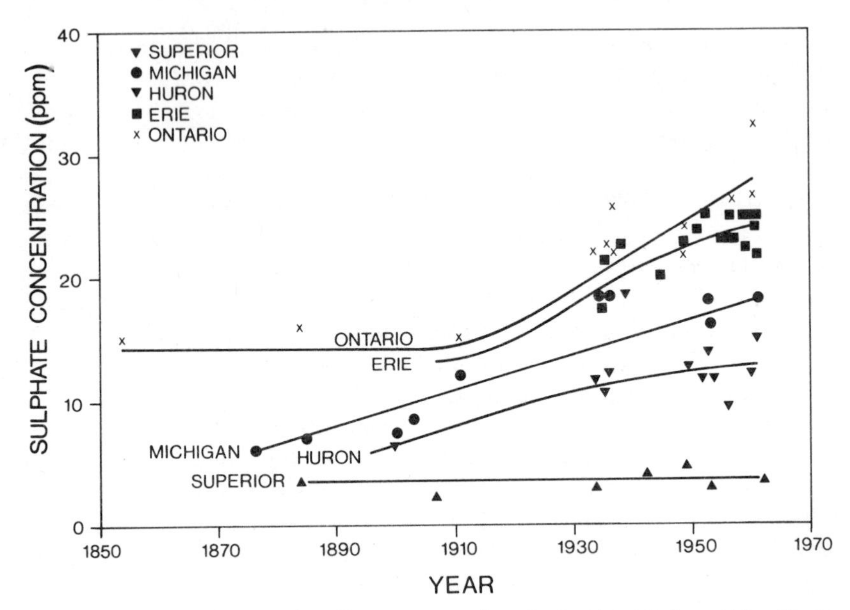

Figure 1. Historical changes in sulfate concentrations in the Great Lakes. (After Beeton, 1965).

pollution is most intense in surface water close to the source, sufficient pollutant sulfur enters the atmosphere to influence the levels in surface waters at very remote locations. Analyses of the Greenland permanent ice sheets (Koide and Goldberg, 1971; Weiss et al., 1975), in fact, showed much higher levels of sulfate in samples accumulated during the past decade (126 to 200 μg/kg water) than in ice samples 60 or more years older (74 to 81 μg/kg water).

3. EQUILIBRIUM CHEMISTRY OF SULFUR

Although some of the important environmental processes involving sulfur cannot be expected always to attain chemical equilibrium, an equilibrium model is useful for showing limiting conditions. It can also serve as a starting point for evaluating the probabilities of occurrence of certain species in solution, and for reviewing other aspects of sulfur aqueous chemistry. Biochemical factors are reviewed later in this chapter.

Chemical thermodynamic data on sulfur species that are significant in aqueous environmental chemistry are given in Table 3. This table is a

Table 3. Chemical Thermodynamic Data

Species	ΔG_f^0 (kcal/mol)	Source of Data
S^{2}(aq)	20.5	Wagman et al. (1968)
S_2^{2-}(aq)	19.0	Wagman et al. (1968)
S_3^{2-}(aq)	17.6	Wagman et al. (1968)
S_4^{2-}(aq)	16.5	Wagman et al. (1968)
S_5^{2-}(aq)	15.7	Wagman et al. (1968
HS^-(aq)	2.88	Wagman et al. (1968)
H_2S(aq)	−6.66	Wagman et al. (1968)
$H_2S_2O_3$(aq)	−125.15	Wagman et al. (1968)
$S_2O_3^{2-}$(aq)	−122.8	Wagman et al. (1968)
$H_2S_2O_4$(aq)	−147.4	Wagman et al. (1968)
$HS_2O_4^-$(aq)	−146.9	Wagman et al. (1968)
$S_2O_4^{2-}$(aq)	−143.5	Wagman et al. (1968)
SO_2(aq)	−71.748	Wagman et al. (1968)
H_2SO_3(aq)	−128.56	Wagman et al. (1968)
HSO_3^-(aq)	−126.16	Wagman et al. (1968)
SO_3^{2-}(aq)	−116.3	Wagman et al. (1968)
HSO_4^-(aq)	−177.97	Wagman et al. (1968)
Ag_2S(c), acanthite	−9.72	Wagman et al. (1969)
Ag_2S(c), argentite	−9.43	Wagman et al. (1969)
CdS(c)	−34.81	Robie and Waldbaum (1968)
CuS(c), covelite	−11.72	Robie and Waldbaum (1968)
Cu_2S(c)	−20.6	Wagman et al. (1969)
FeS(c), pyrrhotite	−24.0	Wagman et al. (1969)
FeS(c), mackinawite	−22.31	Berner (1967)
FeS(c), precipitated	−21.3	Berner (1967)
Fe_3S_4(c), greigite	−70.89	Berner (1967)
FeS_2(c), pyrite	−38.30	Robie and Waldbaum (1968)
HgS(c), red	−12.1	Robie and Waldbaum (1968)
HgS(c), black	−10.34	Robie and Waldbaum (1968)
MnS(c)	−52.14	Robie and Waldbaum (1968)
NiS(c)	−20.6	Robie and Waldbaum (1968)
PbS(c), galena	−22.96	Robie and Waldbaum (1968)
Sb_2S_3(c)	−41.46	Robie and Waldbaum (1968)
ZnS(c), sphalerite	−48.62	Robie and Waldbaum (1968)
ZnS(c), wurtzite	−45.76	Robie and Waldbaum (1968)
$CaSO_4 \cdot 2\ H_2O$(c), gypsum	−429.84	Latimer et al. (1933)
H_2O(l)	−56.687	Wagman et al. (1968)
OH^-(aq)	−37.594	Wagman et al. (1968)
H^+(aq)	0.0	Wagman et al. (1968)
HCO_3^-(aq)	−140.26	Wagman et al. (1968)

compilation of standard Gibbs free energies of formation for the common aqueous sulfur species, and related ions and compounds required for evaluating stabilities in aqueous media. Table 5 is a compilation of equilibrium constants for computing solubilities of sulfates and some related species.

Figure 2 is a stability-field or pH-potential diagram for the system H_2O + S at 25°C and 1 atm. The diagram is based on data in Table 3. Boundaries between solute-species domains represent conditions where the activities of the two species are equal at equilibrium. Boundaries between reduced and oxidized forms were computed in two steps. First,

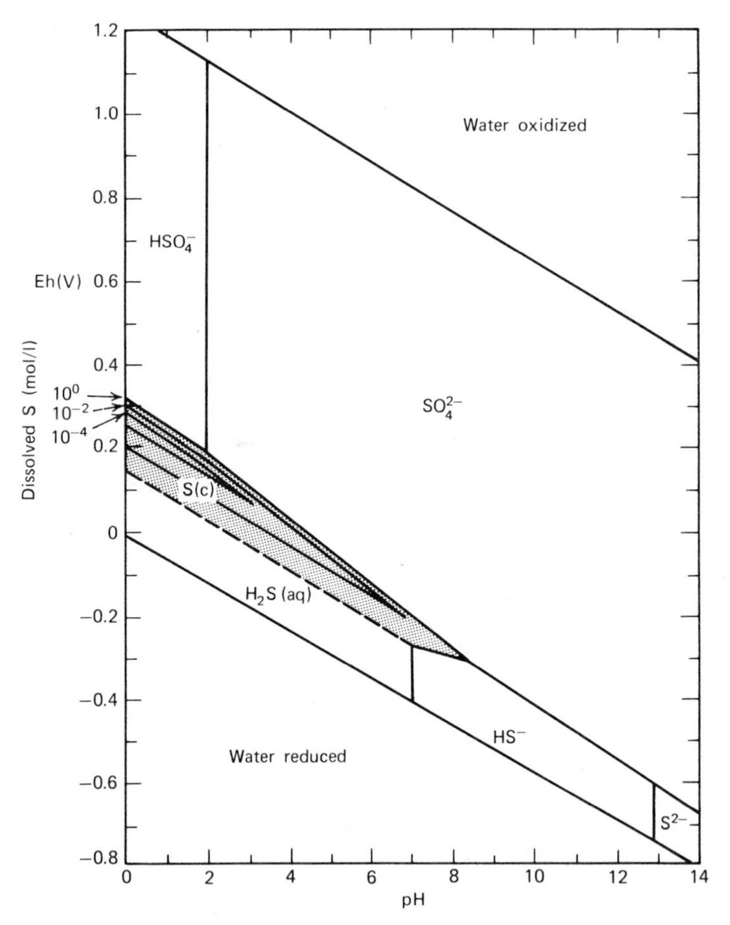

Figure 2. Stability fields of aqueous sulfur species and of elemental sulfur in the system S + H_2O + O_2 at 25°C and 1 atm pressure.

it is necessary to calculate a standard potential utilizing the relationship

$$E^\circ = \Delta G \frac{R}{nF}$$

where ΔG_R is the net standard free energy change in the equilibrium half-reaction, n is the number of electrons gained or lost in that half-reaction, and F is the Faraday constant. When ΔG_R is in kilocalories per mole, F = 23.06 kcal/V·mol. The standard potential is then used in the Nernst equation to compute the redox potential for the half-reaction at specified concentrations of reactants. The form of the Nernst equation used here is

$$Eh = E^\circ + \frac{RT}{nF} \ln \frac{[\text{oxidized}]}{[\text{reduced}]}$$

The symbol Eh represents the redox potential in volts referred to the hydrogen electrode as zero, R is the gas constant, T is the temperature (°K), and the terms in brackets are the thermodynamic activities of the reacting species in the half-reaction. Details of the calculations involved in preparing pH-potential diagrams are given in various papers (Pourbaix, 1949; Garrels and Christ, 1965).

The shaded area in Figure 2 is the region in which elemental sulfur is a stable phase when the solutes are all at unit activity. The boundaries around the sulfur field are shifted when different solute activities are present. Boundaries are also shown for the sulfur field when the total dissolved sulfur species are $10^{-2.00}$, $10^{-4.00}$, and $10^{-6.00}$ M. The boundaries shown between domains of solute species are independent of total sulfur activity. At 0°C and 1 atm the solubility of H_2S given by Weast (1968, p. B204) is about $10^{-1.87}$ M; hence unit activity of H_2S(aq) cannot be reached at equilibrium, and the boundary representing this activity is a dashed line in Figure 2. The sloping upper and lower boundaries represent the limits of equilibrium stability of water toward oxidation or reduction under the specified conditions.

Figure 2 shows that in an aqueous system at equilibrium at 25°C and 1 atm there are five possible dominant species of dissolved sulfur. Over a temperature range of a few tens of degrees and a few atmospheres of pressure, the fundamental thermodynamic relationships are not significantly different from those shown in Figure 2; hence it is reasonable to postulate that this should be applicable to most natural waters at ambient temperatures at or near the land surface.

Compositions of natural water, in general, are in agreement with this postulate: that is, stable oxidation states for sulfur aqueous species that are reported in water analyses are mostly forms either of S^{6+} or of S^{2-}. Intermediate oxidation states such as polysulfide, where the oxidation state may be between −1 and 0, thiosulfate ($S_2O_3^{2-}$), thionates such as $S_2O_4^{2-}$, and sulfite (SO_3^{2-}) are occasionally reported in chemical analyses of hydrothermal or grossly polluted water, but seem never to occur as dominant species. However, small amounts could be widely distributed, as can be shown by calculations based on the data in Table 3.

Conversion of S^{2-} to S^{6+} in a single step in a simple system like that of Figure 2 is obviously very improbable. The oxidation of aqueous sulfide by oxygen, for example, is likely to entail several steps, and some of these may involve substantial energy inputs (see below). The reaction path may be effectively blocked if these are large. Barriers in the reduction of sulfate to sulfide can also be expected, and in addition a rather strong reductant would be needed to reach the low *Eh* required for sulfide stability. Thus it may not be easy to reach equilibrium in the sulfur redox processes, and strictly inorganic mechanisms sometimes do not attain it.

Measurements of *Eh* using standard and inert metal electrodes may encounter difficulties in sulfate-sulfide systems. It may be possible, however, to estimate *Eh* values in the highly reducing region from ionic compositions of aqueous solutions. The computer program for evaluating water-mineral equilibria, SOLMNEQ, provides for this approach (Kharaka and Barnes, 1973).

Biological mediation of sulfur redox processes is the route by which a steady state may be attained in many natural systems (see below). Forms of *Thiobacillus,* for example, may facilitate the conversion of S^0 or H_2S(aq) to SO_4^{2-} in aerated solutions. This is a common process in geothermal areas, where reduced sulfur species in ascending solutions encounter oxygen near the land surface (Ehrlich and Schoen, 1967). Some thiobacilli can thrive in solutions whose pH is less than 2.0, but are not active at temperatures above 55°C. Other species, however, are thermophilic and perform sulfur oxidations in the temperature range of 60 to 94°C (Orr, 1974a).

Using data given by Naumov et al. (1971) and taking into account the effect of temperature in the Nernst equation, it is possible to estimate some effects of increased temperature on the redox boundaries in Figure 2. At temperatures near 100°C, the areas of dominance of oxidized species increase somewhat at the expense of the domains of H_2S and HS^-. Although biota are not active at these high temperatures, the inorganic processes increase in rate as temperature rises, and attainment of equilibrium is more likely than it would be in a completely inorganic system at 25°C.

A somewhat more complete evaluation of possible aqueous sulfur species, both at equilibrium and in certain nonequilibrium conditions, can be made by considering complete redox reactions rather than the half-reactions used in preparing Figure 2. In real systems a chemical reduction cannot take place without an equivalent oxidation of some other reacting species. The wide range of possible oxidation states of sulfur permits various kinds of disproportionation processes, in which some of the sulfur is undergoing reduction while another part of it is being oxidized. From the data in Table 3, it is possible to weigh the thermodynamic feasibility of these processes and to calculate a more complete distribution of equilibrium solute species than is shown in Figure 2, where only two species are considered at a time.

The proportions of polysulfide species that may be present in systems containing $H_2S(aq)$ and free sulfur at equilibrium can be computed from equations such as

$$H_2S(aq) + S^0 = S_2^{2-} + 2H^+$$

giving the relationship

$$\frac{[S_2^{2-}]\,[H^+]^2}{[H_2S]} = 10^{-18.81}$$

In a similar fashion, using the thermodynamic data in Table 3, the second stage of polymerization gives

$$H_2S(aq) + 2S^0 = S_3^{2-} + 2H^+$$

and

$$\frac{[S_3^{2-}]\,[H^+]^2}{[H_2S]} = 10^{-17.79}$$

Other species of this type listed in Table 3 can be similarly evaluated. The equations predict an increasing importance of the polysulfides as pH increases, and also indicate that the more highly polymerized forms are favored in sulfur-rich systems. However, as pH increases, the dissociation process

$$H_2S(aq) = HS^- + H^+$$

becomes increasingly important. The equilibrium constant for this reaction is $10^{-6.99}$. To compute the distribution of the four species one may write an equation for the total dissolved sulfur species in the system, as a summation of all the specific forms considered. When this is combined with the equations for the chemical equilibria relating the species to one another, it is possible to form a mathematical statement, in terms of total dissolved sulfur and pH, of the activity of any single sulfur species of interest, for example

$$[H_2S] = \frac{\text{total } S_{\text{diss}}}{1 + 10^{-6.99}[H]^{-1} + 10^{-15.60}[H]^{-2}}$$

The last term in the denominator of this equation represents the contribution of all the polysulfide species listed in Table 2. At the highest dissolved sulfur activity, shown in Figure 3 at pH 7.00, elemental sulfur can be a stable phase, and the calculation predicts that approximately 1% of the dissolved sulfur activity would be present as polysulfides. At higher pH levels this percentage would rapidly increase, but sulfur is not an equilibrium phase under most conditions in alkaline solutions. The effects of activity coefficients have not been taken into account in this estimate. In solutions of high ionic strength the proportions of polysulfide would tend to be increased somewhat.

A disproportionation process can also involve species more highly oxidized than elemental sulfur. For example, the reaction

$$4S^0 + 2SO_4^{2-} + H_2O = 3S_2O_3^{2-} + 2H^+$$

gives the equilibrium expression

$$\frac{[S_2O_3^{2-}]\,[H^+]^2}{[SO_4^{2-}]} = 10^{-32.43}$$

A computation like the one for the polysulfide equilibria suggests that only small amounts of $S_2O_3^{2-}$ will occur at equilibrium below neutral pH.

Calculations like those above have been made by Boulegue (1976). His results, which indicate that polysulfide may be important above pH 9, are shown in Figure 3.

An autoxidation mechanism for sulfur species may be written:

$$3H_2S(aq) + SO_4^{2-} + 2H^+ = 4S^0 + 4H_2O$$

Thermodynamic data for these species give the mass-law expression:

$$[H_2S]^3\,[SO_4^{2-}]\,[H^+]^2 = 10^{-21.12}$$

Species summations and other related equilibria that can be set up for this system permit calculation of the equilibrium pH for a solution in contact with elemental sulfur when a total of dissolved sulfur species and a ratio of H_2S to SO_4^{2-} activity are specified. This may be considered the equivalent of an open system where waters from more than one

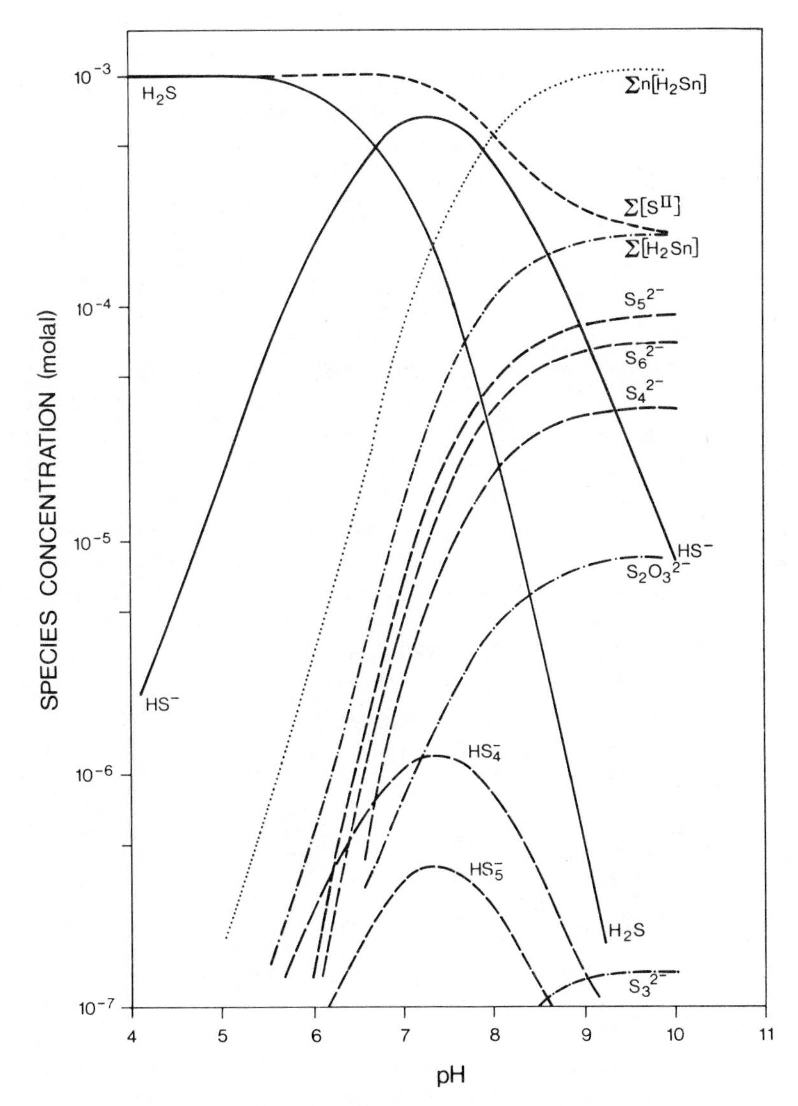

Figure 3. Distribution of sulfur species in the system H_2S-$S_8(\alpha)$-H_2O at 25°C and a total sulfur concentration of 10^{-3} M: $[H_2S_n] =]HS_n^-] + [S_n^{2-}]$; $n[H_2S_n] = n[HS_n^-] + n[S_n^{2-}]$; $[S^{II}] = [H_2S] + [HS^-] + [H_2S_n] + 0.5\,[S_2O_3^{2-}]$. (After Bouleque, 1976.)

source are mixed. The mixture may be out of equilibrium, but the thermodynamic data show the direction in which any reaction might proceed and the amount of energy, or the thermodynamic driving force, that might be present for potential utilization by microorganisms mediating the processes.

This disproportionation equilibrium is another way of evaluating the conditions in solution in the shaded S^0 stability field of Figure 2. At any point within the elemental sulfur field there will be a characteristic ratio of $H_2S(aq)$ species to SO_4^{2-} species and a range of permissible total dissolved sulfur, for any given value of the pH.

These relationships may be indicative of thermodynamic factors in systems where sulfur has been or is being deposited. Water moving past a sulfur deposition site would tend to attain a pH, $H_2S:SO_4^{2-}$ ratio, and total dissolved sulfur activity compatible with the equilibrium condition.

The oxidation of sulfide solutions has been extensively investigated (e.g., see Cline and Richards, 1969; Snavely and Blount, 1969; Chen and Morris, 1972a, 1972b; Almgren and Hagstrom, 1974). As the theoretical calculations predict, the products of the oxidation are a complex function of several factors, including the initial sulfide and dissolved oxygen concentrations, pH, *Eh,* temperature, and presence or absence of catalysts and inhibitors, as well as biological activity. The reported products include elemental sulfur, thiosulfate, sulfite, sulfate, polythionates, and polysulfides. The effects of pH and *Eh* on the distribution of the reaction products should be apparent from Figures 2 and 3. The formation of sulfur oxyanions is reported at high initial sulfide concentrations, the formation of thiosulfate and sulfate at lower initial concentrations of elemental sulfur and polysulfides.

4. FORMATION OF METAL SULFIDES

The maximum equilibrium solubility of sulfur species shown in Figure 3 is unlikely to be encountered in natural systems. This is particularly true of reducing environments, where H_2S species predominate. Many cations form sulfides with low solubilities. Table 3 includes standard Gibbs free energy values for some of the more common metal sulfides.

Because iron is an abundant metal, its effect on sulfur solubility in reduced systems is an important aspect of the geochemistry of both elements. The most common iron sulfides include pyrite and marcasite, both having the formula FeS_2, and several different crystalline forms that have sulfur in a more reduced state, including greigite (Fe_3S_4), mackinawite (FeS), and pyrrhotite (also FeS). Iron is abundant enough that these precipitates can be major sulfur sinks. In some groundwaters where sulfate reduction has occurred, the sulfide content can be very small and the sulfate concentration near zero (Hem, 1970, pp. 148, 168). Probably, after being reduced the sulfur was removed by precipitation as an iron sulfide.

Like other sulfur redox processes, the synthesis and dissolution of pyrite and other iron sulfides may not be readily reversible and may require biological mediation to reach equilibrium.

Figure 4 is a pH-potential diagram showing the solubility of sulfur derived from pyrite in the presence of a fixed activity of dissolved iron. The critical region for sulfur behavior in this system occurs where the equilibrium aqueous form is sulfate. Pyrite is rather easily oxidized in the presence of air and water, but the reverse of the reaction is generally more difficult to accomplish. In some such systems a more reduced sulfide species may precipitate first. Figure 5 shows a similar system in which sulfur solubility is controlled by greigite (Fe_3S_4). In the low-pH region (below pH 4) the solubility of sulfur may be more stringently limited by the sulfur-sulfide redox couple, and the region where sulfur is the controlling solid is indicated by shading.

Figure 6 represents the solubility of sulfur in the presence of mackina-

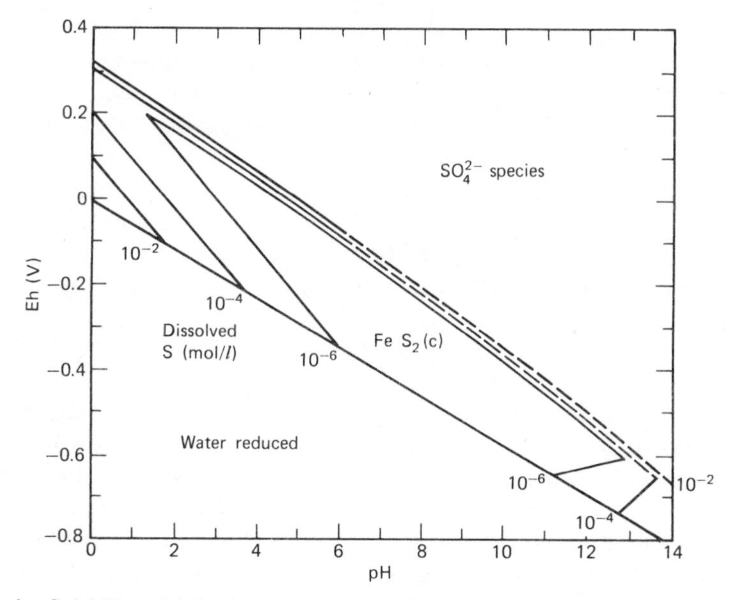

Figure 4. Solubility of sulfur in the system S + H_2O + O_2 + Fe, in the presence of pyrite and a fixed ferrous iron activity at $10^{-4.00}$ mole/l at 25°C and 1 atm pressure.

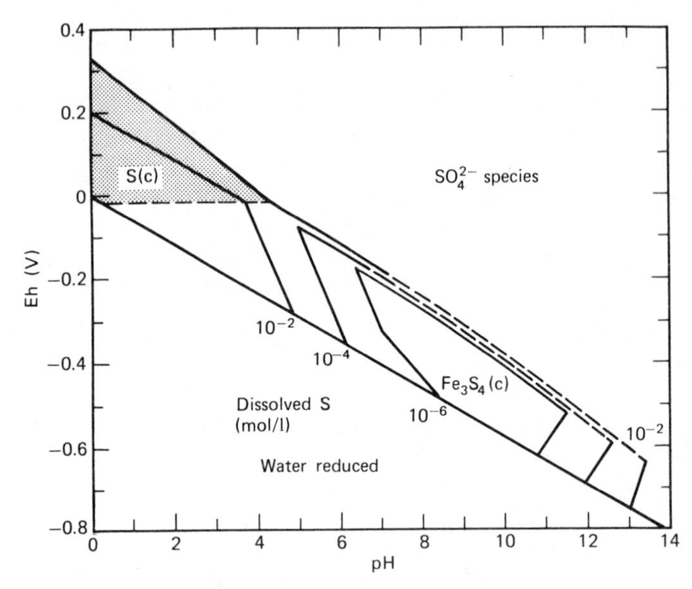

Figure 5. Solubility of sulfur in the system S + H_2O + O_2 + Fe, in the presence of greigite (Fe_3S_4) with a fixed ferrous iron activity of $10^{-4.00}$ mole/l at 25°C and 1 atm pressure.

wite. This mineral may be oxidized to form sulfate ions, but its sulfur is fully reduced and no increase in solubility occurs with decreasing *Eh*. A more stable form of FeS, pyrrhotite, has a somewhat lower solubility. The ferrous sulfides in Figures 5 and 6 are probably synthesized in natural systems at low temperatures; and, although metastable with respect to pyrite, they may persist in some environments for long periods. It should be emphasized that the stability fields for most heavy-metal sulfides generally fall in the stability region covered by pyrite in Figure 4.

Sulfide oxidation inevitably produces large amounts of H^+. When large amounts of reduced sulfide mineral species become available for oxidation, as in coal mines where pyrite accompanies the coal, in waste spoil piles that accumulate near such mines, and in metal mines exploiting sulfide ores, the acidity produced in water draining from the workings and associated debris has a severely detrimental effect on the quality of the water in the receiving streams. Such drainage may have a pH below 3.0 and can carry high concentrations of sulfate and dissolved metals.

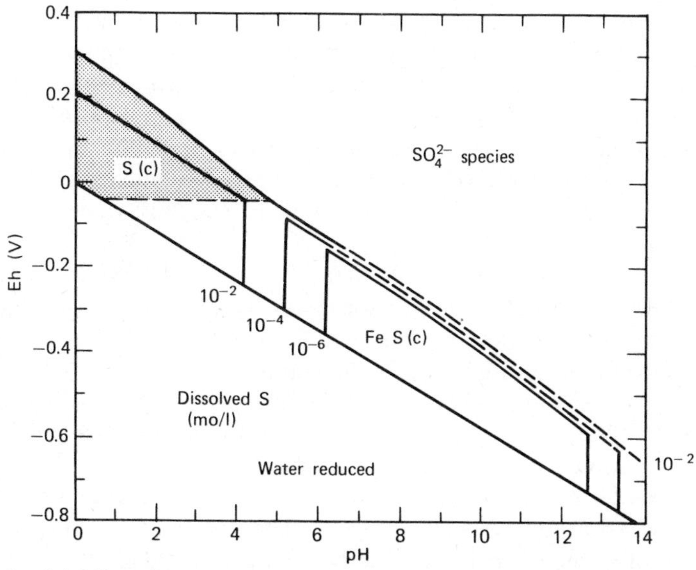

Figure 6. Solubility of sulfur in the system S + H_2O + O_2 + Fe, in the presence of mackinawite (Fes) with a fixed ferrous iron activity of $10^{-4.00}$ mol/l at 25°C and 1 atm pressure.

5. RATES OF AQUEOUS SULFIDE OXIDATION

The rate of oxidation of aqueous sulfide tends to be complicated and not readily amenable to simple kinetic models. For example, Chen and Morris (1972b) reported that in acidic solutions with pH <6 the oxidation rate was very slow. The rate then increased rapidly to a maximum around pH 8, fell to roughly 25% of the maximum value near pH 9, and went again to a maximum at a pH of about 11, decreasing in more alkaline solutions. These authors found the induction period for the sulfide oxidation in fresh water to be 0.2 to 6 hr and suggested that the reaction proceeded through a chain mechanism with a reaction order of 1.34 with respect to sulfide. When the experimental conditions are severely controlled, however, the reaction rate can usually be explained in terms of first-order or pseudo-first-order kinetics with respect to dissolved sulfide. For example, Avrahami and Golding (1968) found that between pH 11 and 14 the oxidation of low levels of sulfide was close to first order, with the rate-determining step believed to be

$$HS^- + 1.5O_2 \rightarrow SO_3^{2-} + H^+$$

Similarly, Skopintsev et al. (1964), Ostlund and Alexander (1963), and Almgren and Hagstrom (1974) found that the oxidation of sulfide in seawater can be treated as a first-order reaction. Table 4 summarizes the available data on sulfide oxidation rates in natural waters. It is clear that there is considerable disagreement in regard to both the reported half-lives and the proposed oxidation mechanisms. Skopintsev et al. (1964) found that, in general, the larger the initial $H_2S:O_2$ ratio, the slower the oxidation rate.

The dramatic catalytic effects of metal ions, first observed by Krebs in 1928, have been examined in greater detail by Chen and Morris (1972a) and by Snavely and Blount (1969). Of the metals studied, the catalytic effects were found to follow the sequence $Ni^{2+} > Co^{2+} > Mn^{2+} > Cu^{2+} > Fe^{2+} > Ca^{2+}$ or Mg^{2+}. These cations increase the sulfide oxidation rate by 5- to >100-fold in the pH range of 7 to 9 and also reduce the induction period dramatically (Chen and Morris, 1972a). The actual mechanisms for the catalysis remain to be determined. Chen and Morris (1972a) suggested two possible schemes. The first mechanism entails the formation of a free radical which activates the autoxidation reaction:

$$M^{n+} + O_2 + H^+ \rightarrow M^{(n+1)+} + HO_2$$

$$M^{(n+1)+} + HS^- \rightarrow M^{n+} + HS$$

The second mechanism involves the formation of metal sulfide complexes which lower the activation energy of autoxidation and result in rapid precipitation of sulfur:

$$M(HS)^{(n-1)+} + O_2 \rightarrow M^{n+} + S^0 + HO_2^-$$

$$HO_2 + HS^- \rightarrow 2OH^- + S^0$$

It is interesting that the metals showing high catalytic effects are able to participate in one-electron exchanges, a notable trait for the production of radical chains.

Organic compounds also influence the oxygenation rate of aqueous sulfide. Some enhance the rate of oxidation (generally by two- to tenfold); these include phenols, aldehydes, aniline, urea, and vanillin. Others that inhibit the oxidation of sulfide (usually by less than fivefold) include EDTA, NTA, cyanide, citrate, peptone, glycerol, and even Roccal, a commercial disinfectant (Chen and Morris, 1972a; Cline and Richards, 1969). Several organic compounds block the oxidation of the intermediary oxyanions; compounds that have been found to inhibit the oxygenation of sulfite include alcohols, resorcinol, ethylamine, phenols, and amines. Only the empirical effects of these compounds have been observed; the actual mechanisms involved have not been investigated.

Table 4. Sulfide Oxidation Rates in Natural Waters

Initial H_2S (μM)	Initial O_2 (μM)	t (°C)	pH	Half-Life (hr)	Order of Reaction[a]	Medium	Reference
40	200	25	8.2	0.4	1	Seawater	Ostlund and Alexander (1963)
32 (15–90)[b]	75 (15–200)[b]	15–22	8–8.5	65 (15–>125)[b]	1	Seawater	Skopintsev et al. (1964)
Saturated	100	—	>11	2.2	1	Fresh water	Avrahami and Golding (1968)
60	60	9.8	7.5–7.8	14.4	1	Seawater	Cline and Richards (1969)
60	120	9.8	7.5–7.8	5.6	1	Seawater	Cline and Richards (1969)
60	240	9.8	7.5–7.8	2.9	1	Seawater	Cline and Richards (1969)
15	Air	18	7.4	15	1	Buffered medium	Sorokin (1970)
160	Air	18	7.4	40	1	Buffered medium	Sorokin (1970)
15	Air	8–9	7.6–7.7	10	1	Seawater	Sorokin (1971)
100	800	25	7.9	50	1.34	Fresh water	Chen and Morris (1972b)
50	200	23–24	8.0	3	1	Seawater	Almgren and Hagstrom (1974)
200	200	23–24	8.0	4.7	1	Seawater	Almgren and Hagstrom (1974)

[a] The order of reaction is with respect to the dissolved sulfide.
[b] The range in data reported is shown in parentheses.

In fact, Chen and Morris (1972a) showed that organic compounds not only modify the reaction rates but may also affect the reaction products. One thing, however, is clear: these observations on the inhibition and catalysis of sulfide oxidation by metal ions and organic ligands have important environmental consequences. The rapid disappearance of dissolved sulfide at the oxic-anoxic interfaces soon after lake overturn may actually be catalyzed by the trace metals in these waters. On the other hand, oxygen and sulfur may coexist for a long time in aquatic systems that receive substantial quantities of organic pollutants. At any rate, the complexity introduced by the effects of metal ions and organic compounds makes it almost impossible to predict the lifetimes of sulfides, even in sterile aquatic environments.

6. AQUEOUS-PHASE OXIDATION OF SULFUR DIOXIDE

The aqueous-phase oxidation of sulfur dioxide is an important process, since each year about 4×10^{10} kg of sulfur is deposited into natural waters as SO_2 (Spedding, 1974). The solutions formed by the introduction of SO_2 into water usually contain free sulfurous acid (H_2SO_3), SO_2(aq), HSO_3^-, SO_3^{2-}, and $S_2O_5^{2-}$ in proportions that vary with pH and total sulfur concentration. Sulfurous acid is unknown as a free acid and is often represented as $H_2O{\cdot}SO_2$. The principal reactions suggested as being responsible for the formation of these species in solution include (Spedding, 1974) the following:

(*a*) $(SO_2)(g) + H_2O \rightleftarrows (SO_2)(aq)$

(*b*) $(SO_2)(aq) + H_2O \rightleftarrows H_2O{\cdot}SO_2$

(*c*) $H_2O{\cdot}SO_2 + H_2O \rightleftarrows H_3O^+ + HSO_3^-; \quad K_c = 1.6 \times 10^{-2}$

(*d*) $HSO_3^- + H_2O \rightleftarrows H_3O^+ + SO_3^{2-}; \quad K_d = 1.0 \times 10^{-7}$

(*e*) $2HSO_3^- \rightleftarrows S_2O_5^{2-} + H_2O; \quad K_e = 7 \times 10^{-2}\,\text{mol}^{-1}$

The subsequent oxidation of the sulfite by atmospheric oxygen is a rather slow process that proceeds via the radical chain mechanism (Haynon et al., 1972; Spedding, 1974):

Initiation:

(*f*) $SO_3^- + O_2 \rightarrow SO_5^-$

(*g*) $SO_3^{2-} + Me^{n+} \rightarrow SO_3^- + Me^{(n-1)+}$

Propagation:

(*h*) $SO_5^- + SO_3^{2-} \rightarrow SO_4^- + SO_4^{2-}$

(*i*) $SO_4^- + SO_3^{2-} \rightarrow SO_4^{2-} + SO_3^-$

Termination:

(*j*) $SO_4^- + SO_4^- \rightarrow S_2O_8^{2-}$

(*k*) $SO_3^- + SO_3^- \rightarrow S_2O_6^{2-}$

(*l*) $SO_5^- + SO_5^- \rightarrow$ products

It is important to note that the rate of SO_2 autoxidation increases rapidly with increasing pH; because H_2SO_4 is a stronger acid than $H_2O{\cdot}SO_2$, the rate of SO_2 autoxidation is further decreased as the pH falls below the pK value for reaction *c*.

Alternative mechanisms for metal-catalyzed oxidation of SO_2 in oxygenated solutions have also been advanced. These reaction schemes commonly invoke a metal sulfite complex as the active intermediate in the catalytic oxidation of sulfurous acid. For example, the following mechanism has been proposed (Freiberg, 1975) for the simultaneous oxidation of SO_2 by O_2 in the presence of Fe^{3+} as a catalyst, and by Fe^{3+} in the presence of O_2:

(*m*) $Fe^{3+} + HSO_3^- \rightleftarrows Fe(HSO_3)^{2+}$

(*n*) $Fe(HSO_3)^{2+} + SO_3^{2-} \rightleftarrows Fe(HSO_3)(SO_3)$

(*o*) $Fe(HSO_3)(SO_3) + O_2 + H_2O \rightarrow Fe(OH)^{2+} + 2HSO_4^-$

(*p*) $Fe(HSO_3)(SO_3) + Fe^{3+} + H_2O \rightleftarrows HSO_4^- + HSO_3^- + H^+ + 2Fe^{2+}$

(*q*) $Fe(OH)^{2+} + H^+ \rightleftarrows Fe^{3+} + H_2O$

Reaction *n* was believed to be the rate-determining step. The oxidation of SO_2 by Fe^{3+} will compete with the catalytic oxidation by O_2 when the Fe^{3+} concentration exceeds 2×10^{-3} M (Freiberg, 1975).

7. SOLUBILITY OF SULFATES

In oxidized systems, where sulfur exists only in the sulfate form, an important set of solubility controls involves the alkaline earth metals. Calcium, the most abundant of these in most aqueous systems, forms gypsum ($CaSO_4{\cdot}2H_2O$), anhydrite ($CaSO_4$), and some other crystalline species of similar composition. These are moderately soluble minerals but occur extensively in evaporite rock formations. Celestite ($SrSO_4$) and barite ($BaSO_4$) are much less soluble but occur more rarely because strontium and barium are not as abundant as calcium. Some other metal sulfates or hydroxysulfates also have low solubilities, but seldom serve as significant controls on sulfur concentrations in natural water.

When present in moderate to high concentrations, sulfate ions participate in forming complexes and ion pairs which must be considered in evaluating sulfate solubilities. Some complexing constants and solubility products for sulfate species are given in Table 5. The complexing constants in this table suggest that in solutions having moderately high activities of sulfate (greater than about $10^{-2.00}$ M) there may be extensive association of sulfate with certain cations, and the geochemical behavior of the cations, as well as the sulfate, may be influenced by this factor (see below). The stronger associations occur with cations of high valence (3+ or greater), and with certain specific metals, lanthanides and actinides. Data for these species are given in Sillen and Martell (1964). As noted below, the sulfate species in seawater include many of these complexes.

The solubility of $CaSO_4$ decreases as temperature increases between 30 and 300°C. In part, this represents a change in the crystal structure of the solid produced. At ordinary earth-surface temperatures the stable form is gypsum ($CaSO_4 \cdot 2H_2O$). Anhydrite is formed at higher temperatures and increased pressure. The transition temperature between the two solids in solutions near saturation with sodium chloride was reported by Ostroff (1964) to be 90.5°C. However, lower temperatures for this transition have been reported by other investigators. The form of the solid is also influenced by pressure and by the ionic strength of the solution.

Limits on the solubility of sulfate in natural water commonly are related to the solubility of gypsum. Because of the effects of ion pairing and of ionic strength on thermodynamic ion activity, however, the calculations needed to evaluate gypsum solubility tend to be rather complicated. For a pure solution containing only calcium and sulfate ions, the equilibrium concentrations for calcium and sulfate at saturation can be calculated by using data in Table 5 and an iteration technique (Garrels and Christ, 1965). The results suggest that at 25.0°C the equilibrium concentrations of the two ions should total about 2100 mg/l as $CaSO_4$, including about 600 mg/l Ca and 1500 mg/l SO_4. Similar results were reported by Tanji and Doneen (1966), who developed a mathematical model.

Natural water in gypsiferous terrains commonly approaches these

Table 5. Chemical Equilibria Involving Sulfate and Related Species

Reaction	Equilibrium Constant, 25°C	Reference
$CaSO_4$, gypsum $= Ca^{2+} + SO_4^{2-}$	$10^{-4.625}$	Latimer et al. (1933)
$SrSO_4(c) = Sr^{2+} + SO_4^{2-}$	$10^{-6.35}$	Calculated from Robie and Waldbaum (1968)
$BaSO_4(c) = Ba^{2+} + SO_4^{2-}$	$10^{-9.97}$	Nancollas and Purdie (1963)
$PbSO_4(c) = PB^{2+} + SO_4^{2-}$	$10^{-7.78}$	Kolthoff et al. (1942)
$KAl_3(OH)_6(SO_4)_2(c) + 6H^+ = 3Al^{3+} + 2SO_4^{2-} + 6H_2O$	$10^{-85.33}$	Calculated from Henley et al. (1969)
$Cu_4(OH)_6SO_4$, brochantite, $+ 6H^+ = 4Cu^{2+} + SO_4^{2-} + 6H_2O$	$10^{-15.37}$	Calculated from Wagman et al. (1969)
$Na^+ + SO_4^{2-} = NaSO_4^-$	$10^{-0.23}$	Truesdell and Jones (1974)
$Ca^{2+} + SO_4^{2-} = CaSO_4(aq)$	$10^{2.31}$	Bell and George (1953)
$Mg^{2+} + SO_4^{2-} = MgSO_4(aq)$	$10^{2.34}$	Hanna et al. (1971)
$Cu^{2+} + SO_4^{2-} = CuSO_4(aq)$	$10^{2.30}$	Calculated from Wagman et al. (1969)
$Cd^{2+} + S_2O_3^{2-} = CdS_2O_3(aq)$	$10^{3.94}$	Gimblett and Monk (1955)
$Ca^{2+} + HCO_3^- = CaCO_3(c) + H^+$	$10^{1.92}$	Jacobson and Langmuir (1974)

concentrations (White et al., 1963, p. F24). The implied equilibrium activity of calcium is rather high, however, and in many systems there will also be an influence related to the solubility of calcite. Waters that are near saturation with both calcite and gypsum are common in some areas. For such a water the activity relationship

$$[SO_4^{2-}]\,[H^+] = 10^{-6.54}[HCO_3^-]$$

can be developed from data in Table 5. In many irrigated areas the increase in concentration of residual water in the root zone of cultivated fields, resulting from evapotranspiration, causes calcite and gypsum to precipitate. Hem (1966) cited data for irrigation return flows in Arizona, New Mexico, and Colorado, which indicated that gypsum saturation was commonly attained or closely approached and calcite saturation appeared to be exceeded to a moderate degree in rivers that carried major amounts of such drainage water.

The residual irrigation drainage and waste water from Imperial Valley, California, a closed basin, drains into Salton Sea, which occupies the lowest part of the valley. Concentration from evaporation has produced a solution that is apparently saturated with respect to gypsum (Hely et al., 1966, p. 25).

As the ionic strength of the solution is increased by other, more soluble ions, such as sodium and chloride, the effective solubility of gypsum is increased. Figure 7, based on calculations made by Cherry (1968), shows how gypsum solubility increases as sodium chloride is added to the solution. Curve *A* represents solubilities in the system gypsum + calcite + NaCl + CO_2 + H_2O at 25°C and 1 atm. Sulfate

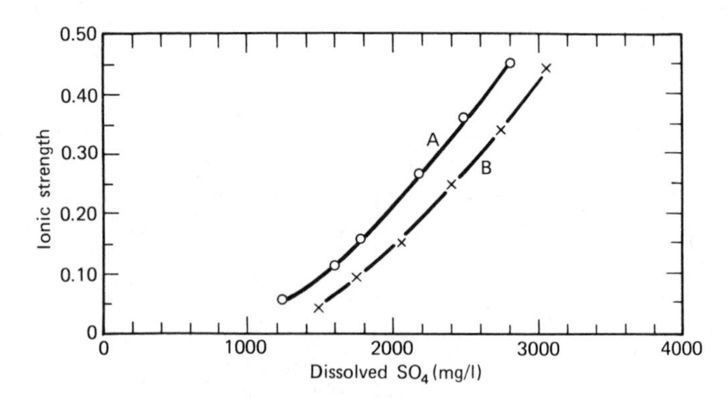

Figure 7. Solubility of sulfate as a function of ionic strength in systems. *A*. Gypsum + calcite + NaCl + CO_2 + H_2O at 25°C and 1 atm pressure. *B*. Gypsum + NaCl + H_2O at 25°C and 1 atm pressure. (Data from CHerry, 1968.)

solubility is somewhat greater in the system gypsum + NaCl + H_2O, given by curve *B*. The difference represents the effect of calcium equilibrated with calcite. This amounts to a substantial concentration at a partial pressure of 1 atm CO_2.

Gypsum has a maximum solubility near 30°C. Data quoted in Sillen and Martell (1971) indicate that the solubility product for gypsum at 100°C is about half as large as the value at 30°C. These constants are corrected to zero ionic strength. Evidently, the effect of temperature on gypsum solubility at higher ionic strengths is not as great.

Solubility products for celestite and barite are given in Table 5. They indicate that solutions equilibrated with these solids alone could have relatively low sulfate concentrations. This is especially true of barite. Generally, however, in natural water the supply of sulfate from other solids tends to overwhelm the available supply of strontium and barium. Therefore these equilibria are more commonly controls of the alkaline earth metal cation concentrations.

Groundwaters from certain aquifers in southeastern Wisconsin seem to contain strontium concentrations in equilibrium with those of celestite (Hem, 1970, p. 196). The very low concentrations of barium that occur in most natural water seem, in general, to be in the range that would be predicted from the barite solubility equilibrium.

8. TRANSPORT OF SULFUR IN AQUATIC ECOSYSTEMS

Sulfur is an essential element in life processes, and an important feature of every aquatic ecosystem is the flow of sulfur within and through it. The geochemical mobility in any given situation is determined by the oxidation state of the sulfur, the biological productivity, and the synergistic effects of the other elemental and nutrient cycles. Because sulfur has several valence states and is an essential element, it is involved in numerous biochemical and inorganic reactions in aquatic systems.

Sulfur is a characteristic constituent of enzymes and proteins and is present also in a wide variety of naturally occurring compounds, many of which play key roles in metabolism. As would be expected, living organisms possess an extensive array of enzymes that can act on a wide range of sulfur compounds. Organisms differ markedly, however, in their ability to assimilate different chemical forms of sulfur. Most plants and many microorganisms are not exacting in their requirements and can use inorganic sulfate as their sole source of sulfur. On the other hand, higher animals require preformed sulfur-containing amino acids and vitamins. The requirements of numerous microorganisms lie between these two extremes.

Living organisms contain highly variable amounts (0.01 to 5%) of sulfur; mean values have been given as 0.5% for plants and decomposers and 1.3% for animals (Orr, 1974a). Many natural organic compounds and most organic detritus also contain sulfur. In spite of the ubiquity of organic sulfur in natural waters, the concentrations are commonly quite small compared to the inorganic sulfur levels. As a result studies of the distribution and cycling of sulfur have emphasized the inorganic species, and little attempt has so far been made to characterize the organosulfur pools in natural waters.

Each aquatic ecosystem thus contains a large inorganic sulfur pool and a small but very active organic sulfur pool. The interconversions of sulfur between the two reservoirs are frequently represented by a cycle (Figure 8). The usual starting point is the sulfate, the most abundant sulfur species in oxic natural waters. Sulfide and sulfite may become the primary sources of sulfur in geothermal environments or because of pollution. The actual microbial transformations fall into four categories: assimilation of sulfur, decomposition of organic sulfur compounds, oxidation of sulfur and its inorganic compounds, and reduction of sulfate and other inorganic sulfur compounds (Starkey, 1956). The metabolic pathways and physiology of the microorganisms involved in the sulfur cycle are the subject of several memoirs and numerous review articles (e.g., see Starkey, 1956; Young and Maw, 1958; Freney and Stevenson, 1966; Postgate, 1968; Trudinger, 1969; Roy and Trudinger, 1970; Kuznetsov, 1970; Sorokin, 1970; Wetzel, 1975; also, Zinder and Brock, this volume); no attempt will be made to review the material here. It is important to note, however, that there are several processes which remove sulfur from the subaqueous biogeochemical cycle: (*a*) the formation of insoluble iron (and other metal) sulfides, as discussed in Section 9.6 (see also Goldhaber and Kaplan, 1975); (*b*) the precipitation of sulfates (particularly gypsum and anhydrite) out of the solution, which may be engendered by extended desiccation; (*c*) the removal of the organic debris and other sulfur-containing suspended particulates; and (*d*) the exchange of sulfur across the air-water and sediment-water interfaces.

It follows that in aerobic surface waters the loss of sulfur should be quite small and entails primarily the sedimentation of suspended particulate and organic matter. Indeed, studies show that $<1.0\%$ of the sulfur entering any of the Great Lakes annually is immobilized in the sediments (Torrey, 1976; Nriagu, 1975; Nriagu and Coker, 1976). It seems unlikely that the leakage of sulfur out of aerobic reservoirs exceeds 5% of the

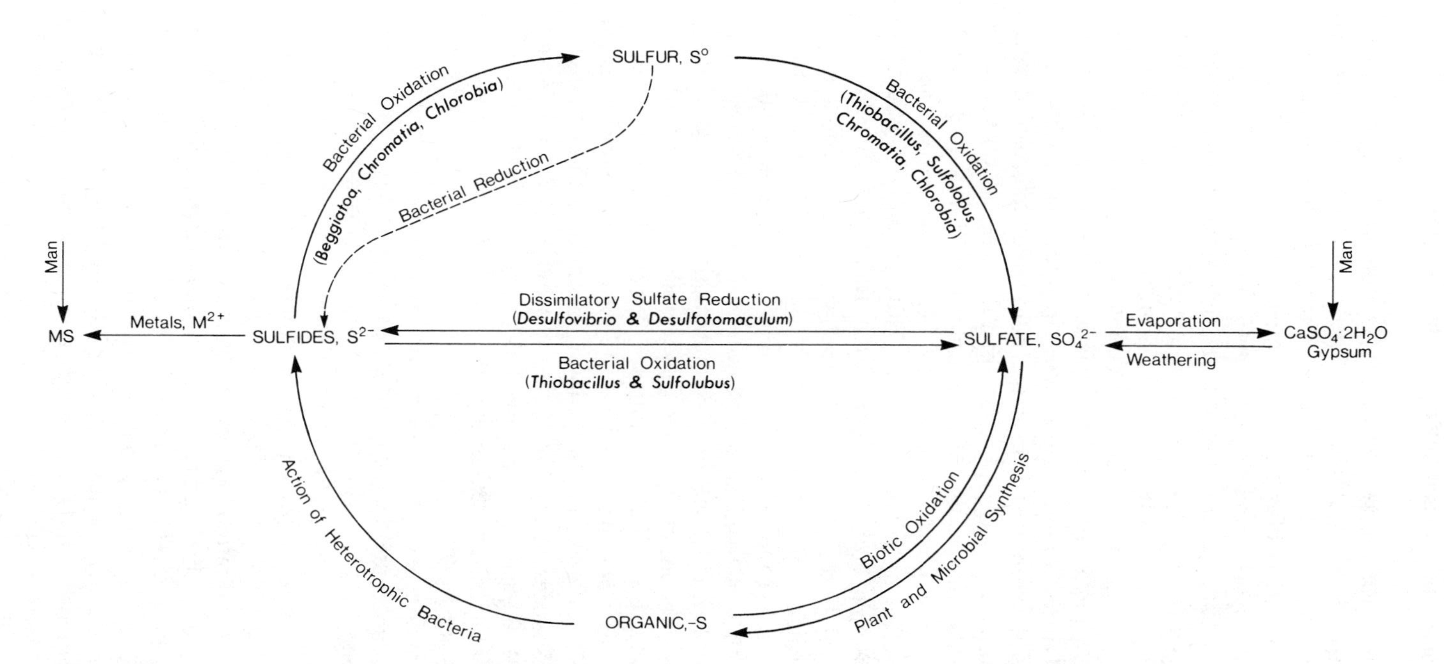

Figure 8. Principal pathways of sulfur transformation in nature.

annual input, with the result that the sulfur concentrations generally reflect changes in the sulfur input. The situation for maritime ecosystems is, however, different; because of the high concentrations of Ca^{2+} and SO_4^{2-} ions, the sulfur may be removed by precipitation of gypsum and anhydrite (Holser and Kaplan, 1966).

Substantial leakage of sulfur out of anaerobic aquatic systems does occur. Microbial sulfate reduction is a very important biogeochemical process which accounts for the sulfides in recent and ancient sediments, and for most epigenetic sulfide ore deposits, as well as for the sulfur in fossil fuels. Stuiver (1967) used ^{35}S as a tracer to study the total sulfur budget of Linsley Pond, Connecticut, in appreciable detail. He found that 520 kg of sulfur was lost at the mud-water interface during the 4-month stratification period; the rate of sulfate reduction varied from 6.6 to 9.0 mg S/day·cm^2 sediment-water interface. Substantial reductions in total sulfur concentrations have also been observed in the anaerobic hypolimnia of many lakes and ponds (Hutchinson, 1957; Sorokin, 1970) and in anaerobic bottom waters of the Black Sea (Skopintsev et al., 1964), several fjords (Richards, 1965), and some polluted coastal and estuarine regions (Richards, 1965; Deuser, 1975). Usually the loss of sulfur by the escape of H_2S to the atmosphere is small, except in marshes with less than 50 cm of water overburden (Bogander, 1977). Transport of sulfur by means of sedimentation of biological material or through production of H_2S by biodegradation of organic matter is quantitatively insignificant (Stuiver, 1967; Goldhaber and Kaplan, 1975); this may not, however, be the case in organic-rich ecosystems where the dissolved sulfate is low. Invariably, though, the organic material influences the sulfur cycle by providing the carbon source for bacterial metabolism (Hutchinson, 1957). Figure 9 depicts the close link between biological processes and the sulfur species in Lake Belovod, USSR (Sorokin, 1970).

The formation and fission of sulfur bonds are also important in the transfer of energy in aquatic ecosystems. This is necessarily so because of the strong link between the biogeochemical cycle of sulfur and the cycles of C, H, O, N, and P. In the biosphere, these six elements are (*a*) combined together as biochemical components of living organisms by photosynthesis and other chemosynthetic processes, (*b*) transformed together by respiration and catabolism or released together as waste products, and (*c*) biodegraded into various organic and inorganic products (Orr, 1974a). The sulfur cycle itself operates with two different groups of bacteria, the aerobic sulfur-oxidizing bacteria and the anaerobic sulfur-reducing bacteria. During the reduction of sulfate by the second group of bacteria, the energy released is either transferred to the resultant H_2S molecules or used for growth and development. The energetics of sulfate reduction is illustrated by the following reactions mediated by *Desulfovibrio* or *Desulfotomaculum* (Wetzel, 1975; Orr, 1974a):

$$2 \text{ Lactate} + SO_4^{2-} \rightarrow 2 \text{ Acetate} + 2HCO_3 + H_2S(\Delta G_r^0 = -40 \text{ kcal/reaction})$$

$$4 \text{ Pyruvate} + SO_4 + 4H_2O \rightarrow 4 \text{ Acetate} + 4HCO_3^- + H_2S + 2H^+(\Delta G_r^0 = -85.2 \text{ kcal/reaction})$$

$$2(CH_2O) + SO_4^{2-} \rightarrow H_2S + 2HCO_3^-$$

where CH_2O is a schematic representation of carbohydrates.

The energy locked up in the H_2S molecules is subsequently transferred to the interface (such as the metalimnion) between the oxic and anoxic waters. Here it may be lost in the form of heat by the chemical oxidation. It can also be used by the sulfur oxidizing bacteria in chemosynthesis and cell development, thereby participating in the formation of primary food resources. The energy flow picture is complicated by the fact that both the biotic and the inorganic oxidation of H_2S results in the formation of several intermediate products, including molecular sulfur, thiosulfate, sulfite, and polythionates (see Roy and Trudinger, 1970). Many sulfur bacteria are, however, capable of making use of these intermediary lower sources of energy.

The rates of sulfate reduction are most intense at the mud-water interface (Kaplan et al., 1963) and below the trophogenic zones (Sorokin, 1964, 1970); these locations receive high inputs of allochthonous and autochthonous organic material. The oxidation and reduction of sulfur thus also entail mobilization of the energy associated with both the natural and the pollutant organic materials in anaerobic water. The little information available suggests that the sulfur bacteria development may represent a significant fraction of the annual photosynthetic productivity of some lakes (see Wetzel, 1975). Recently, the suggestion has been made that sulfate uptake can be used as a measure of planktonic microbial production in freshwater ecosystems (Monheimer, 1974, 1975). Because of the rather specific requirements of the sulfur cycle bacteria (in terms of redox potential, pH conditions, and sulfur species), however, the development of genotype populations generally occurs in rather localized trophic zones.

The foregoing account clearly illustrates the close link between the chemical and biological processes associated with the transport of sulfur in aquatic ecosystems.

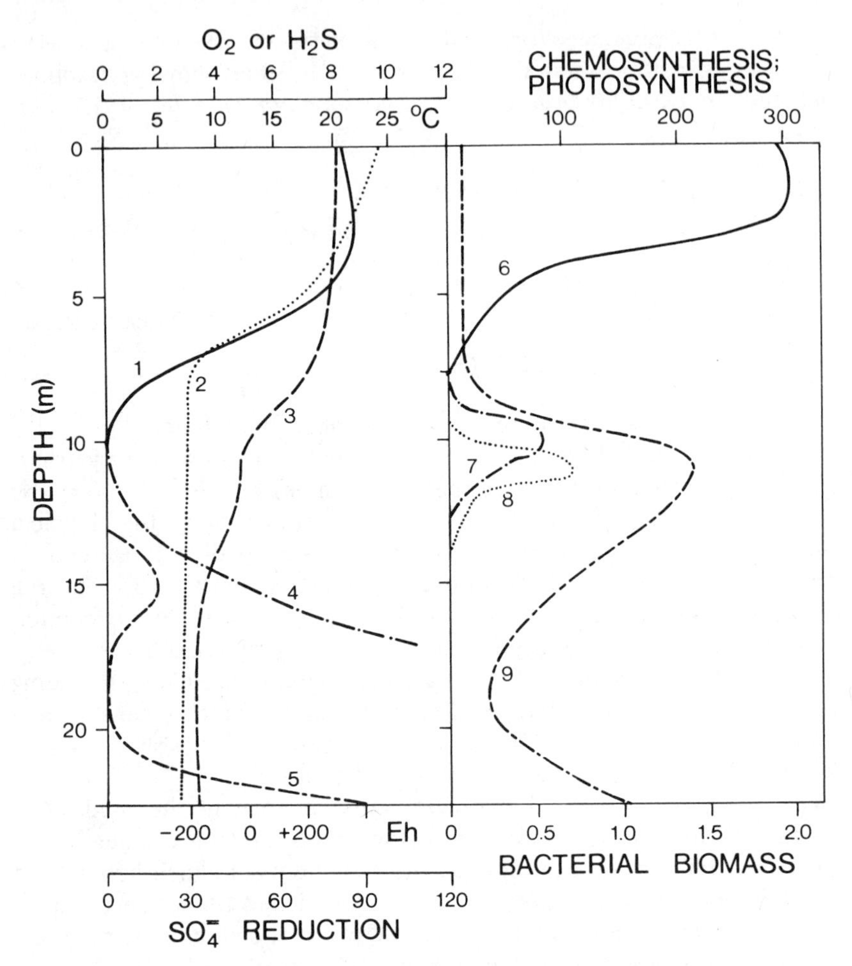

Figure 9. Midsummer distribution of the chemical and biological parameters (pertaining to the sulfur cycle) in meromictic Lake Belovod, USSR. 1, Oxygen (ppm); 2, temperature (°C); 3, *Eh* (mV); 4, sulfide concentration (ppm); 5, rate of sulfate reduction (mg H_2S formed/m^3·day); 6, algal photosynthesis (mg C/m^3·day); 7, chemosynthesis (mg C/m^3·day); 8, purple sulfur bacteria photosynthesis (mg C/m^3·day); 9, bacterial biomass (ppm). (Data plotted from Sorokin, 1970.)

9. CHEMICAL IMPACTS OF SULFUR ON AQUATIC ECOSYSTEMS

Sulfur influences the physical, biological, and chemical characteristics of its environment in many ways. For example, the excess sound absorption at frequencies greater than 10^4 Hz has been attributed to magnesium sulfate association (Fisher, 1967). The availability of sulfur in reduced form causes the accumulation of base metals into large ore bodies. Also, the production of toxic H_2S in anaerobic waters usually entails extensive changes in the fauna and flora of the particular aquatic ecosystems (see above). The brief discussions below will be restricted to chemical impacts, however.

9.1. Metal Complexation

As noted above, a significant fraction of Mg^{2+} and Na^+ in seawater is complexed with the sulfate ion. In river waters it has been estimated that sulfate complexes account for about 5% and 2%, respectively, of the Ca^{2+} and Mg^{2+} in solution (Millero, 1975). Kester et al. (1975) considered the speciation of iron in marine systems and showed that $FeSO_4^0$ may account for 2 to 8% of total dissolved iron(II), depending on pH and alkalinity. The percentage of Fe^{3+} ions in association with SO_4^{2-} was shown to be slightly higher (5 to 10%). In fresh water, $FeSO_4^+$ comprises about 1% of the dissolved iron(III) in solution (Kester et al., 1975). A small fraction (<5%) of the zinc in seawater is complexed by sulfate ions (Millero, 1975). Evaluation of the extent of metal association with dissolved sulfur oxides has been hampered by lack of accurate thermochemical data. The strong association of H^+ with sulfate ions in seawater is noteworthy. Apparently the formation of significant amounts of HSO_4^- is chiefly responsible for the discrepancy between the pH scales for seawater based on measurements of total and of free (subscripts T and F, respectively) hydrogen ion concentrations:

$$(H^+)_T = (H^+)_F + (HSO_4^-) + (HF)$$

A discussion of the pH scales based on buffers prepared in seawater is given in Hansson (1973) and Pytkowicz et al. (1975).

Several studies have noted in both H_2S-rich interstitial waters and anaerobic bottom waters concentrations of trace metals that greatly exceed the values expected to be in equilibrium with the metal sulfides (e.g., see Presley et al., 1972; Harris et al., 1975; Leckie and Nelson, 1975). Gardner (1974) considered the organic versus inorganic trace-metal complexes in such sulfidic marine waters and concluded that bisulfide and polysulfide complexes may be important agents of trace-metal solubility in such anaerobic environments. Gardner's (1974) model, however, considered only simple amino acids and hydroxycarboxylic acids, and some of the constants (for the thiocomplexes) used in the calculations will probably be revised. A more cogent observation is that the sulfhydryl (—SH) groups of organic molecules generally have a strong affinity for trace metals. Indeed, these thioorganic molecules have been suggested as the principal ligands involved in the complexation of lead in natural waters (Rickard and Nriagu, 1977).

The information available, though meager, clearly suggests that thioligands play an important role in the speciation of metals in natural waters.

9.2. Effects on pH

The influence of sulfur and its derivates on the pH (and *Eh*; see below) of an aquatic environment is inescapably linked to the C-N-S-H-O system. Furthermore, the impact is linked to biological activity, which commonly determines the rates and pathways of sulfur transformation. In aerobic waters the pH is invariably buffered by the carbonate and/or silicate system. In sulfur-rich anaerobic waters, however, the pH will probably be modified by sulfate reduction, as depicted in the schematic reactions given above. Thorstenson (1970), Nissenbaum et al. (1972), and Gardner (1974) have presented detailed discussions of chemical models for sulfate reduction in closed anaerobic environments. They show that, if all the sulfate in an enclosed parcel of seawater is reduced, the pH of the solution will essentially be buffered at a value of about 6.6 (see Figure 10). As indicated by the schematic reaction, a decrease in SO_4^{2-} levels is balanced by increases in HS^- and HCO_3^- concentrations; below a pH value of 6,6, the formation of neutral H_2CO_3 and H_2S molecules would create a charge imbalance in the system. The extent of pH reduction obviously depends on the sulfate content of the water (Ben-Yaakov, 1973). For example, model calculations show that for sulfate concentrations <100 ppm (typical of most fresh waters) the decrease in pH that may be effected by complete sulfate reduction will generally be <0.5 pH unit. Very often, the H^+ produced is consumed in other side reactions, and the overall effect of sulfate reduction on the pH of natural waters is considerably decreased. Figure 10 clearly illustrates the fact that calcium carbonate and iron oxides generally act to prevent the lowering of the pH. More general discussions of pH buffering by sulfate reduction are presented by Thorstenson (1970), Berner (1971b), Ben-Yaakov (1973), Gardner (1974), and Goldhaber and Kaplan (1974). Microbial sulfate reduction is believed to be the principal pH buffering process in sediments (Ben-Yaakov, 1973).

Sulfuric acid is now widely implicated as the major cause of acidity in precipitation (Kramer, Part I; Oden, 1976). As noted above, inland waters located in igneous and metamorphic areas tend to be character-

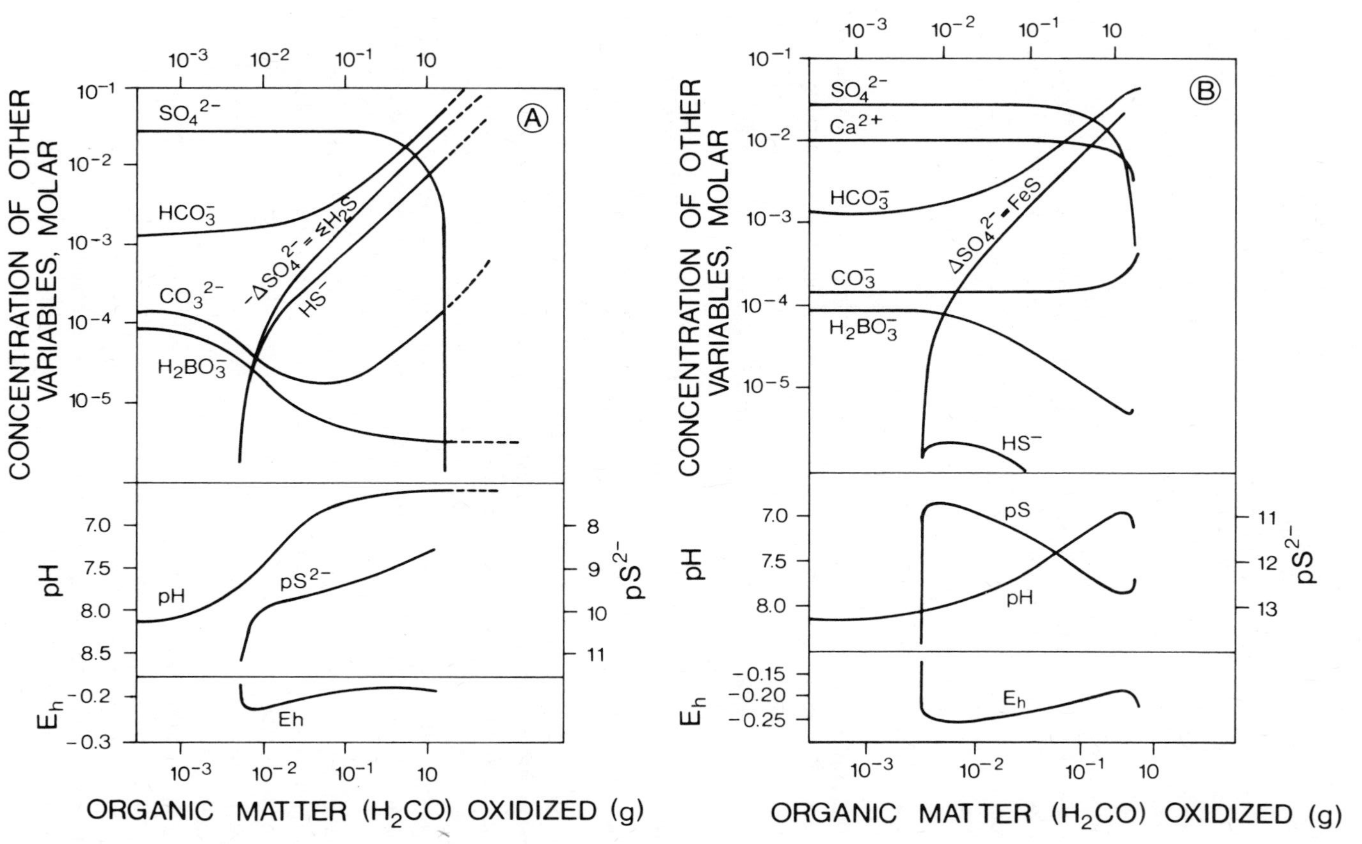

Figure 10. Chemical models for sulfate reduction in *(A)* inert sediments, and *(B)* sediments with $CaCO_3$ and Fe_2O_3. Model *B* also assumes metastable formation of FeS. (After Gardner, 1973.)

ized by low ionic concentrations and very low buffer capacities and hence are vulnerable to inputs of acids from the atmosphere. (Buffer capacity is usually defined as the incremental change in total dissolved protolytic species required to shift the pH by 1 unit.) Lakes whose pH values have been reduced to very low levels (intolerable to the biota and flora) by acid precipitation have been reported in shield areas of the United States, Canada, and Scandinavia. Excellent reviews of the impact of acid precipitation on the chemistry of surface waters are presented by Gorham (1976), Oden (1976), Hornbeck et al. (1977), and Almer et al. (this volume); an annotated bibliography on the subject has been published by Wright (1977).

9.3. Effects on Alkalinity

Bacterial reduction of sulfate results in the formation of bicarbonate ion (Hem, 1970). This is usually necessary to maintain electroneutrality after the microbial conversion of divalent SO_4^{2-} ion to monovalent HS^- or neutral H_2S species. A more generalized respiratory reaction may be written (Richards, 1965):

$$\tfrac{1}{53}(CH_2O)_{106}(NH_3)_{16}H_3PO_4 + SO_4^{2-} \rightarrow 2HCO_3 + H_2S + \tfrac{16}{53}NH_3 + \tfrac{1}{53}H_3PO_4$$

It is seen that the reduction of 1 mmol of sulfate will (theoretically) increase the alkalinity by 2 meq/kg solution. Indeed, laboratory studies in synthetic media have found the ratios of alkalinity produced to SO_4^{2-} reduced to be 1.85 to 2.45 (Abd-el-Malek and Rizk, 1963). Model calculations of the changes in alkalinity associated with sulfate reduction are shown in Figure 10. In nature, marked increases in dissolved carbon dioxide attributable to biological sulfate reduction have been reported in anaerobic marine basins (Knull and Richards, 1969), anoxic lake hypolimnia (Gaines and Pilson, 1972), and pore water of reduced sediments (Presley and Kaplan, 1968; Berner et al., 1970). It is interesting that increasing bacteriological and geochemical evidence now suggests that microbial reduction of Na_2SO_4 is the main factor responsible for the major Na_2CO_3 deposits of the world (Abd-el-Malek and Rizk, 1963; Postgate, 1968). Although sulfate reduction is likely to be the predominant alkalinity-modifying process during early diagenesis, several other processes also affect alkalinity in anaerobic environments. These include precipitation or dissolution of $CaCO_3$, ammonia formation by bacterial hydrolysis of basic nitrogen compounds, authigenic silicate formation and respiration of benthic organisms (see Berner et al., 1970).

The oxidation of H_2S may lead to a decrease in alkalinity, as demonstrated in the Black Sea by Skopintsev et al. (1964).

9.4. Effects on *Eh*

Figure 10 suggests that large changes in *Eh* would occur during sulfate reduction in closed anaerobic marine environments if equilibrium were maintained between the important redox couples in the C-N-S-H-O system. Actual measurements of *Eh,* however, are commonly not related to equilibria involving combinations of the dissolved species SO_4^{2-}, HS^-, S^-, N_2, NH_4, CH_4, and HCO_3^- (of the C-N-S-H-O system) because CH_4, N_2, HCO_3^-, and SO_4^{2-} do not react rapidly at an electrode surface (i.e., are not electroactive). In fact, the actual electroactive substances that determine the *Eh* of natural waters are unknown (Stumm, 1966; Doyle, 1968). In many sulfidic sediments, however, Berner (1963) found that the measured *Eh* is the same as that predicted for the couple

$$HS^- \rightleftharpoons S^0 + H^+ + 2e$$

The electroactive species are believed to be HS^- and a series of polysulfide ions formed by the reaction of HS^- with S^0 (Berner, 1971b).

Although *Eh* values obtained with a platinum electrode are not affected by redox couples involving the major anions, the distribution of dissolved species in some anoxic environments does, in fact, closely approximate the predictions based on equilibrium sulfate reduction models. Thorstenson (1970) calculated the *Eh* values for the N_2-NH_4^+, SO_4^{2-}-HS^-, and HCO_3^--CH_4 redox pairs, using published data for the Cariaco Trench, Black Sea, and Lake Nitinat, as well as the pore waters from Bermuda and Santa Barbara basin sediments. His results (Figure 11) fit the equilibrium model quite closely, as shown by the clustering of field measurements around the theoretical broken line. The implication is clear that somehow, the oxidation of organic matter during sulfate reduction has an important effect on the chemistry of anoxic waters.

9.5. Effects on the Release of Metals and Nutrients from Sediments

By dominating the pH and *Eh* buffering processes during early diagenesis, sulfate reduction plays a major role in the regeneration of material

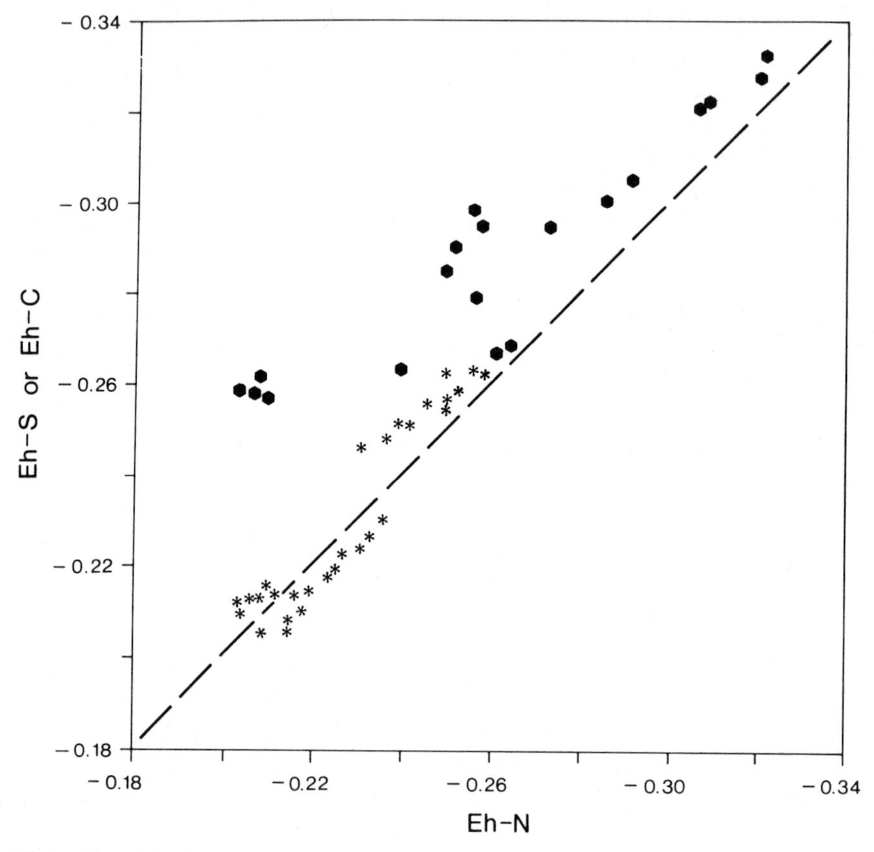

Figure 11. Calculated versus measured *Eh* in anaerobic environments. The calculations assumed equilibria between the redox couples N_2/NH_4^+ (*Eh*-N), SO_4^{2-}/HS^- (*Eh*-S), HCO_3^-/CH_4 (*Eh*-C). (After Thorstenson, 1970, and Berner, 1971b.)

from recent sediments. Ferromanganese oxyhydroxides are well known to be highly effective scavengers for dissolved organic and inorganic species. When these oxyhydroxides are reduced in the sediments, they release most of the material culled from the water; the liberated compounds may subsequently be trapped at the oxidized mud-water interface or be released to the oxygen-depleted hypolimnatic waters. Massive releases of phosphorus, ammonia, and silica, as well as trace metals, from sediments have now been documented in many anaerobic marine basins and lake hypolimnia (Mortimer 1941, 1942; Holmes et al., 1974). For example, Burns and Ross (1972) estimated that about 2.4×10^6 kg of phosphorus was released by the sediments in the central basin of Lake Erie during 4 summer months of 1970. The mixing of the phosphorus-enriched bottom waters with the surface waters was further shown to have stimulated a massive algal bloom (Burns and Ross, 1972). It is clear that because of the synergism that exists in aquatic environments sulfur may be regarded as a catalyst for many nutrient cycles.

9.6. Effects on Heavy-Metal Concentrations

Heavy metals may be transported to the sediments by ferromanganese oxyhydroxides and particulate organic matter. They may also be introduced in various forms as industrial and domestic wastes. In anaerobic environments these metals may be mobilized by the reduction of the oxyhydroxides or the oxidation of organic matter. Subsequent reaction of the released or pollutant metals with sulfides may then involve (*a*) precipitation of pure sulfide phases, (*b*) formation of solid solutions with other, more abundant metal sulfides, and (*c*) sorption into the neoformed iron sulfides (Leckie and Nelson, 1975). These heavy-metal sulfides are noted for their great insolubility, and their formation may be an important buffer mechanism controlling the levels of dissolved heavy metals in both pore waters and sulfide-rich waters (Hutchinson, 1957; Lawrence and McCarty, 1965). Indeed, Loring (1976) observed that 70 to 90% of the copper, zinc, and lead in Saguenay fjord sediments is associated with discrete, diagenetic sulfide minerals. The role of natural heterogeneous sulfide systems in controlling the concentration and distribution of heavy metals in aquatic environments has been discussed by Holmes et al. (1974) and Leckie and Nelson (1975).

9.7. Effects on Mineral Associations

Sulfidization of terrigenous sediments influences the stability of the mineral assemblages, with important consequences for the surrounding aquatic environments. In addition to the precipitation of metal sulfides, other minerals (especially those susceptible to changes in *Eh*) may be altered partially or completely. For example, Drever (1971) proposed that, in the presence of dissolved sulfide, iron may be extracted from clay minerals to form a sulfide, the charge deficit left by the iron being filled by Mg^{2} from the surrounding water. The reaction may be shown schematically by the conversion of nontronite (Fe^{3+} montmorillonite) to saponite (Mg^{2+} montmorillonite):

$$\text{nontronite} + 6Mg^{2+} + 8SO_4^{2-} + 15[C] + 2H_2O \rightarrow \text{saponite} + 4FeS_2 + 11CO_2 + 4HCO_3^-$$

Evidence in support of such reactions is discussed by Drever (1971) and Manheim and Sayles (1974).

It is also true that the oxygenation of an anoxic sulfuretum affects both the mineral associations and the composition of the aqueous matrix. The best known example is the formation of acid waters associated with the subaerial oxidation of pyrite and other metal sulfides. In many instances, the reactions may be mediated by microorganisms such as *Thiobacillus ferroxidans* (see Barton, this volume).

10. DISTRIBUTION OF SULFUR IN NATURAL WATERS

Sulfur in its various forms is a major component of natural waters, being exceeded only by chloride in seawater and by bicarbonate and, in some cases, silica in fresh waters. In acid lakes and/or because of pollution, sulfate may become the most abundant anion. Because of anthropogenic emissions of sulfur, the $SO_4^{2-}:HCO_3^-$ ratios in rain and many freshwaters have been extensively modified (Berner, 1971a; Oden, 1976; Gorman, 1976).

The results of the chemical and biochemical processes outlined in preceding sections are manifested in the composition ranges and chemical characteristics of natural waters.

The reported concentrations of sulfur range from <0.1 ppm, reported in thermal springs of Tuscany, Italy (Bencini et al., 1977), to over 222,000 ppm, found in subsurface brine (see White et al., 1963). Precambrian shield lakes in Africa are noted for their low sulfate contents (<0.5 ppm), which may be a factor limiting the algal productivity of these lakes (see Talling and Talling, 1965). Representative sulfur concentrations of the various types of natural waters are shown in Tables 2 and 6.

The sulfate (g/kg) to chloride ratio in seawater is constant, 0.1400 (Morris and Riley, 1966), suggesting that the mixing time of the oceans is fast compared to the input rates and reactivities of the two constituents. Studies of isotopic variations (Holser and Kaplan, 1966) indicate that neither the $SO_4^{2-}:Cl^-$ ratio nor the mean concentration of sulfate in open oceans, which is 2650 ppm or 27.6 mmol/l (Garrels and Mackenzie, 1971), has changed since the Miocene. The sulfate in seawater is strongly complexed, the percentage concentrations of the various spe-

Table 6. Representative Sulfate and Sulfide Concentrations in Natural Waters[a]

Locality	Sulfate (ppm)	Dissolved Sulfide (ppm)	Source
	Thermal Springs		
Japan	1650 0.11 ($S_2O_3^{2-}$)	3.2	Iwasaki and Yamaya (1973), Sugawara, (1967)
New Zealand	77 (12–>1000)	4.8 (0–74)	Ellis (1973), Mahon (1973)
Tuscany, Italy	19.6 (0.1–39)	—	Bencini et al. (1977)
Oregon	216 (13–434)	—	Mariner et al. (1974a)
Nevada	142 (7–597)	—	Mariner et al. (1974b)
Wyoming	276 (14–600)	—	Cox (1973)
Virginia	179 (25–387)	—	Helz and Sinex (1974)
USSR	65	1–17	Ivanov (1967), Pinneker and Lomonosov (1973)
Vichy Basin (France)	164 (110–228)	—	Michard et al. (1976)
Yellowstone National Park	118 (28–1928) 5.4 (elem. S) (0–651)	0.29 (0–2.7)	Zinder and Brock (1977)
Lake Magadi basin, Kenya	186 (73–249)	—	Jones et al. (1977)
	Subsurface Brines		
Israel	361 (6–2754)	—	Bentor (1969)
Saline County, Mo.	242 (56–1337)	4.8 (0.1–19)	Carpenter and Miller (1969)
Western Canada	489 (2–4777)	—	Billings et al. (1969)
Paradox Basin, U.S.	1704 (47–8400)	—	Henshaw and Hill (1969)
Kansas	1264 (2–8500)	—	Dingman and Angino (1969)
Illinois	990	—	White (1968)
Wind River basin, Wyo.	615 (35–6770)	1.9 (0.3–6.5)	Vredenburgh and Cheney (1971)
Southwestern Louisiana	53 (0–407)	—	Dickey et al. (1972)
Sacramento Valley, Calif.	133	—	Berry (1973)

Table 6. (Continued)

Locality	Sulfate (ppm)	Dissolved Sulfide (ppm)	Source
Kern County, Calif.	29 (1–142)	—	Chave (1960)
Sinai desert	433 (2–1687)	—	Gat and Issar (1974)
Gulf of Suez, Egypt	1014 (346–4800)	—	Issar et al. (1971)
Salado brine, N.M.	6970–222,000	—	White et al. (1963)
	Saline Lakes		
Antarctica	2,243 (1,1110–3,810)	—	Yoshida et al. (1973)
Gulf of Aquaba	5,510	—	Friedman et al. (1973)
Searles Lake, Calif.	47,700	—	White et al. (1963)
Great Salt Lake, Utah	3,680	—	White et al. (1963)
Lake Magadi, Kenya	1,788 (598–2,580)	—	Jones et al. (1977)
Owens Lake, Calif.	24,600 (12,500–39,600)	—	Friedman et al. (1976)
Eyre, Australia	1,940	—	Livingstone (1963)
Devils Lake, N.D.	6,600 (194–13,600)	—	Livingstone (1963)
Salt lakes, Saskatchewan	9,810 (361–51,720)	—	Livingstone (1963)
Salt lakes, B.C.	98,000 (6,240–203,900)	—	Livingstone (1963)
Hot Lake, Wash.	103,680		
Crinean salt lakes, USSR	24,380 (3,850–74,200)	—	Livingstone (1963)
	Lakes		
Superior (1973)	3.0 (1.2–7.2)	—	Original
Victoria (1961)	1.8	—	Talling and Talling (1965)
Huron (1975)	15 (3.4–27)	—	Original
Georgian Bay (1974)	16 (14–17.1)	—	Original
Michigan (1971–1974)	21 (11–35)	—	Torrey (1975)
Tanganyika	3	—	Talling and Talling (1965)
Baikal (1945–1960)	5.9	—	Kozhov (1963)
Kozhov (1963)			

Malawi	5.3	—	Talling and Talling (1965)
Erie (1975)	25 (17–34)	—	Original
Great Slave (1969)	19.0	—	Reeder et al. (1972)
Winnipeg (1937)	205	—	Livingstone (1963)
Ontario (1976)	29 (24–42)	—	Original
Chad (1967)	4.8	—	Beadle (1974)
Kivu (1959)	15	—	Talling and Talling (1965)
Rudolph (1954)	64	—	Talling and Talling (1965)
Albert (1961)	45	—	Talling and Talling (1965)
Mweru (1961)	3.7	—	Talling and Talling (1965)
Kyoga	3.8	—	Livingstone (1963)
Gluboka, Antarctica	6.3	—	MacNamara (1973)
Lagernoye, Antarctica	3.1	—	MacNamara (1973)
West Greenland lakes	2.5	—	Livingstone (1963)
English lakes	6.4	—	Macan (1970)
African lakes	13.5	—	Beadle (1974)
New Zealand lakes	15.8	—	Livingstone (1963)
Northern Indiana lakes	32	—	Wetzel (1965)
Northern Wisconsin lakes	4.0	—	Juday et al. (1938)
Southern Wisconsin lakes	10.6	—	Birge and Juday (1911)
	Rivers		
North America	20	—	Livingstone (1963)
South America	4.8	—	Livingstone (1963)
Europe	24	—	Livingstone (1963)
Asia	8.4	—	Livingstone (1963)
Africa	13.5	—	Livingstone (1963)
Australia	2.6	—	Livingstone (1963)
World average	11.2	—	Livingstone (1963)

[a] Ranges in reported concentrations are shown in parentheses.
[b] The sampling year is indicated after each lake.

cies being as follows: 39 to 46%, free SO_4^{2-}; 26 to 40%, $NaSO_4^-$; 15 to 22%, $MgSO_4^0$; 3 to 5%, $CaSO_4^0$; and 0.3 to 2%, KSO_4^- (Pytkowicz and Kester, 1971; Elgquist and Wedborg, 1974; Whitfield, 1975; Morel et al., 1976). In basins and coastal areas that develop bottom anoxia, the reduction of sulfate and the precipitation of metal sulfides may deplete the $SO_4^{2-}:Cl^-$ ratio by as much as 4% (see Wilson, 1975). On the other hand, the ratio may be increased in volcano-prone basins.

Of the continental waters, the sulfur contents of subsurface brines rank among the highest (Table 6). Some of the samples are believed to represent fossil water, which may be contemporaneous with the enclosing rocks (White, 1957). Others have been extensively diluted with meteoric water, whereas a few consist almost entirely of meteoric waters. The sulfate concentrations therefore vary from <1.0 to well over 8000 ppm, as a consequence of the differences in origin and past histories of the brines. Although only a few of the brines have been analyzed for H_2S, high levels of dissolved sulfide should be expected, considering that these waters are generally in contact with oil and gas reservoirs which are noted for their high reduced-sulfur contents (see Orr, 1974b; Thode et al., 1954). In fact, these brines have been implicated as a source and carrier of reduced sulfides species involved in the formation of many epigenetic mineral deposits (Jackson and Beales, 1967). The important thing, though, is that these brines may be flushed out by deep circulating groundwaters and thus may affect the quality of surface waters. Detailed discussions of the classification, origin, chemistry, and diagenesis of these waters may be found in White (1957), White et al. (1963), and a special issue of *Chemical Geology* (Vol. 4, pp. 7–370, 1969).

The sulfate contents of groundwaters (Table 2) reflect both the edaphic effects and the original composition of the water in the catchment area. Groundwaters from igneous rocks normally have low levels (<10 ppm) of sulfate, although values as high as 200 ppm may be associated with the weathering of magmatic sulfide deposits (White et al., 1963). Abnormally high levels of sulfate (>1000 ppm) generally suggest that the groundwater has passed through gypsiferous sedimentary beds. Otherwise, the groundwaters in sedimentary rocks usually contain <100 ppm sulfate, which may be derived from the flushing of remnants of connate brines trapped by the sediments and from the oxidation of sedimentary sulfide minerals. The impact of human activities on groundwater quality is now a matter of concern (Pettyjohn, 1972). The sulfur concentrations in groundwater may be (adversely) affected by the disposal of solid wastes, sewage, and wastewaters on land, by agricultural use of fertilizers, and by various mining operations.

For example, groundwater draining an industrial waste disposal site near Derby, Colorado, has been shown to contain 4 to 30 times more sulfate than the uncontaminated waters from the same aquifer (see Pettyjohn, 1972). Pettyjohn also describes an interesting case of groundwater contamination (sulfate content >800 ppm) caused by the leaching of surficial material near a former horse stable.

The data of Livingstone (1963) for the mean sulfate contents of rivers in the various continents are shown in Table 6. Rainwater, as already noted, is a significant source of sulfur in rivers and can account for most of the sulfur in headwater streams draining igneous rock terrains. Because of increasing inputs via the atmosphere, the sulfate contents of rivers, particularly in industrialized regions, have risen steadily. Superimposed on such long-term trends are short-term changes in concentration which may be due to (*a*) variations in chemistry of rainfall, (*b*) changes in amount and composition of runoff additions, (*c*) evaporation losses from the drainage basin, and (*d*) varying amounts of sewage and wastewater inputs from domestic and industrial sources. Large rivers may also show regional variability and/or local irregularities, which reflect (1) differences in the nature of the rocks in various parts of the drainage basin, and (2) the pickup of increasing amounts of pollutant sulfur. Thus, after flowing through the highly industrialized parts of Germany, the sulfate content of the Rhine River increases from 20 to 130 ppm (Forstner and Muller, 1974). It is equally impressive that between 1837 and 1971 the sulfate concentration in the lower reaches of this river increased by approximately 600% (Forstner and Muller, 1974). Some rivers that are polluted with sulfur-rich organic wastes may develop anaerobic stretches where the sulfur may become immobilized as metal sulfides. Otherwise, the sulfate is transported mostly as a conservative species.

The sulfur concentrations in the major lakes of the world and in representative saline lakes are shown in Table 6. Lakes, in general, are more stable than streams and do not show such striking changes in the amounts and proportions of the principal dissolved components (Livingstone, 1963). As noted above, however, the sulfate contents of lakes in many parts of the world (see Figure 1) have increased substantially during the past 100 years. The literature on sulfur in limnetic environments is voluminous because of the influence of the sulfur cycle on redox potential, biological productivity, and other key element and nutrient cycles (see Hutchinson, 1957). The most detailed studies of the biogeochemical cycling of sulfur in lakes have been made by Ohle (1934, 1954), Juday et al. (1938), Mortimer (1941, 1942), Stuiver (1967), Kuznetsov (1970), and Sorokin (1970). In aerobic oligotrophic and

Table 7. Chemical Analyses of Water from Various Sources, Showing Sulfur Species and Concentrations, and Related Characteristics

Analysis number[a]	1	2	3	4	5
Date of sample	—	Oct. 1, 1937	May 13, 1952	—	Oct. 1, 1954
pH	—	7.2	8.3	4.20	1.7
Constituent (mg/l)					
SiO_2	13.1	20	16	49	274
Fe	0.67	—	0.15	830	112
Ca	15	495	3.0	386	79
Mg	4.1	441	7.4	278	8.7
Na	6.3	9,640	857	305	455
K	2.3	441	2.4	—	9.6
CO_3	—	—	57	—	—
HCO_3	58.4	1,010	2,080	—	—
HSO_4	—	—	—	—	1,590
SO_4	11.2	1.1	1.6	4,530	2,020
H_2S	—	279	—	—	—
HS	—	139	—	—	—
Cl	7.8	16,900	71	10	93
F	—	—	2.0	—	0.9
NO_3	1	—	0.2	—	—

Analysis number[a]	6	7	8	9	10
Date of sample	Nov. 25, 1949	Mar. 3, 1969	Feb. 2, 1939	—	May 28, 1964
pH	—	8.0	—	7.3	—
Constituent (mg/l)					
SiO_2	29	22	—	5.9	—
Fe	—	—	—	4.4	—
Ca	636	385	97	1,340	776
Mg	43	177	38,300	368	928
Na } K }	17	1430			
Na			43,900	8,010	9,540
K			2,090	96	—
				Sr 41	
				Ba 66	
CO_3	—	—	—	—	—
HCO_3	143	342	1,380	120	180
HSO_4	—	—	—	—	—
SO_4	1570	1540	222,000	0	7,010
H_2S	—	—	—	—	—
HS	—	—	—	—	—
Cl	24	2040	16,800	15,500	13,500
F	—	—	—	—	—
NO_3	18	47	—	—	—
Diss. solids	2410	5820	325,000	25,700	31,800

[a] 1. Average river water (Livingstone, 1963). 2. Resort well, Chokrak, Crimea, USSR (White et al., 1963). 3. Well in Fort Union Formation, Richland County, Mont. (Hem, 1970). 4. Water from coal mine shaft, Wilkes-Barre, Pa. (Barnes et al., 1964). 5. Lower Mendeleev Spring, Kunashir Island, USSR (White et al., 1963). 6. ''Jumping Springs'' from gypsum, Eddy County, N.M. (Hem, 1970). 7. Gila River at Gillespie Dam, Ariz. (U.S. Geological Survey, 1974). 8. Brine, Salado Formation, Eddy County, N.M. (White et al., 1963). 9. Brine well, Pennsylvanian sandstone, Paintsville, Ky. (White et al., 1963). 10. Salton Sea, Imperial County, Calif. composite sample (Hely et al., 1966).

mesotrophic lakes, sulfate is the dominant form of sulfur in the water. In eutrophic lakes, the deoxygenation of the hypolimnetic waters may result in extensive reductions of the sulfur oxides to the sulfide species. Complete reduction of the sulfate is rare but has been reported in, for instance, Lake Vechten (Cappenberg, 1972). The dichotomy in the vertical distribution of principal sulfur species and sulfur-cycle microorganisms in stratified lakes may be illustrated using the data from Lake Belovod (Figure 9). After the spring and fall overturns (of stratified lakes), the H_2S is quickly oxidized to sulfate, and the oxidation of the iron sulfides in the sediment may result in the oxic regeneration of sulfate, as reported by Mortimer (1941, 1942) and Wetzel (1975).

Table 7 presents chemical analyses for 10 samples of natural or polluted waters which illustrate the results of processes that determine the chemical state and concentration of dissolved sulfur, and the way in which other dissolved constituents relate to these effects.

Analysis 1 in the table is the computed "average" river water as reported by Livingstone (1963). This analysis is heavily weighted by the compositions of large rivers of the world, especially the Amazon. In many areas the dissolved sulfate concentrations of river water are higher than the concentration given in analysis 1. However, most streams contain sulfate concentrations in the range of $10^{-3.00}$ to $10^{-4.00}$ mol/l (about 10 to 100 ppm as SO_4) most of the time.

Analysis 2 represents a groundwater in the Black Sea region of the USSR, in which sulfate reduction has strongly influenced the form of the dissolved sulfur species. Here the sulfide concentration is near saturation at 25°C and 1 atm pressure. The sulfate reduction process tends to add to the dissolved CO_2 species concentration, and this effect is shown in the high bicarbonate content of this water.

Analysis 3 represents water from a deep well in the Tertiary Fort Union Formation in eastern Montana, where sulfate reduction has depleted the sulfate content of the water nearly to extinction. The bicarbonate content has been increased, but sulfide species are reported to be absent, probably because they have been precipitated as iron sulfide in the aquifer.

Analyses 4 and 5 are representative of waters affected by sulfide oxidation. The first is for water from a flooded anthracite coal mine in eastern Pennsylvania, where pyrite associated with the coal has been oxidized to produce sulfate, ferrous iron, and H^+. At higher temperatures, in geothermal systems, the H_2S flux moving into the oxidizing regime near the land surface commonly gives rise to acid waters containing major amounts of sulfate, as demonstrated in analysis 5 for water from a volcanic region in the eastern USSR.

Analysis 6 represents a spring issuing from gypsum in southeastern New Mexico. This water is essentially at saturation with respect to gypsum at 25°C.

Analysis 7 is for water from the Gila River downstream from the intensively irrigated area around Phoenix, Arizona. The composition of this water is close to what would be predicted for saturation with calcite and gypsum, with a considerable admixture of sodium and chloride. This water is essentially all residual drainage from irrigated soils in which calcite and gypsum are likely to be precipitated as the result of evapotranspiration.

Sample 8 represents a highly concentrated brine of unusual composition, dominated by magnesium and sulfate, from an exploratory well in evaporite strata in southeastern New Mexico.

Sample 9 is a brine from a well in Kentucky, where sulfate concentration is low and strontium and barium concentrations are unusually high. In this solution the sulfate concentration is probably at a low value because of limitations on the solubility of barium sulfate.

Sample 10 represents water from Salton Sea, in Imperial County, California. This water is near saturation with gypsum and calcite, because of the effects of evaporation and inflows of relatively highly mineralized irrigation drainage.

REFERENCES

Abd-el-Malek, Y. and Rizk, S. R. (1963). "Bacterial Sulfate Reduction and the Development of Alkalinity," *J. Appl. Bacteriol.* **26,** 7–34.

Ackermann, W. C., Harmeson, R. H., and Sinclair, R. A. (1970). "Some Long-Term Trends in Water Quality of Rivers and Lakes," *EOS Trans. Am. Geophys. Union* **51,** 516–520.

Almgren, T. and Hagstrom, I. (1974). "The Oxidation Rate of Sulfide in Seawater," *Water Res.* **8,** 395–400.

Avrahami, M. and Golding, R. M. (1968). "The Oxidation of the Sulfide Ion at Very Low Concentrations in Aqueous Solutions," *J. Chem. Soc.* (A), pp. 647–651.

Barnes, I., Stuart, W. T., and Fisher, D. W. (1964). *Field Investigation of Mine Waters in the Northern Anthracite Field, Pennsylvania.* U.S. Geological Survey, Prof. Paper 473B, 8 pp.

Beadle, L. D. (1974). *The Inland Waters of Tropical Africa.* Longmans, Green, New York, 365 pp.

Beeton, A. M. (1965). "Eutrophication of the St. Lawrence Great Lakes," *Limnol. Oceanogr.* **10,** 240–254.

Bell, R. P. and George, J. H. B. (1953). "The Incomplete Dissociation of Some Thallous and Calcium Salts at Different Temperatures," *Faraday Soc. Trans.* **49,** 619–627.

Bencini, A., Duchi, V., and Martini, M. (1977). "Geochemistry of Thermal Springs of Tuscany (Italy)," *Chem. Geol.* **19,** 229–252.

Bentor, Y. K. (1969). "On the Evolution of Subsurface Brines in Israel," *Chem. Geol.* **4,** 83–110.

Ben-Yaakov, S. (1973). "pH Buffering of Pore Water of Recent Anoxic Marine Sediments," *Limnol. Oceanogr.* **18,** 86–94.

Berner, R. A. (1963). "Electrode Studies of Hydrogen Sulfide in Marine Sediments," *Geochim. Cosmochim. Acta* **27,** 563–575.

Berner, R. A. (1967). "Thermodynamic Stability of Iron Sulfides, *Am. J. Sci.* **265,** 773–785.

Berner, R. A. (1971a). "Worldwide Sulfur Pollution of Rivers," *J. Geophys. Res.* **76,** 6597–6600.

Berner, R. A. (1971b). *Principles of Chemical Sedimentology.* McGraw-Hill Book Co., New York, 240 pp.

Berner, R. A., Scott, M. R., and Thomlinson, C. (1970). "Carbonate Alkalinity in the Pore Waters of Anoxic Marine Sediments," *Limnol. Oceanogr.* **15,** 544–549.

Berry, F. A. F. (1973). *High Fluid Potentials in California Coast Ranges and Their Tectonic Significance.* American Association of Petroleum Geologists, Bull. **57,** pp. 1219–1249.

Billings, G. K., Hitchon, B., and Shaw, D. R. (1969). "Geochemistry and Origin of Formation Waters in the Western Canada Sedimentary Basin," *Chem. Geol.* **4,** 211–223.

Birge, E. A. and Juday, C. (1911). *The Inland Lakes of Wisconsin: The Dissolved Gases of the Water and Their Biological Significance.* Wisconsin Geological and Natural History Survey, Bull. **22,** Sci. Series 7, 259 pp.

Bogander, L. E. (1977). "*In Situ* Studies of Sulfur Cycling at the Sediment-Water Interface," paper presented at the 3rd International Symposium on Environmental Biogeochemistry, Wolfenbuttel, West Germany, Mar. 27–Apr. 3.

Boulegue, J. (1976). "Equilibria in the System H_2S-S_8 Colloid-H_2O," *C.R. Acad. Sci. Paris,* Series D, **283,** 591–594.

Burns, N. M. and Ross, C. (1972). *Project Hypo—An Intensive Study of the Lake Erie Central Basin Hypolimnion and Related Surface Water Phenomena.* U.S. Environmental Protection Agency, Tech. Rept. TS-05-71-208-24, 182 pp.

Cappenberg, Th. E. (1972). "Ecological Observations on Heterotrophic, Methane Oxidizing and Sulfate Reducing Bacteria in a Pond," *Hydrobiologia* **40,** 471–485.

Carpenter, A. B. and Miller, J. C. (1969). "Geochemistry of Saline Subsurface Water, Saline County (Missouri)," *Chem. Geol.* **4,** 135–167.

Chave, K. E. (1960). "Evidence on the History of Sea Water from Chemistry of Deeper Subsurface Waters of Ancient Basins," *Bull. Amer. Assoc. Petrol. Geol.,* **44,** 357–370.

Chen, K. Y. and Morris, J. C. (1972a). "Oxidation of Sulfide by O_2: Catalysis and Inhibition," *Proc., Am. Soc. Civil Eng.,* Sanitary Eng. Div. **SA1,** 215–227.

Chen, K. Y. and Morris, J. C. (1972b). "Oxidation of Aqueous Sulfide by O_2," *Environ. Sci. Technol.* **6,** 529–537.

Cherry, J. A. (1968). "Chemical Equilibrium between Gypsum and Brackish and Slightly Saline Waters at Low Temperatures and Pressures," *Chem. Geol.* **3,** 239–247.

Chukhrov, F. V., Churikov, V. S., Yermilova, L. P., and Nosik, L. P. (1975). "On the Variation of Sulfur Isotope Composition in Some Natural Waters," *Geochem. Intl.* **12,** 20–36.

Cline, J. D. and Richards, F. A. (1969). "Oxygenation of Hydrogen Sulfide in Seawater at Constant Salinity, Temperature and pH," *Environ. Sci. Technol.* **3,** 838–843.

Cox, E. R. (1973). *Water Resources of Yellowstone National Park, Wyoming, Montana and Idaho.* U.S. Geol. Survey Open-File Report, 161 p.

Deuser, W. G. (1975). "Reducing Environments." In J. P. Riley and G. Skirrow, Eds., *Chemical Oceanography,* Vol. 3, 2nd ed., Academic Press, London, pp. 1–37.

Dickey, P. A., Collins, A. G., and Fajardo, I. (1972). *Chemical Composition of Deep Formation Waters in Southwestern Louisiana.* American Association of Petroleum Geologists, Bull. 56, pp. 1530–1570.

Dingman, R. J. and Angino, E. E. (1969). "Chemical Composition of Selected Kansas Brines as an Aid to Interpreting Change in Water Chemistry with Depth," *Chem. Geol.* **4,** 325–339.

Dochinger, L. S. and Seliga, T. A., Eds. (1976). *Acid Precipitation and the Forest Ecosystem.* U.S. Department of Agriculture, Forest Service, Gen. Tech. Rept. NE-23, Northeast Forestry Experiment Station, Upper Darby, Pa., 1974 pp.

Doyle, R. W. (1968). "The Origin of the Ferrous Ion-Ferric Oxide Nernst Potential in Environments Containing Dissolved Ferrous Iron," *Am. J. Sci.* **266,** 840–859.

Drever, J. I. (1971). Magnesium-Iron Replacements in Clay Minerals in Anoxic Marine Sediments," *Science* **172,** 1334–1336.

Ehrlich, G. G. and Schoen, R. (1967). *Possible Role of Sulfur Oxidizing Bacteria in Surficial Acid Alteration near Hot Springs.* U.S. Geological Survey, Prof. Paper 575C, p. 110.

Elgquist, B. and Wedborg, M. (1974). "Sulfate Complexation in Seawater," *Marine Chem.* **2,** 1–15.

Ellis, A. J. (1973). "Chemical Processes in Hydrothermal Systems—A Review." In E. Ingerson, Ed., *Hydrogeochemistry,* Vol. I. The Clarke Co., Washington, D.C., pp. 1–26.

Field, C. W. (1972). "Sulfur: Element and Geochemistry." In R. W. Fairbridge, Ed., *Encyclopedia of Geochemistry and Environmental Sciences.* Van Nostrand Reinhold, New York, pp. 1142–1148.

Fisher, F. H. (1967). "Ion-Pairing of Magnesium Sulfate in Seawater Determined by Ultrasonic Absorption," *Science* **157,** 823.

Fleetwood, A. (1969). "The Chemical Composition of the Precipitation and Surface Water and Its Relation to Evaporation on Bjornoya Island, Norway," *Tellus* **21,** 113–126.

Forstner, U. and Muller, G. (1974). *Schwermetalle in Flüssen und Seen.* Springer-Verlag, New York, 225 pp.

Freiberg, J. (1975). "The Mechanism of Iron Catalyzed Oxidation of SO_2 in Oxygenated Solutions," *Atmos. Environ.* **9,** 661–672.

Freney, J. R. and Stevenson, F. J. (1966). "Organic Sulfur Transformations in Soils," *Soil Sci.* **101,** 307–316.

Friedman, G. M., Amiel, A. J., Braun, M. and Miller, D. S. (1973). *Generation of Carbonate Particles and Laminites in Algal Mats—Example from Sea-Marginal Hypersaline Pool, Gulf of Aquaba, Red Sea.* American Association of Petroleum Geologists, Bull. 57, pp. 541–557.

Friedman, I., Smith, G. I., and Hardcastle, K. G. (1976). "Studies of Quarternary Saline Lakes. II: Isotopic and Compositional Changes during Desiccation of the Brines in Owens Lake, California, 1969–1971." *Geochim. Cosmochim. Acta* **40,** 501–511.

Friend, J. P. (1973). "The Global Sulfur Cycle." In S. I. Rasool, Ed., *Chemistry of the Lower Atmosphere.* Plenum, New York, pp. 177–201.

Gaines, A. G. and Pilson, M. E. Q. (1972). "Anoxic Water in the Pettaquamscutt River," *Limnol. Oceanogr.* **17,** 42–49.

Gardner, L. R. (1973). "Chemical Models for Sulfate Reduction in Closed Anaerobic Marine Environments," *Geochim. Cosmochim. Acta* **37,** 53–68.

Gardner, L. R. (1974). "Organic versus Inorganic Metal Complexes in Sulfidic Marine Waters—Some Speculative Calculations Based on Available Stability Constants," *Geochim. Cosmochim. Acta* **38,** 1297–1302.

Garrels, R. M. and Christ, C. L. (1965). *Solutions, Minerals and Equilibria.* Harper and Row, New York, 450 pp.

Garrels, R. M. and MacKenzie, F. T. (1967). "Origin of the Chemical Compositions of Some Springs and Lakes." In *Equilibrium Concepts in Natural Waters.* American Chemical Society, Washington, D.C., pp. 222–242.

Garrels, R. M. and MacKenzie, F. T. (1971). *Evolution of Sedimentary Rocks.* W. W. Norton & Co., New York, p. 101.

Gat, J. R. and Issar, A. (1974). "Desert Isotope Hydrology: Water Sources of the Sinai Desert," *Geochim. Cosmochim. Acta* **38,** 1117–1131.

Gimblett, F. G. R. and Monk, C. B. (1955). "Spectrophotometric Studies of Electrolytic Dissociation. I: Some Thiosulfates in Water," *Faraday Soc. Trans.* **51,** 793–802.

Goldhaber, M. B. and Kaplan, I. R. (1974). "The Sulfur Cycle." In E. D. Goldberg, Ed., *The Sea,* Vol. 5. Wiley-Interscience, New York, pp. 569–655.

Gorham, G. (1976). "Acid Precipitation and Its Influence upon Aquatic Ecosystems—an Overview," *Water, Air Soil Pollut.* **6,** 457–481.

Grey, D. C. and Jensen, M. L. (1972). "Bacteriogenic Sulfur in Air Pollution," *Science* **177,** 1099–1100.

Haar, T. E. and Coffey, P. E. (1975). *Acid Precipitation in New York State.* New York State Department of Environmental Conservation, Tech. Paper 43, 52 pp.

Hanna, E. M., Pethybridge, A. D., and Prue, J. E. (1971). "Ion Association and the Analysis of Precise Conductimetric Data," *Electrochim. Acta* **16,** 677–686.

Hanshaw, B. B. and Hill, G. A. (1969). "Geochemistry and Hydrodynamics of the Paradox Basin Region, Utah, Colorado and New Mexico," *Chem. Geol.* **4,** 263–294.

Hansson, I. (1973). "A New pH Scale and Standard Buffers for Sea Water," *Deep-Sea Res.* **20,** 479–491.

Harris, R. L., Helz, G. R., and Cory, R. L. (1975). "Processes Affecting the Vertical Distribution of Trace Components in the Chesapeake Bay." In T. M. Church, Ed., *Marine Chemistry in the Coastal Environment.* American Chemical Society, Symposium Series, 18, pp. 176–185.

Haynon, E., Treinin, A., and Wilf, J. (1972). "Electronic Spectra, Photochemistry and Autoxidation Mechanism of the Sulfite-Bisulfite-Pyrosulfite System," *J. Am. Chem. Soc.* **94,** 47–57.

Hely, A. G., Hughes, G. H., and Irelan, B. (1966). *Hydrologic Regimen of Salton Sea, California.* U.S. Geological Survey, Prof. Paper 486C, 32 pp.

Helz, G. R. and Sinex, S. A. (1974). "Chemical Equilibria in Thermal Spring Waters of Virginia," *Geochim. Cosmochim. Acta* **38,** 1807–1820.

Hem, J. D. (1966). "Chemical Controls of Irrigation Drainage Water Composition. In *Proceedings of the 2nd American Water Resources Conference.* American Water Resources Association, Chicago, Nov. 20–22, 1966, pp. 64–67.

Hem, J. D. (1970). *Study and Interpretation of the Chemical Characteristics of Natural Water.* U.S. Geological Survey, Water Supply Paper 1473, 363 pp.

Hemley, J. J., Hostetler, P. B., Gude, A. J., and Mountjoy, W. T. (1969). "Some Stability Relations of Alunite," *Econ. Geol.* **64,** 599–612.

Hitchon, B. and Krouse, H. R. (1972). "Hydrogeochemistry of the Surface

Waters of the Mackenzie River Drainage Basin, Canada—III. Stable Isotopes of Oxygen, Carbon and Sulfur," *Geochim. Cosmochim Acta,* **36,** 1337–1357.

Holland, H. D. (1973). "Systematics of the Isotopic Composition of Sulfur in the Oceans during the Phanerozoic and Its Implications for Atmospheric Oxygen," *Geochim. Cosmochim. Acta* **37,** 2605–2616.

Holmes, C. W., Slade, E. A., and McLerran, C. J. (1974). "Migration and Redistribution of Zinc and Cadmium in Marine Estuarine System," *Environ. Sci. Technol.* **8,** 255–259.

Holser, W. T. and Kaplan, I. R. (1966). "Isotope Geochemistry of Sedimentary Sulfates," *Chem. Geol.* **1,** 93–135.

Hornbeck, J. W., Likens, G. E., and Eaton, J. S. (1977). Seasonal Patterns in Acidity of Precipitation and Their Implications for Forest Stream Ecosystems," *Water, Air Soil Pollut.* **7** (in press).

Hutchinson, G. E. (1957). *A Treatise on Limnology,* Vol. 1. John Wiley and Sons, New York, pp. 753–786.

Issar, A., Rosenthal, E., Eckstein, Y., and Bogoch, R. (1971). *Formation Waters, Hot Springs and Mineralization Phenomena along the Eastern Shore of the Gulf of Suez, Egypt.* International Association of Scientific Hydrology, Bull. 26, pp. 25–44.

Ivanov, V. V. (1967). "Principal Geochemical Environments and Processes of the Formation of Hydrothermal Waters in Regions of Recent Volcanic Activity." In A. P. Vinogradov, Ed., *Chemistry of the Earth's Crust,* Vol. 2, Israel Program of Scientific Translations, Jerusalem, pp. 260–281.

Iwasaki, I. and Yamaya, K. (1973). "Chemical Composition of Hot Springs in Japan." In E. Ingerson, Ed., *Hydrogeochemistry,* Vol. I. The Clarke Co., Washington, D.C., pp. 93–103.

Jackson, S. A. and Beales, F. W. (1967). "An Aspect of Sedimentary Basin Evolution: The Concentration of Mississippi Valley-Type Ores during the Late Stages of Diagenesis." *Can. Pet. Geol. Bull.* **15,** 383–433.

Jacobson, R. L. and Langmuir, D. (1974). "Dissociation Constants of Calcite and $CaHCO_3^+$ from 0° to 50°C," *Geochim. Cosmochim. Acta* **38,** p. 301–318.

Jones, B. F., Eugster, H. P., and Rettig, S. L. (1977). "Hydrogeochemistry of the Lake Magadi Basin, Kenya," *Geochim. Cosmochim. Acta* **41,** 53–72.

Juday, C. E., Birge, E. A., and Meloche, V. W. (1938). "Mineral Content of the Lake Waters of Northeastern Wisconsin," *Trans. Wisc. Acad. Sci. Arts Lett.* **31,** 223–276.

Kaplan, I. R., Emery, K. O., and Rittenberg, S. C. (1963). "The Distribution and Isotopic Abundance of Sulfur in Recent Marine Sediments of Southern California," *Geochim. Cosmochim. Acta* **27,** 27–331.

Kester, D. R., Byrne, R. H., and Liang, Yu-J. (1975). "Redox Reactions and Solution Complexes of Iron in Marine Systems," In T. M. Church, Ed.,

Marine Chemistry in the Coastal Environment. American Chemical Society, Symposium Series, 18, pp. 5–79.

Kharaka, Y. K. and Barnes, I. (1973). SOLMNEQ: *Solution-Mineral Equilibrium Computations*. U.S. Geological Survey Computer Contribution, Menlo Park, Calif., 82 pp. (Available from National Technical Information Service, Springfield, Va. 22151, as Rept. PB-215 899.)

Knull, J. R. and Richards, F. A. (1969). "A Note on the Sources of Excess Alkalinity in Anoxic Waters," *Deep-Sea Res.* **16,** 205–212.

Koide, M. and Goldberg, E. D. (1971). "Atmospheric Sulfur and Fossil Fuel Combustion," *J. Geophys. Res.* **76,** 6589–6596.

Kolthoff, I. M., Perlich, R. W., and Weiblen, D. (1942). "Solubility of Lead Sulfate and Lead Oxalate in Various Media," *J. Phys. Chem.* **46,** 561.

Kozhov, M. (1963). *Lake Baikal and Its Life*. W. Junk, The Hague, 344 pp.

Krebs, H. A. (1929). "Uber die Wirkung der Schwermetalle auf die Autoxydation der Alkalisulfide und des Schwefelwasserstoffs," *Biochim. Z.* **204,** 343–346.

Kuznetsov, S. I. (1970). *Microflora of Lakes and Their Geochemical Activities*. Izdatel'stvo Nauka, Leningrad, 440 pp.

Latimer, W. M., Hicks, J. F. G., Jr., and Schutz, P. W. (1933). "Heat Capacity and Entropy of Ca and Ba Sulfates—Entropy and Free Energy of the Sulfate Ion," *J. Chem. Phys.* **1,** 620.

Lawrence, A. W. and McCarty, P. L. (1965). "The Role of Sulfide in Preventing Heavy Metal Toxicity in Anaerobic Treatment," *J. Water Pollut. Control Assoc.* **37,** 392–406.

Leckie, J. O. and Nelson, M. B. (1975). "Role of Natural Heterogeneous Sulfide Systems in Controlling the Concentration and Distribution of Heavy Metals," paper presented at the 2nd International Symposium on Environmental Biogeochemistry, Burlington, Ontario, Apr. 8–12.

Livingstone, D. A. (1963). *Chemical Composition of Rivers and Lakes*. U.S. Geological Survey, Prof. Paper 440G, 64 pp.

Loring, D. H. (1976). "The Distribution and Partition of Zinc, Copper, and Lead in the Sediments of the Saguenay Fjord," *Can. J. Earth Sci.* **13,** 960–971.

Macan, T. T. (1970). *Biological Studies of the English Lakes*. Elsevier, New York, 260 pp.

MacNamara, E. E. (1973). "Freshwater Lakes and Their Dynamics, Coastal Enderby Land, Antarctica." In E. Ingerson, Ed., *Hydrogeochemistry,* Vol. I. The Clarke Co., Washington, D.C., pp. 633–646.

Mahon, W. A. J. (1973). "The Chemical Composition of Natural Thermal Waters." In E. Ingerson, Ed., *Hydrogeochemistry,* Vol. I. The Clarke Co., Washington, D.C., pp. 196–210.

Makarenko, F. A., Zverev, V. P., and Kononov, V. I. (1973). Subsurface Chemical Runoff in the USSR area," In E. Ingerson, Ed., *Hydrogeochemistry,* Vol. I. The Clarke Co., Washington, D.C., pp. 567–573.

Malmer, N. (1974). *On the Effects on Water, Soil and Vegetation of an Increasing Atmospheric Supply of Sulfur.* National Swedish Environmental Protection Board, Rept. SNV PM 402E, Solna, 98 pp.

Manheim, F. T. and Sayles, F. L. (1974). "Composition and Origin of Interstitial Waters of Marine Sediments, Based on Deep Sea Drill Cores." In E. D. Goldberg, Ed., *The Sea,* Vol. 5. Wiley-Interscience, New York, pp. 527–568.

Mariner, R. H., Rapp, J. B., Willey, L. M. and Presser, T. M. (1974a). *The Chemical Composition and Estimated Minimum Thermal Reservoir Temperatures of the Principal Hot Springs of Northern and Central Nevada.* U.S. Geol. Survey Open-File Report, 32 p.

Mariner, R. H., Rapp, J. B., Willey, L. M. and Presser, T. M. (1974b). *The Chemical Composition and Estimated Minimum Thermal Reservoir Temperatures of Selected Hot Springs in Oregon.* U.S. Geol. Survey Open-File Report, 27 p.

Michard, G., Stettler, A., Fouillac, C., Ouzounian, G., and Mandeville, D. (1976). "Subsuperficial Changes in Chemical Composition of the Thermomineral Waters of Vichy Basin: Geothermal Implication," *Geochem. J.* **10,** 155–161.

Millero, F. J. (1975). "The Physical Chemistry of Estuaries." In T. M. Church, Ed., *Marine Chemistry in the Coastal Environment.* American Chemical Society, Symposium Series 18, pp. 25–55.

Monheimer, R. H. (1974). "Sulfate Uptake as a Measure of Planktonic Production in Freshwater Ecosystems," *Can. J. Microbiol.* **20,** 825–831.

Monheimer, R. H. (1975). "Sulfate Uptake by Microplankton Communities in Western Lake St. Clair," *Limnol. Oceanogr.* **20,** 183–190.

Morel, F., McDuff, R. E., and Morgan, J. J. (1976). "Theory of Interaction Intensities, Buffer Capacities and pH Stability in Aqueous Systems with Application to the pH of Seawater and a Heterogeneous Model Ocean System," *Marine Chem.* **4,** 1–28.

Morris, A. W. and Riley, J. P. (1966). "The Bromide/Chlorinity and Sulfate/Chlorinity Ratios in Sea Water," *Deep-Sea* Res. **13,** 699–705.

Mortimer, C. H. (1941, 1942). "The Exchange of Dissolved Substances between Mud and Water in Lakes," *J. Ecol.* **29,** 280–329; **30,** 147–201.

Nancollas, G. H. and Purdie, N. (1963). "Crystallization of Barium Sulfate in Aqueous Solution," *Faraday Soc. Trans.* **59,** 735–740.

Naumov, G. B., Ryzhenko, B. N., and Khodakovsky, I. L. (1971). *Handbook of Thermodynamic Data* (translated from the Russian by G. J. Soleimani, 1974). (Available from U.S. Department of Commerce, National Technical Information Service, Springfield, Va. 22151, as Rept. PB-226 722.)

Nissenbaum, A., Presley, B. J., and Kaplan, I. R. (1972). "Early Diagenesis in a Reduced Fjord, Saanich Inlet, British Columba. I: Chemical and Isotopic Changes in Major Components of Interstitial Water," *Geochim. Cosmochim. Acta* **36,** 1007–1027.

Nriagu, J. O. (1975). "Sulfur Isotopic Ratio Variations in Relation to Sulfur Pollution of Lake Erie." In *Isotope Ratios as Pollutant Source and Behavior Indicators.* International Atomic Energy Agency, Vienna, pp. 77–93.

Nriagu, J. O. and Coker, R. D. 1976). "Emission of Sulfur from Lake Ontario Sediments," *Limnol. Oceanogr.* **21,** 485–489.

Nriagu, J. O. and Coker, R. D. (1977). "Isotopic Composition of Sulfur in Precipitation within the Great Lakes Basin," *Tellus* **29** (in press).

Oden, S. (1976). "The Acidity Problem—an Outline of Concepts," *Water, Air Soil Pollut.* **6,** 137–166.

Ohle, W. (1934). "Chemische und Physikalische Untersuchungen norddeutscher Seen," *Arch. Hydrobiol.* **26,** 386–464, 584–658.

Ohle, W. (1954). "Sulfat als "Katalysator" des limnischen Stoffkreislaufes." *Vom Wasser* **21,** 13–32.

Orr, W. (1974a). "Biogeochemistry of Sulfur." In Vol. II/4. Springer-Verlag, New York, 23 pp.

Orr, W. L. (1974b). *Changes in Sulfur Content and Isotope Ratios of Sulfur during Petroleum Maturation—Study of Big Horn Paleozoic oils.* American Association of Petroleum Geologists, Bull. 58, pp. 2295–2318.

Ostlund, G. H. and Alexander, J. (1963). "Oxidation Rate of Sulfide in Sea Water: A Preliminary Study," *J. Geophys. Res.* **68,** 3995–3997.

Ostroff, A. G. (1964). "Conversion of Gypsum to Anhydrite in Aqueous Salt Solutions," *Geochim. Cosmochim. Acta* **28,** 1363–1372.

Palache, C., Berman, H., and Frondel, C. (1951). *Dana's System of Mineralogy,* Vol. 2. John Wiley and Sons, New York, 1124 pp.

Pauling, L. (1940). *The Nature of the Chemical Bond, 2nd ed.* Cornell University Press, Ithaca, N.Y.

Pettyjohn, W. A., Ed. (1972). *Water Quality in a Stressed Environment.* Burgess, Minneapolis, 309 pp.

Pinneker, E. V. and Lomonosov, I. S. (1973). "On the Genesis of Thermal Waters in the Sayan-Baikal Highland." In E. Ingerson, Ed., *Hydrobiogeochemistry,* Vol. 1. The Clarke Co., Washington, D.C., pp. 246–254.

Postgate, J. R. (1968). "The Sulfur Cycle." In G. Nickless, Ed., *Inorganic Sulfur Chemistry.* Elsevier, Amsterdam, pp. 259–279.

Pourbaix, M. J. N. (1949). *Thermodynamics of Dilute Aqueous Solutions.* Edward Arnold and Co., London, 136 pp.

Presley, B. J. and Kaplan, I. R. (1968). "Changes in Dissolved Sulfate, Calcium and Carbonate from Interstitial Water of Near-Shore Sediments," *Geochim. Cosmochim. Acta,* **32,** 1037–1048.

Presley, B. J., Kolodny, Y., Nissenbaum, A., and Kaplan, I. R. (1972). "Early Diagenesis in a Reducing Fjord, Saanich Inlet, British Columbia. II: Trace Elements Distribution in Interstitial Water and Sediments," *Geochim. Cosmochim. Acta* **36,** 1073–1090.

Pytkowicz, R. M. and Kester, D. R. (1971). "The Physical Chemistry of Sea Water," *Oceanogr. Mar. Biol. Ann. Rev.* **9,** 11–60.

Pytkowicz, R. M., Atlas, E., and Culberson, C. H. (1975). Chemical Equilibrium in Seawater." In T. M. Church, Ed., *Marine Chemistry in the Coastal Environment.* American Chemical Society, Symposium Series, 18, pp. 1–24.

Reeder, S. W., Hitchon, B., and Levinson, A. A. (1972). "Hydrogeochemistry of the Surface Waters of the Mackenzie River Drainage Basin, Canada. I: Factors Controlling Inorganic Composition," *Geochim. Cosmochim. Acta* **36,** 825–865.

Richards, F. A. (1965). "Anoxic Basins and Fjords." In J. P. Riley and G. Skirrow, Eds., *Chemical Oceanography,* Vol. 2. Academic Press, London, pp. 611–645.

Rickard, D. T. and Nriagu, J. O. (1977). "Aqueous Environmental Chemistry of Lead." In J. O. Nriagu, Ed., *Biogeochemistry of Lead.* Elsevier, Amsterdam (in press).

Robie, R. A. and Waldbaum, D. R. (1968). *Thermodynamic Properties of Minerals and Related Substances at 298.15°K (25.0°C) at 1 Atmosphere (1.013 Bars) Pressure, and at Higher Temperatures.* U.S. Geological Survey, Bull. 1259, 256 pp.

Roy, A. B. and Trudinger, P. A. (1970). *The Biochemistry of Inorganic Compounds of Sulfur.* Cambridge University Press, 400 pp.

Sillen, L. G. and Martell, A. E. (1964). *Stability Constants of Metal-Ion Complexes.* Chemical Society (London, Spec. Publ. 17, 754 pp.

Sillen, L. G. and Martell, A. E. (1971). *Stability Constants,* Suppl. 1. Chemical Society (London), Spec. Publ. 25, 865 pp.

Skopintsev, B. A., Karpov, A. V., and Vershinina, O. A. (1964). "Study of the Dynamics of Some Sulfur Compounds in the Black Sea under Experimental Conditions," *Soviet Oceanogr.* **4,** 55–72.

Snavely, E. S. and Blount, F. E. (1969). "Rates of Reaction of Dissolved Oxygen with Scanvengers in Sweet and Sour Brines," *Corrosion* **25,** 397–402.

Sorokin, Yu. I. (1964). "On the Trophic Role of Chemosynthesis in Water Bodies," *Int. Rev. Ges. Hydrobiol.* **49,** 307–324.

Sorokin, Yu. I. (1970). "Inter-relations between Sulfur and Carbon Turnover in Meromictic Lakes," *Arch. Hydrobiol.* **66,** 391–446.

Sorokin, Yu. I. (1971). "Experimental Data on Oxidation Rate of Hydrogen Sulfide in the Black Sea," *Okeanologija* **11,** 423–431.

Spedding, D. J. (1974). *Air Pollution.* Clarendon Press, Oxford, pp. 33–46.

Staples, L. W. (1972). "Sulfates, Sulfides and Sulfosalts." In R. W. Fairbridge, Ed., *Encyclopedia of Geochemistry and Environmental Sciences.* Van Nostrand Reinhold, New York, pp. 1123–1142.

Starkey, R. L. (1956). "Transformations of Sulfur by Microorganisms," *Ind. Eng. Chem.* **48,** 1429–1437.

Starkey, R. L. (1966). "Oxidation and Reduction of Sulfur Compounds in Soils," *Soil Sci.* **101,** 297–306.

Stuiver, M. (1967). "The Sulfur Cycle in Waters during Thermal Stratification," *Geochim. Cosmochim. Acta* **31,** 2151–2167.

Stumm, W. (1966). "Redox Potential as an Environmental Parameter; Conceptual Significance and Operational Limitations." In *Proceedings of the 3rd International Conference on Water Pollution Research, Munich,* pp. 1–16.

Sugawara, K. (1967). "Migration of Elements through Phases of Hydrosphere and Atmosphere." In A. P. Vinogradov, Ed., *Chemistry of the Earth's Crust,* Vol. 2. Israel Program for Scientific Translations, Jerusalem, pp. 501–510.

Talling, J. F. and Talling, I. B. (1965). "The Chemical Composition of African Lake Waters," *Int. Rev. Ges. Hydrobiol.* **50,** 421–463.

Tanji, K. K. and Doneen, L. D. (1966). "Predictions on the Solubility of Gypsum in Aqueous Salt Solutions," *Water Resources Res.* **2,** 543–548.

Thode, H. G., Wanless, R. K., and Wallouch, R. (1954). "The Origin of Native Sulfur Deposits from Isotope Fractionation Studies," *Geochim. Cosmochim. Acta* **5,** 286–298.

Thorstenson, D. C. (1970). "Equilibrium Distribution of Small Organic Molecules in Natural Waters," *Geochim. Cosmochim. Acta* **34,** 745–770.

Torrey, M. S. (1976). *Environmental Status of the Lake Michigan Region.* Vol. 3: *Chemistry of Lake Michigan.* Argonne National Laboratory, Rept. ANL/ES-40, Argonne, Ill., 418 pp.

Trudinger, P. A. (1969). "Assimilatory and Dissimilatory Metabolism of Inorganic Sulfur Compounds by Micro-organisms," *Adv. Microbiol. Enzymol.* **3,** 111–158.

Truesdell, A. H. and Jones, B. F. (1974). "WATEQ, a computer program for calculating chemical equilibria of natural waters," *U.S. Geol. Survey J. Res.,* **2,** 233–248.

U.S. Geological Survey (1974). *Quality of Surface Waters of the United States 1969, Parts 9 and 10.* U.S. Geological Survey, Water Supply Paper 2148, p. 270.

Vredenburgh, L. D. and Cheney, E. S. (1971). *Sulfur and Carbon Isotope Investigation of Petroleum, Wind River Basin, Wyoming.* American Association of Petroleum Geologists, Bull. 55, pp. 1954–1975.

Wagman, D. D., Evans, W. H., Parker, V. B., Halow, I., Bailey, S. M., and Schumm, R. H. (1968). *Selected Values of Chemical Thermodynamic Properties.* U.S. National Bureau of Standards, Tech. Note 270-3, 264 pp.

Wagman, D. D., Evans, W. H., Parker, V. R., Halow, I., Bailey, S. M., and Schumm, R. H. (1969). *Selected Values of Chemical Thermodynamic Properties.* U.S. National Bureau of Standards. Tech. Note 270-4, 141 pp.

Weast, R. D., Ed. (1968). *Handbook of Chemistry and Physics,* 49th ed. Chemical Rubber Co., Cleveland, Ohio, Pa-1-F-286.

Weiss, H., Bertine, K., Koide, K., and Goldberg, E. D. (1975). "The Chemical Composition of Greenland Glacier," *Geochim. Cosmochim. Acta* **39,** 1–10.

Wetzel, R. G. (1965). "Productivity and Nutrient Relationships in Marl Lakes of Northern Indiana," *Verh. Int. Ver. Limnol.* **16,** 321–332.

Wetzel, R. G. (1975). *Limnology.* W. B. Saunders Co., Philadelphia, pp. 246–286.

White, D. E. (1957). "Magmatic, Connate and Metamorphic Waters," *Geol. Soc. Am. Bull.* **68,** 1659–1682.

White, D. E. (1968). "Environments of Generation of Some Base-Metal Ore Deposits," *Econ. Geol.* **63,** 301–335.

White, D. E., Hem, J. D., and Waring, G. A. (1963). *Chemical Composition of Subsurface Waters.* U.S. Geological Survey, Prof. Paper 440-F, 67 pp.

Whitfield, M. (1975). "Sea Water as an Electrolyte Solution." In J. P. Riley and G. Skirrow, Eds., *Chemical Oceanography,* Vol. 1, 2nd ed. Academic Press, London, pp. 43–171.

Wilson, T. R. S. (1975). "Salinity and the Major Elements of Sea water." In J. P. Riley and G. Skirrow, Eds., *Chemical Oceanography,* Vol. 1. Academic Press, London, pp. 365–413.

Wright, R. F. (1977). "Acid Precipitation and Its Effects on Freshwater Ecosystems: An Annotated Bibliography," *Water, Air Soil Pollut.* **7** (in press).

Wuensch, B. J. (1972). "Crystal Chemistry of Sulfur." In *Handbook of Geochemistry,* Vol. II/3. Springer-Verlag, New York, 23 pp.

Yoshida, Y., Torii, T., and Yamagata, N. (1973). "Antarctic Saline Lakes." In E. Ingerson, Ed., *Hydrogeochemistry,* Vol. I. The Clarke Co., Washington, D.C., pp. 652–660.

Young, L. and Maw, G. A. (1958). *The Metabolism of Sulfur Compounds.* Methuen, London.

Zinder, S. and Brock, T. D. (1977). "Sulfur Dioxide in Geothermal Waters and Gases," *Geochim. Cosmochim. Acta* **41,** 73–79.

7

SULFUR POLLUTION AND THE AQUATIC ECOSYSTEM

Brodde Almer

Institute of Freshwater Research, Drottningholm, Sweden

William Dickson
Christina Ekström
Einar Hörnström

National Swedish Environment Protection Board, Solna, Sweden

"When we had got some distance away from the resting-place we felt a strong smell of sulphur, which was eventually seen rising to the west of the city of Falun, getting so intense closer to the city that it all but asphyxiated anyone who was not used to it.

"Out of this mine there ascended a steady pillar of smoke, which, together with the physical features of the mine, made us understand that the description of hell given to us by "theologists," to impress on man's mind, was taken from this or similar mines. Never has a poet described *Styx, Regnum Subterraneum* and *Plutonis,* or a "theologist", hell, with such horror as it appears here. For outside, there ascends a poisonous, pungent sulphur smoke, poisoning the air wide around so that no one will get there without hardship. This corrodes the earth so that no herbs can grow around it.

"*The ground* was completely naked ¼ mile (2.5 km) away, consisting of large loose rocks, as if thrown on sterile gravel ground. The reason for this, in particular, seems to be the large amount of sulphurous, vitriolic, cold and corrosive copper smoke that Vulcani copper boys light at Falun; all the more as the mountains nearest to Falun or to other cold roasters are more naked, whereas the dells, whither the smoke to a lesser extent expands and where it is more easily washed away, are greener."

This description is from Carl von Linné's *Iter Dalekarlicum* (1734), dealing with the 500-year-old copper smelter at Falun in the province of Dalarna, Sweden (1).

1. INTRODUCTION

Sulfur problems are old, and the effects of sulfur on the environment have been known for centuries. They may be the result of discharges to waters, to soil, or to the air. As emissions of sulfur dioxide (SO_2) from tall stacks increase, the effects will become a concern of the many. In addition to the toxicity of SO_2 to organisms, other effects include increased leaching of cations and metals, acidification of soils and waters, and changes in the rates of nutrient cycling.

Most soils reflect thousands of years of atmospheric input and weathering, while waters reflect what has happened in the very last years. Therefore the effects of an increased deposition of sulfur will be first noticed in the water ecosystem, but they certainly reflect what is happening in the drainage area too.

The effects of dissolved SO_2 are similar to those of nitric or hydrochloric acid or of acidifying ammonium compounds. In large areas sulfur is still the major acidic substance, although nitrogen compounds have recently received due attention. In Scandinavia, sulfur effluents account for about 70% of the total acidification.

Water will be acidified if the watershed is poorly buffered. Most sensitive are waters on low-weathering siliceous bedrock, quartzites, sandstones, granites, and felsic gneisses, where the contribution of alkaline substances to the water is small. Although low-weathering rocks are found all over the world, acidified waters have been noticed only in connection with acid deposition, acid outlets or sulfuric soils, ditching, or other contaminated situations. The effects of atmospheric acidification on water have been studied at different places, for example, in the Adirondack Mountains of the United States, in the La Cloche Mountains of Canada, and in Scandinavia.

This research has been well documented in various publications (2–5). The major observations will be summarized in this chapter.

2. LOADING AND CONCENTRATIONS OF SULFUR IN LAKES AND RIVER WATERS

2.1. Natural Loading to Waters on Hard Rocks

In unpolluted areas the loading of "excess" sulfur into waters is usually very small. A load of 0.2 g S/m^2 of lake or river water each year (or even less) may be a realistic "natural" value, giving a sulfur content of 0.02 to 0.06 meq/l (0.3 to 1.0 mg/l) at 700 to 200 mm runoff/year. Though sulfur is a nutrient essential to water organisms, it is probably not a limiting factor even at these concentrations. Today these low levels in lake waters are found only in very remote areas.

The concentrations in polluted areas may be five times higher than the levels in pristine environments. In Europe alone, the deposition of atmospheric sulfur today is likely to be 10 times the preindustrial rate (6).

2.2 Sulfur Deposition and Acidification

The emission rate in Europe is about 60 million tonnes SO_2/year (1973), a large portion of which is emitted from the continent and the United Kingdom, with only 1.2 million tonnes coming from Scandinavia. A gradient in the deposition picture is to be expected, along with distance from the major sources.

The deposition models for sulfur have been considered in detail by others in this work. Here it should only be mentioned that near the emission source dry deposition will usually comprise the major portion of the total sulfur deposition, whereas wet deposition accounts for the largest portion at long distances from the source. Only a part of the total amount of sulfur deposited will escape the soils and reach the water. The rest will be bound to the soil or reemitted into the air.

An attempt at sulfur budgets for three different regions of Sweden is shown in Table 1. The total sulfur deposition is two or three times larger on the West Coast of Sweden than in central Sweden. Most of it is in the form of dry deposition, wheras further north the main removal mechanism is wet deposition. There is a strong gradient from the West Coast, making the deposition in south-central Sweden more like that in central

Table 1. Sulfur Budgets for Three Parts of Sweden[a]

Source	Sulfur (kg/ha·year)					
	West Coast		South Central Sweden		Central Sweden	
Wet deposition	11.5	(1973–1975)	10.3	(1968–1973)	8.2	(1968–1973)
Dry deposition	20	(1971–1972)	7.5	(1971–1972)	4.5	(1971–1972)
Lake water	25	(1973–1976)	9.4	(1968–1973)	3.7	(1968–1973)

[a] Budgets for south central and middle Sweden from Eriksson (7). Budget for the West Coast from Dickson (12). Dry deposition calculated by Högström, who used a deposition velocity of 0.8 cm/sec (8). For the West Coast 1–3 kg is sea salt sulfur.

Sweden, 400 km to the north, than that on the West Coast, 200 km to the south. In central and northern Sweden the amount of wet deposition alone is larger than what is found in lake waters.

Figure 1 shows the total sulfur deposition (excess sulfur), that is, precipitation plus calculated dry deposition, for Sweden. Figure 2 shows the amounts of sulfur found in poorly buffered lake waters, expressed as grams of sulfur per square meter per year. The decrease northward in both deposition and leaching is evident.

As mentioned earlier, the character of the watershed is of crucial importance for any acidification to take place. In Figure 3 some lakes with varying sulfur loadings in different parts of Sweden, considered to be located in the most sensitive bedrock are compared with respect to their present pHs and sulfur influxes, (expressed as grams of sulfur per square meter per year). The dots follow a typical acid-base titration graph (curve 1). It can be seen that these lakes, the most sensitive ones that can be found in Sweden, may resist a load of about 0.3 g S/m^2 lake water each year, but no more. At 1 g S/m^2 they will be strongly affected, and the pH will probably decrease below 5.0. Curve 2 illustrates that lakes on slightly less sensitive ground will resist acidification at considerably higher loading rates.

Acidification of fresh waters was noticed in southern Norway as early as the 1920s, being the cause of lower yields of salmon in some large rivers. By then, these rivers occasionally showed pH values below 5.5. Present values are often below 5.0 (9). Today the sulfur deposition in this area is probably three or four times larger than it was at the end of last century; compare the estimated emissions in Europe from fossil fuel in Figure 4 (10). On the assumptions that curve 1 in Figure 3 is typical of the most sensitive waters, and that one fourth of the current load was

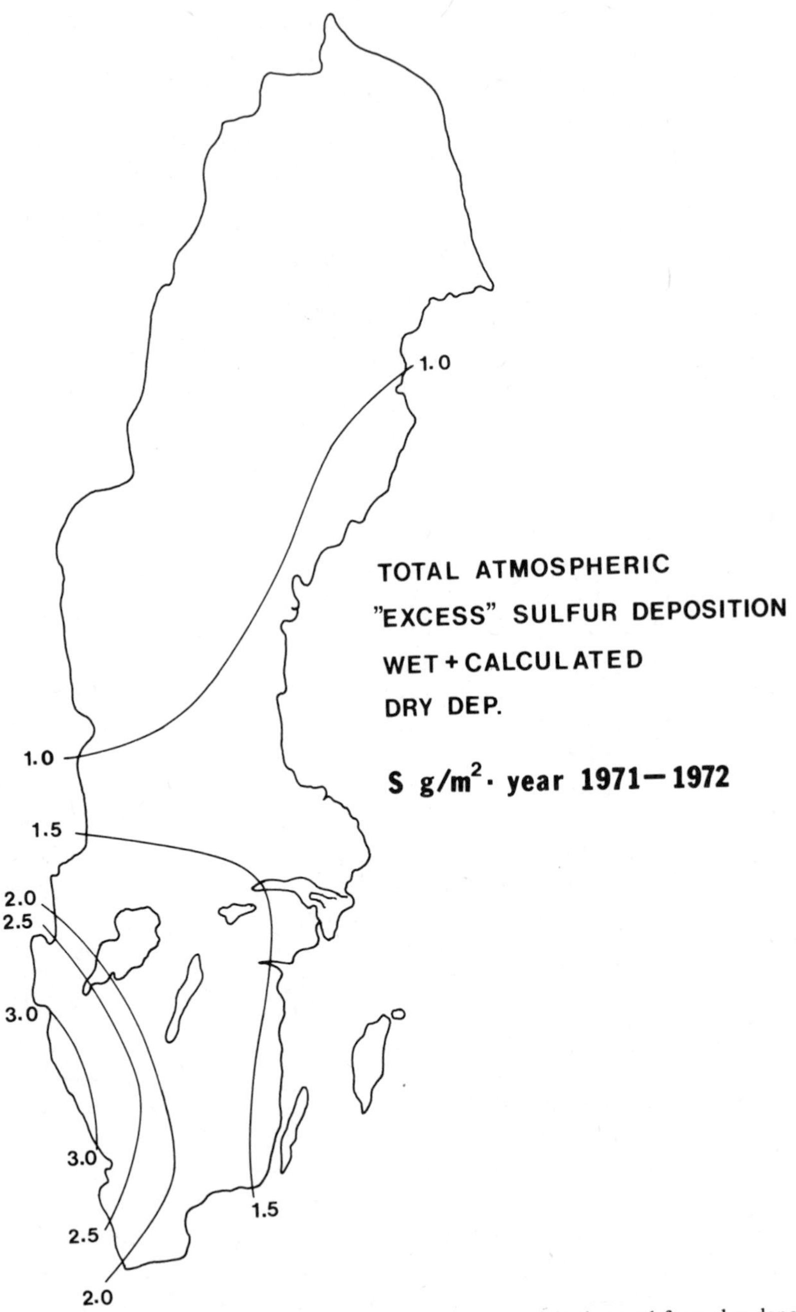

Figure 1. Atmospheric load of "excess" sulfur from precipitation and from dry deposition, 1971–1972 (g s/m² year). Dry deposition calculated from a deposition velocity of 0.8 cm/sec (8).

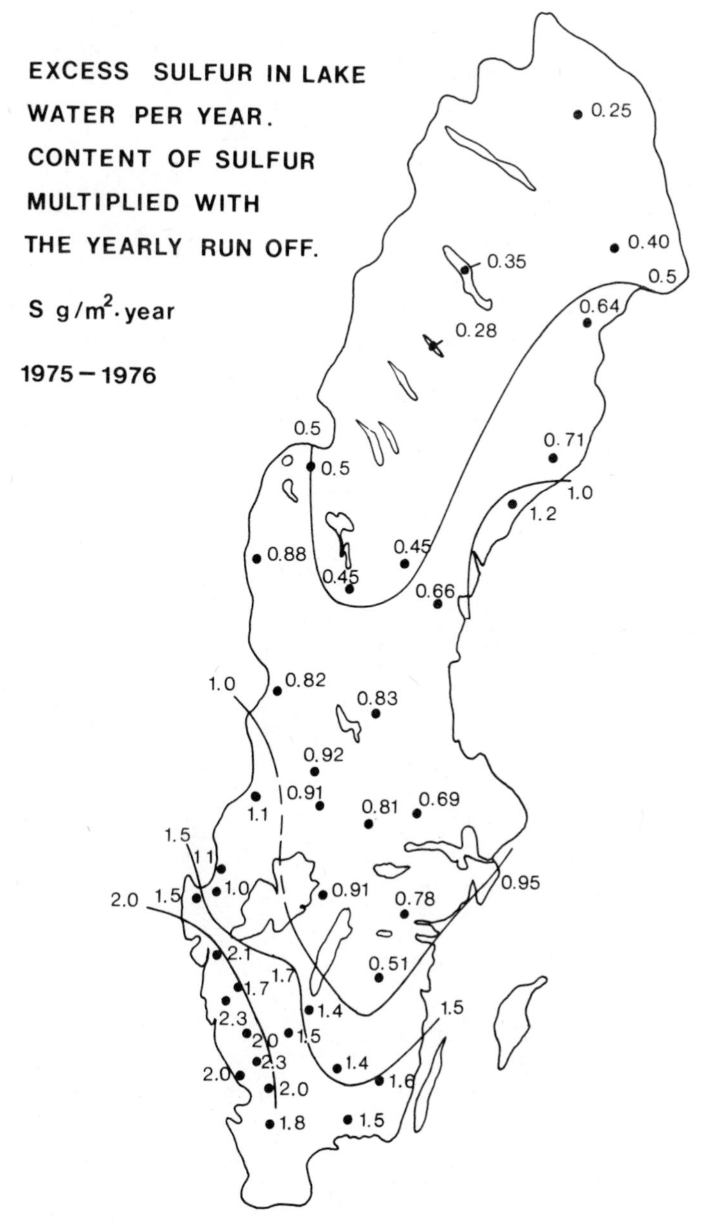

Figure 2. "Excess" sulfur in lake water per year (g s/m² year). (Concentration of "excess" sulfur multiplied by the yearly runoff.)

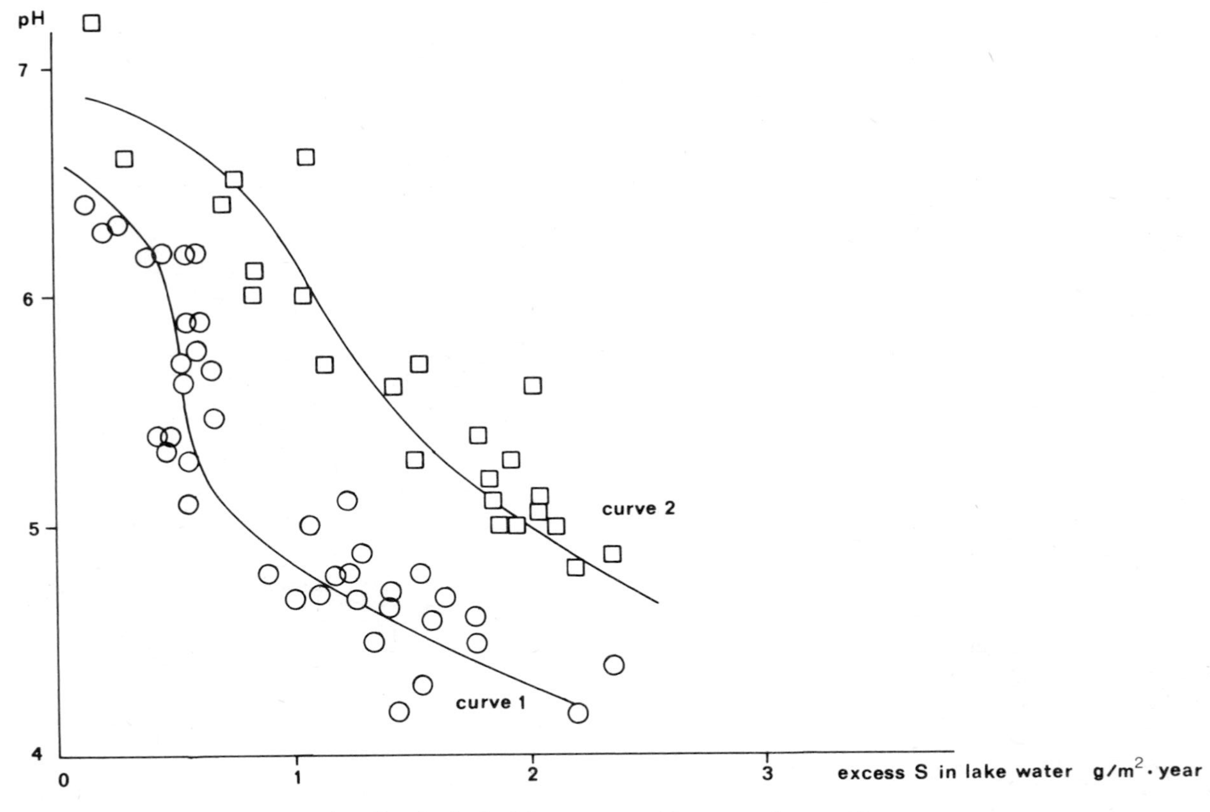

Figure 3. The pH values and sulfur loads in lake waters with extremely sensitive surroundings (curve 1) and with slightly less sensitive surroundings (curve 2). (Load = concentration of ''excess'' sulfur multiplied by the yearly runoff.)

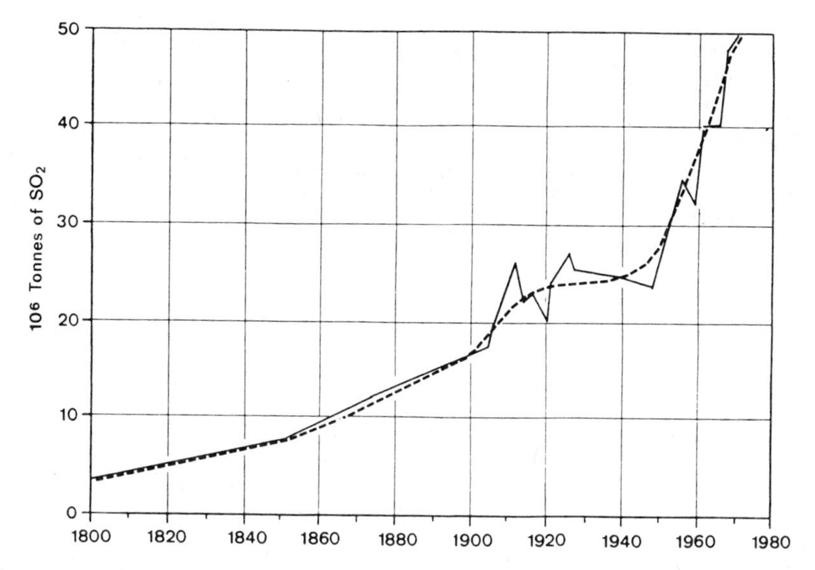

Figure 4. Estimated emissions of sulfur dioxide from combustion of fossil fuels in Europe (10).

discharged at the end of the nineteenth century, the pH of the waters was probably around 6 in those days. But very small increases in the loading rate will drastically change the pH. In southern Scandinavia now, large areas are suffering from the acid precipitation (11,12).

In Sweden the first biological symptoms of acidification were noticed in the 1920s and 1930s, with a decrease in number of the most sensitive fish species (roach) in some very poorly buffered lakes with pH values below 6 during summer. Today these lakes are the most acid in Sweden, with pH values of 4.0 to 4.5 or even less during the time of snowmelt. A pH decrease of up to 2 units can be noticed in some lakes.

3. INCREASED CATION LEACHING

3.1. Calcium and Magnesium

The alkalinity of water derives from the weathering of calcareous and siliceous minerals and from other chemical processes in the soil or the water, such as denitrification, nitrate transformation, and sulfide formation. Usually there should be good correlation (about 1:1) between the content of nonmarine calcium and magnesium and the concentration of bicarbonate in waters. However, acid stress will diminish the level of bicarbonate and also will increase the leaching of calicum and magnesium.

Figure 5 shows two diagrams of alkalinity and nonmarine calcium and magnesium in lakes. One diagram represents a rather unpolluted part of

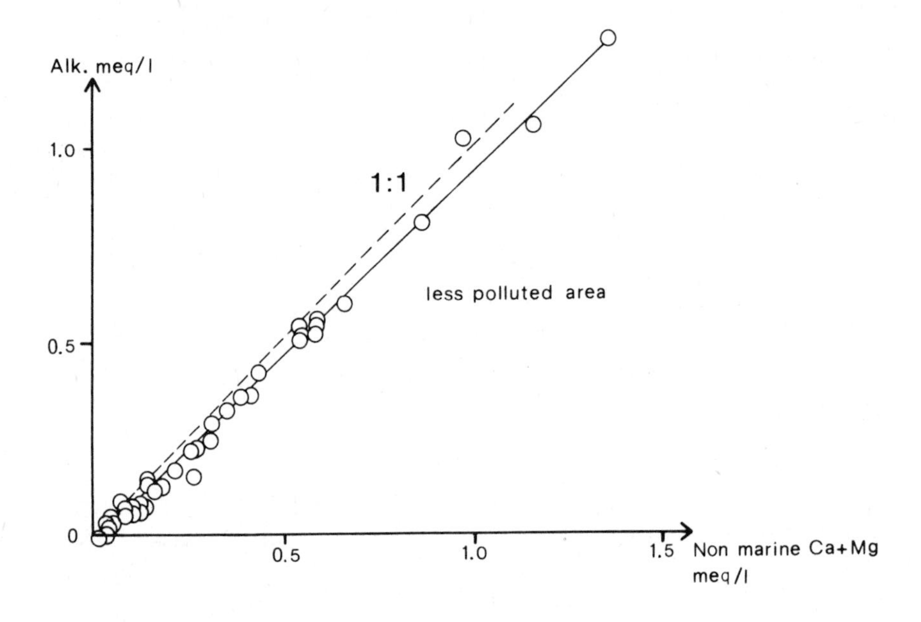

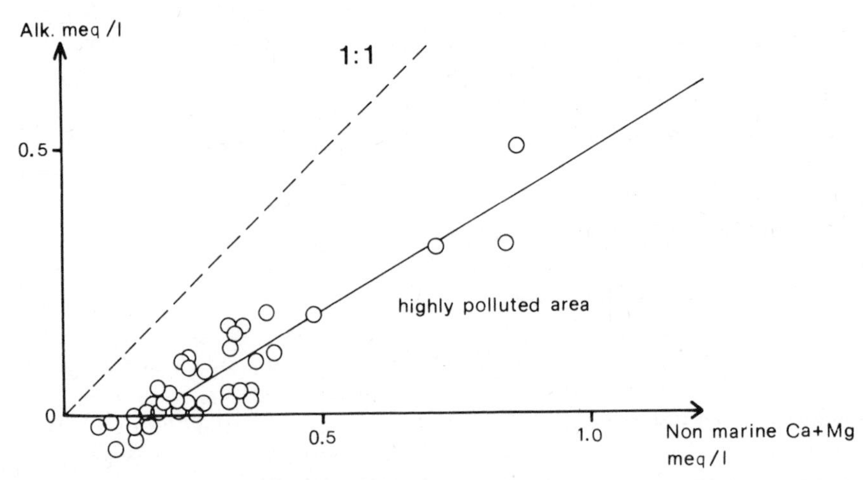

Figure 5. Alkalinities and contents of calcium and magnesium of nonmarine origin in lakes in two regions in Sweden with different sulfur loads.

Sweden; the other, the highly polluted West Coast. In the northern, less polluted area, the increased leaching of calcium and magnesium is about 0.04 meq/l, but in the West Coast region it is 0.15 to 0.4 meq/l. The actual concentrations of "excess" sulfate in the areas are 0.06 and 0.19 to 0.45 meq/l, respectively. The different slope of the line in Figure 5 from the most polluted area depends upon the larger leaching of sulfur from the best buffered soils; it may also be due to the fact that the sulfur deposition within the area shows large gradients. Some lakes situated near large emission sources receive two times more "excess" sulfur than do the more remote lakes within the area.

3.2. Sodium and Potassium

Increased leaching of sodium may also occur when the sulfur load is large. Comparison of lakes on low-weathering bedrocks in regions with different acid stresses shows that leaching of sodium in the polluted area may be about 0.05 meq/l, whereas it is 0.01 to 0.02 meq/l in unpolluted areas.

In highly polluted areas about 0.01 meq/l of potassium may be leached, probably the result of leaching from (*a*) needles on trees and (*b*) the bedrocks.

3.3. Aluminum

Aluminum is the most common metal on the earth. The distribution of water-soluble aluminum has been studied by several authors. In aqueous solutions with pH below 5, $Al(H_2O)_6^{3+}$, $AlOH(H_2O)_5^{2+}$, and $Al_8(OH)_{20}^{4+}$ are said to be the principal ionic species. At higher pH, when hydroxide is more abundant, polymerization leads to the formation of colloidal particles [$Al(OH)_3$]. Aging increases the size, accompanied by a decrease in the pH of the solution. At higher pH of 7.5 to 10, the predominant dissolved form is $Al(OH)_4^-$ (13,14).

In natural waters other anions will successfully compete with the hydroxide in complexing aluminum. Thus the solubility and colloidal stability of aluminum hydroxide precipitates are highly dependent on sulfate in the water. The aluminum hydroxide hydrosol, which may be rather stable in a nitrate system over a wide pH range, is rapidly destabilized in the presence of sulfate. Sulfate is also incorporated in the resulting precipitate, forming a mixed salt (14). In natural water much of the aluminum found probably is complexed to humic substances.

In soil water of low acidity, the concentration of aluminum is rather small, and on reaching surface waters with pH above 5.5 the aluminum will be precipitated out. Lake sediments, therefore, have large contents of hydromorphic aluminum.

The concentration of soluble aluminum will increase drastically, however, if the soil is acidified, for example, by sulfur. At pH values below 5, large amounts of aluminum will be present in the incoming solution, and more can be dissolved from the lake sediment. Acid waters, therefore, are highly enriched in aluminum.

Figure 6 shows lakes from the Swedish West Coast with pH values varying from 3.95 to 7.4 and the concentration of aluminum varying according to pH from 670 to 10 μg/l. Because of its buffering capacity and complexing capacity with regard to the sulfate, phosphate, and humus, aluminum has a great influence on many processes in the soil and the water.

Buffering Capacity of Aluminum

When the soil or the water is acidified, the dissolution of aluminum silicate buffers the solutions. But when dissolved in acid water, aluminum acts as a week acid. In Figure 7 water from an acid lake of pH 4.1 and 0.45 mg Al/l, and two synthetic lake waters acidified with hydrochloric acid to pH 4.1, one of them with 0.5 mg Al/l added, have been titrated with a base, and the titration curves are compared. For instance, with an addition of 0.11 meq/l of the base, the "blank" with hydrochloric solution reached a pH of 6.7, whereas the acid lake and the 0.5 mg/l aluminum solution reached only a pH of 5.1.

Then the waters were titrated back to the original pH. As can be seen from Figure 8, the acid lake water and the aluminum solution needed less acid than the "blank" to reach the original pH, 4.1. The "extra need of base" exists in the pH range 4.5 to 5.5 and is almost equivalent to the amount of Al present in the waters. This amount apparently becomes bound as stable hydroxides and is not affected by the acid added—at any rate, not instantly.

Apparently acid lakes may be regarded as weak acids of aluminum salts, and they may be much more acid than is indicated by the pH value. One becomes aware of this when liming them. The amount of aluminum needed is often twice that required to buffer strongly acid water.

Aluminum as Precipitant for Humus

In the same pH range where it exerts its buffering effects, aluminum also acts as a very efficient precipitant for humic substances. Figure 9 shows its effect on the color of brown water. Humic substances are readily

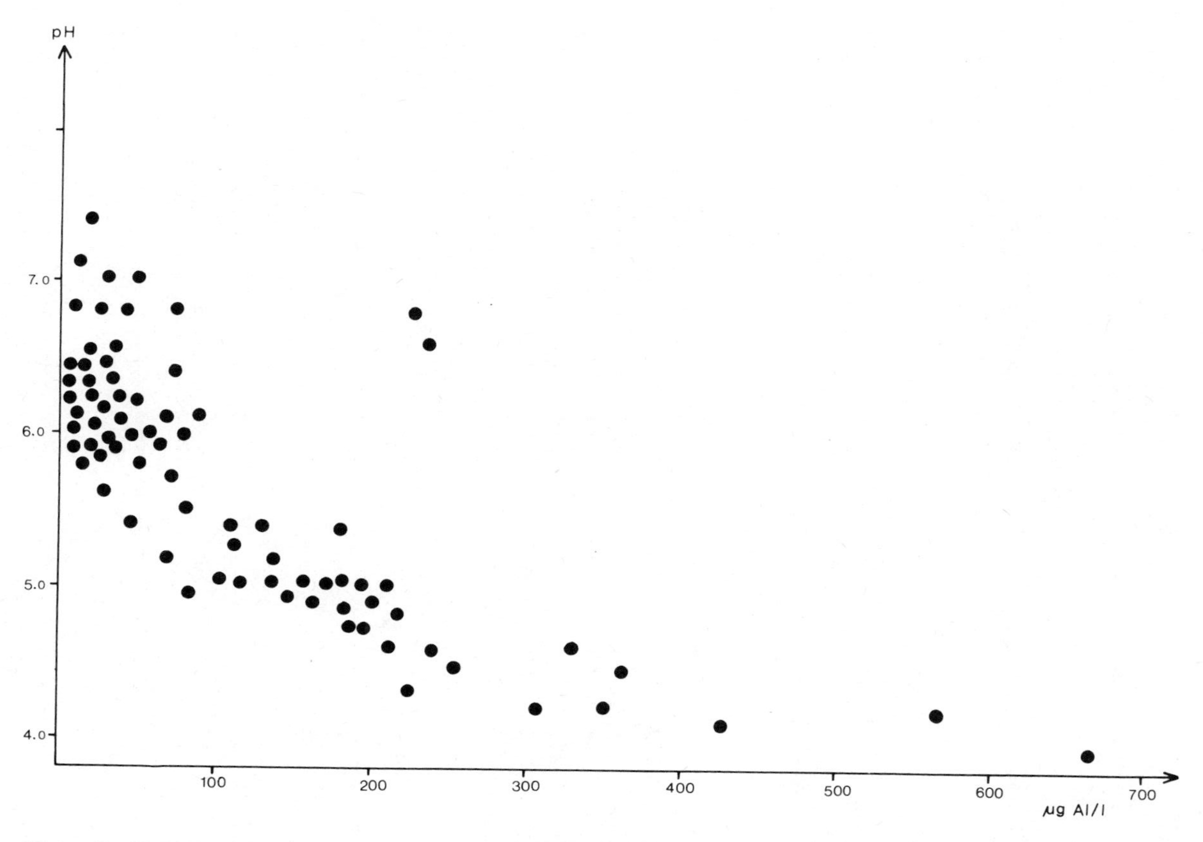

Figure 6. The pH values and aluminum contents in lakes on the Swedish West Coast.

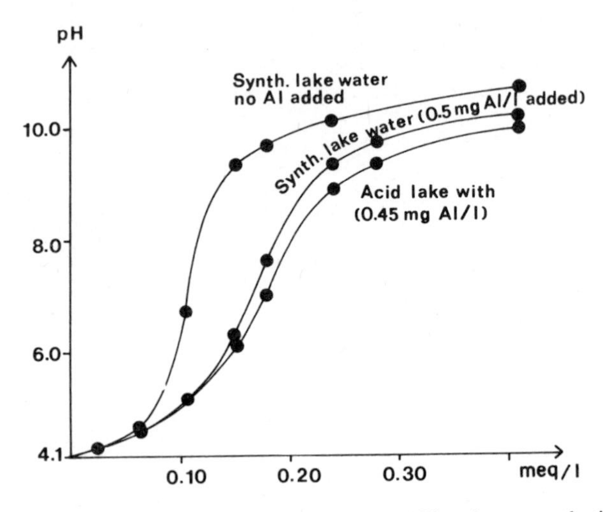

Figure 7. One acid lake of pH 4.1 containing 0.45 mg Al/l and two synthetic lake waters of pH 4.1, one with 0.5 mg Al/l added, titrated with base to different pHs.

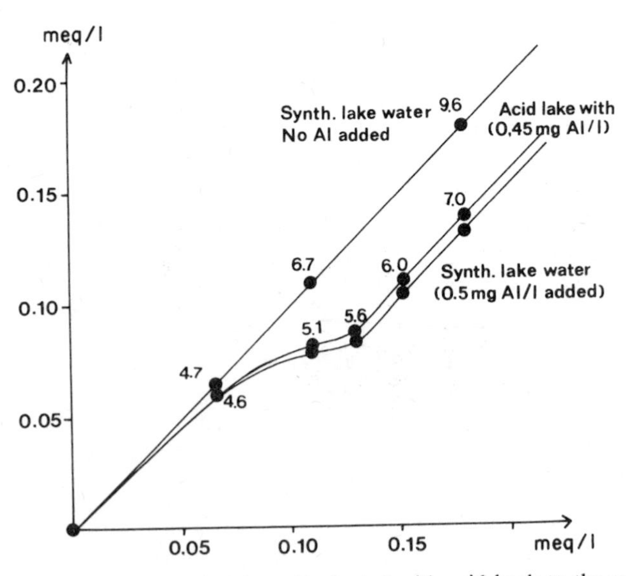

Figure 8. The same waters as in Figure 7, titrated with acid back to the original pH of 4.1.

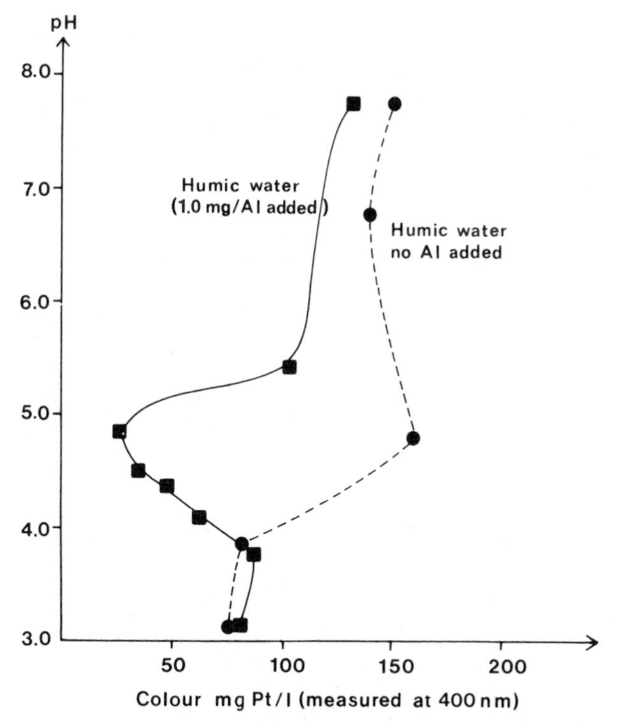

Figure 9. Color of humic water at different pHs with no aluminum added and with 1.0 mg Al/l added. A rapid precipitation of the humus occurred at pH 4–~5 when aluminum was added. There was no precipitation when only acid or base was added.

precipitated in the pH range of 4.0 to 5.0. Probably this is a contributing reason why acid lakes become so clear. Acidity alone gives a slight discoloration, but not a precipitate. The transparency may increase by 10 to 15 m after the acidification. Figure 10 shows one example. From 1958 to 1973 the $KMnO_4$ demand decreased from 24 to 8 mg/l, and the transparency increased more then 7 m over that at the beginning of the century. In 1973 the content of aluminum was 0.37 mg/l. The humic substances either are bound in the forest soil or are precipitated to the bottom of lakes.

Figure 11 shows the color and the concentrations of aluminum in some Swedish lakes during summer. A high content of aluminum is coupled with very clear water. Extremely colored water, however, may keep large amounts of aluminum complexed where the humus is not precipitated.

The decreasing content of humus in lakes exposed to aluminum

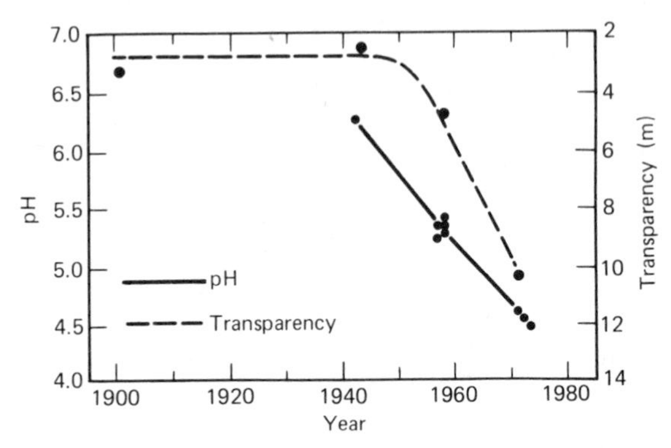

Figure 10. Changes in transparency since 1900 and in pH since 1943 in one lake on the Swedish West Coast.

additions is also important because humus not only is a carrier of several minor elements but also may chelate poisonous metals. Therefore metals are probably more toxic in clear, dehumidified water than in the original, colored lake.

Aluminum as Precipitant for Phosphorus

Aluminum also easily removes dissolved phosphorus, which is an element of crucial importance to the trophic state of the waters. This means that waters within a wide pH range are "oligotrophicated" because of increased removal of phosphorus by aluminum.

The fluocculating and precipitating ability is strongest at the pH interval of 5 to 6 (Figure 12). This also means that waters of pH 4.5 or less may contain more soluble phosphorus than does less acid water.

Aluminum as Toxic Metal to Fish

In acid waters the concentration of aluminum can reach toxic levels. Jones (15) sets the lethal concentration limit for aluminum as 0.07 mg/l. Acid waters generally contain 0.2 to 0.6 mg/l (cf. Figure 6). At these levels, the pH of 4 to 5 is already quite harmful, though the toxicity of aluminum increases the ill effects. When the pH of the waters is raised (e.g., by liming), much of the aluminum is removed to the bottom as hydroxide. Nevertheless, trout introduced just after the liming, when the concentration was still high (0.3 mg/l), died, though the pH was "safe," at about 6 (16). Also, in less acid waters concentrations of aluminum should not exceed 0.1 mg/l if the sensitive fish species are to survive (17).

Since aluminum always seems to be present in water of low pH,

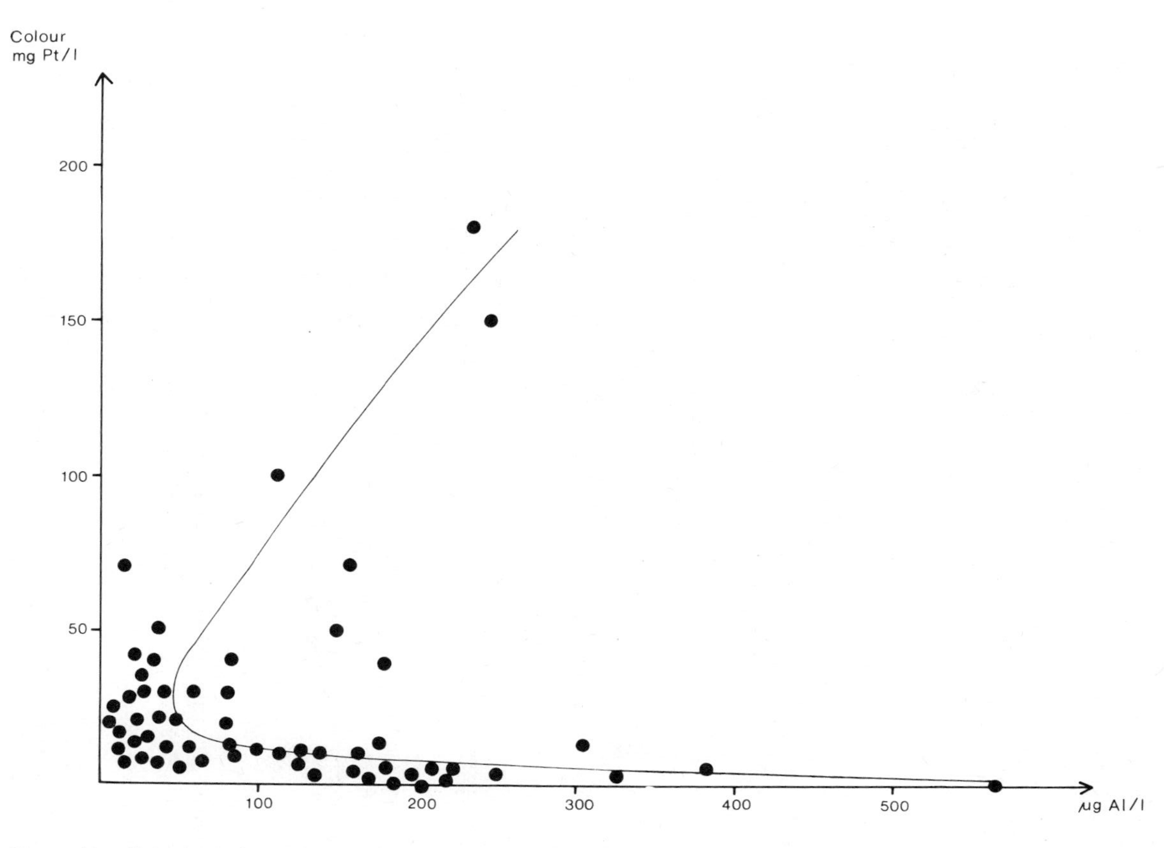

Figure 11. Color and aluminum contents during summer in some Swedish lakes.

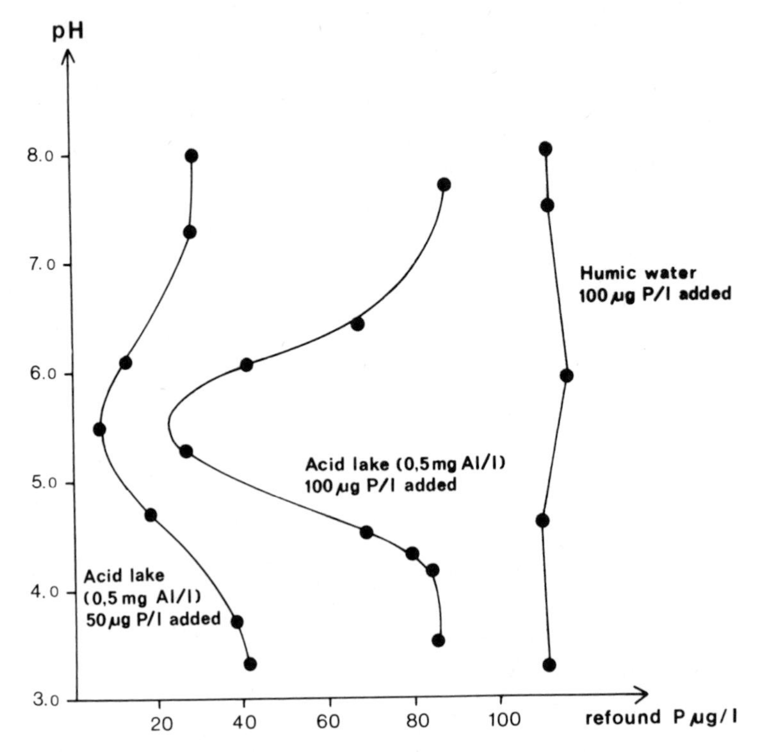

Figure 12. Refound phosphorus after additions of 50 and 100 μg P/l to water from an acid lake of pH 4.1 and 0.5 mg Al/l. Also, addition of 100 μg P/l to humic water of pH 6.2 containing 0.2 mg/l of aluminum complexed to humus. The pH was adjusted to different values. The water samples were stored for 5 days before they were analyzed.

analysis for this element should also be included in the bioassay tests concerned with the effects of pH on aquatic biota. At present this is seldom done.

4. INCREASED DISSOLUTION OF OTHER METALS

4.1. Manganese

Manganese will dissolve in acid milieu (Figure 13). Below pH 5.0 considerable amounts are dissolved in some lakes. One of the lakes in Figure 13 is used for domestic/industrial water supply. Along with the increase in acidity from pH 6.3 to 4.5, the content of manganese

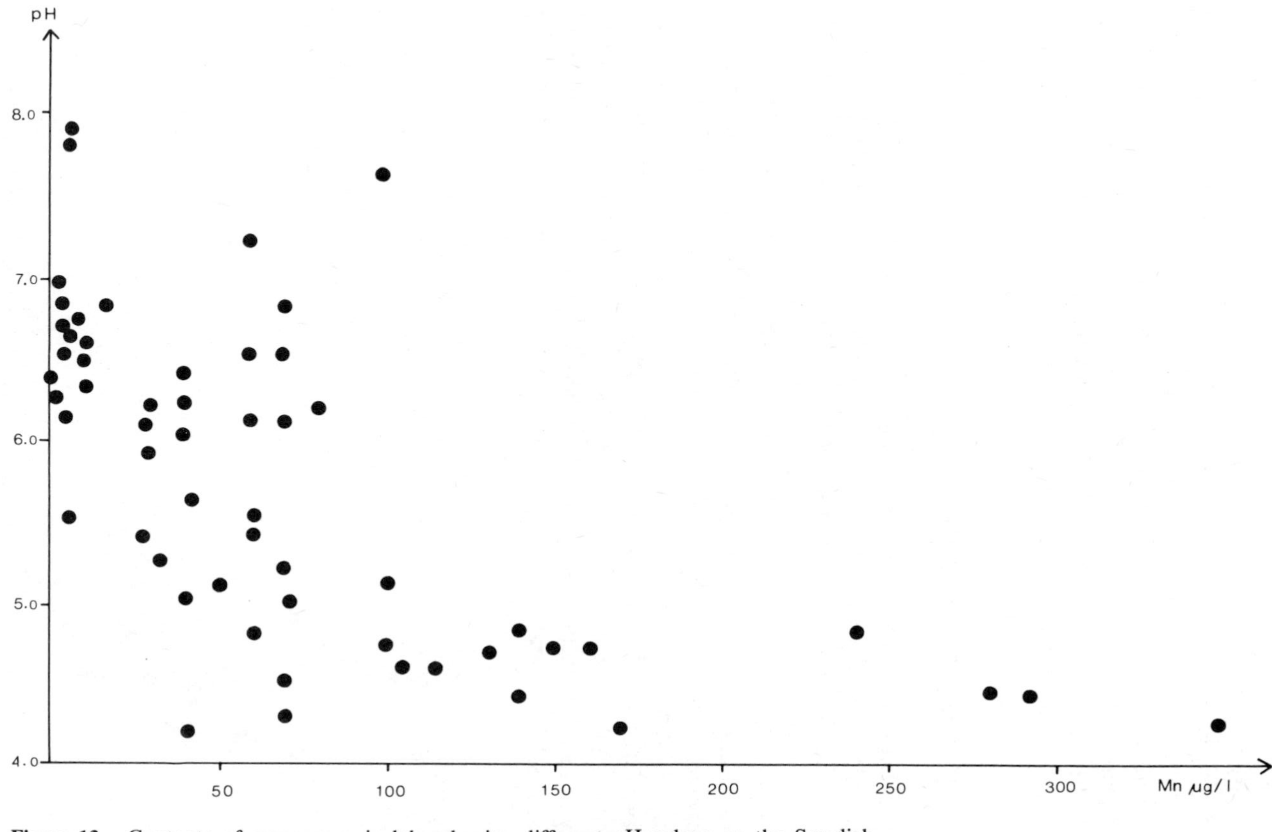

Figure 13. Contents of manganese in lakes having different pH values on the Swedish West Coast.

increased from 0.05 to 0.35 mg/l.

Some acid lakes, though, have low manganese concentrations. A condition for the level to increase is, of course, that there is some manganese to dissolve. Compared with the concentration of iron in sediments of acid lakes, the content of manganese is low. For example, the Mn:Fe ratio was 0.002 to 0.03 in three acid lakes but 0.06 to 0.16 in two nonacidified lakes.

4.2. Cadmium and Mercury

At neutral pH, cadmium is bound to other metal compounds, for example, iron and manganese oxyhydroxides. At low pH values, however, the dissolved fraction may increase (Figure 14), resulting in high-level accumulations in organisms.

The same may be true for mercury (Figure 15). The absence of young fish in many acid lakes, a new, accumulating flora (18), and possibly a delayed decomposition (see Section 6.5), however, may also check the accumulation of fish in some lakes. When the acid waters are limed, however, the pH increases, and one theory is that the new conditions in water and sediment caused by this treatment may increase the availability of mercury to plankton and newly hatched fish. At any rate, in newly limed lakes the mercury content is often higher in fish liver than in the

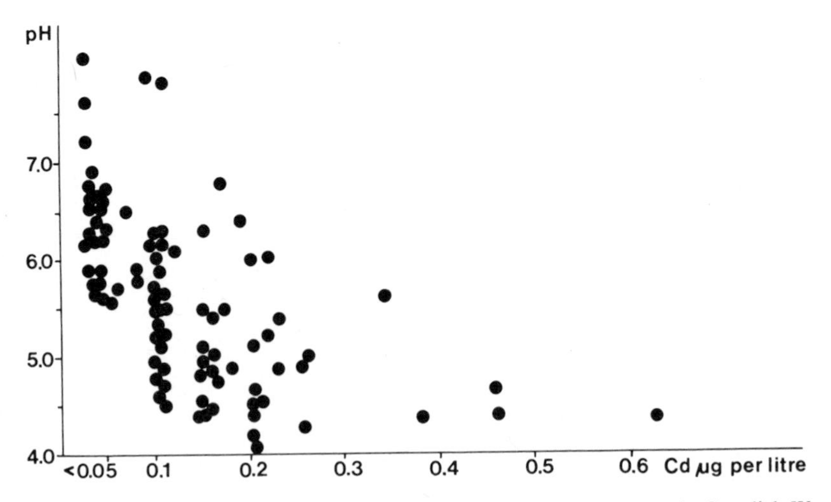

Figure 14. Contents of cadmium in lakes having different pH values on the Swedish West Coast.

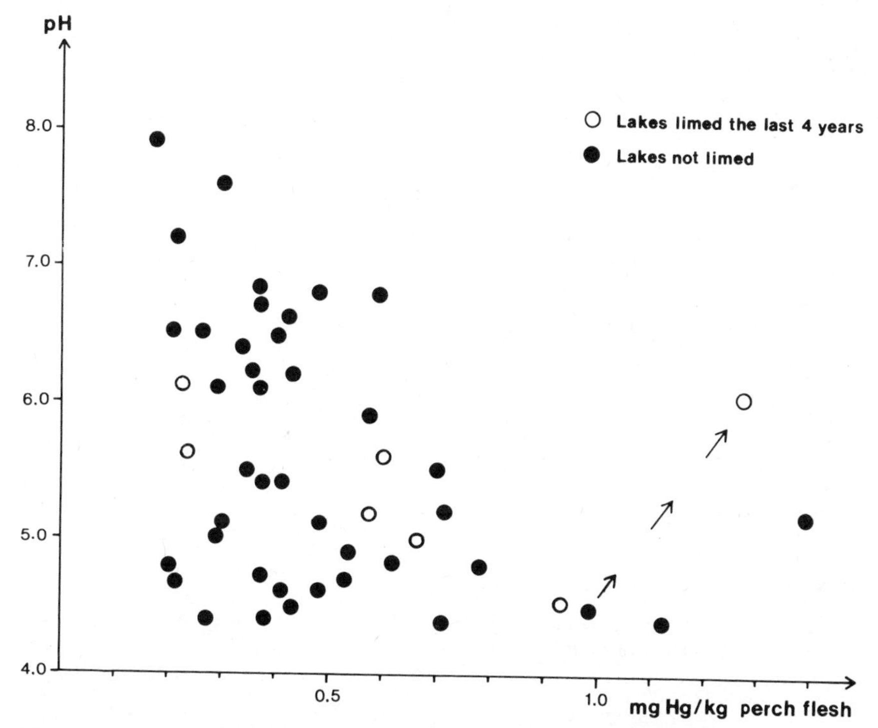

Figure 15. Contents of mercury in flesh of perch (*Perca fluviatilis*) from lakes having different pH values. Each dot represents the mean value of five fish.

flesh, whereas the opposite situation is usually found, especially in pH-normal lakes (Figure 16) (16).

5. CONCLUSION

The increasing sulfur influx into natural waters is usually accompanied by increased leaching of different cations and metals into these aquatic ecosystems. Had there been no addition of cations to the waters, the present loading rate of sulfur in large areas would have resulted in a pH of 3.5 in many lakes and rivers.

6. BIOLOGICAL EFFECTS OF AN INCREASED SULFUR LOAD

Large inputs of sulfate to lake water facilitates the formation of hydrogen sulfide; but when additions are in the form of acid sulfate,

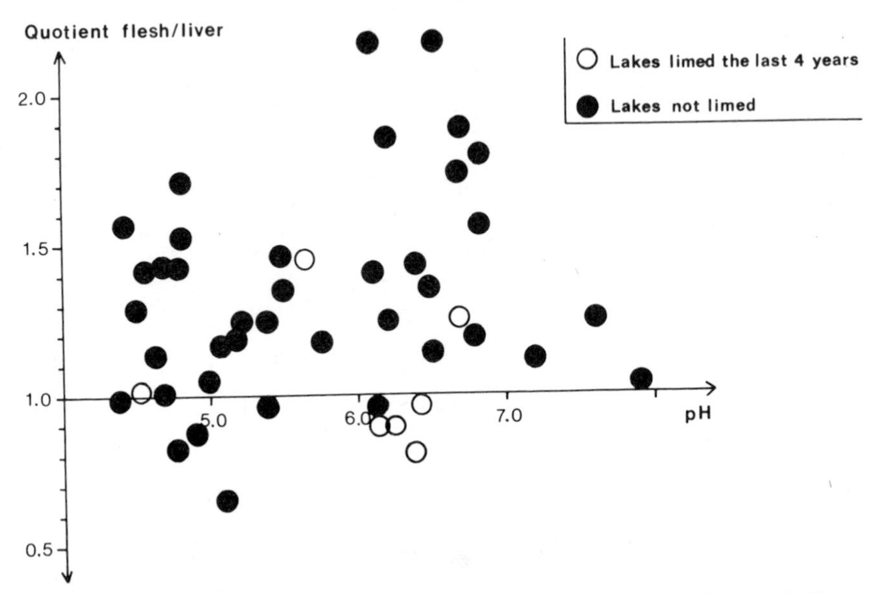

Figure 16. Quotients of amounts of mercury in flesh and in liver for perch (*Perca fluviatilis*) in lakes having different pH values. Each dot represents the mean value of five fish.

these kinds of problems are rare. Instead the ecosystem is affected by the acidity of the water. The effects are strikingly similar in pit waters of extremely high ionic strength and in almost "distilled" water polluted by atmospheric sulfur.

6.1. Primary Producers, Phytoplankton

Generally, at normal pH values of 6 to 8, lakes in the West Coast region of Sweden contain 30 to 80 species of phytoplankton per 100-ml sample in the middle of August. On the other hand, lakes with pH below 5 have about a dozen species (Figure 17), and in some very acid lakes (pH <4) only three species have been noticed. The distribution of algal species and other quantitatively important groups also has a clear connection with pH (Figure 18). Thus the phytoplankton community in acid lakes is of a typical kind. In late summer it consists mainly of peridinians, and in spring of small chrysophyceans. However, there are also acid lakes dominated by the genus *Oocystis* (Chlorophyceae).

The dominant species in acid lakes are *Peridinium inconspicuum* and

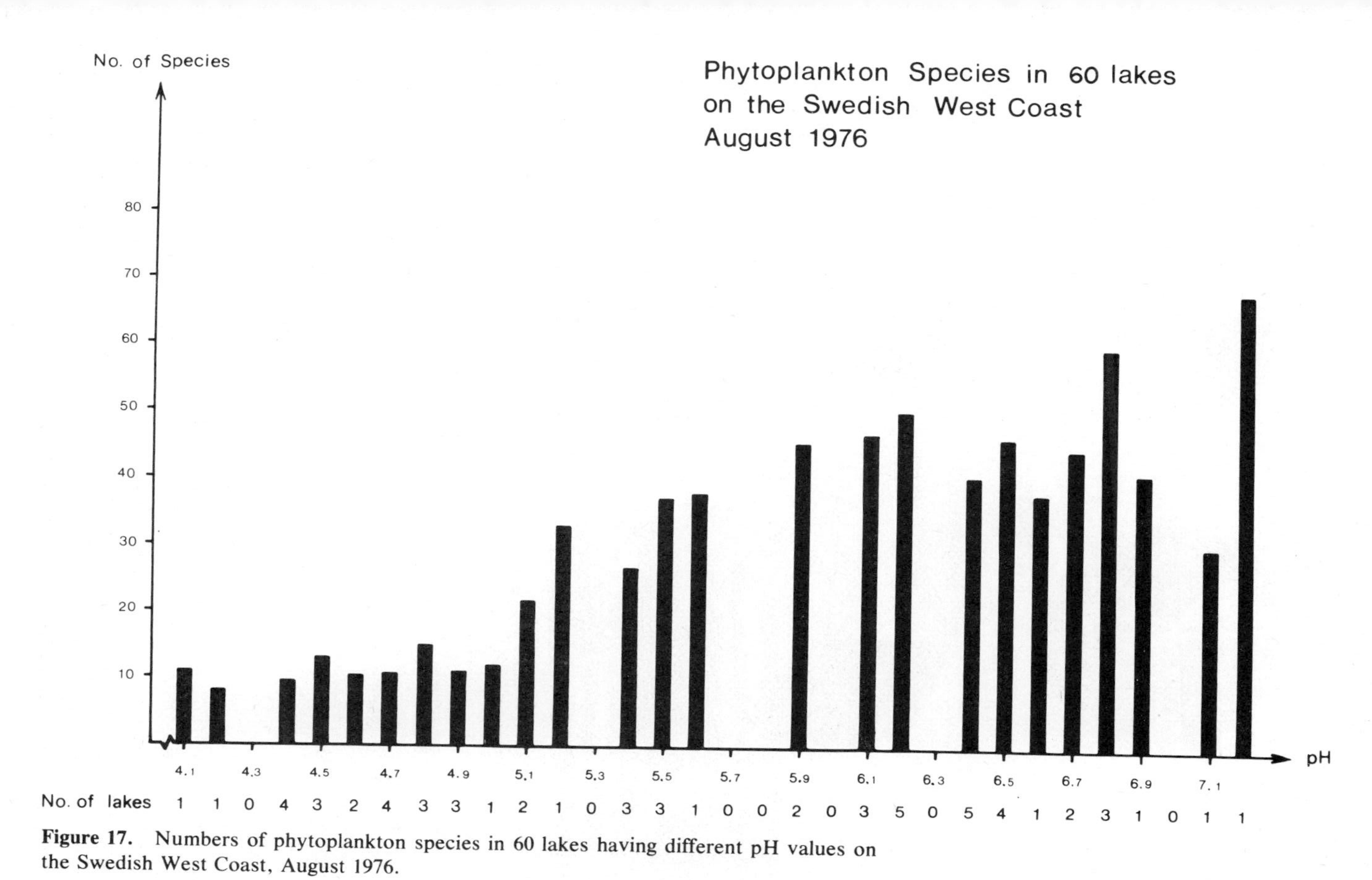

Figure 17. Numbers of phytoplankton species in 60 lakes having different pH values on the Swedish West Coast, August 1976.

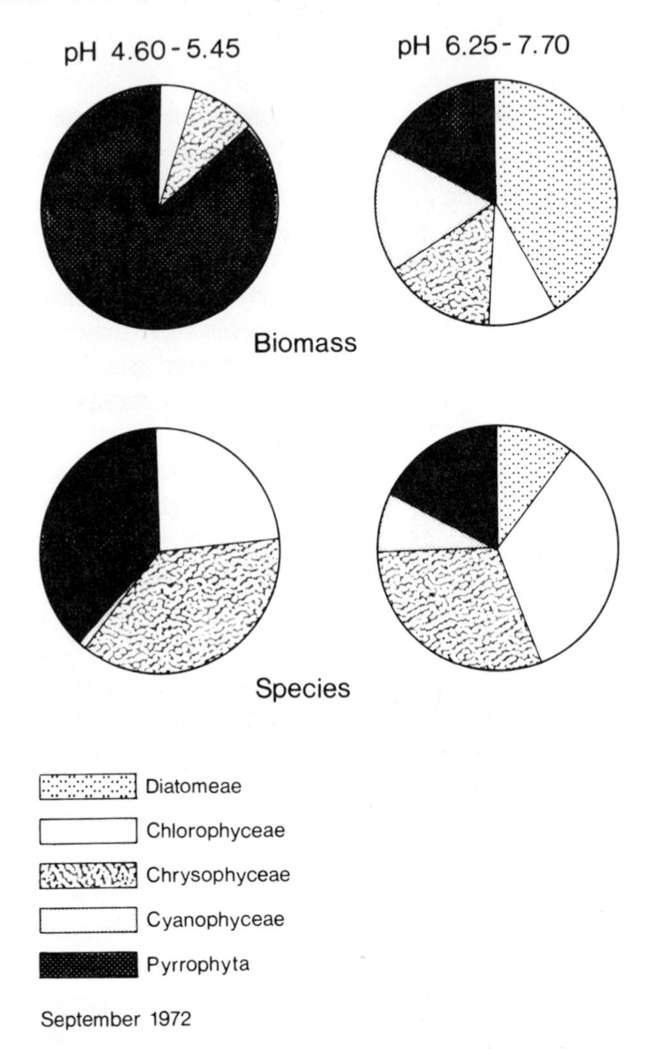

Figure 18. Percentage distribution of phytoplankton species and their biomasses, September 1972.

Gymnodinium cf. *uberrimum* (Peridineae). In particular, the former generally constitutes the bulk of the biomass in the most acidified lakes. Other common species, most of them of much less quantitative importance, are small chrysophyceans and members of the genus *Chlamydomonas* (Chlorophyceae).

The families Chlorophyceae and Chrysophyceae show greatly reduced numbers of species in acidified lakes. The most striking change, however, is the disappearance of diatoms and blue-green algae (5).

Another special feature of many acidic rivers or lakes is the appearance of the filiform alga, *Mougeotia* (Conjugatae). It is frequently observed floating free in enormous masses in the pelagic region, and after death it covers the lake bottoms (5,19,20). The same mass occurrence of *Mougeotia* is observed in drainage water deriving from mining activities (21).

6.2. Diatoms as pH Indicators

Valves from dead diatoms resist decomposition in the sediment and provide a means for studying changes in the aquatic ecosystem.

The sediment from one lake on the Swedish West Coast was analyzed for diatoms. As possible consequences of the acidification, the following changes in the diatom flora were noted in the uppermost sediment of the lake.

The frequency of diatom valves decreases strikingly from the older to the more recent layers of the sediment. Planktonic diatoms show a distinct decrease as compared to the remaining flora of bottom forms and epiphytes. On the other hand, diatoms characteristic of extremely acid milieu and acidobionts (pH optimum less than 5.5) *Eunotia* spp., *Tabellaria binalis,* and *Amphicampa hemicyclus* increase in number when going from the lower to the uppermost parts of the sediment.

The change in the composition of acid species (acidobiontic and acidiphilous) shows a pH lowering from about 6 to 4.5 when expressed as an index: ω = (acid units)/(number of acid species). As this drop has actually occurred (Figure 10), index ω seems to be a good indicator of the acidity of the sedimentation milieu (5,22).

6.3. Algal Biomass and Nutrient Levels

It is easy to find a correlation between biomasses and pH value, indicating that acid lakes have the smallest biomasses (23). This will be the result obtained if the lakes studied have shifting nutrient levels, because pH generally increases with increasing nutrient content.

An investigation made in August 1976 of 58 nutrient-poor lakes in the Swedish West Coast region, however, showed the largest mean biomass in the most acid lakes (pH <4.5) (Figure 19). It was impossible to distinguish differences in phosphorus levels because of the overall low

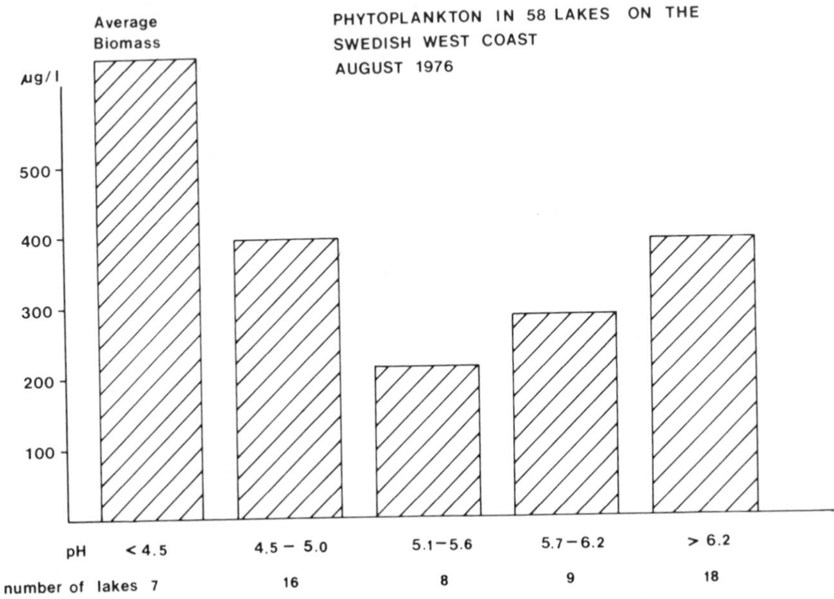

Figure 19. Average biomass of phytoplankton in 58 lakes having different pH values on the Swedish West Coast, August 1976.

concentrations, (<10 μg P/l). As to the possibility of aluminum removing the phosphorus (cf. Figure 12), it is not unlikely that the most acid lakes were actually richer in phosphorus and therefore had the largest biomasses. In view also of the masses of free-floating *Mougeotia,* the algal biomasses could be much higher in acid lakes than in lakes with higher pH.

As the summer plankton of acid lakes normally consist of large forms that do not interfere greatly with the transparency, acid lakes still have the highest transparency, 10 to 20 m.

In addition to the sulfur pollution the deposition of nitrogen and phosphorus is increasing. Thus the atmospheric precipitation often contains 1.5 mg N/l and 20 μg P/l. The concentration of total nitrogen in lakes on the Swedish West Coast is 0.4 to 1 mg N/l before the production season and 0.2 to 0.7 mg N/l during summer. In summertime, acid lakes usually have the lowest nitrogen values, and most of the nitrogen appears as nitrate, in the range 0.1 to 0.3 mg/l. Biological denitrification occurs preferentially as circumneutral pH and is probably disturbed in acid waters. Apparently nitrogen is not a limiting factor for algae in these waters.

6.4. Primary Production of Algae

Production studies using the radioactive carbon technique may be less suitable for very poorly buffered lakes, and it is necessary to employ very dilute ^{14}C-bicarbonate solutions. A study of three lakes of different pH levels serves to illustrate the production in acidified lakes (Figure 20). The acid lake of blue-green color and a transparency of 10–20 m has a more even production throughout the water column, with the maximum at a greater depth, than do other lakes. Because of the thickness of the trophogenic layer, therefore, the production per square meter may be as large as in lakes of higher pH, or larger. The investigation in question gave no indication that the acidification should affect the photosynthesis. What has happened is just that the original phytoplankton community has been replaced by more acid-tolerant species.

6.5. Sediment and Macrophytes

In acid lakes, benthic algae, leaves, and organid débris are frequently found on the bottom, indicating that acidity lowers the rate of decomposition and mineralization. Experimental studies also indicate that this is the case (24). This will be of crucial importance to the nutrient regeneration in the lakes and may also intensify the choking up of the lakes.

Evident is an invasion by peat moss (*Sphagnum* spp.) from land to the bottom of several acid waters. These effects are well known from pit water and sulfate soil lakes (21,25,26). In acid lakes where conditions are suitable, peat moss may cover more than 50% of the bottom above the 4-m depth, and may also grow at much lower depths. The invasion starts just below pH 6 (27). The ion exchange capacity of *Sphagnum,* such as calcium uptake and hydrogen release, is well known (28) and will probably increase the acidification of the lake ecosystem as long as the white mosses are not decomposed.

In most oligotrophic lakes, species of *Lobelia, Isoëtes,* and *Litorella* are common. Gradually they become overgrown by white mosses and covered by dense felts of *Mougeotia.* Even the isoëtides may meet with this competition when invading the bottoms of deep, clear lakes.

The felts of white mosses (benthic algae), leaves, and nondecomposed organic material settle on formerly firm bottoms of the lake, thus diminishing the buffering contact with the sediment. These soft bottoms are colonized by species such as *Juncus bulbosus, J. filiformis, Spar-*

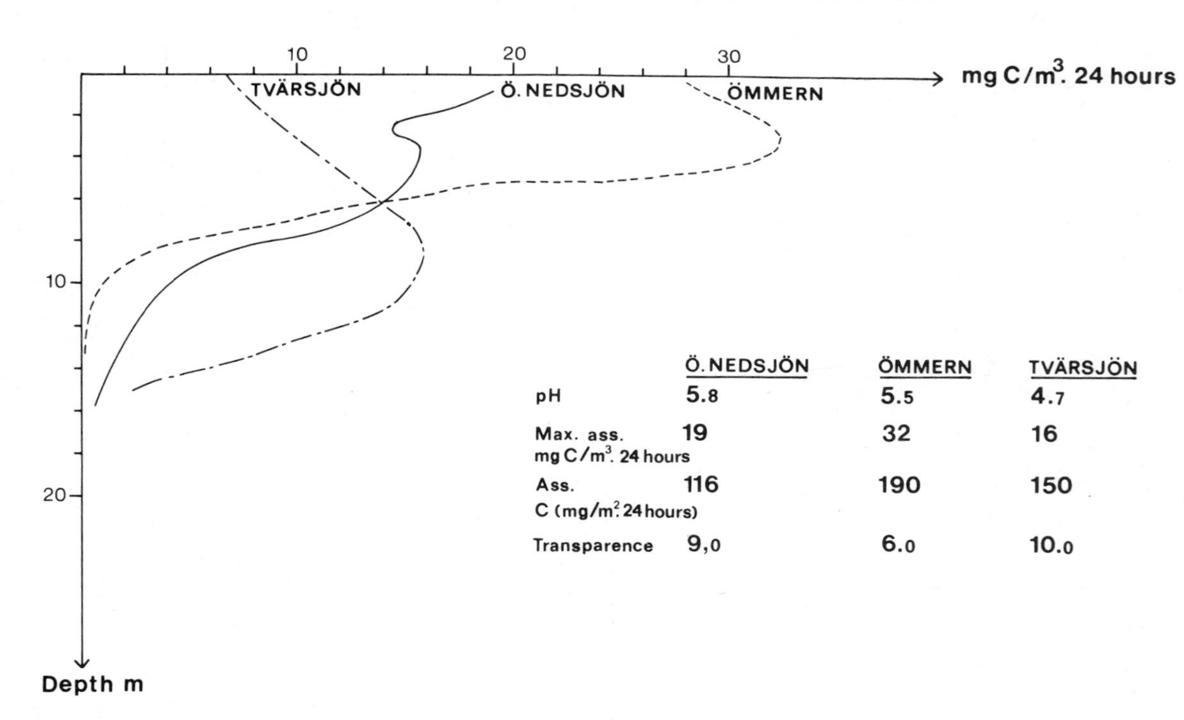

Figure 20. Production studies by ^{14}C technique in three lakes having different pH values on the Swedish West Coast, May 13–15, 1975.

ganium, and *Utricularia. Nuphar* and *Nymphaea* are also well adapted and may occur in dense felts in very acid waters. Thus the production of macrophytes in acid lakes with suitable bottoms may be very large. In view of the large production of benthic algae and possibly also of phytoplankton in acid waters as compared to normal oligotrophic waters, a better characterization of the process would be "acidotrophication," rather than "oligotrophication."

7. ZOOPLANKTON

The number of zooplankton, too, decreases with increasing acidity of the water (5,20,29). Certain species (e.g., of the genera *Bosmina, Cyclops, Diaptomus,* and rotatorians, of the genera *Polyarthra, Keratella,* and *Kellicottia*) have a high tolerance and can be found anywhere in the pH interval 4.4 to 7.9. Others, such as cladocerans of the *Daphnia* genus, are more sensitive; they have only rarely been found at pH <6 (Figure 21).

In highly acidified lakes (pH <5) *Polyarthra remata, Bosmina coregoni,* and *Diaptomus gracilis* often dominate. At pH 5 to 5.5 the

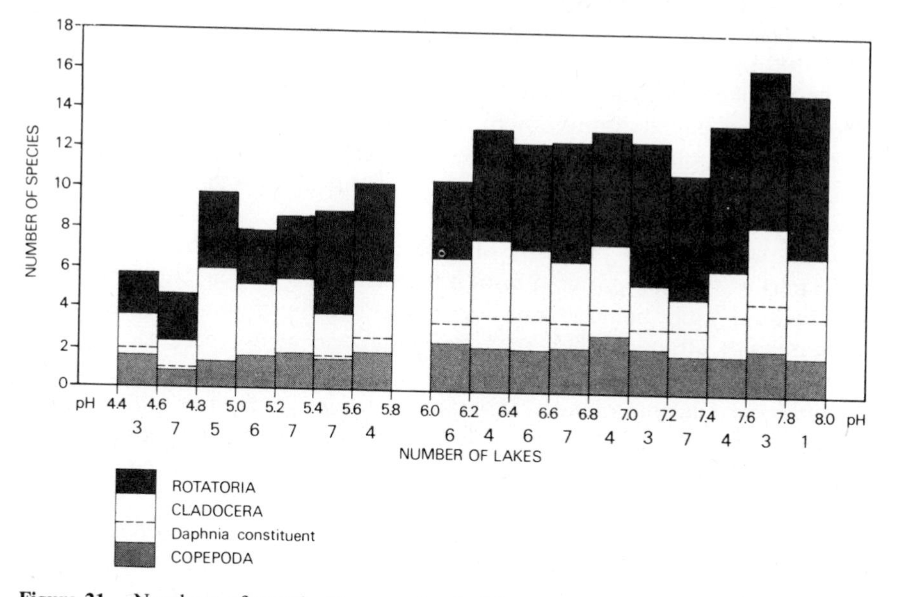

Figure 21. Numbers of zooplankton species in lakes having different pH values on the Swedish West Coast, August 1971.

occurrence of *Holopedium, Keratella,* and *Kellicottia* increases. At pH closer to 6 the species composition is nearly "normal."

Bythotrephes longimanus, a relatively large cladoceran, is difficult to catch by net sampling and is fairly seldom found in plankton samples from acid lakes. In stomachs of fish from acid lakes, however, it is very common (30). At higher values of pH it is less frequently found in the fish diet, probably because predation in these circumstances is more intensive. The same is generally true of the larger plankton forms (31,32).

8. BOTTOM FAUNA

Benthic organisms, too, are impoverished in acid lakes, even if the pH often is higher in the sediments than in the superjacent water. Investigations made in North America, Norway, and Sweden (23,33–35) have shown that the number of species is lower in an acidified lake than in a lake with normal pH. Species with great demands on their environment are replaced by more eurytopic species (34), although the number of individuals may increase. Mass developments of whirligig beetles (Gyrinidae), water boatmen (Corixidae), and damsel-flies (*Zygoptera*) commonly occur in acid waters, probably as a consequence of the lower predation by fish (36).

The change in bottom fauna is caused not only by the drop in pH but also by the changed conditions in the surface layers of the lake bottom, where the periphyton has increased and decomposition decreased, as mentioned above. Thus the living environment of bottom organisms has changed in several respects.

Selective predation probably also affects the benthic fauna. When a great part of the fish population of a lake is wiped out, as happens upon acidification (see below), one or more benthic fauna species may predominate, although they may be subject to heavy predation under normal circumstances (30).

Larvae of Chironomidae, *Trichoptera, Odonata, Hemiptera, Megaloptera,* and sometimes the alderfly *Sialis lutaria,* are very common in acid lakes. The aquatic sow bug (*Asellus aquaticus)* and aquatic earthworms *(Oligochaeta)* are other common organisms.

Many species of Mayflies (*Ephemeroptera*) and stoneflies (*Plecoptera*) are reduced by acidification (35), but others are evident at low pH values, notably *Leptophlebia vespertina* (29,34,35,37) and Siphlonuridae at pH 5.2 (33).

9. OTHER INVERTEBRATES

One of the freshwater shrimps, *Gammarus lacustris,* an important element in the diet of fish in Scandinavian lakes, is not found at a pH below 6.0 (38). Experimental studies have shown that adults of this species have a low tolerance for pH values below 5.5. At pH 5.0, more than 80% died within 2 days (39). Another important prey for the fish is *Lepidurus arcticus*. This euphyllopod, which occurs in the highlands of Scandinavia, has not been found in lakes with pH below 6.1 (39).

Molluscs are very sensitive to acid water. In 832 Norwegian lakes (40) no snails were found at pH values below 5.2; snails were rare in the pH range 5.2 to 5.8 and occurred less frequently at 5.8 to 6.6 than in more neutral or alkaline waters. About the same is true for the mussels, *Lamellibranchiata*.

The crayfish *(Astacus astacus)*, which is regarded as a delicacy in Scandinavia, has been seriously affected by the acidification of lakes. It seems to prefer pH values above 6.5 (41), and the species is rare in lakes where the pH is below 6.0 in summer.

10. FISH

Sörlandet, located in southernmost Norway, is an area with high precipitation and very sensitive bodies of water. During the last few decades, acidification has caused very severe damage to the fish stocks in this region. Within the framework of the Norwegian SNSF (Acid Precipitation—Effects on Forest and Fish) Project great efforts have been made to investigate and map the effects of acidification on fish (42).

Of 2083 lakes investigated (8), 642 had lost their populations of brown trout *(Salmo trutta)*. For instance, in 1950 the brown trout was lacking in 44 of 266 lakes in the Tovdal River catchment area. In 1975 the number of stocks that had been wiped out was as high as 169, and in many lakes the populations were decreasing. At the turn of the century the Tovdal was one of the most important salmon rivers in southern Norway—about five tonnes of Atlantic salmon *(Salmo salar)* and sea trout *(Salmo trutta)* were caught each year. Since 1970, however, the catch has been negligible, and the same gloomy situation exists in regard to several other rivers in southern Norway. Figure 22 compares the yield of salmon in seven acid rivers with the yield in 68 less acid rivers.

In Sweden the effects on fish have been studied mainly in the West Coast region, which is the most exposed area of the country. In some lakes, stocks of the very sensitive roach *(Rutilus rutilus)* were wiped out as early as the 1920s and 1930s.

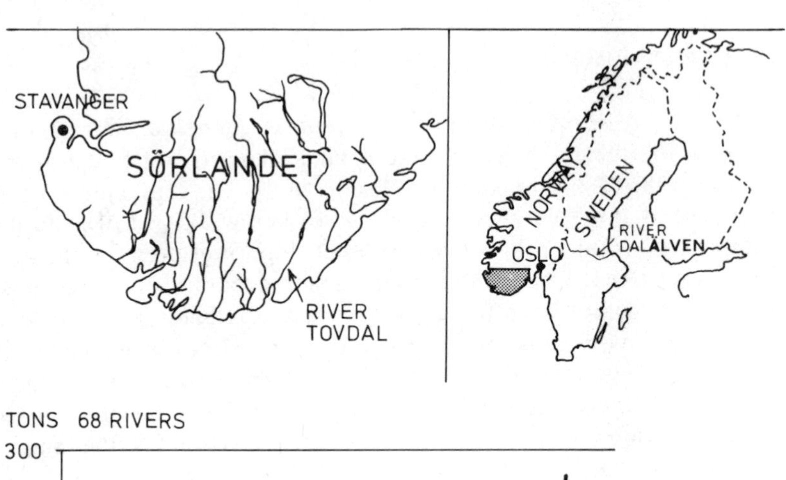

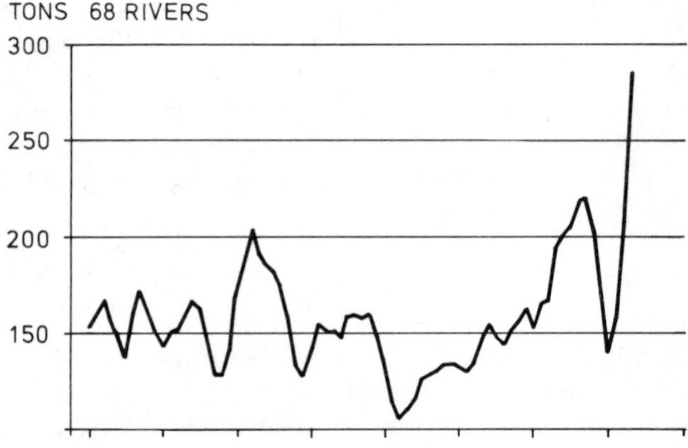

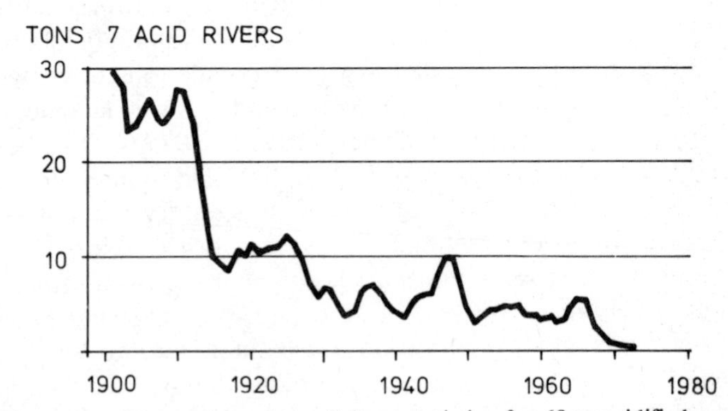

Figure 22. Norwegian salmon fishery statistics for 68 unacidified and 7 acidified rivers (42).

In southern Sweden test fishing and other investigations have been carried out in 100 lakes with pHs in the 4.4 to 7.5 range (Table 2). Biologically and statistically, the two cyprinids, the minnow *(Phoxinus phoxinus)* and the roach, seem to be the most sensitive species. Salmon and sea trout, which live in fresh water only during spawning and at the start of the period of growth, are not quite so sensitive to acid water. The eel *(Anguilla anguilla)* seems to be the least sensitive fish in lakes, probably because it spawns in the sea. In one lake with pH 4.4 some eels were found dead in fyke nets.

The first reports in North America linking lake acidification and the extinction of fish populations to acid precipitation were derived from studies of lakes in the vicinity of Sudbury, Ontario (46). Populations of lake trout *(Salvelinus namaycush)*, lake herring *(Coregonus artedii)*, white suckers *(Catostomus commersoni)*, and other species disappeared rapidly during the 1960s from a group of remote lakes in the La Cloche Mountain region, some 65 km from Sudbury. The rapid lake acidification and fish extinction in this sensitive region were attributed to the spread of acid fallout from the metal smelters in Sudbury, which emit more than 1 million tonnes of sulfur dioxide annually. In a survey of 150 lakes, 33 were classified as "critically acidic" (pH less than 4.5) and, 37 were described as "endangered" (pH 4.5 to 5.5). Subsequent consideration of the pH levels observed to affect fish reproduction in the lakes of this region indicated that even lower acidity levels should be ascribed to these categories for the maintenance of successful reproduction of the most sensitive species.

Intensive studies of acid precipitation effects on fish populations have been conducted in the Adirondack Mountains of New York State, which is one of the major dilute lake districts in the eastern United States (47). A survey of the higher elevation lakes in this region revealed that 51% had pH values less than 5, and 90% of these lakes were devoid of fish life. Comparable data from the period 1929 to 1937 indicated that at that time only 4% of these lakes had pH values below 5 and were devoid of fish. Entire fish communities consisting of brook trout *(Salvelinus fontinalis)*, lake trout, white sucker, brown bullhead *(Ictalurus nebulosus)*, and several cyprinid species were eliminated over a period of 40 years in association with decreased pH.

10.1. Reproduction and Survival

The most sensitive phases in the life of fish occur during the roe and fry stages. Even if the roe hatches in acid water, the fry are often malformed and will die later on.

Table 2. Effects of Acidification in 100 Swedish Lakes (pH Interval 4.3 to 7.5) South of Dalälven River (36,37,43–45)

Family	Species	Number of Previous Stocks	Number of Remaining Stocks	Stocks Exterminated by Acidification No.	Stocks Exterminated by Acidification %	pH Critical to Reproduction
Cyprinidae	Minnow (*Phoxinus phoxinus*)	28	16	12	43	≤5.5
Cyprinidae	Roach (*Rutilus rutilus*)	77	52	25	32	≤5.5
Salmonidae	Arctic char (*Salvelinus alp.*)[a]	36	25	7	19	≤5.2
Salmonidae	Brown trout (*Salmo trutta*)	28	24	4	14	<5.0
Coregonidae	Cisco (*Coregonus albula*)	21	19	2	10	<5.0
Esocidae	Pike (*Esox lucius*)	79	72	7	9	4.4–4.9
Percidae	Perch (*Perca fluviatilis*)	99	95	4	4	4.4–4.9
Anguillidae	Eel (*Anguilla anguilla*)[b]	76	73	2	4	~ 4.5[b]

[a] Four stocks of Arctic char and one of eel have been exterminated through factors other than acidification. In many of the acidified lakes the stocks of different fish species are thinning and are on the way to vanishing.
[b] The eel spawns in the Sargasso Sea. Probably elvers and young eels avoid waters with pH around 4.5.

In Swedish lakes it has been observed that the reproduction of minnow and roach may be disturbed severely at pH 5.5. Field tests showing the sensitivity of roach eggs (Figure 23) were performed in three lakes close to Stockholm. The eggs of perch *(Perca fluviatilis)* that hatched in the most acid lake (Lake Trehörningen) were malformed (45). In Sweden the roach is used as an indicator species that gives information about the pH situation in a particular waterbody. The first sign of change occurs when lakes display mostly big and old roach. When reproduction fails, the number of individuals decreases and food competition slackens. As a result the remaining fish grow bigger than before (Figure 24). This thinning effect has also been observed for other species, such as the Arctic char *(Salvelinus alpinus)*, the brown trout, and the perch.

Table 3 shows the development in three acidified lakes (pH at spawning time in 1976: 4.95, 4.8, and 4.4, respectively) from 1971 to 1976. The stocks have been thinned, and the perch have increased their

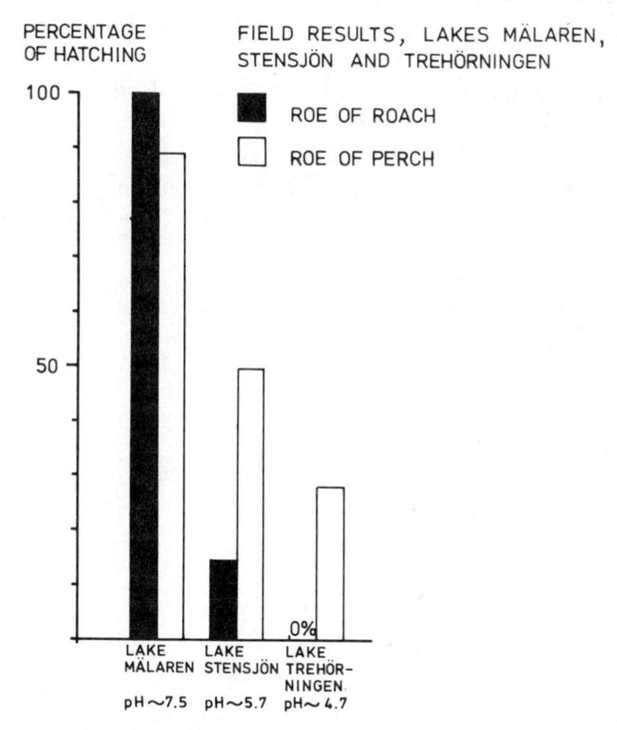

Figure 23. The hatching percentage of roe of roach (*Rutilus rutilus*) and perch (*Perca fluviatils*) in waters of different pH levels (45).

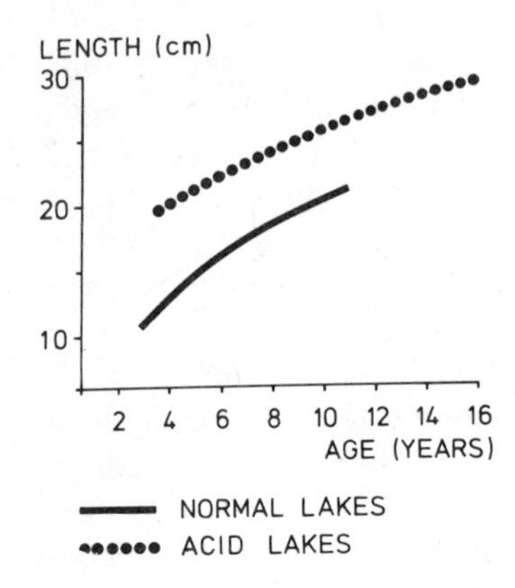

Figure 24. Increased size and growth of roach (*Rutilus rutilus*) in acidified lakes with pH values at spawning time of 4.6 to 5.5, compared to the size in normal lakes of pH 6.3 to 6.8.

weight about three times. Lake Ålevatten and Lake Västersjön have become only slightly acidified, whereas in Lake Rishagerödvatten the pH has decreased by 0.6 unit. Adult minnow and roach are also sensitive to acid water. In laboratory tests, minnows died before pH of 5.0 was reached. Roach tend to avoid water having a pH below 5.6, and salmon parr avoids water in which the pH is below 5.3 (48).

After very heavy rains in Sweden in the autumn of 1974, many stocks

Table 3. Thinning Effects in Stocks of Perch *(Perca fluviatilis)*[a]

Length of Fish (cm)	Ålevatten 1971	Ålevatten 1976	Västersjön 1971	Västersjön 1976	Rishagerödvatten 1971	Rishagerödvatten 1976
<15	10	0	3	0	10	0
15.0–19.9	20	1	37	5	58	0
20.0–24.9	61	1	94	5	3	30
25.0–29.9	5	7	24	12	0	9
30.0–34.9	0	22	1	17	0	0
35.0–39.9	0	3	0	1	0	0
Total number	96	34	159	40	71	39
Mean weight (g)	103	353	122	303	49	167

[a] Fishing surveys in the Swedish West Coast region, 1971 and 1976.

of roach were severely affected by acid water in the spring of 1975. In some lakes (pH ~ 4.5) the fins (mainly the tailfins) of perch are often deformed; this is probably an effect of acid water.

10.2. Body Salt Regulation

The acidified waterbodies of southern Norway are very low in salt. In 1975 a fish kill was reported in the Tovdal River at the time of snowmelt. The river was partly covered by ice, and the bottom was littered with thousands of dead brown trout over a stretch of at least 30 km. All age groups were represented, and it was estimated that several tonnes of trout had been killed. Both an investigation in the field and some tank experiments were made by SNSF to confirm the cause of death. Physiological studies of the fish strongly indicated that failure in body salt regulation was the primary cause of death. Both chloride and sodium in the blood plasma had decreased as a result of acid stress. Therefore it is evident that the concentration of dissolved salts in the water influences acid tolerance. Field surveys show that fish populations tend to disappear even at higher pH in lakes with extremely low ion contents (49).

10.3. Food

In a small acid lake, the pike, *Esox lucius,* was forced to eat insects because its normal prey had been killed through acidification (50). Examination of the food composition of 825 perch from the West Coast region of Sweden (30) showed that fish, which normally comprise the bulk of food for adult perch, was replaced by various invertebrates at decreasing pH (Figure 25). The food of the small perch, however, was nearly the same throughout the whole pH range (4.4 to 7.45). With a moderate decrease in pH, gnats and the water-louse *(Asellus aquaticus)* appear to be of major importance to larger perch. In the most acidified lakes, the largest perch eat mainly *Corixa* bugs. They become exceptionally fat and grow very fast because of sparse fish population and the reduced competition for food.

11. OTHER ANIMALS

There is reason for fear that frogs, toads, and newts, too, will disappear when the water becomes acid.

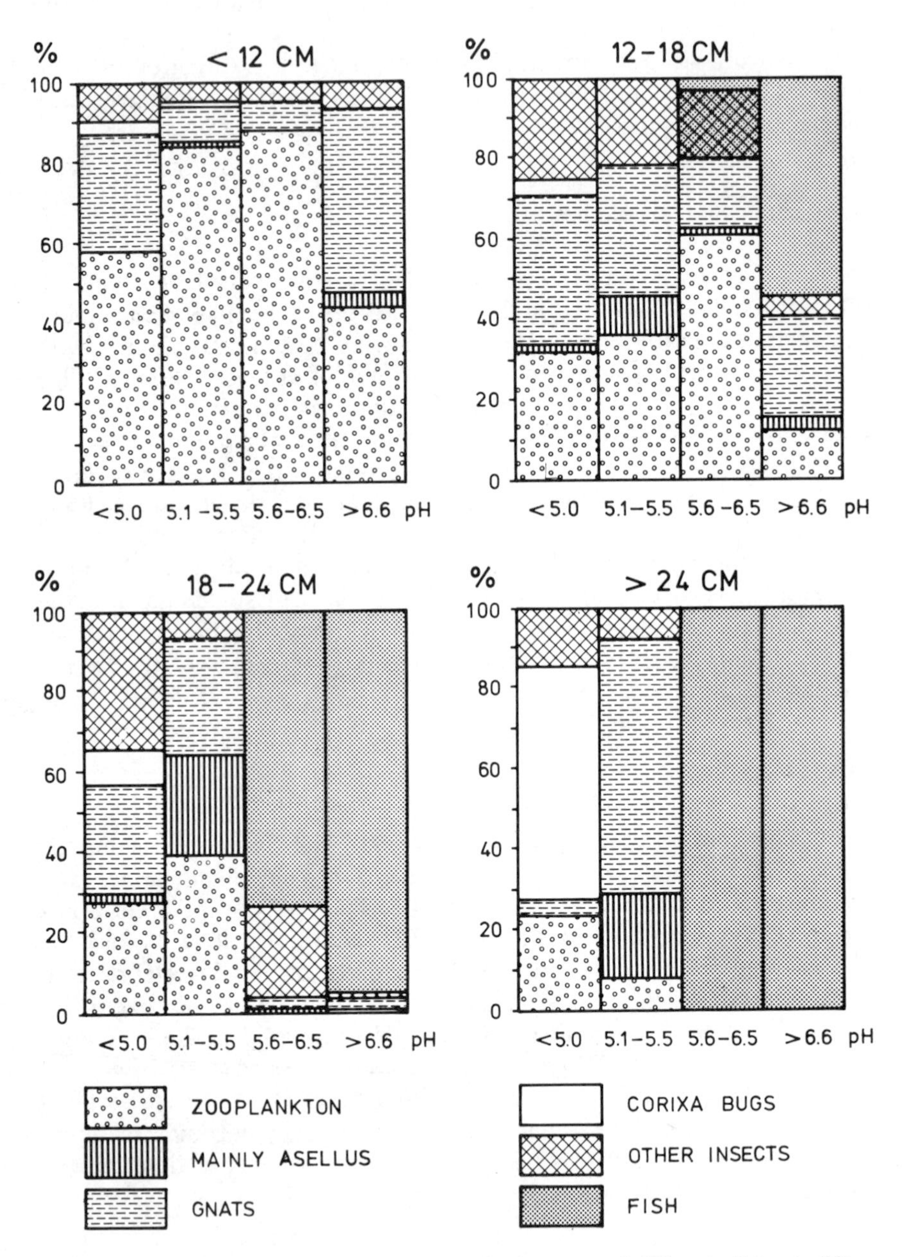

Figure 25. The food of perch (*Perch fluviatilis*) in waters of different pH levels (30).

In the West Coast region of Sweden fish-eating birds such as mergansers and loons have been forced to migrate from several acid lakes with decreasing fish stocks to new lakes with ample food supply. In this way many territories will become vacant, and this will lead to decreasing stocks.

12. CONCLUSION

Atmospheric sulfur deposition affects thousands of poorly buffered waters in Scandinavia and North America. Many of these are situated in virgin forests, in national parks, and in mountain areas. The biological response to acidification through sulfur deposition is very similar to that observed in acid effluents from mining operations and drainage from sulfur soils. Large areas are losing their unpolluted waters and much of the habitat for normal aquatic life.

In the future, large efforts will be made to compensate for this acidification by treatment with lime and other basic substances. This may improve the situation significantly but certainly will not solve the sulfur problem.

REFERENCES

1. von Linné, Carl, *Dalaresa, Iter Dalekarlicum,* 1734. *Natur och Kultur,* Stockholm (1964) (in Swedish).
2. *Water, Air Soil Pollut.* Vols. 6–8 (1976, 1977).
3. *Proceedings of the First Specialty Symposium on Atmosperic Contribution to the Chemistry of Lake Waters.* J. Great Lakes Research, *2*(Suplement 1) (1976).
4. *AMBIO,* **5**(5–6) (1976).
5. Almer, B., Dickson, W., Ekström, C., Hörnström, E., and Miller, U. (1974). *AMBIO* **3,** 30.
6. Svensson, B. H. and Söderlund, R. Eds. (1976). *Nitrogen, Phosphorus and Sulphur—Global Cycles.* Ecol. Bull., Swedish Natural Science Research Council, 22, and Scope, Rept 7, Stockholm.
7. Eriksson, E. (1974). *Forsk. Framsteg* **4,** 41 (in Swedish).
8. Högström, U. (1977). *The Characteristic Scale of Wet Fallout of Sulfur and the Budget of Atmospheric Sulfur for Sweden.* Meteorological Institute of Uppsala, Rept. 48.
9. Jensen, K. W. and Snekvik, E. (1972). *AMBIO* **1,** 223.

10. Fjeld, B. (1976). *Forbruk av Fossilt Brensel i Europa og utsläpp av SO_2 i Perioden 1900–1972*. NILU, Keller, Norway, T.N. 1 (in Norwegian).
11. Wright, R. F. and Henriksen, A. (1977). *Limnol. Oceanogr.* (in press).
12. Dickson, W. (1975). *Rept. Inst. Freshwater Res. Drottningholm* **54,** 8.
13. Hem, J. D. and Roberson, C. E. (1967, 1968, 1969). *Chemistry of Aluminum in Natural Water.* Geological Survey, Water Supply Paper 1827 A-C, U.S. Government Printing Office, Washington, D.C.
14. Hayden, P. L. and Rubin, Alan J. (1974). In A. J. Rubin, Ed., *Aqueous-Environmental Chemistry of Metals*. Ann Arbor.
15. Jones, J. R. (1964). *Fish and River Pollution*. Butterworths, London.
16. Dickson, W. (1977). "Effects of Liming," unpublished manuscript.
17. Everhart, W. H. and Freeman, R. A. (1973). U.S. Environmental Protection Agency, Rept. R-3-73-011b.
18. Grahn, O. and Hultberg, H. (1977). Personal communication.
19. Dannevig, G. (1968). *Zool. Revy* **30,** 53.
20. Hendrey, G. R., Baalsrud, K., Traaen, T. S., Laake, M., and Raddum, G. (1976). *AMBIO* **5,** 224.
21. Harrison, A. D. (1958). *Verh. Int. Ver. Limnol.* **12,** 603.
22. Nygaard, G. (1956). *Folia Limnol. Scand.* **8,** 32.
23. Conroy, N., Hawley, K., Keller, W., and Lafrance, C. (1976). *Proceedings of the First Specialty Symposium on Atmospheric Contribution to the Chemistry of Lake Waters.* J. Great Lakes Research *2*(Supplement 1) p. 146.
24. Laake, M. (1976). *Effekter av lav pH på Produksjon, Nedbrytning og Stoffkretsløp i Littoralsonen.* SNSF, Project IR 29, 1432 Aas, Norway (in Norwegian, summary in English).
25. Vallin, S. (1953). *Rept. Inst. Freshwater Res. Drottningholm* **34,** 167.
26. Ohle, W. (1936). *Arch. Hydrobiol.* **30,** 604.
27. Hultberg, H. and Grahn, O. (1976). *Proceedings of the First Specialty Symposium on Atmospheric Contributions to the Chemistry of Lake Waters.* J. Great Lakes Research *2*(Supplement 1) p. 208.
28. Anschütz, J. and Gessner, F. (1954). *Flora* **141,** 178.
29. Sprules, W. C. (1975). *Ver. Int. Ver. Limnol.* **19,** 635.
30. Andersson, B. (1972). "Food Selectivity of Perch *(Perca fluviatilis* L.) in Acidified Lakes on the West Coast of Sweden," *Inform. Inst. Freshwater Res. Drottningholm* **17,** (In Swedish, summary in English).
31. Stenson, J. (1972). *Rept. Inst. Freshwater Res. Drottningholm* **52,** 132.
32. Nilsson, N-A. and Pejler, B. (1973). *Rept. Inst. Freshwater Res.* Drottningholm **53,** 51.
33. Hansson, M. (1976). "The Biology of an Acid Mountain Lake as Illustrated by the Food of Arctic Char," *Inform. Inst. Freshwater Res. Drottningholm* **5** (in Swedish, summary in English).

34. Mossberg, P. and Nyberg, P. (1976). "Effects of Acidification on Bottom Fauna and Fish in Lake Västra Skälsjön," *Inform. Inst. Freshwater Res. Drottningholm* **9** (in Swedish, summary in English).

35. Borgström, R., Brittain, J., and Lillehammer, A. (1976). *Invertebrates and Acid Water; Survey of Sampling Sites.* SNSF, Project IR 21, 1432 Aas, Norway (in Norwegian, summary in English).

36. Grahn, O., Hultberg, H., and Landner, L. (1974). *AMBIO* **3,** 93.

37. Dickson, W., Hörnström, E., Ekström, C., and Almer, B. (1975). "Char Lakes South of River Dalälven," *Inform. Inst. Freshwater Res. Drottningholm* **7** (in Swedish, summary in English).

38. Ökland, K. A. (1969). *Norw. J. Zool.* **17,** 111.

39. Borgström, R. and Hendrey, G. R. (1976). *pH Tolerance of the First Larval Stages of* Lepidurus Arcticus *(Pallas) and Adult* Gammarus Lacustris *G.O. Sars.* SNSF, Project IR 22/76, 1432 Aas, Norway.

40. Ökland, J. (1969). *Malacologia* **9,** 143.

41. Svärdson, G. (1974). "Review of the Laboratory's Activity and the Projects of the Year 1974," *Inform. Inst. Freshwater Res. Drottningholm* **1,** (in Swedish, summary in English).

42. Braekke, F. H. Ed. (1976). *Impact of Acid Precipitation on Forest and Freshwater Ecosystems in Norway.* SNSF-Projekt, Research Rept. 6/76, 1432 Aas, Norway.

43. Almer, B. (1972). "The Effect of Acidification on Fish Stocks in Lakes on the West Coast of Sweden," *Inform. Inst. Freshwater Res. Drottningholm* **12** (in Swedish, summary in English).

44. Lessmark, O. (1976). "The Effect of Acidification on the Fish Fauna of Some Lakes in the Province of Småland," *Inform. Inst. Freshwater Res. Drottningholm* **7** (in Swedish, summary in English).

45. Milbrink, G. (1975). *Fauna Flora* **2,** 41 (in Swedish, summary in English).

46. Beamish, R. J., Lockhart, W. L., Van Loon, J. C., and Harvey, H. H. (1975). *AMBIO* **4,** 98.

47. Schofield, C. L. (1976). "Acid Precipitation: Effects on Fish, *AMBIO* **5,** 228.

48. European Inland Fisheries Advisory Commission (EIFAC) (1969). "Water Quality Criteria for European Freshwater Fish—Extreme pH Values and Inland Fisheries," *Water Research* **3,** 593.

49. Leivestad, H. and Muniz, I. P. (1976). *Nature* **259,** 391.

50. Hultberg, H. and Stenson, J. (1970). *Flora Fauna* **1,** 11 (in Swedish, summary in English).

8

THE ACID MINE DRAINAGE

Paul Barton

Chemical Engineering Department, The Pennsylvania State University, University Park, Pennsylvania

1. SOURCES, MECHANISMS, AND EFFECTS OF POLLUTION

1.1. Sources

Chemical pollution resulting from mining activities is caused by exposing minerals to oxidation and leaching, producing drainage waters that often contain undesirable concentrations of dissolved minerals. Sulfur-acid pollution can occur in mining organic materials (coal), metal ores (uranium, silver, zinc, lead, and copper), pyritic sulfur, and geothermal fluids. Acid drainage is also produced in mineral processing and converting facilities (coal, copper, zinc). Other industrial activities that produce acid waste water include the pickling of steel, metal etching and plating, and petroleum refining (acid sludge). Unlike industrial waste pollution, mine drainage can be an indefinitely continuing source of pollution long after completion of mining.

Acid mine drainage is produced whenever water containing oxygen comes into contact with sulfur present as sulfides (particularly pyrites) in the ore, country rock, or mine tailings. Some of the more important metal sulfides are pyrite, FeS_2; pyrrohotite, $Fe_{11}S_{12}$; chalcopyrite, $CuFeS_2$; galena, PbS; sphalerite, ZnS; arsenopyrite, $FeS_2 \cdot FeAs$; and pentlandite, $(Fe, Ni)_9S_8$. The acid mine drainage associated with coal mining is caused by the iron disulfide associated with the enclosing rock strata and that present in the coal seam itself.

Analyses of selected wastewater effluents from several mining activities involved with the production of coal, zinc, copper, and uranium are given in Table 1. Note the variability in acidity (expressed as ppm $CaCO_3$) for a relatively small change in pH, because of the presence of hydrolyzable heavy-metal salts. For neutralization treatment, acidity is an important parameter in determining titrant requirements. In coal mine drainage, ferrous and ferric iron and aluminum are important pollutants. In heavy-metal production, other metal ions such as arsenic, cadmium, copper, lead, manganese, and zinc add to the above species. The sulfate levels are orders of magnitude greater than the 250 mg/l drinking water quality recommendation of the U.S. Environmental Protection Agency (5).

The land area affected by mining activity in the United States exceeds 13 million acres. Of this total, 7 million have been undercut, 3 million disturbed by surface mining, and 3 million given over to mining-related waste accumulations (6). It has been estimated that 10,000 miles of streams and 29,000 surface acres of impoundments and reservoirs are seriously affected by mine drainage. About 40% of this drainage comes

Table 1. Characteristics of Selected Acid Drainage Waters

Constituent	Reference and Source of Effluent			
	1 Coal Mine Drainage, Tarrs, Pennsylvania pH 2.7 (ppm)	2 Zinc Mine Drainage, Kellog, Idaho pH 2.2 (ppm)	3 Uranium Tailings Pile, Elliot Lake, Ontario pH 2.0 (ppm)	4 Copper Smelting (Acid Plant Scrubber) pH1.8 (mg/l)
Acidity as $CaCO_3$	752	—	14,600	—
Aluminum	55	347	588	—
Arsenic	—	—	—	8.2
Cadmium	—	22.5	0.05	0.09
Calcium	194	31.6	416	—
Copper	—	13.4	3.6	0.12
Total iron	122	16,250	3,200	0.1
Ferrous iron	96	—	1,750	—
Lead	—	0.8	0.67	0.91
Magnesium	83	1,500	106	—
Manganese	4	2,625	5.6	—
Nickel	—	4.8	3.2	<0.001
Silicon	41	—	—	—
Sodium	—	0.5	920	—
Uranium	—	—	7.2	—
Zinc	—	14,560	11.4	13.7
Cyanide	—	—	—	<0.005
Nitrate	—	77.5	—	—
Sulfate	1474	63,000	7,440	490

from active mines; the remainder is from abandoned surface mines (25%) and shaft and drift mines (35%) (7).

About 75% of the coal mine drainage in the United States occurs in the Appalachian region. The pH, sulfuric acid loadings, and sulfate concentrations in rivers in this region, as measured by the U.S. Geological Survey (8), are reported in Figure 1. The free mineral acid loads exceed 4000 tons (of H_2SO_4) per day and are highest in the West Branch Susquehanna River, Monongahela River, and Kiskiminetas River basins, where the streams have pH values less than 4.5. Sulfate concentrations in excess of 50 ppm are found throughout the coal mining areas of the Appalachian region, the median concentration being 160 ppm in Pennsylvania and Ohio.

The mixture of alkaline streams with mine drainage waters eventually neutralizes all acid streams in Appalachia (Figure 1). Bicarbonate concentrations are less than 50 ppm in the polluted streams and in the 50- to 290-ppm range in the downstream areas.

1.2. Mechanisms

The qualitative mechanism for acid water formation when water and air come into contact with the iron disulfide associated with a coal seam is generally accepted to be as follows:

$$2FeS_2 + 2H_2O + 7O_2 \rightarrow 2FeSO_4 + 2H_2SO_4 \tag{1}$$

$$4FeSO_4 + 2H_2SO_4 + O_2 \rightarrow 2Fe_2(SO_4)_3 + 2H_2O \tag{2}$$

$$Fe_2(SO_4)_3 + 6H_2O \rightarrow 2Fe(OH)_3 + 3H_2SO_4 \tag{3}$$

$$FeS_2 + 14Fe^{3+} + 8H_2O \rightarrow 15Fe^{2+} + 2SO_4^{2-} + 16H^+ \tag{4}$$

In equation 1 the sulfur in the iron disulfide is oxidized to the sulfate ion by oxygen; in equation 2 ferrous iron is oxidized to ferric iron; in equation 3 ferric iron is hydrolzyed to ferric hydroxide; in equation 4 ferric iron oxidizes the pyrite. There is no typical acid mine water, which is a more complex solution than these equations indicate. The ferrous to ferric iron ratio varies, and pH is not a true indication of total acidity. Additional ions such as silica, aluminum, manganese, calcium, and magnesium are present in significant quantities.

While coal-associated iron disulfide usually is isolated from oxygen and water in its natural environment, deep-mining the coal seam removes support from the overlying strata and induces caving, while strip mining physically disrupts the strata. Influx of water and air, together with exposure of acid-producing materials, provides the necessary conditions for the above reactions to occur. Caved areas and backfilled areas provide for easier movement of water, and overflow or through-flow becomes the major contributor of acid mine water. Cessation of mining activity may aggravate the problem, since the water is no longer pumped or drained regularly and exposure time for the acid-producing reactions is increased.

McKay and Halpern (9) report that the rate-determining step in the oxidation of pyrite is a heterogeneous process on the pyrite surface that involves only an oxygen molecule. This is supported by the observation that the rate is first order in O_2, and independent of the solution composition. They suggest that, from a common intermediate formed in the reaction between FeS_2 and O_2, competing reactions for the formation of S^0 and H_2SO_4 occur and that the latter predominates. This results in the path shown in equation 1. The reaction expressed by equation 2 determines the ratio of Fe^{3+} to Fe^{2+} in solution. McKay and Halpern claim that this reaction has no effect on the oxidation of pyrite, but their experiments were performed at elevated temperatures (100 to 130°C).

In contrast to the above, Singer and Stumm (10) state that the rate-limiting step in the oxidation of iron pyrite and the formation of acidity in streams associated with coal and copper mines is the oxidation of ferrous iron (equation 2). Their experiments were made at 25°C. They claim that it is irrelevant whether pyrite or marcasite, the orthorhombic polymorph of FeS_2, is considered as the sulfide source. In nonbacterial solution they found that the partial pressure of oxygen had no effect on the rate of reaction, but that ferric iron was the prime oxidant (equation 4). The ferrous ions generated are then reoxidized (equation 2) to propogate the pyrite oxidation cycle. These authors also found that oxidation of ferrous iron to the ferric state is not influenced by pH below 4.5, but above this pH value the oxidation rate is second order with respect to the acid concentration.

Hanna et al. (11) express the opinion that the kinetics of the sulfide to sulfate reaction can be greatly influenced by the mineralogy of the system, in contrast to the results obtained with crystalline samples of pyrite and marcasite. Other minerals present may act as catalysts or supply the galvanic couple. It is implied from laboratory observations that the reaction rate is directly related to the surface area of a solid reactant and that subsequent or intermediate reactions, which probably occur as liquid-phase ionic reactions, are relatively rapid (12,13). Evidence for the latter point is the minor quantities of intermediate reactants present in the reaction system.

A number of variables, in addition to exposed surface area, are reported to influence the overall reaction rate (11). These include water, oxygen partial pressure, pH, ferric ion concentration, catalytic agents such as bacteria or specific ions, and inhibitors such as bactericides or alkaline materials. Intermediate reactions may shift product concentrations of ferric and hydrogen ions to alter the overall kinetics. It is conceivable that capillary diffusion is one of the rate-controlling mechanisms.

The iron-oxidizing bacteria of the genus *Thiobacillus* are widely distributed in acid waters associated with deposits of metal sulfides and sulfide-bearing coals. These are indigenous to environments containing metal sulfides, oxygen (either atmospheric or in aqueous solution), and acid conditions. A mechanism by which *Thiobacillus ferrooxidans* can accelerate the oxidation is as follows. The ferrous to ferric oxidation reaction in equation 2 proceeds slowly below pH 4.5 in the absence of a catalyst, but rapidly in the presence of iron-oxidizing bacteria. *Thiobacillus ferrooxidans* has been shown to accelerate this reaction by a factor of more than 10^6 at pH values less than 3.5 (14). In the pH range 3.5 to 4.5, filamentous iron bacteria of the genus *Metallogenium* significantly catalyze iron oxidation (15). The ferric ions generated then oxidize pyrite as in equation 4. The ferrous ions generated by the reaction in equation 4 are then reoxidized (equation 2), and the pyrite degradation cycle is propagated.

Thiobacillus ferrooxidans can also oxidize sulfur and sulfur-containing compounds to sulfuric acid. In the biogeochemical sulfur cycle in nature, the sulfur-oxidizing bacteria have the important oxidative role of converting the reduced sulfur compounds (H_2S and S^0) to oxidized sulfur compounds (SO_3^{2-}, $S_2O_3^{2-}$, SO_4^{2-}). Anaerobic sulfate-reducing bacteria complete the cycle by reducing the oxidized sulfur compounds back to the reduced state.

In a pilot-scale study (16) made at the McDaniels Test Mine in Ohio with the specific objectives of defining better the mechanism and kinetics of acid water production, the following conclusions were reached. The major sites of pyrite oxidation lie above the groundwater table, in regions where gas rather than liquid is the continuous phase. Oxygen transport to the pyrite oxidation sites occurs by gas-phase convection and diffusion. There was no evidence of significant bacterial catalysis of pyrite oxidation. Although autotrophic iron-oxidizing bacteria are active in the drainage after it has left the pyrite oxidation sites and has been diluted with groundwater, bacteria are not present in sufficient numbers at the oxidation sites to affect significantly the rate of pyrite oxidation. Since pyrite oxidation sites are located above the water table, the transport of oxidation products away from the pyrite is a slow process, limited by molecular diffusion in water films, seepage by wetting, percolation, and flushing due to changes in water table elevations.

It becomes apparent that a consensus agreement has not been reached on the quantitative nature of acid mine water production. If the reaction mechanism and kinetics of the conversion of pyrites to sulfuric acid could be clearly defined, the pertinent variables would be known, and characterization of a mine site with regard to potential acid formation and suggestions for abatement could be more effectively accomplished.

1.3. Effects

Upon emerging from the ground, the iron hydroxide (hydrated ferric oxides) entrained in the acid mine water settles into the stream beds. The dissolved ferric hydroxide precipitates as the pH is raised by neutralization with stream bed minerals or dilution with more basic waters. The ferrous iron present reacts with atmospheric oxygen and hydrolyzes according to the reaction

$$4Fe^{2+} + O_2 + 10H_2O \rightarrow 4Fe(OH)_3 + 8H^+ \quad (5)$$

to produce ferric hydroxide and even more acidity (200 ppm of Fe^{2+} will lower a neutral pH to 2). Ferrous iron is thus seen to raise neutralization chemical demand as well as chemical oxygen demand (COD). In the absence of acid, ferrous sulfate is oxidized and hydrolyzed to form a basic ferrous sulfate precipitate according to the reaction

$$4FeSO_4 + O_2 + 2H_2O \rightarrow 4Fe(OH)SO_4 \quad (6)$$

The colloidal ferric hydroxide and basic ferric sulfate result in the characteristic red or brown sludge deposits in acid mine drainage stream beds and reservoirs. These deposits may contain significant aluminum hydroxide precipitate.

Sulfuric acid in mine drainage can be neutralized or reduced by passage through calcium- and magnesium-containing formations, or by reaction with mica, feldspar, or clay in which the acid is affected by potassium, aluminum, and silicon oxides in complex association (6). Potassium oxides will neutralize some acid. So will aluminum oxide, but aluminum sulfate hydolyzes to restore some acid.

The devastation wrought to plant life by acid mine wash contamination of soils is exemplifed in Figure 2. High hydrogen ion concentrations favor the increased solubility of elements such as Fe, Al, Mn, Zn, Pb, and As that can be toxic to plants and animals. Exchangeable aluminum constitutes most of the total exchangeable acidity in soils. Blevens et al. (17) found pH values of 3.1 to 4.2 and aluminum concentrations of 3.2 to 5.0 meq/100 g soil in surface soil samples from the drainage area shown in Figure 3. These exceed the acidity (pH 5.5) and aluminum content of 1.0 meq/100 g soil reported to inhibit root growth. Soluble salts were present in quantities sufficient to limit the growth of salt-sensitive plants, but not enough to be considered harmful to most agronomic species. The contaminated soils required approximately 18,000 kg/ha of agricultural lime ($CaCO_3$ equivalent of 100%) to raise the pH from 3.9 to 6.5. This neutralized the acidity due to Al^{3+}. With the addition of both lime and

Figure 2. Area of Clear Creek Floodplain, Hopkins County, Kentucky, contaminated by acid mine wash. (Reprinted by permission from R. L. Blevins, H. H. Bailey, and G. E. Ballard, *Soil Sci.* **110,** 191–196. © 1970 by the Williams & Wilkins Co., Baltimore, Maryland 21202.)

phosphorus to samples of the contaminated soils in greenhouse studies, high corn yields were attained. Thus the reclamation of acid-mine-wash-contaminated soils can be accomplished.

The iron and aluminum hydroxides in acid mine drainage smother biological life natural to waterways. Even if the drainage is neutralized, the stream may not provide optimal conditions for life support. For example, Updegraff and Sykora (18) have shown in a laboratory study that coho salmon avoid lime-neutralized iron hydroxide suspensions in the range of 4.25 to 6.45 mg Fe/l. Their experiment simulated an aquatic environment with neutral pH and a high oxygen content, similar to shallow, fast-flowing streams where aeration is maximized. Under these conditions, ferric hydroxide floc formation would probably eliminate food sources and would cause fish to avoid this stretch of water. A worse fate is encountered in slightly acidic waters with low oxygen content as they often contain large quantities of dissolved Fe^{2+}. Fish present in this environment would be exposed to the insidious deposition of iron on their gills and on the perivitelline membrane covering their eggs.

Sykor et al. (19) measured the effects of ferric hydroxide, as found in neutralized acid mine water, on aquatic life. Iron concentrations ranged from 50 to 1.5 mg/l. The highest concentration of suspended iron that does not seem to affect the survival and growth of fathead minnows appears to be below 12 mg Fe/l. Hatchability was reduced at 1.5 mg Fe/l, probably because of small particles of suspended iron clogging the egg pores. Data on the growth of juvenile brook trout kept for a full generation in different concentrations of iron revealed a definite trend toward slower growth with increasing concentration of suspended ferric hydroxide, with larger trout in 6 mg Fe/l and in the control. The favorability of the 6-mg-Fe/l situation over the control may be due to the importance of iron as an inorganic nutrient supplied to younger, growing fish without the excessive turbidity occurring in high concentrations, which impairs visibility and feeding. Tests conducted with invertebrates have revealed low susceptibility of caddis-fly larvae to suspended ferric hydroxide up to 20 mg Fe/l. With freshwater shrimp the safe concentration for reproduction and growth seems to be less than 3 mg Fe/l.

2. POLLUTION ABATEMENT

The antipollution techniques that can be applied to acid mine drainage are watershed management, disposal methods, and treatment methods. An abatement program is essential for active mines and is required by law in the United States. Wide-scale abatement of acid pollution caused by inactive mines would present water collection problems of enormous magnitude, even though the neutralization treatment per se of the collected waters could be in the realm of current economics and technology. For comparison, the drainage from active and abandoned mines amounts to over 4 million tons acidity/year in the United States, or about 13% of the commercial sulfuric acid production.

2.1. Watershed Management

Watershed management has some potential for abating acid mine pollution from surface mines and tailing piles, but has limited application to deep mines.

The techniques applicable to surface mines include interception and diversion to reduce water contact with acid-forming materials, improved overburden segregation and backfilling for water infiltration control, and low-wall mineral barriers to retain water. Long-wall strip mining is considered (20) a possible means of reducing water pollution. In this technique, coal is cut by machine and transported to the outcrop by a conveyor belt. The mine roof is held up by hydraulic jacks that progress forward with the cutting equipment, allowing the roof to collapse behind the miner with minimum disturbance of the overburden.

Covering a refuse pile (21) with a few feet of soil and planting a grass cover reduced the production of acid mine water by 91% to 16 lb acid as $CaCO_3$/acre·day, at a cost of about $10,000/acre.

Water management techniques applicable to deep mining include grouting off or intercepting groundwater to minimize inflow to the mine, arranging the workings so that the inflow can be removed from the mine as quickly as possible, partial extraction of coal, roof fracture control and backfilling of waste materials to avoid caving, placement of mine openings for either complete inundation or zero discharge on completion of mining, and planned flooding and sealing of the mine.

Mine sealing and inundation of underground mine workings is employed to reduce the oxidation of pyritic materials. Seals have also been used to prevent the entrance of air or water to the underground mine. Seals can slow the exchange of air within the mine as the barometric pressure fluctuates. Sealing involves construction of a physical barrier in the mine opening and the maintenance of mineral (coal) barriers within the mine. The barriers must be designed to withstand the maximum expected pressure head of water that will be exerted against them. Sealing underground mines is somewhat analogous to creating a surface water impoundment; a major portion of the dam structure is already in place, and the seal merely closes the opening. The perimeter of the mine forms most of the impoundment and often is not able to withstand any significant pressure. Underground mine seals have seldom been 100% successful because of leakage rates and weak points (22).

In the Moraine State Park project (23) in Pennsylvania, underground mine hydraulic sealing and grouting work resulted in an overall reduction in discharge flow rates from 146 to 57 gal/min, an overall decrease in net acidity from 501 to 160 lb/day, and an overall increase in iron from 34 to 42 lb/day. The cost of the 65 seals needed and the associated grouting ranged from $8300 to $58,400 per seal, for an average of $19,500.

Permeable aggregate underground mine sealing involves the use of ungrouted alkaline aggregate material that will neutralize acid water passing through it. Precipitates form, and the ability to neutralize decreases with time. The goal is to have the precipitate eventually plug the barrier and seal the mine. Penrose and Holubec (24) found in laboratory tests that limestone, $\frac{3}{8}$-in.-to-dust size, at high placment densities was the most satisfactory natural material tested. Limestone plugs will perform best on ferric mine waters and poorest on ferrous mine waters. Significant stone volume losses can occur when limestone plugs are exposed to acid mine water flow because of settling of the stone upon being wetted, erosion, and consumption of the stone. This situation generally causes seal failure at the top of the shaft opening (22). The addition of hydrated lime or powdered limestone to the water behind a rubble barrier to form a gelatinous precipitate provided only temporary sealing of the outflow from an abandoned mine (25).

2.2. Disposal Methods

Disposal methods for unneutralized acid mine drainage involve injection wells, drainage fields, and lagooning with dilution-flow control.

The rate at which a disposal well can accept fluid depends upon the fluid properties, the rock properties, the size of the well, and the pressure gradient in the system. The necessary injection pressure plus the pressure of the fluid column in the well must balance the well bottom formation pressure and friction loss in the injection tube. The formation pressure can be estimated by multiplying the well depth by the geological pressure gradient. The generally accepted pressure gradient is 1 psi pressure/ft depth. To increase the flow rate for a given injection pressure, the formation properties may be altered. Several methods of well stimulation are practiced in the petroleum industry to achieve a more efficient well, such as hydraulic fracturing, nitroshooting, well perforating, and acidizing. Hydraulic fracturing is the most popular of these methods.

The main concern in injecting wastes into a porous formation is the possibility of plugging the disposal zone. Of particular concern is the presence of suspended solids, bacteria, dissolved gases, and concentrations of iron.

The bituminous coal fields of central and western Pennsylvania lie in a region of well-stratified sedimentary rock. The principal rock types are sandstones, limestones, shales, siltstones, and claystones, all varying in areal extent and thickness. Shales, siltstones, and claystones are very impermeable rocks and are generally ruled out as well disposal reservoirs. Sandstones often have high porosity and permeability values. The silica content is comparatively nonreactive with acid mine water. Limestones generally exhibit lower permeability and porosity than sandstones and are often partially dissolved by groundwater, forming a series of subterranean channels. The fluid contact of the acid water with the reactive limestone is minimal.

Stefanko et al. (26) tested the injection of fresh water into a 130-ft-thick "salt sand" formation covered by about 1400 ft of overburden, in southwestern Pennsylvania. The tests were disappointing in that the pumping pressures required were high and the amount of sandstone penetrated was small, even with hydrofracturing. This points out the need for reliable geologic data and extensive core drilling information regarding any proposed injection site.

The use of drainage fields for the disposal of acid mine water can have an adverse effect on the biological life associated with the soil. Growth of plants is generally retarded or restricted in soil media of pH less than 4.0. Exchangeable aluminum is usually the toxic constituent in such acid media. However, germination and growth may be limited by the hydrogen ion concentration per se (27).

The pH of soils contaminated by acid mine wash can be raised to levels compatible with crop growth by the addition of lime (17). However, this will not lower the concentrations of sodium or soluble salts that can result in additional long-term soil degradation. Barley was found to be adaptable to lime-neutralized sludge, presumably because of its tolerance to high levels of salts (27). Grass cover can be established directly on acid refuse piles by application of sufficient agricultural limestone to raise the pH of the material to above 5.0, followed by proper addition of fertilizer (28).

Growth of trees is generally inversely proportional to the level of acidity and to the concentration of aluminum. Growth, especially root growth, is poor and mortality is high with a pH of 2.5 or an aluminum concentration of 240 ppm (29). Birches were found to be somewhat resistant to aluminum, and pines to acid. The symptoms displayed by plants growing under high aluminum concentrations are similar to those found under low phosphorus conditions, indicating that aluminum may cause a phosphorus deficiency in plants.

Lagooning with dilution-flow control involves the purposeful discharge of polluted water into a receiving stream during periods of high flow. Lagoons are used for the settling of tailings from mineral processing facilities, as well as for storage of polluted mine water. Negative features include inadequate dilution, erosion of fines, air pollution caused by blowing fines, and revegetation problems. Groundwater pollution due to seepage does occur, as in the case where a strip mine in eastern Ohio was used for the disposal of acid pickling liquors and serious water pollution became evident in surrounding areas (30).

Laws to regulate abatement and treatment actions for wastewaters from the mining, milling, and metallurgical industries are mandated if we are to reverse the pollution damage already inflicted on our environment. In the United States, the passage by Congress of Public Law 92-500, The Federal Water Pollution Control Act Amendments of 1972, established a comprehensive set of new requirements that are directed ultimately at a goal of zero discharge of water pollutants by 1985. Although attainment of this goal seems unlikely, the law gives the Environmental Protection Agency the authority to ascertain that an all-out effort is made. Guidelines have been presented for effluent quality limitations, based on the application of control technology now considered "practicable" and "currently available," as well as minimum acceptable interim effluent levels for the mining and milling industries. The effluent limitations shown in Table 2 for the mining and milling industries are expressed only in concentrations. Because of the wide variations in ores processed, the EPA has concluded that a pound-per-day limitation is impractical, though this policy may change for the milling industry. The amendments stipulate that the intent of Congress is to control the quality of point discharges or groups of point discharges. The definition of "point" becomes important inasmuch as most wastewater impoundments lose water by seepage. The law may be interpreted to include quality control of groundwater as well as of surface water.

Table 2. Wastewater Effluent Limitations Recommended by the U. S. Environmental Protection Agency

Constituent	Minimum Interim pH 6.0 to 8.5 (mg/l)	"Practicable" pH 6.0 to 8.5 (mg/l)
Iron (dissolved)	7.0	0.5
Manganese	1.0	0.1
Cyanide	0.03	0.02
Suspended solids	30	20
Sulfate		
Heavy metals and toxic materials	Determined regionally	

3. TREATMENT OF POLLUTED WATERS

3.1. Methods

The many variable factors associated with acid mine drainage often render the planning of a treatment facility especially complex. Instability occurs in regard to changing concentration resulting from surface water seepage, variations in rainfall, caving conditions within mines, and the intersection of long-abandoned deep mines with later strip operations. In addition, the concentrations of pollutants are constantly changing because of the oxidation of ferrous iron by atmospheric oxygen to form various insoluble iron compounds that coat the bottom of any receiving stream. Gathering the waters to deliver them to a common destination may possibly require extensive pipe lining and pumping costs. Decisions must establish whether small treatment plants or a centralized treatment plant after water collection should be employed. The overall problem may be complicated by whether the treatment is to be carried out by an individual company, a governmental agency, or an interagency authority.

Treatment processes fall into two main areas: (1) all the soluble constituents are removed from the water by rejection as a high-salt-content brine, and (2) pollutants are removed from the system as an insoluble precipitate. The first area is generally characterized by high-quality water products but yields waste brines which probably will have to be neutralized before disposal. The second area is characterized by water products high in soluble salts, though usually of an innocuous nature. Both areas are characterized by the need to dispose of significant quantities of waste products that are difficult to dewater and dry.

The unit operations that have received considerable attention for

Table 3. Mine Drainage Treatment Methods

Neutralization
Reverse osmosis
Reverse osmosis-neutralization combination
Ion exchange
Flash distillation
Freezing
Electrodialysis
Solvent extraction

treating acid mine drainage are listed in Table 3. Some of these techniques (namely, flash distillation, freezing, electrodialysis, solvent extraction, and reverse osmosis) have been extensively developed for the production of desalted water from seawater and inland brines, and extension of the available technologies to the treatment of acid mine water was investigated by various agencies in the United States.

Multistage flash distillation is technically capable of producing very pure water along with a highly concentrated brine. The acid water can be neutralized either before or after distillation, in order to provide a neutral, concentrated waste for disposal. Prior neutralization provides more solids precipitation problems in the evaporators than does neutralization afterwards. Materials of construction are a problem when distilling unneutralized acid, exotic materials such as titanium being required for good corrosion resistance. The advantage of the combination prior neutralization-distillation process over neutralization itself is the potential for producing pure water and, possibly, solidified wastes.

The Commonwealth of Pennsylvania was in the process of contracting for a $14 million distillation plant for acid mine drainage. The plant, which was to be built by Westinghouse Electric Company, was expected to produce 5 million gal/day of purified water containing less than 10 ppm dissolved solids. Rising costs, however, dictated a canceling of the plant. In view of the recent inflation in the cost of energy, it is doubtful that distillation can ever become an attractive means of treating a significant fraction of acid mine drainage. An efficient multistage flash distillation requires 124 Btu/lb water produced (31). At an energy cost of $2.00/million Btu, the energy cost alone would amount to $2.18/1000 gal. This is a high continuing cost to pay for a pollution treatment operation. Consider, for example, that some deep mine owners in the anthracite region of Pennsylvania pumped as much as 14,000 gal of water for every ton of coal mined. The Blue Coal Corporation pumped 50 million gal of water out of its deep mines each day.

Schroeder and Marchello (32) compared the costs of various saline water conversion processes applied to acid mine drainage. The freezing and hydrate processes cost more than multistage flash distillation. Difficulty in separating the pure water (ice crystals) from the acid brine makes the technical feasibility of freezing processes questionable. Just 1% entrainment of pH 3 brine into neutral water will lower the pH of the product water to 4.

The cost of the electrodialysis process for treating acid mine water was estimated to be about the same as that for flash distillation (32). Electrodialysis has been rejected by the EPA as a technically feasible treatment method, however, because of poisoning of the membranes through iron precipitation (33).

Powell (34) has investigated the extraction of purified water from acid mine brine by hydrocarbons. The hydrocarbon solvent, *n*-heptane, was chosen on the basis of the solubility of water in the hydrocarbon phase at high temperature and the selectivity of the solvent for water over salt (35), as well as the resistance of the hydrocarbon to reaction or emulsification with the acid mine water. In this process the liquid solvent and acid water are mixed and heated; water is extracted and then vaporized into the hydrocarbon layer; the hydrocarbon layer is decanted from the brine layer; the phases are cooled separately by heat exchange with the feed; and purified water precipitates from the hydrocarbon phase and is decanted. At a temperature (514°F) just 2.5°F above the critical point of *n*-heptane, the water content in the *n*-heptane phase was 43 mol %. The ratio of the distribution factor for water to that for acid* was 330. The extracted water had 0.5 ppm or less of iron, aluminum, or calcium, a pH of 4.35, 12 ppm acidity, and negligible dissolved solids, from a simulated acid mine water containing 100 ppm Fe, 50 ppm Al, 50 ppm Ca, 730 ppm sulfate, and 780 ppm acidity. Because of the high temperature needed, an exotic metal such as Hastelloy C is required, and the vessel wall thicknesses need to be high to contain the vapor pressure exerted (1350 psig). Consequently, plant costs are expected to be noncompetitive with the costs of other separational processes.

It has been proposed (36) that liquid extraction be used to extract the iron and aluminum ions from acid mine water before neutralization. A solvent with very low solubility in water and with the ability to form complexes with large metal ions is needed; an example is alkylphosphoric acid with long hydrocarbon chains. This process offers several advantages. (1) Liquid extraction is noted as a low energy consumer when compared to distillation. (2) Inexpensive crushed limestone can be effectively used for the neutralization step, providing a more compact sludge than that resulting from other bases. (3) The unwanted ions in the water are removed, not simply exchanged as in some ion exchange processes. (4) The process is continuous; it does not require the use of cyclic operation through packed beds as with solid ion exchange resins. (5) It has the potential for metals recovery by precipitation from solution via reduction with hydrogen.

* $$\frac{K_{H_2O}}{K_{H_2SO_4}} = \frac{\text{wt } H_2O \text{ in hydrocarbon phase/wt } H_2O \text{ in brine phase}}{\text{wt } H_2SO_4 \text{ in hydrocarbon phase/wt } H_2SO_4 \text{ in brine phase}}$$

3.2. Ion Exchange

Ion exchange with solid resins can be used to remove both the anions and the cations from mine drainage waters. Potable water was produced at Burgettstown, Pennsylvania, in a 500,000-gal/day plant costing $800,-000 (37). This facility purified high-sulfate water, containing over 1500 ppm of dissolved solids, by passing it successively through a strongly acidic cation exchange resin and a strongly basic anion exchange resin and then filtering and chlorinating. The effluent contained 150 ppm dissolved solids. The ion exchange system was supplied by The Dow Chemical Company. The cation resin was regenerated with sulfuric acid, and the anion one with a lime slurry. By-product cake from the filtration step, consisting mainly of gypsum, was used as excavation fill. Two major problems with strongly ionized resins are the regeneration costs and the consumption of strong acid and strong alkali.

One method that utilizes a weak electrolyte ion exchange resin is the modified Desal Process of the Rohm and Haas Company, tested in an 800,000-gal/day plant near Hawk Run, Pennsylvania (38). In this process coal mine drainage is passed through AMBERLITE IRA-93 resin in the bicarbonate form, and the sulfate ions in the mine drainage are exchanged for bicarbonates. In addition to removing the sulfates, this step also converts the metal sulfates into soluble metal bicarbonates. In the next stage, aeration releases carbon dioxide which raises the pH, and oxidizes the ferrous ions to the ferric state. At the increased pH, iron, aluminum, and manganese precipitate as insoluble hydrous oxides. Some calcium and magnesium carbonates also precipitate. Lime treatment precipitates more calcium and magnesium by converting the bicarbonates into less soluble carbonates. The resin is regenerated by passing a dilute solution of ammonium hydroxide through the resin bed. This converts the anion exchange resin back to the free base form. Introduction of carbon dioxide at 50 psig with the acid mine water then converts the resin back to the bicarbonate form.

The exchange reaction is as follows:

$$\{2\{R—N(H_3O)^+ + HCO_3^-\}\} + M^{2+} + SO_4^{2-} \rightarrow \{\{R—NH_3O)^+\}_2 + SO_4^{2-}\} + M(HCO_3)_2 \quad (7)$$

where M represents the cations found in the mine drainage, and the species in the brackets are in the resin phase. Feiler (39) was able to model the kinetics of this exchange reaction using a two-film model, yielding mass transfer coefficients in both the resin and the surrounding fluid. The regeneration reaction proceeds as follows:

$$\{\{R—N(H_3O)^+\}_2 + SO_4^{2-}\}\} + 2NH_4^+ + 2OH^- \rightarrow \{2\{R—N(H_3O)^+ + OH^-\}\} + (NH_4)_2SO_4 \quad (8)$$

The amount of ammonia used is 120% of the stoichiometric quantity. The reaction that converts the resin from the free base form to the bicarbonate form is

$$\{R—N(H_3O)^+ + OH^-\} + H_3O^+ + HCO_3^- \rightarrow \{R—N(H_3O)^+ + HCO_3^-\} + 2H_2O \quad (9)$$

The bicarbonate becomes associated with the resin only when accompanied by acidity, which neutralizes the hydroxide ion. These resins have a low base strength, and, being weakly ionized, bicarbonate can be taken up only if acidity is present for neutralization. Resin utilization increases as the amount of carbon dioxide increases, up to a plateau at about twice the stoichiometric amount of carbon dioxide (39).

Two pairs of 11-ft-diameter, 6.5-ft-deep ion exchange beds were employed. Upflow operation was used to prevent precipitates from depositing in the beds. The flow rate, based on empty tower cross-sectional area, was 2.5 gal/min·ft². The beds were operated for 18 hr between regenerations.

A flow diagram of the process is shown in Figure 3. Regenerant and rinse waters are treated with calcium hydroxide, filtered, and distilled to

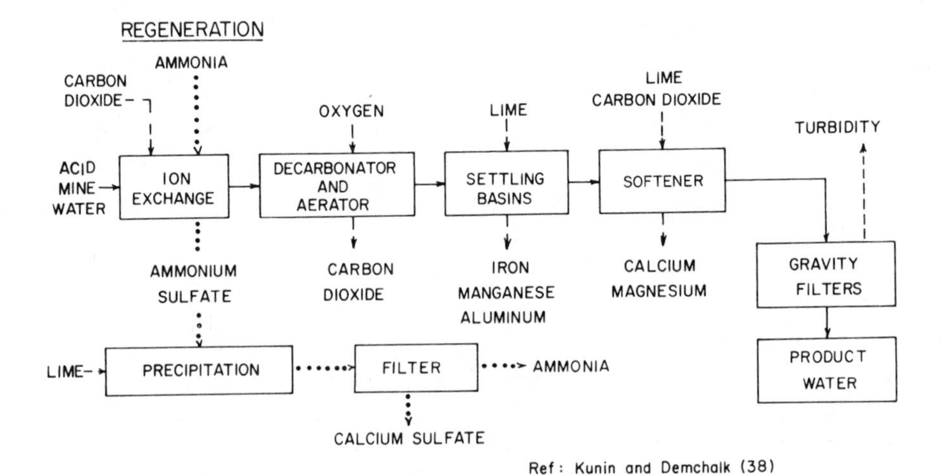

Figure 3. Flow diagram of ion exchange process for treating acid mine drainage.

recover ammonia. Total loss of ammonia from the system is 5%/cycle. Solid plant wastes are iron oxide, calcium carbonate, and a calcium sulfate. Liquid wastes include lagoon overflow and ammonia stripper still bottoms. Typical feed and product water analyses are presented in Table 4. The pH of the product water is decreased from 9.5 to 7.8–8.0 and the hardness to 70 ppm by lime softening. Subsequent addition of carbon dioxide (recarbonation) neutralizes any excess lime or alkalinity and converts carbonates to soluble bicarbonates. The product water meets the 250-ppm-sulfate and 0.3-ppm-iron specifications of the EPA (5). Operating costs, excluding amortization, were 50 cents/1000 gal.

3.3. Reverse Osmosis

Osmosis occurs when two solutions of different concentrations in the same solvent are separated from one another by a membrane. If the membrane is more permeable to the solvent than to the solute, solvent flow occurs from the more dilute to the more concentrated solution. This solvent flow continues until the concentrations are equalized or the pressure on the more concentrated side of the membrane rises to the osmotic level. If a pressure in excess of the osmotic pressure is applied to the more concentrated solution, the solvent can be caused to flow into the more dilute solution. This reverse osmosis process can be used to separate water from acid and salts in acid mine water without a change in phase. Acid mine water flows over a membrane at high pressure; partly desalted water flows through the membrane; a concentrated brine is the other exiting stream.

Table 4. Typical Water Analyses for Ion Exchange Acid Mine Drainage Treatment Plant (38)

Item	Acid Mine Water	Product Water
pH	3.7	9.5
Acidity (ppm $CaCO_3$)	362	—
Iron (ppm)	101	0.2
Calcium (ppm $CaCO_3$)	295	85
Magnesium (ppm $CaCO_3$)	100	99
Total hardness (ppm $CaCO_3$)	395	184
Sulfate (ppm)	648	192
Total dissolved solids (ppm)	1084	284

The rate of flow of water through the membrane can be expressed by

$$J = A(\Delta P - \Delta\pi) \tag{10}$$

where J = product water flux (l/sec·m²)
A = membrane constant (1/sec·m²)/(kN/m²)
ΔP = differential pressure across the membrane (kN/m²)
$\Delta\pi$ = osmotic pressure (kN/m²)

The rate at which salt leaks through the membrane can be expressed by

$$J_s = B(C) \tag{11}$$

where J_s = salt flux (g/cm²·sec)
B = salt flux coefficient (cm/sec)
ΔC = difference in salt concentration between the brine and the product water (g/cm³)

The osmotic pressure of a representative acid mine water (40) with a conductivity of 1040 μmho/cm is 76 kN/m² (kilonewtons per square meter) (19 psi). Typical values for operating pressure drops are in the range of 2800 to 4100 kN/m² (400 to 600 psi). Salt rejection measures the ability of a membrane to reject specific ions, and is expressed as the percent of salt in the feed that does not exit with the desalted water product. Values of 92 to 99.5% are common (41). Loss in flux occurs with operating time in reverse osmosis membranes, even with high-purity water, because of compaction of the membrane and/or its porous support structure. In systems free of fouling, this flux loss follows a linear log-log plot in respect to time. Fouling becomes apparent by a downward curvature on this plot.

Reverse osmosis systems were tested by the EPA (41) on acid mine drainage discharges at four different locations (Norton, West Virginia; Morgantown, West Virginia; Ebensburg, Pennsylvania; and Mocanaqua, Pennsylvania) whose water quality characteristics were quite varied. The three basic types of modules employed were tubular, spiral wound, and hollow fiber. The combination neutralization-reverse osmosis process was also tested. The thrust of the test program was toward increasing recoveries in order to produce the smallest volume of waste brine.

The tubular reverse osmosis system used in the tests consisted of sixty 7.6 cm × 2.39 m plastic tubular modules. Each module contained eighteen 1.27-cm-i.d. porous fiberglass tubes that were lined with cellulose acetate membrane and connected in series. The membrane area was 1.57 m²/module for a total system membrane area of 94.4 m².

The spiral-wound reverse osmosis module, shown in Figure 4, con-

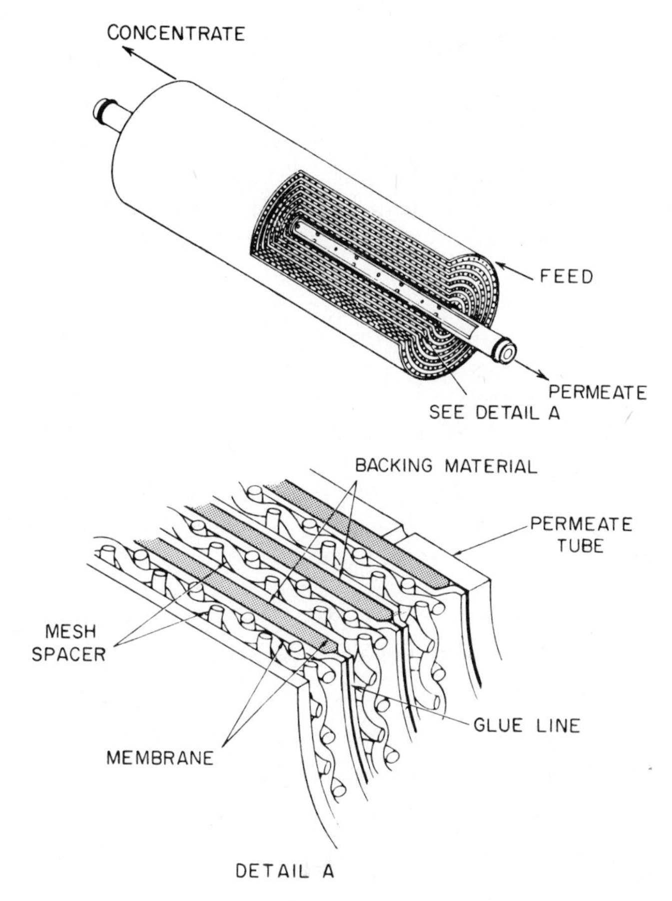

Figure 4. Spiral-wound reverse osmosis module. (This material originally appeared in *Chem. Eng. Progress,* **64,** No. 12, 1968. It appears here courtesy of the copyright owners, the American Institute of Chemical Engineers.)

sisted of two sheet membranes laminated to a sheet of porous, flexible material rolled upon a tubular manifold concentrically with a plastic sheet mesh that provides a spacer between successive turns. The three-layer membrane laminate was connected to the central manifold and inserted into a pressure vessel. Pressurized feed is forced to flow axially along the cylinder between successive membrane layers; permeate flows spirally through the core layer of the laminate and out the central tube. The brine flows out the end of the pressure vessel opposite the feed pipe. The 10.2 cm × 3.05 m vessel contained three modules in series, each with 4.65 m^2 (50 ft^2) of modified cellulose acetate membrane.

The hollow-fiber reverse osmosis module, shown in Figure 5, resembles a shell-and-tube heat exchanger in construction. Approximately 140 m^2 (1500 ft^2) of modified nylon hollow-fiber membrane was packed in each 15.2 cm × 1.22 m stainless steel pressure vessel. The ends of the tube bundle were fixed into a plug of resin that serves as a header. Pressurized liquid is supplied to the shell side of the bundle through a porous central tube; desalted water permeates through the walls of the hollow-fiber tubes and exits through the ends of the tube bundle; brine concentrate is removed at the shell wall.

A flow diagram (41) of a reverse osmosis plant for treating acid mine drainage is shown in Figure 6. Included is the neutralization step used to convert acid brine waste to a less voluminous and neutral sludge. The feed water must be filtered to remove particles as small as 10 μm and thus avoid plugging the membranes. Several reverse osmosis units are placed in series to attain higher brine concentrations in the effluent. To keep recovery at or beyond 70% and yet maintain sufficient brine flow velocity and turbulence in each module to prevent boundary layer precipitation, a portion of the brine is recycled into the feed.

The selectivities of the three reverse osmosis modules in treating acid mine water (41) are compared in Table 5. The ion rejection performance

Table 5. Selectivity of Reverse Osmosis Modules in Treating Acid Mine Water at 75% Recovery (41)

		Amount (mg/l)[a]				
Sample Designation	pH	Acidity	Ca	Fe	Mn	SO_4
Tubular module						
Feed	3.4	250	125	78	14	660
Brine	3.1	560	330	230	39	1650
Product	4.2	46	2.2	0.9	0.31	4.4
Rejection (%)		81.6	98.2	98.8	97.8	99.3
Spiral-wound module						
Feed	3.4	240	130	77	(17)	750
Brine	2.9	810	490	330	(43)	2300
Product	4.4	38	0.4	0.4	(0.5)	0.9
Rejection (%)		91.7	99.8	99.8	(98.8)	99.9
Hollow-fiber module						
Feed	3.4	(210)	110	65	14	740
Brine	2.9	(720)	420	260	53	2700
Product	4.5	(32)	0.55	0.54	0.08	2.2
Rejection (%)		84.8	99.5	99.2	99.4	99.7

[a] Values in parentheses are estimated.

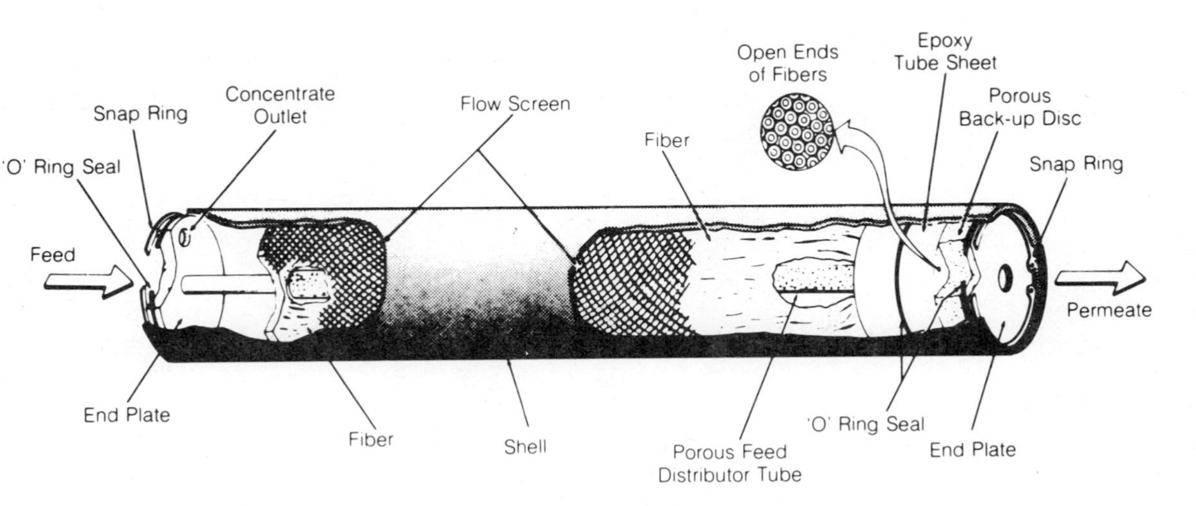

Figure 5. Hollow-fiber reverse osmosis module. (Copyright © E. I. duPont de Nemours & Co. 1976.)

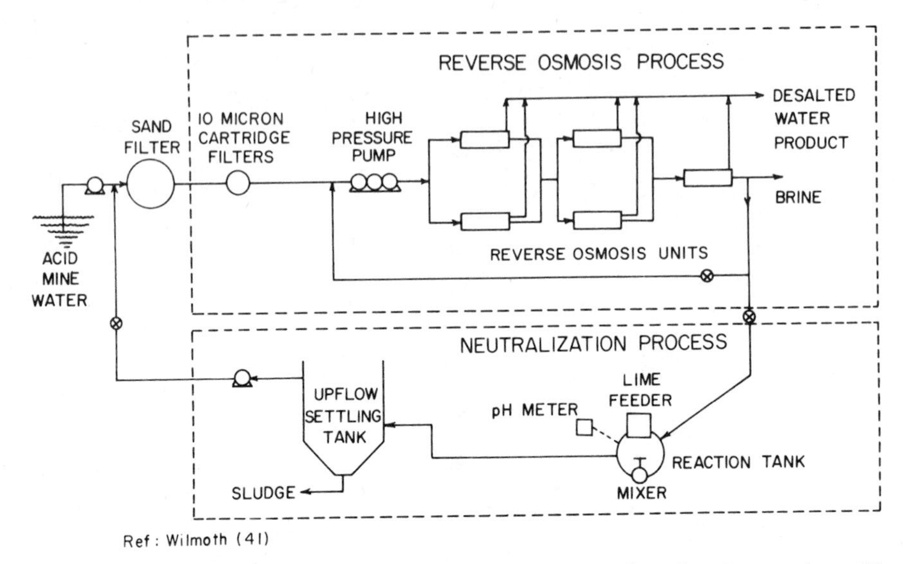

Figure 6. Flow diagram for reverse osmosis plus neutralization plant for treating acid mine drainage.

of each membrane is seen to be in the range of 98 to 100% except for acidity. The product water would require neutralization to increase the pH to an acceptable 6.0 to 8.5 and, for many acid mine waters, to precipitate iron and manganese to meet EPA drinking water standards (0.3 mg Fe/l, 0.05 mg Mn/l). For manganese, the pH would have to be increased to 9 to 10 to precipitate the element, followed by acidification to acceptable pH limits.

The combination reverse osmosis-neutralization process in Figure 7 also did not produce a neutral product water, as shown in Table 6. The

Table 6. Chemical Analyses for Combination Reverse Osmosis-Neutralization Treatment of Acid Mine Water at 91% Recovery (41)

		Amount (mg/l)			
Sample Designation	pH	Acidity	Ca	Fe	SO_4^{2-}
Hollow-fiber module					
Feed	2.7	660	100	120	980
Brine	2.0	6000	1000	1200	11,000
Product	3.4	130	3.0	1.6	21
Rejection (%)		95.1	99.4	99.7	99.6

Table 7. Comparison of Average Reverse Osmosis Fluxes during Treatment of Mocanaqua, Pa., Acid Mine Water (41)

	Module		
	Tubular	Spiral Wound	Hollow Fiber
Output per vessel volume at 323°K ($m^3/m^3 \cdot day$)			
At 2760 kN/m^2	—	2026	5723
At 4140 kN/m^2	418	2559	—
Membrane flux at 323°K ($m^3/m^2 \cdot day$)			
At 2760 kN/m^2	—	500	101
At 4140 kN/m^2	634	784	—

yield of product water in the reverse osmosis step could be increased to 91%, compared to 75% for reverse osmosis alone.

Comparison of the water production capabilities shown in Table 7 for treatment of acid mine waters by reverse osmosis shows that the spiral-wound modules excel when throughput is expressed on the basis of membrane area, and that the hollow-fiber modules excel on the basis of output per unit volume of pressure vessel. These values are based on run times on the order of 800 to 2500 hr at 75% recovery of product water at the Mocanaqua site. The operating life of the membranes is shortened considerably at higher recoveries or at other sites where acid mine water contains higher concentrations of calcium, iron, or sulfate. At Norton the hollow-fiber system experienced major colloidal and iron fouling problems. Calcium sulfate is usually the species causing fouling. In general, calcium sulfate fouling occurred when $PMC/(2.16 \times 10^{-4})$ was greater than 2.0, where PMC = product of molar concentrations of calcium and sulfate in the brine stream.

To extend permeator life, flushing with sodium hydrosulfite was employed. Ten percent failure of the modules was reported after 813 hr of operation (42). The cost of the permeators was reported (41) to be in the range of $290 to $1000/1000 gal/day capacity.

3.4. Neutralization

The use of chemical engineering operations that tend to concentrate all the ions in the brine discharge and to deionize the product water is not warranted for most acid mine drainages because the treated water would not be in demand directly as a public water supply. Probably the product water would be returned to the stream, where less stringent pollution criteria are in effect. Consequently, the virtually unanimous choice of mine operators for treating acid mine drainage, from both the technological and the economic viewpoint, is neutralization. It raises the pH, precipitates the problematical iron and aluminum, and under some conditions reduces the sulfate, magnesium, and manganese concentrations. The reactions can be considered as cation exchange reactions in which calcium, magnesium, or sodium ions replace the hydrogen, iron, and aluminum ions in the drainage. The product waters usually show no appreciable decrease in total dissolved solids.

For neutralization reactions, the hydroxyl ion is really the only ion species feasible to form a weakly ionized, polar compound with the hydrogen ion. Carbonate ion reacts with hydrogen ion to form carbonic acid, which will dissociate to form water and carbon dioxide. Carbon dioxide is easily removed from solution because of its low solubility. Carbonates also form insoluble compounds with iron and aluminum. These two reagent anions are available for mine drainage treatment as compounds of calcium, magnesium, sodium, and ammonia at feasible costs. Ammonium hydroxide is not recommended since it is a nutrient for plants and organisms, nor is magnesium ion always desirable since it is a cathartic. The reagents that can be considered include calcium carbonate (high-calcium limestone), calcium oxide (calcined, quick, or pebble lime), calcium hydroxide (hydrated lime), calcium carbonate-magnesium carbonate (dolomite or dolomitic limestone), calcium oxide-magnesium oxide (burnt or calcined dolomite), calcium hydroxide-magnesium hydroxide (hydrated dolomite), sodium carbonate (soda ash), and sodium hydroxide (caustic soda). The oxides listed are anhydrides that react with water to form hydroxides.

Lovell (43) compared the costs of these neutralization agents, at the source and upon delivery to a 500,000-gal/day acid mine treatment plant at Hollywood, Pennsylvania. This plant was designed specifically to study and compare the various neutralization agents. In Table 8 the delivered costs of the compounds are compared, based on equal neutralization potentials, and are seen to increase from crushed limestone to caustic soda. The least expensive reagent is limestone, at a cost of 1 cent/1000 gal water per each 500 mg/l acidity (equivalent to pH 3) on the 1970 basis of $1.5/ton limestone FOB at the quarry. On a 1976 basis, crushed limestone at the quarry costs about $2.3/ton. Calcium carbonate is also available in the form of clear crystals of calcite; it is more expensive at $17/ton No. 2 grit ($\frac{3}{16}$-in. screened crystal) in bags.

Table 8. Cost of Neutralization Reagents (43)

Reagent	Cost (\$/ton FOB) at Source	Delivered Cost, Hollywood, Pa. (\$/ton) (1968–1971)	Reagent Cost,[a] (¢/1000 gal per 500 mg/l acidity)
Limestone			
Stone sizes	1.5	4.5	1
Pulverized	—	8.5	1.8
Calcined lime			
Bulk	17	21	2.5
Bagged	19	23	2.8
Hydrated lime			
Bulk	20	24	3.7
Bagged	23	28	4.4
Dolomite			
Stone sizes	2.5	10	1.9
Pulverized	7	15	2.9
Calcined dolomite			
Bulk	16	24	2.4
Bagged	21	29	2.9
Dolomite, pressure hydrated			
Bagged	23	31	4.3
Soda ash, powder			
Bulk	—	40	9
Bagged	—	76	17
Caustic soda			
50% solution			
Tanker	—	66	11
Drum	—	268	45
Flake			
Drums, bulk	—	78	13
Solid, small lots	—	110	18

[a] Assumes 100% reagent purity.

Other considerations become basic in choosing among reagents. The selection is established by drainage water analyses and volume, plant location, and reagent performance potential. The following factors should be considered in the optimization: source geographical distribution, availability, transportation, handleability, cost, reactivity, and chemical and physical properties and mechanistic action in the treatment process, as these affect the quality of the water product and the characteristics of the sludges.

High-calcium and dolomitic limestones are naturally occurring minerals. The fact that the other reagents in Table 8 are manufactured products partially accounts for their higher cost. The quality and the reactivity of limestones and their products vary widely with their origin and manufacturing process. Calcined materials contain 5% or more of inert constituents. Differences in chemistry between high-calcium and dolomitic materials must be considered. The physical form of caustic soda and soda ash (solid, flake, solution) has an impact in determining reaction conditions, ambient storage and reaction temperatures, transportation, storage, and safety. The demand for these reagents for mine drainage treatment is not market controlling; therefore consumption should be planned around prevailing commercial products.

Pulverized limestone and lime products are commonly transported in bulk, using pneumatic handling, in loads ranging from 15 to 28 tons. Air pollution by dust release during unloading can be controlled. Reagent, shipment, and handling costs for bags of up to 100 lb are higher. Larger particle sizes of limestone ($>\frac{3}{8}$ in.) that do not contain large amounts of fines are easiest to handle, do not require protected storage facilities, and are not affected by the weather. Pulverized stone products, calcined lime, and hydrated lime are conveniently stored in bins. Since bin storage of hydrated lime can result in extensive water and carbon dioxide absorption within several weeks, long-term storage in large bins should be avoided. Pneumatically handled hydrated lime may be slurried (10 to 50 wt %) directly from the delivery tanker and stored as a suspension.

The chemical characteristics of the reagents used to react with acid mine drainage are well known. Each of the reagents dissociates extensively in aqueous solution, providing reactive ionic species. The hydroxyl and carbonate anions neutralize the acid and precipitate the metals; typical stoichiometry is indicated by the following equations:

$$Ca(OH)_2 + H_2SO_4 \rightarrow CaSO_4 \cdot 2H_2O \tag{12}$$

$$3Ca(OH)_2 + Fe_2(SO_4)_3 + 6H_2O \rightarrow 3CaSO_4 \cdot 2H_2O + 2Fe(OH)_3 \downarrow \tag{13}$$

$$CaCO_3 + H_2SO_4 + H_2O \rightarrow CaSO_4 \cdot 2H_2O + CO_2 \uparrow \tag{14}$$

$$3CaCO_3 + Fe_2(SO_4)_3 + 9H_2O \rightarrow 3CaSO_4 \cdot 2H_2O + 2Fe(OH)_3 \downarrow + 3CO_2 \uparrow \tag{15}$$

These cation exchange reactions occur essentially instantaneously when the reactive species come into contact. The reaction rate is controlled primarily by physical availability and contact of the ionic species. Reagent and product solubility, mechanical aspects (mixing), and diffusion become the controlling factors. Many different equilibria are involved in the neutralization of acid mine water, and they play a role in defining the direction and extent of the system reactions.

Lovell (43) has shown that the relative rate variations among neutralization reagents are quite extensive, as illustrated in Figure 7. Sodium hydroxide is more responsive than hydrated lime, while soda ash can be expected to react somewhat more slowly, as controlled by the bicarbonate equilibria. The final pH levels shown in Figure 7 are a function of the equilibria developed in a multitude of reactions. The upper pH limit in the carbonate systems is controlled by mechanical factors such as particle size and conditions affecting the hydroxide-bicarbonate-carbonate equilibria. Calcium carbonate reagents offer the advantage that, when present in excess of the stoichiometric amount, they automatically buffer the effluent to a pH no higher than 8.3. Large-particle-size

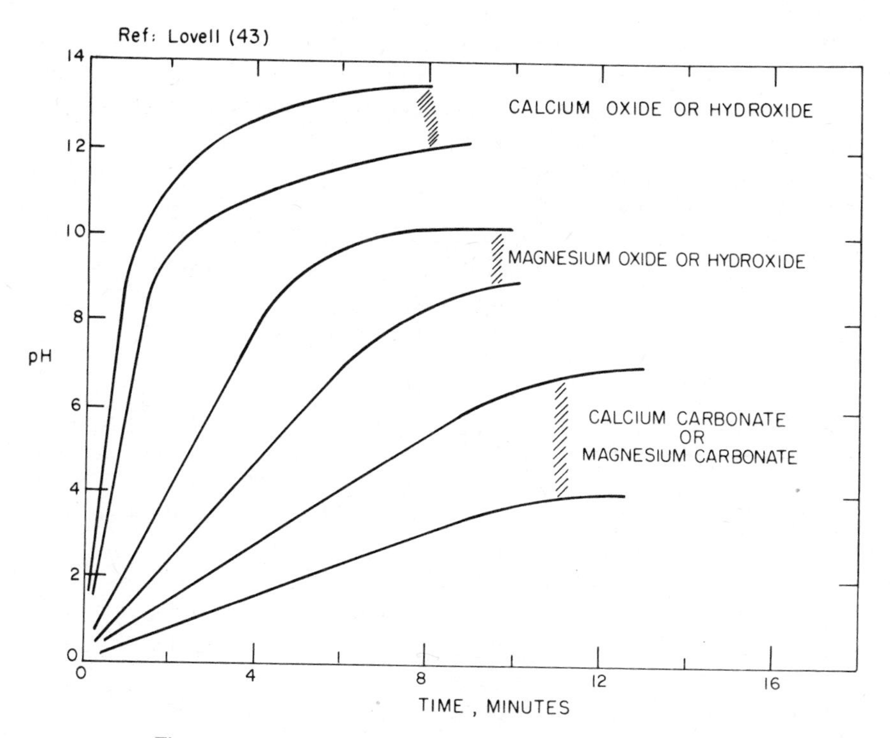

Figure 7. Relative reaction rates for various alkali reagents.

limestone can be utilized in a rotary cement-mixer-type reactor to increase consumption efficiency. With strong bases automatic process control units are generally required to avoid overtitration. The time available for reaction to a neutral pH is a pragmatic decision; some reagents require very large reactors and power inputs as the reaction time exceeds 10 min.

Lovell (43) found that utilization of neutralization reagent commonly was on the order of 50%. Excess reagent consumption is highest when mine waters are strongly acid and when the pH of the effluent exceeds 7.0. It is suspected that reactions involving aluminum and magnesium are the cause. The reaction of base with aluminum hydroxide beyond its minimum solubility level (about pH 6.2) appears to lead to its resolubilization as the aluminate. This means less removal of aluminum from waters and higher reagent consumption. Despite theoretical considerations, magnesium appears to precipitate at unexpectedly low pH levels, consuming excess base. About 95% of stoichiometric strong-base reagent consumption can be attained with sufficient iron removal by maintaining the effluent pH near 6.0. Limestone process efficiencies are about 80%, but are most easily maintained at 50 to 60%.

When heavy metals are present in the mine drainage waters, sulfide may be used to precipitate them (44). Lime is employed in the first stage to precipitate ferric and aluminum hydroxides, and barium sulfide in the second stage to precipitate sulfides such as CuS and ZnS.

Some metal ions require either oxidation or reduction in order to be precipitated at an acceptable pH level. Ferrous iron should be oxidized to the ferric state. Hexavalent chromium must be reduced to the trivalent state, usually at a pH of 2.5 to 3, and then precipitated as chromic hydroxide by increasing the pH to 8.5. Sodium hydrosulfite, sulfur dioxide, or ferrous hydroxide can be used as the reductant (45).

Limestone Neutralization

Barton and Vatanatham (46) investigated the rates of reaction of limestone particles with sulfuric acid and simulated acid mine waters in an aerated, stirred batch reactor and a flow-through bed reactor. The pH and CO_2 concentrations were monitored as functions of reaction time. To devise a reactor design model based on rate-limiting mechanisms, it is necessary to employ mechanistic chemical reactions and associated equilibrium expressions to fit experimental results. To restrict the complexity of the system, only data for pure sulfuric acid with a high-calcium limestone were modeled in this fashion.

In the bulk liquid phase it is assumed that all reactions are in equilibirum and that dissolved CO_2 is in equilibrium with the gas bubbles in the solution. With these assumptions, the following equilibirum equations as presented by Butler (47) can be applied:

$$HSO_4^- \leftrightharpoons H^+ + SO_4^{2-} \tag{16}$$

$$HCO_3^- \leftrightharpoons H^+ + CO_3^{2-} \tag{17}$$

$$HCO_3^- \leftrightharpoons OH^- + CO_2(aq) \tag{18}$$

$$H_2O \leftrightharpoons H^+ + OH^- \tag{19}$$

$$Ca^{2+} + HCO_3^- \leftrightharpoons CaHCO_3^+ \tag{20}$$

$$Ca^{2+} + OH^- \leftrightharpoons CaOH^+ \tag{21}$$

The following chemical reactions can be applied also. These three reactions are first tested to determine whether they are applicable and whether they are at equilibrium.

$$CaCO_3(s) \leftrightharpoons Ca^{2+} + CO_3^{2-} \tag{22}$$

$$Ca(OH)_2(s) \leftrightharpoons Ca^{2+} + 2OH^- \tag{23}$$

$$CaSO_4 \cdot 2H_2O(s) \leftrightharpoons Ca^{2+} + SO_4^{2-} + 2H_2O \tag{24}$$

At equilibrium the chemical activity of each species can be related through the equilibrium constants of the above reactions. In addition, an electrical neutrality equation and a mass balance on sulfate groups are applied.

Consider the case where no calcium hydroxide or calcium sulfate precipitates, and equilibrium is not established with the solid calcium carbonate. Since all the reaction equilibrium constants, as well as $[H^+]$, $[CO_2(aq)]$, and total sulfate, are known, eight unknowns with eight relationship equations can be applied. Thus the activity of each species can be calculated with a minimum of experimental measurements at any given time during a neutralization reaction. The concentration of each ionic species can be related to the activities by means of activity coefficients correlated with ionic strength, using the modified Davies equation presented by Butler (47). On the assumption that aqueous CO_2 is in equilibrium with the gas evolved from the solution, the concentration of aqueous CO_2 can be related to the partial pressure of CO_2 in the gas by using Henry's law.

From data based on varying limestone particle size, temperature, and carbon dioxide partial pressure, the following kinetic equation was developed, based on a rate model that approaches first order with respect to hydrogen diffusion:

$$-\frac{d[CaCO_3(s)]}{[CaCO_3(s)]^{2/3}\,dt} = \frac{6kM[CaCO_3(s)_0]^{1/3}}{\rho_p D_0}([H^+]_t - [H^+]_e) \tag{25}$$

where $[CaCO_3(s)]$ = moles of solid $CaCO_3$ at time t
$[CaCO_3(s)_0]$ = moles of initial solid $CaCO_3$
D_0 = initial particle diameter
$[H^+]_e$ = molar hydrogen ion concentration in bulk solution at equilibrium
$[H^+]_t$ = molar hydrogen ion concentration in bulk solution at time t
k = rate constant
M = molecular weight of $CaCO_3$
ρ_p = density of limestone

This rate expression was shown to fit data for the neutralization of sulfuric acid with a limestone containing 85 wt % $CaCO_3$ for particles in the range of 0.037–0.044 to 0.500–0.595 mm, temperatures from 274 to 348°K, and carbon dioxide partial pressures of 0 to 188 mm Hg over the pH range of 2 to 6. The value of k (units of l/cm²·sec) is given as a function of temperature for the stirred, aerated batch reactor by

$$k = \exp\left[\frac{-1833}{T(°K)} - 6.769\right] \tag{26}$$

The apparent activation energy calculated from these rate constants is 1.5×10^4 J/mol, which is rather small for chemical-reaction-controlled situations and corroborates a hydrogen-diffusion-controlled model in the pH range noted. One explanation is that the hydrogen ion, despite its small size, is the only species that diffuses toward the solid surface, while other species such as CO_2, Ca^{2+}, CO_3^{2-}, and HCO_3^- are all diffusing out. The hydrogen ion is in essence diffusing upstream against the bulk convective flow of materials.

The hydrogen-diffusion-controlled model was corroborated by the work of Wentzler and Aplan (48), who investigated the dissolution of limestone in nitric acid at constant pH, using the rotating disk method. Pearson and McDonnel (49), in testing the reaction rate of sulfuric acid with 2½- to 4-in.-diameter crushed limestone at ambient temperature, indicated that a carbonate surface reaction mechanism should be coupled with hydrogen diffision above a pH of about 5. Lund et al. (50) reported that, in studies with the rotating disk method, dolomite reacts approximately 100 times slower than calcite as marble and limestone, and that its dissolution rate in hydrochloric acid is controlled by the reaction between hydrogen adsorbed on the surface and the solid dolomite.

The effect of particle size on the reaction time of sulfuric acid with twice the stoichiometric amount of limestone at 298°K is shown in Figure 8 for the stirred, aerated batch reactor of Barton and Vatanatham (46). Starting with a solution with a pH of 1.9 (acidity 961 ppm $CaCO_3$ equivalent), a neutral solution (pH = 7) is reached in 13 min with 37 to 44-μm particles, but 3 hr is required with 0.5- to 0.6-mm particles.

The reaction rate was reduced by a factor of 3 when the flow-through bed reactor, in which the limestone particles are not agitated, was used. Lowering the amount of limestone from twice stoichiometric to stoichiometric lowered the reaction rate toward the end by a factor of 10, and lowered the final pH from the usual 7.5–8.0 value to 3.2 (starting at 1.8 to 2.0). This finding might be attributed to interference of diffusion by the 12 wt % solid inerts in the limestone. There was no calculated or visible evidence of precipitation of gypsum ($CaSO_4 \cdot 2H_2O$) on the limestone in any of the experiments, though this could happen with severely loaded acid mine waters. Gypsum coating should not be anticipated with waters containing less than 1400 ppm SO_4^{2-}. In this case dolomitic limestone can be used to reduce the calcium content of the solution as it is being neutralized.

The effect of various metal ions in sulfuric acid solution on the

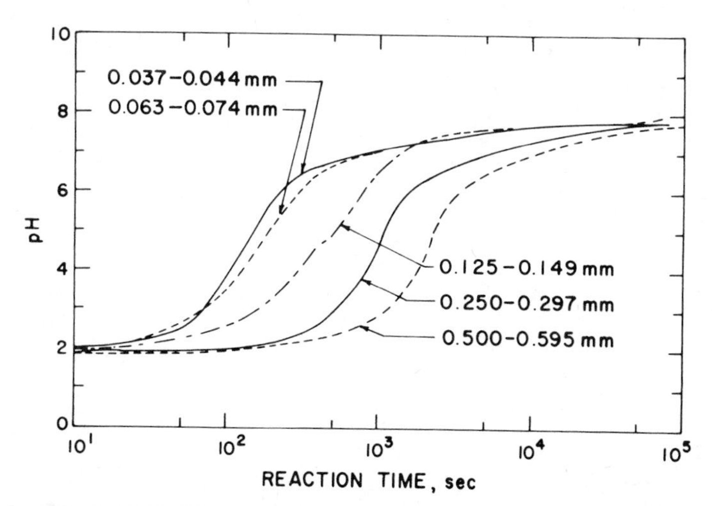

Figure 8. Effect of initial limestone size on neutralization rate of sulfuric acid. (Reprinted with permission from *Environ. Sci. Technol.* **10,** No. 3, 262–266, March 1956. Copyright by the American Chemical Society.)

neutralization rate with limestone particles in the stirred, aerated batch reactor can be seen in Figure 9 (46), starting with solutions of comparable acidity. The mixture containing Fe^{2+}, Fe^{3+}, and Al^{3+} is typical of strong acid coal mine drainage. The metals are seen to slow down the neutralization rate by a factor of 10; additional work is needed to describe quantitatively how this happens. As the pH rises above 4.5, the equilibrium that yields hydroxyl ions becomes important. The solubility products of the hydrated ferric and aluminum compounds are exceeded, and their precipitation occurs. For the mixture containing Fe^{2+}, the pH rises sharply to above 6, drops down, and then rises again to above 6 toward the end point. This can be explained by considering the oxidation of ferrous to ferric ion by oxygen in the purge gas according to equation 5. The oxidation proceeds faster at higher pH,

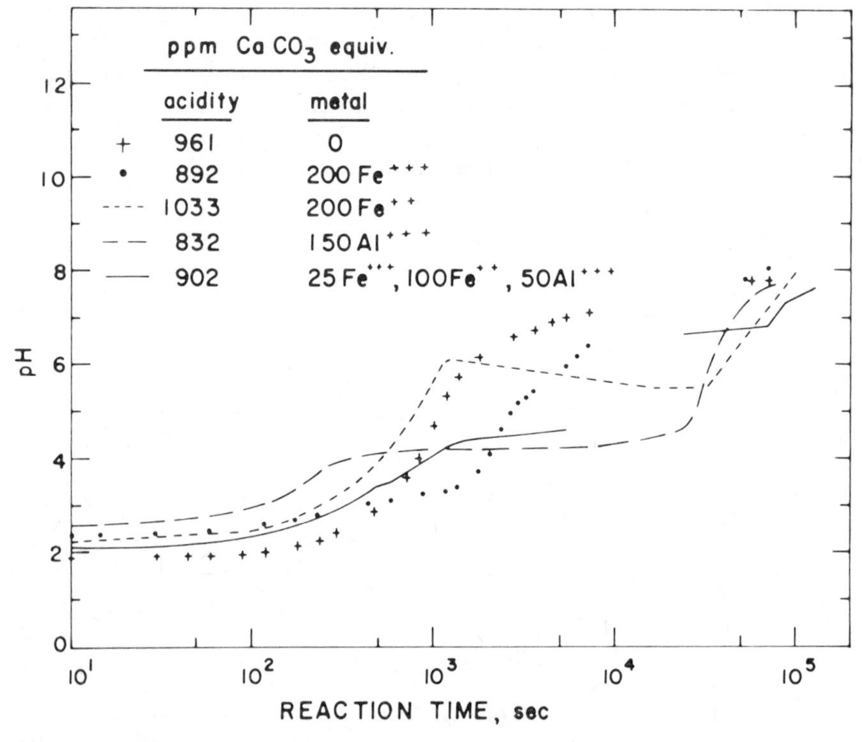

Figure 9. Effect of metal ions on neutralization rate of sulfuric acid with 0.250 to 0.297 mm limestone at 298.2°K. (Reprinted with permission from *Environ. Sci. Technol.* **10**, No. 3, 262–266, March 1956. Copyright by the American Chemical Society.)

producing hydrogen ions which require additional base for neutralization.

The effect of the compositions of various limestones on the neutralization of acid mine waters was investigated by Bituminous Coal Research (51). The reaction rate was proportional to the CaO content, with neutral pH being reached in 0.5 hr or less with limestones containing >70 wt % CaO at twice the stoichiometric amount of 37 to 44 μm limestone with dilute sulfuric acid (pH = 3.2). Under similar conditions dolomitic limestones (30 to 45 wt % MgO) neutralized only to pH values of 5.1 to 6.4 in 0.5 hr. Magnesite with 80 wt % MgO was nonreactive.

Ferrous Iron Oxidation

In the treatment of coal mine drainage with alkalies, it is desirable to convert the iron to the ferric state during the neutralization process. The fact that the solubility product of ferric hydroxide (1.1×10^{-36} at 18°C) is lower than that of ferrous hydroxide (1.64×10^{-14} at 18°C) means that the desired iron concentrations can be obtained at lower pH levels, thus expanding the potential use of limestone for neutralization. Elimination of the ferrous iron eliminates the potential for additional acid formation downstream or in the sludge. Reagent savings may be possible by not increasing the pH to the level required to precipitate unoxidized ferrous ion, by the avoidance of side reactions probably involving aluminum and magnesium (43).

The oxidation of ferrous iron may be accomplished with atmospheric oxygen, activated-carbon-catalyzed aeration (52), peroxides, or chlorine, or performed electrochemically or biologically. The use of the more exotic reagents involves intolerably high reagent costs for the flow rates involved. Electrochemical oxidation has been performed experimentally only on a small scale (53).

Biological oxidation of Fe^{2+} can be performed under acid conditions before the neutralization step. This would be especially applicable to limestone neutralization of waters containing more than 100 mg Fe^{2+}/l, where lengthy aeration of limestone suspensions during neutralization would otherwise be required. The biochemical oxidation system requires aeration as well as consideration of cell concentrations and growth rates as controlled by cell metabolism and interferences. A trickling-filter-type biochemical reactor provided oxidation rates of 6000 mg/l·hr, as compared to 50 mg/l·hr for a deep vat reactor (43), using *Ferrobacillus* and *Thiobacillus* bacteria. This can be compared to conventional aeration values on the order of several hundred milligrams per liter per hour. Sludges resulting from hydrolysis after biochemical iron oxidation of drainage waters are poorly crystallized, amorphous, ferric oxyhydroxides. Further study is needed on the effects of temperature and sludge deposition in biochemical reactors.

A controlled process of Fe^{2+} oxidation by oxygen from air is nearly universally utilized. The process design for this oxidation is controlled by the rate of transfer of oxygen from air to the reacting system and by the rate of Fe^{2+} oxidation with oxygen. The parameters that have been established to govern the reaction step are temperature, pH, Fe^{2+} concentration, and Al^{3+} concentration (43). The rate increases proportionately with the first three but shows a reciprocal relationship to Al^{3+} concentration. Catalytic and inhibiting agents may also be involved.

The overall reactions may be described as follows:

$$4FeSO_4 + O_2 + 2H_2SO_4 \rightarrow 2Fe_2(SO_4)_3 + 2H_2O \tag{27}$$

$$Fe_2(SO_4)_3 + 6H_2O \rightarrow 2Fe(OH)_3 + 3H_2SO_4 \tag{28}$$

These are probably oversimplifications of the actual mechanistic steps. One mole of sulfuric acid is formed per mole of Fe^{2+} oxidized and hydrolyzed. Thus the equilibrium pH that develops during the oxidation becomes the parameter which controls the oxidation rate. Mihok (54) proposed a design basis for Fe^{2+} oxidation in a limestone reactor, ranging proportionately from 200 mg/l·hr at an Fe^{2+} concentration of 50 mg/l to 750 mg/l·hr at 2000 mg/l for the pH range of 6.5 to 7.5 at 0 to 10°C.

Sludge Properties

Sludge control for acid mine water neutralization processes may incorporate three stages: solids-fluid separation of the sludge, sludge dewatering, and sludge disposal. Two techniques may be employed satisfactorily for the solid-fluid separation step: settling lagoons and mechanical clarifier-thickeners. Lagoons are most commonly employed and may provide limited sludge storage or permit permanent disposal. Mechanical clarifier-thickeners require the least land area and provide the least sludge storage, but permit continuous, automatic sludge removal.

Further dewatering of settled sludge before disposal is desirable because of its voluminous nature, low percentage of solids, and tendency to gel. This may involve *in situ* evaporative drying or direct water removal by a drying basin, a filter, or a centrifuge. The settled sludge can be pumped through pipes to the disposal area or to tanker trucks for haulage transfer. After dewatering, the solids may be transferred by dump truck. Ultimate disposal is usually made into abandoned deep or surface mines or as landfill. Removal of settled sludge from a lagoon bottom is difficult and costly because of its poor flow properties. Some sludges tend to gel, while others achieve a plastic, claylike character.

Important physical properties of the sludge are settling rate, final settled volume after compression, bulk density, and flowability. All are related to the hydrated character of the sludge. The sludges settle to a solids content between 0.5 and 14 wt % (43), with most in the lower range. When dried or frozen, the hydrated character of the sludge is changed and cannot be restored upon rewetting.

The properties of the sludges vary widely with the neutralization reagent, as shown in Table 9. Limestone sludges settle more rapidly, are much less voluminous, and contain higher solids contents than those produced by other reagents. In contrast to other reagents, limestone treatment yields turbid supernatant liquor. The volume of settled sludge increases with the iron-aluminum concentration of the raw water and can exceed 30% v/v of the total effluent. The sludge settling rate seemed to approach a maximum at iron-aluminum levels in the 100 to 200-ppm range. Laboratory tests indicated maximum settling rates near pH 6.8, though this trend was not found in plant tests.

The sludges were found to be primarily iron oxyhydroxides (43). Natural *yellowboy* is poorly crystallized goethite (orthorhombic $Fe_2O_2 \cdot H_2O$) and amorphorous ferric oxyhydroxides. Sludges prepared by synthetic neutralization yield a mixture of poorly crystallized goethite and lepidocrocite [FeO(OH)]. Neutralization sludges also contain calcium carbonate, basic magnesium carbonate, and gypsum. No feasible

Table 9. Sludge Settling Data for Several Neutralization Reagents at the Hollywood, Pa., Plant (43)

Source Water						
Total Acidity (mg/l)	Fe (mg/l)	Al (mg/l)	Alkali Used	Sludge Settling Rate (ft/hr)	Sludge Settled Volume (%)	Solids in Settled Sludge (wt %)
114	4	15	NaOH	—	2.2	0.23
206	35	17	CaO	—	4.5	0.21
1707	420	210	CaO	1.02	29.6	0.46
1305	250	191	$Ca(OH)_2$	1.47	18.0	0.76
1398	241	205	Soda ash	1.78	11.0	2.11
1380	317	191	Limestone	5.40	3.0	5.63
1401	322	191	Pulverized limestone	4.10	2.0	6.78

use for the sludge solids has been developed, nor is there any practical method for the recovery of by-products.

4. TREATMENT PLANTS

All but a few facilities for treating acid water drainage from operating mines use lime as the neutralization agent. The basis of choice between hydrated lime and calcined lime is basically one of economics as related to plant capacity. Calcined lime costs about two thirds as much as hydrated lime on a neutralization equivalent basis and is in favorable supply. However, calcined lime must be slaked, and this process requires additional capital and operating money, including that for labor to dispose of unreactive grit. Effective slaking should be a continuous operation to maintain proper reaction temperature. The minimum-sized commercial slaker processes 25 tons hydrated lime/week; this establishes the lower level of calcined lime neutralization plant capacity.

An example of a commerical acid mine neutralization facility is presented in the flow plan in Figure 10. This shows the original plant, which treats 3 million gal/day, and a new plant, which treats 6 million gal/day of water from a bore hole in the Pittsburgh Seam Mine with hydrated lime (55). The raw mine water is discharged from four 1600-gal/min mine pumps into a 4-million-gal equilibration pond, lined with over 3 ft of compacted, impervious clay to prevent leakage.

Lime is delivered to the plant lime bin in pneumatic tank trucks of approximately 22-ton capacity. When the pH probe signals for lime, the rotary and screw feeders under the bin start to feed lime into the slurry tank. In the new plant the bin is placed directly over the slurry tank. The water pump starts to put water into the same tank, and the lime slurry starts to pump milk of lime from the tank into the mine water as it enters the reaction and aeration tank. In the new plant a triple weir box is used to meter the exact volume of water required into the lime slurry tank. The raw water is limed and aerated for 20 min in the original plant and for 40 min in a 55-ft-diameter by 27-ft-high tank in the new plant. When the pH of the exit water rises to the preset level, the control shuts off both lime feeders and pumps.

The limed and aerated mine water discharges into a flume, passes by a pH control probe, and discharges into the center well of a thickener. The overflow from the thickener is collected in a flume near the periphery and discharged either directly to the stream or to a polishing pond, which retains the water for 12 hr before discharging it to the stream. The underflow of the thickener is pumped into an abandoned mine for

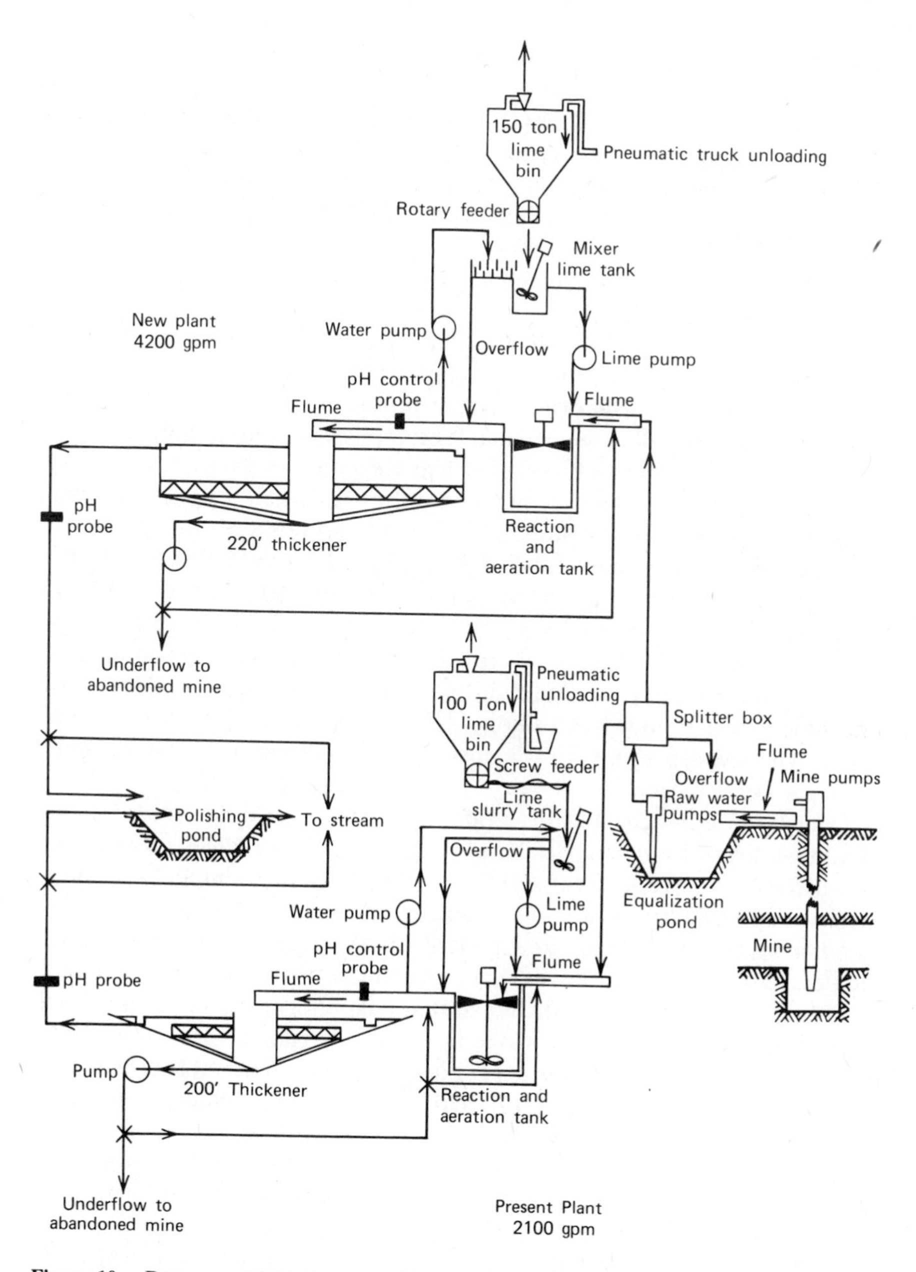

Figure 10. Duquesne Light Company Warwick Mine Portal No. 2 mine water treatment plant flowsheet (Reprinted with permission from "Mine Drainage Experience," 4th Symposium on Coal Mine Drainage Research, Mellon Institute, Pittsburgh, Pennsylvania, April 26–27, 1972.)

disposal into a seam that percolates below the coal seam from which the raw water is pumped.

In the original plant a 200-ft-diameter earthen wall thickener was employed. This thickener produced a turbid effluent on windy days because the wind stirred up the precipitate in the shallow water between the end of the rakes and the overflow trough. The new plant has a 220-ft-diameter by 23-ft-deep thickener with an 8-ft depth of water along its outside wall. This has eliminated the turbidity problem and provided for better sludge compaction.

From 1969 through 1971, the raw mine water had an average pH of 4.0, acidity of 1150 ppm, total iron content of 510 ppm, and ferrous iron content of 350 ppm. The original plant processed 3.4×10^9 gal of water with 12,800 tons of lime, which corresponds to the stoichiometric requirement. The thickener effluent had a volume of 2.7×10^9 gal, with an average pH of 7.9 and an iron content of less than 7 ppm. The plant underflow sludge had a volume of 5.4×10^8 gal (16% of the feed), with an average solids content of 0.78% and an iron concentration of 2890 ppm. Increasing the solids content in the underflow of the original earthen wall thickener to 2.4 wt % by letting the solids build up in the thickener, and to 5.0 wt % by recycling of sludge to the reaction tank, resulted in turbid overflow that did not meet stream requirements. With the new, deeper thickener, sludge solids contents of 3 to 4 wt % and clear overflow are attained. The cost of the lime treatment was 20 cents/1000 gal (1971 basis).

Limestone could be the preferable choice for treating nearly all but the most severely loaded discharges. It has the advantages of availability, lower cost, reduced hazards, ease of application, simplicity of plant design, impossibility of water overtreatment, ease of storage, and higher solids concentration of the precipitated sludge. It has not been accepted, however, because of its lower reactivity in contrast to other reagents.

Mihok et al. (56) have demonstrated the effectiveness of limestone neutralization of acid mine water in a 580,000-gal/day pilot plant. The process is described in Figure 11, and the facility is pictured in Figure 12. The process consists of producing a very fine (minus 400 mesh) limestone slurry, mixing the slurry with the mine water, aerating the resulting mixture to remove carbon dioxide and precipitate iron, and separating the solids from the liquid by sedimentation. A mine discharge of pH 2.8, containing 1700 ppm acidity, 36 ppm ferrous iron, and 360 ppm total iron, was treated. The treated water ultimately reached a pH of 7.4 and contained no detectable iron. The resulting sludge compacted to approximately one third of the volume of sludge from lime-neutralized mine water. Mihok (54) tabulated the capital and operating costs for this

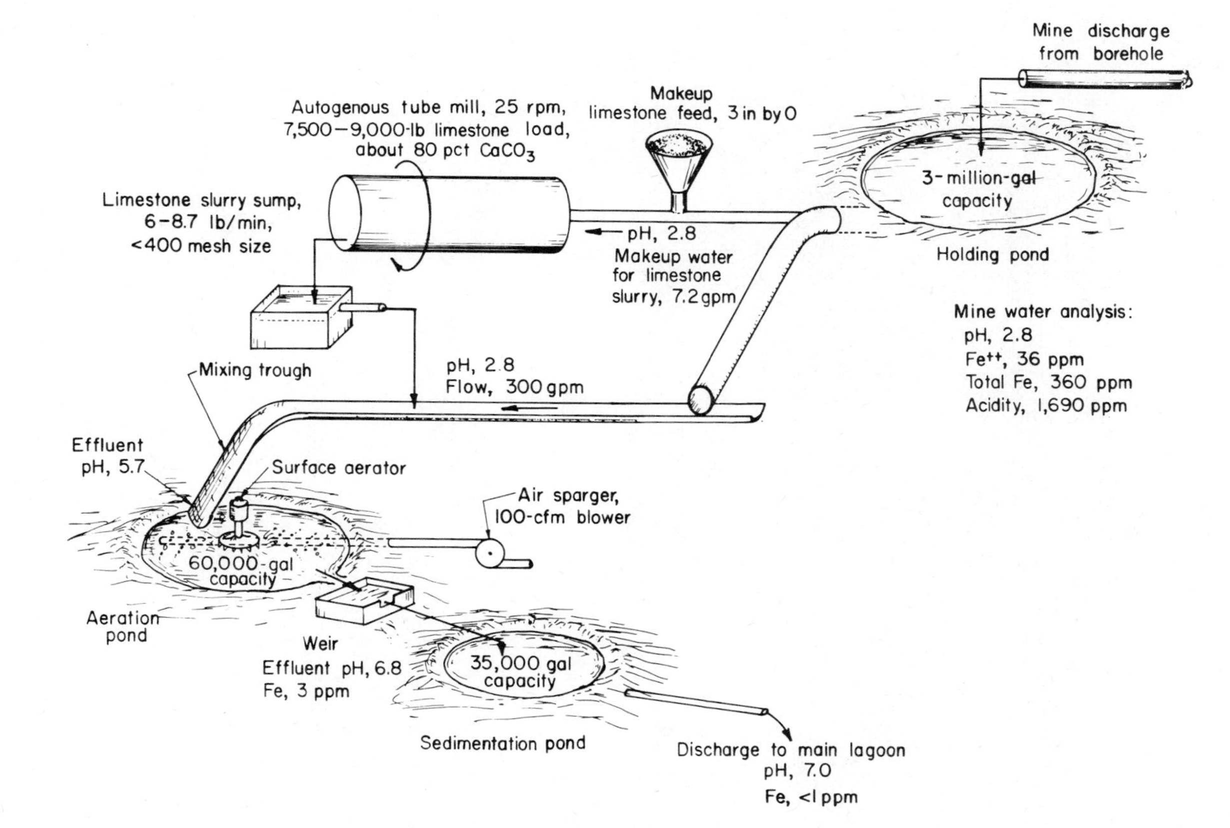

Figure 11. Flowsheet of the limestone neutralization process. (U. S. Bureau of Mines.)

Figure 12. Overall view of limestone treatment plant from edge of holding basin, showing tube mill and aeration pond. (U. S. Bureau of Mines.)

type of plant as a function of plant size, acidity, and ferrous iron. Estimated capital costs ranged from $55,000 to $766,000 in 1970 for 100,000- to 6-million-gal/day plant capacities, respectively. Estimated operating cost, including amortization, ranged from 44 to 2 cents (1970)/ 1000 gal treated, respectively.

Stirred aerated reactors containing suspensions of larger particles of limestone (200 mesh, 0.074 mm) should be considered in future acid mine water neutralization plants. Fixed bed reactors, possibly with upflow and aeration, containing large particles of limestone should be considered for acid mine waters not too heavily loaded with iron; however, there is the possibility that precipitates will eventually plug the void spaces between the particles in the reactor. This type of reactor could be the most attractive means of treating acid mine water in large streams.

REFERENCES

1. Bituminous Coal Research, Inc. (1970). *Studies on Limestone Treatment of Acid Mine Drainage*. Federal Water Pollution Control Administration, U.S. Department of the Interior, DAST-33, 14010 EIZ.

2. Williams, R. E. (1975). *Waste Production and Disposal in Mining, Milling and Metallurgical Industries,* Miller Freeman Publications, San Francisco, Calif.

3. Hawley, J. R. and Shikaze, K. H. (1971). *The Problem of Acid Mine Drainage in Ontario.* Proceedings, Third Annual Meeting of Canadian Mineral Processors, Mines Branch, Department of Energy, Mines and Resources, Ottawa, Ontario, Canada.

4. Battelle Columbus Laboratories (1974). *Development Document for Proposed Effluent Limitations Guidelines and New Source Performance Standards for the Primary Copper, Secondary Copper, Primary Lead, and Primary Zinc Subcategories of the Copper, Lead, Zinc Segment of the Nonferrous Metals Manufacturing Point Source Category.* Draft Report, Contract 68-01-1518, U.S. Environmental Protection Agency.

5. U.S. Environmental Protection Agency (1973). *Water Quality Criteria 1972.* EPA Rept. R3-73-033, March.

6. *Environ. Sci. Technol.* **8**(2), 110 (1974).

7. Browning, J. E. (1970). *Chem. Eng.* Jan. 12, p. 40.

8. Biesecker, J. E. and George, J. R. (1966). *Stream Quality in Appalachia as Related to Coal-Mine Drainage, 1965.* Geological Survey, U.S. Department of the Interior, Circ. 526.

9. McKay, D. R. and Halpern, J. (1958). *Trans. Met. Soc. AIME* **212,** 301.

10. Singer, P. C. and Stumm, W. (1970). *Science* **167,** 1121.

11. Hanna, G. P., Jr., Lucas, J. R., Randles, C. I., Smith, E. E., and Brant, R. A. (1963). *J. Water Pollut. Control Fed.* **35**(3), 275.

12. Nelson, H. W., Snow, R. D., and Keyes, D. B. (1933). *Ind. Eng. Chem.* **25,** 1335.

13. Hayase, Otsuka, and Maridko (1957). *Nippon Kogyo Karshi* **73,** 141.

14. Silverman, M. P. and Ehrlich, H. L. (1964). *Adv. Appl. Microbiol.* **6,** 153.

15. Walsh, F. and Mitchell, R. (1972). *Environ. Sci. Technol.* **10**(9), 809.

16. The Ohio State University Research Foundation (1971). *Pilot Scale Study of Acid Mine Drainage.* U.S. Environmental Protection Agency, 14010 EXA 03/71.

17. Blevins, R. L., Bailey, H. H., and Ballard, G. E. (1970). *Soil Sci.* **110,** 191.

18. Updegraff, K. F. and Sykora, J. L. (1976). *Environ. Sci. Tehcnol.* **10**(1), 51.

19. Sykora, J. L., Smith, E. J., Shapiro, M. A., and Synak, M. (1972). "Chronic Effect of Ferric Hydroxide on Certain Species of Aquatic Animals," Fourth Symposium on Coal Mine Drainage, Mellon Institute, Pittsburgh, Pa., Apr. 26–27.

20. Potomac Engineering and Surveying, Petersburg, W. Va. (1972). *Study of a New Surface Mining Method.* U.S. Environmental Protection Agency, Project 68-01-0763.

21. Kosowaski, Z. V. (1973). *Control of Mine Drainage from Coal Mine Mineral Wastes.* U.S. Environmental Protection Agency, EPA R2-73-230.

22. U.S. Environmental Protection Agency (1973). *Processes, Procedures, and Methods to Control Pollution from Mining Activities*. EPA-430/9-73-011, October.
23. Foreman, J. W. and McLean, D. C. (1973). *Evaluation of Pollution Abatement Procedures, Moraine State Park*. U.S. Environmental Protection Agency, EPA-R2-73-140.
24. Penrose, R. G., Jr. and Holubec, I. (1973). *Laboratory Study of Self-Sealing Limestone Plugs for Mine Openings*. U.S. Environmental Protection Agency, EPA-670/2-73-081.
25. Stoddard, C. K. (1973). *Abatement of Mine Drainage Pollution by Underground Precipitation*, U.S. Environmental Protection Agency, EPA-670/2-73-092.
26. Stefanko, R. K., Linden, V., and Tilton, J. G. (1966). *Development and Testing of an Injection Well for the Subsurface Disposal of Acid Mine Water*. The Pennsylvania State University, Spec. Rept. SR-60 to the Pennsylvania Coal Research Board.
27. Ryan, J. and Stroehlein, J. L. (1976). *World Mining*, February, p. 38.
28. Consolidation Coal Company (1971). *Control of Mine Drainage from Coal Mine Mineral Wastes*. U.S. Environmental Protection Agency, 14010 DDH 08/71, August.
29. Beyer, L. E. and Hutnik, R. J. (1969). *Acid and Aluminum Toxicity as Related to Strip Mine Spoil Banks in Western Pennsylvania*. The Pennsylvania State University, Spec. Res. Rept. SR-72 to the Pennsylvania Coal Research Board, May 1.
30. Pettyjohn, W. A. (1975). *Ground Water* **13**(1), 4.
31. Office of Saline Water (1964). *1964 Saline Water Conversion Report*. U.S. Department of the Interior.
32. Schroeder, W. C. and Marchello, J. M. (1966). *Study and Analysis of the Application of Saline Water Conversion Processes to Acid Mine Waters*. U.S. Department of the Interior, Research and Development Progress Rept. 199, August.
33. Shackelford, J. M. (1972). "Treatment and Recycle of Mine Waste Water," 72-PID-9, ASME Meeting, New Orleans, La., Mar. 28–30.
34. Powell, R. W. (1972). "Hydrocarbon Extraction of Acid Mine Drainage," M.S. thesis, The Pennsylvania State University, University Park, Pa.
35. Barton, P. and Fenske, M. R. (1970). *Ind. Eng. Chem. Process Des. Dev.* **9**(1), 18.
36. Barton, P. (1972). *Purification of Acid Mine Water by Liquid Extraction*. Grant Application from The Institute for Research on Land and Water Resources, The Pennsylvania State University, to the Office of Water Resources Research, U.S. Department of the Interior, October.
37. *Chem. Eng*. May 17, p. 77, 1971.
38. Kunin, R. and Demchalk, J. J. (1974). "The Use of AMBERLITE® Ion

Exchange Resins in Treating Acid Mine Waters at Philipsburg, Pennsylvania," 5th Symposium on Coal Mine Drainage Research, Louisville, Ky., Oct. 22–24.

39. Feiler, H. D. (1973). "Ion Exchange Removal of Sulfate from Coal Mine Drainage," M.S. thesis, The Pennsylvania State University, University Park, Pa.
40. Riedinger, A. and Schultz, J. (1966). *Acid Mine Water Reverse Osmosis Test at Kittanning, Pennsylvania.* U.S. Department of the Interior, Research and Development Progress Rept. 217, October.
41. Wilmoth, R. C. (1973). *Applications of Reverse Osmosis to Acid Mine Drainage Treatment.* U.S. Environmental Protection Agency, EPA-670/2-73-100, December.
42. Rex Chainbelt, Inc. (1970). *Treatment of Acid Mine Drainage by Reverse Osmosis.* Federal Water Quality Administration, U.S. Department of the Interior, 14010 DYK 03/70.
43. Lovell, H. L. (1973). *An Appraisal of Neutralization Processes to Treat Coal Mine Drainage.* U.S. Environmental Protection Agency, EPA-670/2-73-093.
44. Ross, L. W. (1973). *Removal of Heavy Metals from Mine Drainage by Precipitation,* U.S. Environmental Protection Agency, EPA-670/2-73-080, September.
45. Onstott, E. I., Gregory, W. S., and Thode, E. F. (1973). *Environ. Sci. Technol.* **7**(4) 333.
46. Barton, P. and Vatanatham, T. (1976). *Environ. Sci. Technol.* **10**(3), 262.
47. Butler, J. N. (1964). *Ionic Equilibrium, a Mathematical Approach.* Addison Wesley Publishing Co., Reading, Mass.
48. Wentzler, T. H. and Aplan, F. F. (1972). "Kinetics of Limestone Dissolution by Acid Waste Waters," Environmental Control, *Proceedings of the AIME Symposium 1972.* The Metallurgical Society, AIME, New York.
49. Pearson, F. H. and McDonnell, A. J. (1974). *Neutralization of Acidic Wastes by Crushed Limestone.* Institute for Land and Water Resources, The Pennsylvania State University, Res. Publ. 79, University Park, Pa.
50. Lund, K., Fogler, H. S., and McCune, C. C. (1973). *Chem. Eng. Sci.* **28,** 691.
51. Bituminous Coal Research, Inc. (1971). *Studies of Limestone Treatment of Acid Mine Drainage-Part II.* U.S. Environmental Protection Agency, 14010 EIZ 12/71, December.
52. Ford, C. T. and Boyer, J. F., Jr. (1973). *Treatment of Ferrous Acid Mine Drainage with Activated Carbon.* U.S. Environmental Protection Agency, EPA-R2-73-150, January.
53. Tyco Laboratories, Inc. (1972). *Electrochemical Treatment of Acid Mine Waters.* U.S. Environmental Protection Agency, 14010 FNQ 02/72, February.

54. Mihok, E. A. (1970). *Mine Water Research-Plant Design and Cost Estimates for Limestone Treatment*. Bureau of Mines, U.S. Department of the Interior, Rept. of Investigation 7368, April.
55. Draper, J. C. (1972). "Mine Drainage Experinece," 4th Symposium on Coal Mine Drainage Research, Mellon Institute, Pittsburgh, Pa., Apr. 26–27.
56. Mihok, E. A., Deul, M., Chamberlain, C. E., and Selmeczi, J. G. (1968). *Mine Water Research—the Limestone Neutralization Process*. Bureau of Mines, U.S. Department of the Interior, Report of Investigation 7191, September.

9

SULFUR POLLUTION AND SOILS

Marvin Nyborg

Department of Soil Science,
University of Alberta,
Edmonton, Alberta, Canada

1. INTRODUCTION

To consider the many ways in which sulfur could act as a pollutant of soils, and the outcome of this pollution, would be a useful but formidable task. Instead, this chapter will deal only with the obvious and simple ways by which soils become polluted so that their productivity is impaired.

Sulfur pollution in its most serious form includes the acidification of soils, whether from sulfur dioxide (SO_2), sulfur fertilizers, or elemental sulfur. Most, but not all, soils are susceptible to acidification, and the growth of both agricultural crops and forest trees can be reduced. The emissions of SO_2 from industry can acidify large areas or regions of land, whereas the acidity produced from fertilizers or from windblown elemental sulfur is a local phenomenon. Acidification by SO_2 is a slow process occurring over many years; acidification by fertilizers or windblown elemental sulfur can easily take place within several years.

The complicated nature of soils will only be touched on here. One should bear in mind that the various sulfur compounds added to soils as "pollutants" can have many effects on the organic and inorganic compounds in the soils and their complexes. Furthermore, the different sulfur compounds can have various effects on biological transformations in soils; that is, soils are complex bodies that contain sulfur naturally, and sulfur is one of the nutrients needed by plants, animals, and soil microorganisms. Many valid subjects for discussion are involved in the sulfur "pollution" of soils (as we all know, soils are the main "sinks" of anthropogenic sulfur). Nevertheless, in this chapter an attempt will be made to isolate the several subjects that are obviously important to sulfur "pollution".

The general term "sulfur pollution of soils" is perhaps unusual, though it seems fitting enough if kept undefined. "Sulfur pollution of soils" will be used here in the restricted and practical sense of changing the composition of soils so that the productivity of agricultural crops or forests is reduced, or there is a conversion to other species of plants that can better tolerate the change in soil composition. The main emphasis in this chapter will be on the amounts of sulfur that are added to soils, the compounds of sulfur that are acid forming in soils, some of the different species of plants that will accommodate themselves to acidified soils, these soils that are resistant to acidification, and methods of counteracting or correcting acidification. In addition, the importance of sulfur deficiency is compared to that of sulfur pollution in soils; also discussed are the inefficiency of sulfur pollution as a means of correcting sulfur deficiency, and, in the extreme, the fact that are both sulfur deficiency and sulfur pollution of soils may occur simultaneously.

2. FORMS OF SULFUR IN SOILS

In the majority of soils most of the sulfur is held in the organic matter; in many cases well over 95% is found in the organic matter (see Fitzgerald, this volume). For example, 64 surface soils representing the main soil types in Iowa had from 95 to 98% of their total sulfur in the organic form (Tabatabai and Bremner, 1972), and 54 soils from the five major soil zones in Saskatchewan had a similar content, from 96.4 to 99.5%, of organic sulfur (Bettany et al., 1973). It is also of interest that Tabatabai and Bremner (1972) found their 64 Iowa soils to have a range of 57 to 618 ppm of total sulfur with a mean of 294 ppm, and Bettany et al. (1973) reported similar results for their Saskatchewan soils. In some soils, however, the majority of the total sulfur is held as salts (e.g., $CaSO_4$ and Na_2SO_4). In any case, plants can feed only on the sulfate-S from the soil. The organic matter of soil has a fairly constant ratio of nitrogen to sulfur of approximately 10: 1.0 to 10: 1.5 (Walker and Adams, 1958; Williams et al., 1960; Harward et al., 1962). For agricultural soils the organic matter of the plow layer contains from about 0.005 to 0.05% sulfur and from 0.04 to 0.4% nitrogen. The inorganic sulfur in the soil profile may vary from only several parts per million or less to several percent in soils high in sulfates.

The organic sulfur in soils can be divided into three fractions which Williams (1975) summarized as follows: hydriodic acid-reducible sulfur, carbon-bonded sulfur, and Raney-nickel-reducible sulfur. The three fractions, in order, consist of ester sulfate, sulfur that is bonded to carbon, and a subfraction of carbon-bonded sulfur that represents the cystine and methionine in the organic matter of soil. As pointed out by Williams (1975), the proportions of organic sulfur in each fraction may average approximately 50, 50, and 25%. However, the percentage of each fraction is seen to vary widely from one kind of soil to another if one examines the results of different researchers (Lowe, 1964; Freney et al., 1970; Tabatabai and Bremner, 1972). Quite apart from the fractions, there is a host of compounds or complexes of sulfur in the organic matter of soils, both in the more resistant fraction of the organic matter and in the transitory forms (Freney and Stevenson, 1966). Sulfur behaves in a manner similar to nitrogen in that approximately 1 to 3% of the organic sulfur in soils is normally converted into inorganic sulfate each year (mineralization), and a like amount of the sulfate is returned to the soil organic matter (immobilization). In most studies mineralizable sulfur is not closely correlated to the total sulfur content of soils (e.g., Tabatabai and Bremner, 1972). The amount of sulfur mineralized apparently depends more on the N:S ratio of fresh organic matter added to the soil than on the N:S ratio of organic matter in the soil as a whole (Freney et al., 1962).

The inorganic sulfur in soils is more easily related to the availability of sulfur to plants, especially the amount of easily extractable sulfate. This sulfate, extractable with water or with a dilute solution of chloride (usually $CaCl_2$), is used by agricultural soil testing laboratories to determine whether sulfur fertilizer is required. Less than 2 or 3 ppm of sulfate-S in the soil shows that a fertilizer may be needed. The work of Tabatabai and Bremner (1972) and that of Bettany et al. (1973), with soils from Iowa and Saskatchewan, both showed an average content of slightly less than 10 ppm of easily extractable sulfate. Apart from the easily soluble inorganic sulfate, most soils also have inorganic adsorbed sulfate, which is only slowly available to plants. The adsorbed fraction is largest in soils that are acid, fine-textured, and high in aluminum and iron oxides. Neptune et al. (1975) demonstrated the difference among soils in their adsorption of sulfate, showing adsorption in Brazilian soils, and none in Iowa soils. The topic of adsorbed sulfate and its relation to the movement of sulfates in soils is well presented in the paper of Harward and Reisenauer (1966). More recent information is covered by Barrow (1975).

Inorganic sulfate may also occur as $CaSO_4$ in soils, often as a gypsum precipitate at the top of the C-horizon of the soil, but the solubility and the availability to plants (usually very low) depend on the nature of the gypsum crystals. In addition, Williams (1975) has pointed out that insoluble calcium sulfate usually exists in calcareous soils, where it occurs as a cocrystallized impurity in the calcium carbonate. Furthermore, this form may comprise most of the total sulfur in calcareous soils. In saline soils, sulfate occurs in part as Na_2SO_4, and so is mobile within the soil profile with the movement of water. Soluble sulfate is an inherent part of saline soils and may be present in large amounts up to several percent of the soil. Submerged or anaerobic soils, if high in inorganic sulfur content, accumulate large amounts of FeS and may also give off H_2S (Starkey, 1966). When the soils are drained, the sulfides form H_2SO_4 and so may make the soils acid.

Therefore, whereas the soil content of organic sulfur is fairly predictable simply from the amount of soil organic matter, and the organic sulfur content does not change quickly, inorganic sulfates in soils vary greatly (from as little as several parts per million to several percent), and their reactions can be fast.

3. SOME REACTIONS OF SULFUR IN SOILS

The following reactions of sulfur are probably the most pertinent to sulfur pollution of soils: the oxidation and reduction of sulfur, the mineralization-immobilization of soil organic sulfur, the sorption of SO_2, the formation of H_2SO_4 by some sulfur fertilizers and by SO_2 emissions, and the retention and leaching of sulfates in soils. Each of these will be dealt with briefly.

An account of the oxidation and reduction of sulfur in soils is given by Starkey (1966). In aerobic soils (and most soils are aerobic), the inorganic sulfur is held as sulfate, the most highly oxidized form. In either premanently or temporarily anaerobic soils, sulfates are reduced mostly to sulfides, the most reduced form of sulfur, through the action of two genera of sulfate-reducing bacteria: *Desulfovibrio* and *Desulfotomaculum*. The usual product is FeS (and FeS_2), and H_2S, as well, may be evolved. In addition, elemental sulfur can be another product of the reduction of sulfate. When soils are drained (or dried), the oxidation of sulfide to sulfate proceeds quickly (usually in a few days upon soil drying), and the generation of sulfates is both chemical and microbiological. When soils are high in inorganic sulfur content, they produce H_2SO_4 as they are dried; and if they do not have sufficient bases (especially calcium) to counteract the H_2SO_4 (to form neutral salts), the soils become acid as the result of oxidation.

The mineralization of sulfur (conversion from organic to inorganic) from the soil matter, as far as is known, takes place in a manner similar to the mineralization of nitrogen. Soil organic matter has different fractions which decompose at different rates. For example, Campbell et al. (1967), through carbon dating, found that one fraction decomposed at a rate of one half per 25 years, and another fraction at a rate of one half per 1200 years. One can assume that organic sulfur and nitrogen in soil matter decompose at similar rates, that is, apparently some decomposes fairly quickly, and some very slowly. The rates of mineralization of nitrogen and sulfur from soil organic matter are generally in the ratio of 10:1, but in particular studies (e.g., Williams, 1967) variations in this ratio were found from soil to soil. Also, generally both nitrogen and sulfur mineralized annually in the field at a rate of 1 to 3% of the totals found in soil organic matter. The immobilization of sulfur into the soil organic matter is accomplished by, first, the uptake of inorganic sulfur by plants and the formation of organic sulfur in plants, followed by the decomposition of fresh plant residues which form soil organic matter. The rate depends (among other things) on the contents of nitrogen and sulfur in the fresh residues. However, immobilization of sulfur into soil organic matter will approximately keep pace with mineralization of sulfur from soil organic matter.

Soils have a large capacity to sorb SO_2 and to sorb it quickly (Abeles et al., 1971). For example, Terraglio and Manganelli (1966) treated 10 g of a dry loam for less than a day with a total of 3618 ppm of SO_2, and the soil retained 1134 ppm. Johansson (1959) set out potted soil in the field at different distances downwind from an oil-burning plant that emitted SO_2, and concluded that direct adsorption of SO_2 by the soil was the important way in which sulfur was added to soil. Cox (1975) used an SO_2 concentration as low as 0.03 ppm for 48 hr and found that three soils gained about 5 ppm of sulfate-S when dry, and about 10 ppm when moist. As shown by Terraglio and Manganelli (1966), Cox (1975), and Ghiorse and Alexander (1976), the factors that increase uptake of SO_2 by soils are fine texture, high soil matter content, high pH, presence of free $CaCO_3$, high soil moisture content, and presence of soil microorganisms. The effect of most of the factors, however, was minimal.

Most of the SO_2 sorbed on soil first forms sulfite, and then is quickly converted to sulfate. Ghiorse and Alexander (1976) found that in a soil treated with SO_2 sulfate formation was almost complete within 10 hrs, unless the soil was first sterilized and then sulfate formation took about 48 hrs. Unless soils treated with SO_2 have free calcium (calcareous soils), the sulfate formed is, in effect, H_2SO_4. In other words, SO_2 produces H_2SO_4 and a lowering of pH in SO_2-treated soils, as has been shown by Cox (1975) and by Nyborg et al. (1977).

Another reaction of SO_2 is to form nonextractable, apparently organic sulfur by the reaction with soil organic matter. Ghiorse and Alexander (1976) reported that about one quarter of the 54 ppm of SO_2, added to sterile and nonsterile soil, formed organic sulfur. Nyborg et al. (1977) apparently found that one half of the SO_2 formed organic sulfur after exposure of two soils to an atmosphere of 0.024 ppm of SO_2. The importance of the direct reaction of soil organic matter with SO_2 has not been assessed, but particularly in areas where there are large concentrations of man-made SO_2 emissions this reaction may well alter the nature of soil organic matter.

When elemental sulfur is added to aerobic soils, either as a fertilizer or an amendment in agriculture or as a pollutant, the sulfur will be oxidized to H_2SO_4 by soil bacteria. The oxidation is carried out by several species, all of the genus *Thiobacillus*, and one species (*T. thioxidans*) continues to operate under extreme acidity (pH 1 to 2) as H_2SO_4 is generated. Tisdale and Nelson (1975) have listed the important factors governing the rate of oxidation of elemental sulfur as follows: the fineness of the sulfur, the population of *Thiobacillus* in the soil, and the soil temperature and moisture content. At a fineness of about 100 meshes per inch sulfur is oxidized quickly, although sulfur finer than 100 meshes oxidizes even more quickly. When the particle size is 5 to 10 meshes per inch, oxidation is very slow. (In addition, observations in the field indicate that lumps of sulfur several centimeters in diameter will lie on the soil with no apparent degradation for several years.) The rate of sulfur oxidation increases with temperature to approximately 40°C, and the optimum soil moisture content is field capacity. When elemental sulfur is used as a fertilizer (and the rate of application is, say, 20 to 40 kg/ha), on many soils the elemental sulfur is not oxidized completely in the first season. Tisdale and Nelson (1975) discuss results showing increases in rate of oxidation for 10 soils when the soils were first inoculated with a soil plentiful in sulfur oxidation bacteria. However, some soils can oxidize sulfur quickly and in large amounts; and when Bertrand (1973) added 1% of sulfur by weight to a normal agricultural soil, 8400 ppm of sulfur was oxidized during 12 weeks of incubation and soil pH dropped from 5.9 to 2.5. In addition, it is characteristic of soils that have received heavy applications of elemental sulfur to oxidize newly applied elemental sulfur at rates of at least several tonnes per hectare per year (Nyborg, 1974).

Ammonium sulfate, whether as a fertilizer or as a pollutant, generates H_2SO_4, but in a different way than does elemental sulfur. Ammonium in the soil is nitrified through the two-step process by two genera of soil bacteria (*Nitrosomonas* and *Nitrobacter*), and essentially the base ammonia is converted to nitric acid. As a consequence, the sulfate from the $(NH_4)_2SO_4$ is no longer counterbalanced by a base, and the sulfate then becomes H_2SO_4. The process of nitrification takes place very quickly in most cultivated soils; an application of $(NH_4)_2SO_4$ as a fertilizer will be nitrified within 2 or 3 weeks in a warm, moist soil. In noncultivated land, and especially forests, nitrification takes place at a much slower rate (Chase et al., 1968), or not at all.

Hence elemental sulfur, $(NH_4)_2SO_4$, and SO_2 all produce H_2SO_4 in soils, but by different processes. Sulfate added to soils leaches downward very quickly to very slowly, depending on the amount of rainfall, the soil texture, and the ability of the soils to retain adsorbed sulfate. Harward and Reisenauer (1966), in their review, reported that leaching losses in a number of investigations with lysimeters varied from insignificantly small amounts to more than 300 kg/ha yearly. Sulfates move downward in soils until they reach the salt-accumulation layer in the subsoils at a depth of 1 or 2 m, and sulfates may reach groundwater in some kinds of soils. The adsorption of sulfate, which is most common in acid soils, slows the downward movement of sulfate before it is leached more deeply into the subsoil. Field experiments in New Zealand showed a 30- to 60-cm movement of sulfate after 25 cm of rain (Walker and Gregg, 1975). However, crops can make use of sulfate that is well into the subsoil; for example, forage crops in New Zealand drew on sulfate from depths of 50 or 100 cm.

When H_2SO_4 is formed in soils, from elemental sulfur, $(NH_4)_2SO_4$, or

SO_2, for example, the sulfate ion will tend to move downward, but this does not mean that the acidity formed can be leached out of the soil. Although some of the acidity will be distributed between the topsoil and the soil immediately below it, the sulfate ion moves only when accompanied by a cation (usually calcium), and thus the base is lost from the soil and the hydrogen ion is left in its place.

4. SULFUR POLLUTION OF SOILS

Here will be summarized the most important ways in which anthropogenic sulfur is added to soils, the amounts of anthropogenic sulfur that may be added, and the conditions under which the sulfur added to soils can be considered a "pollutant."

The emission of SO_2 into the atmosphere and the deposition of much of this sulfur on the land constitute the type of soil pollution that currently receives the most attention. Approximately 60 million tonnes of SO_2 is emitted annually the world over, and through this mechanism the largest quantity of sulfur is added to the soils as a pollutant. It is difficult to determine the exact amount of sulfur and its form (i.e., H_2SO_4 as opposed to sulfate salts) added to the soils in particular localities. The ways in which emitted SO_2 is returned to the earth include the following: H_2SO_4 and neutral salts from SO_2 are carried down in rain or snow (Barrett and Brodin, 1955); SO_2 may be adsorbed on trees, which then intercept rainfall and the H_2SO_4 formed is carried into the soil (Baker et al., 1973); deposition of particulates containing H_2SO_4 and salts takes place (Brosset, 1973); and SO_2 is adsorbed directly by soils (Johansson, 1959; Terraglio and Manganelli, 1966). The amount of acidification caused by each of these mechanisms depends on the rate and composition of the sulfur deposition. If the depositions consisted only of H_2SO_4, there would be acidification of soils, whatever the rate; but if the depositions were neutral salts, soils would not be acidified. However, SO_2 emissions usually are deposited as mixtures of H_2SO_4 and various salts, with the compositions of these mixtures continuously varying with time and place. The direct sorption of SO_2 by soils is the one mechanism that is simply only acidifying (Cox, 1975; Nyborg et al. 1977). Acidification of soils depends greatly on the kind of soil. For example, a soil well stocked with free $CaCO_3$ (calcareous soils may consist of 20% free carbonate) is essentially immune to acidification by H_2SO_4 simply because the acid reacts with the free $CaCO_3$ to form $CaSO_4$. However, most soils are not calcareous, and exposure to SO_2

emission causes them to become slowly acidified, although the rate of acidification varies greatly among soils (see also Kramer, Part I).

Sulfur dioxide is therefore a "pollutant" insofar as it acidifies soils, particularly those that are poorly buffered, and those that are already slightly acid, so that any further lowering of pH may affect the soil productivity.

The amount of sulfur fertilizers currently used throughout the world is estimated as 10 million tonnes of sulfur annually (D. W. Bixby, 1977, personal communication). Many kinds of sulfur fertilizers are manufactured, but the most important are $(NH_4)_2SO_4$ and superphosphate. Examples of sulfur-based fertilizers include $CaSO_4$, Na_2SO_4, K_2SO_4, elemental sulfur, and $(NH_4)_2SO_4$. Of those, only the last two are acidifying; the others are neutral salts. The fertilizer superphosphate contains 14% sulfur in the form of $CaSO_4$. In agriculture, ground limestone (mostly $CaCO_3$) is used where needed to counteract the acidification by fertilizers; 1 kg of sulfur as elemental sulfur requires 3 kg of $CaCO_3$, and 1 kg of sulfur as $(NH_4)_2SO_4$ requires 6.2 kg of $CaCO_3$. In the same way as acidification by SO_2, some soils are quickly acidified by fertilizers, whereas others are not. However, sulfur fertilizers are as yet little used in forests, although sulfur deficiency in forests has been shown in Australasia and North America (Humphreys et al., 1975). The application of ground limestone is not used in practice for forests, because of the difficulty of spreading the limestone.

Acid-forming sulfur fertilizers can be thought of as "pollutants" on soils that become acidified through the application of sulfur fertilizers. Sulfur fertilizers are needed on sulfur-deficient soils, but such soils are in the minority. The rate of sulfur fertilization is fairly low (in comparison to nitrogen fertilization), usually not more than 20 to 30 kg S/year.

Elemental sulfur and $CaSO_4$ are used as "amendments" on salt-affected soils in amounts as high as several tonnes per hectare. These two sulfur-containing amendments (and others) are added to "reclaim" only soils that are high in sodium content and often have a high pH. While large amounts of sulfur are added, the purpose is to replace the sodium with calcium and, in some cases, to bring the pH of the soil down to neutrality. Treatments with amendments can produce marked improvements in soil structure and increased soil productivity. Here the amount of sulfur added is carefully controlled (Richards, 1969), and the applications of sulfur represent, not a pollution, but rather a benefit to soils.

In anaerobic soils that are high in inorganic sulfur, bacterial activity reduces sulfate to sulfides. As a result, evolution of H_2S can harm or kill plants (Jacq and Dommerques, 1970; Ford, 1973). This is a well-known difficulty with paddy rice, and Vamos (1964) gives details on the problem in Hungary. These soils, which generate H_2S when wet, can be expected to become more acid as they are dried and the sulfides are oxidized to H_2SO_4. A similar reaction in soils reclaimed from the sea in Holland causes the extreme acidity produced when soils high in sulfide generate H_2SO_4 as they dry, until ground limestone is applied to complete the reclamation. The same phenomenon takes place throughout the world when wet soils high in inorganic sulfur are intermittently dried. The process is natural, but it will also occur where man's activity creates periodically wet soils and makes them high in inorganic sulfur.

Pollution of soils by sulfur may take place when industrially produced elemental sulfur is stored in the open. This is probably a small problem, but it is dramatic. In Alberta, Canada, elemental sulfur is produced at sour gas-processing plants, and the molten elemental sulfur is poured and solidified for storage as "blocks" or "slates." The wind erodes some fine dust from the stored elemental sulfur, and more when the sulfur is handled for shipment. The fine elemental sulfur is deposited on the soil within 2 or 4 km downwind, and depositions range from 1 to more than 100 tonnes/ha. Over several years the soils can become acidified to the extent that they have a pH as low as 1, and are completely barren of plant growth (Nyborg, 1974). This is an extreme case of pollution of the soil by sulfur.

5. SOIL ACIDITY AND ITS INFLUENCE ON PLANT GROWTH

Some soils are naturally acid, but the acidity has developed very slowly. These naturally acid soils have been formed through the depletion of bases (mostly calcium) as the soils were leached by high precipitation. In addition, residues from some kinds of vegetation (e.g., coniferous trees) form acid leachate; and, as well, the parent material of soils may originally be low in bases. As a generality, acid soils are found most frequently where precipitation is high and vegetation is coniferous, whereas neutral (and alkaline) soils develop as a result of low precipitation and grassland vegetation. Under agriculture, soils tend to become acidified. Removal of crops from the land slowly depletes the bases from the soil, and the use of ammonium-based nitrogen fertilizers will acidify soils very quickly. The kinds of sulfur fertilizers that acidify soils are as potent as nitrogen fertilizer in making soils acid; however, sulfur fertilizers are used at much lower rates and most soils do not need these fertilizers. Sulfur released as acidifying pollutants from industry (particularly SO_2) can probably result in soil acidification as important as that caused by nitrogen fertilizer. In fact, the acidification by industry may be more difficult to deal with because it affects both agricultural and forest soils.

Since acidification of soils is the main effect of SO_2, elemental sulfur, and some sulfur fertilizers, some details will be given here on the different kinds of effects on soils of acidity. Soils naturally vary in pH from the extremes of about 3.0 to 10, and soils outside this range are usually barren of vegetation. Any particular species of plant is unable to grow over this entire range of soil pH, but is suited only to a portion of the range. Thus the type of vegetation or crops that are supported by a soil changes with the pH of that soil. A soil pH of 5 to 8.5 is the range within which almost all agricultural crops can grow, and between pH 6 to 8 the greatest diversity of agricultural crops will grow with the highest yields. For soil pH below 7, various tables have been established, based on observation or experimental work, to show minimum pH values for good growth of various crops (Woodruff, 1967). In a table given by Jackson (1962), the minimum pH for alfalfa is 6.2; barley, 5.6; wheat and corn, 5.2; oats, 5.0; and grasses, between 4.6 and 5.0. It is of interest that the crops that will grow at the lowest pH values are blueberries at 4.2, and cranberries at 4.0. Consequently, we see that the pH of a soil governs indirectly its suitability for various crops.

Most commercially important forest trees can tolerate quite a low soil pH (at least in North America and northern Europe), with pines, spruces, and firs apparently growing in soils with a pH as low as approximately 3.5. However, the plant species comprising the forest undergrowth have a gradiency of tolerance to soil acidity. In Alberta, measurements on forest soils acidified by windblown elemental sulfur showed that when soil pH was 5 to 4 the number of species in the undergrowth was reduced, when pH was 4 to 3 only a few species grew, and when pH was less than 3 there was no longer any undergrowth (Nyborg, 1976, unpublished study). No direct information was found on relative tolerances among species of trees to soil acidity, but it seems likely that such tolerances occur.

Soil acidity affects the chemistry of soils and consequently the plants that are able to grow well on a particular soil. Soil acidity influences plants in a number of ways. Depending on species, they may suffer from toxic levels of soluble Al, Mn, and H, and from deficiency of Ca, Mg, and Mo. The soil pH, or the H^+-ion concentration, can be directly toxic to plant roots when the pH is about 4 to 3 (Jackson, 1967). When the pH is higher, the hydronium ion has little effect on plants, with the exception of legumes. Legumes can be affected by the H^+-ion concentration in the formation of nodules with symbiotic bacteria (*Rhizobium*) and in the actual symbiotic fixation of nitrogen in the nodules. Alfalfa is one of the plants most sensitive to soil acidity, and nitrogen fixation is curtailed primarily through the H^+-ion concentration in soil, even though the concentration is fairly low. Both nodulation and nitrogen fixation are retarded when the soil pH is less than 6.0 (Rice, et al., 1977). Although other cultivated legumes, and associated *Rhizobium,* are probably not as sensitive to soil acidity as alfalfa, they are recognized as the more sensitive among crops (Holding and Lowe, 1971).

In the last 15 years it has been widely recognized that the main way in which soil acidity impedes the growth of most agricultural crops is through aluminum toxicity (Jackson, 1967). When soils are of pH 5.5 to 5.0, soluble aluminum appears, and the solubility of aluminum increases as the pH decreases further (Hoyt and Nyborg, 1972). Toxicity due to manganese is another factor in soil acidity, although only some crops are sensitive to this element. Soluble manganese may be found at toxic levels in soils slightly above pH 5.0, and solubility increases with decrease in soil pH. One of the important characteristics of aluminum and manganese toxicity is that only slight acidification will increase the solubility of aluminum and manganese, and thereby will increase the toxicity from these two elements.

Calcium (and magnesium) deficiency was thought until recently to be the prime cause of damage to plants growing on acid soils, but it is now seen as only one of the causes. The calcium in soils is usually expressed as percent base saturation, which is the percent of the cation exchange capacity satisfied by calcium (and magnesium). Under field conditions with agricultural soils calcium deficiency is seldom seen, except in some specialty crops (tomatoes, potatoes, peanuts). Calcium is normally not deficient until it shows, let us say, less than 20% base saturation, and for magnesium the value is less than 5% base saturation; only soils that are too acid for agriculture usually have percent base saturations that low. Forest soils can easily have very low percent base saturation values, and Oden (1976) shows that many of the forest soils in Norway and Sweden have less than 20% base saturation. Acid sulfur fallout in Norway has been shown to leach calcium from acid soils (Overrein, 1972), and the growth of forest trees in Sweden has apparently been decreased (Jonsson, 1977).

Generally, the availability of phosphorus to plants is lower in acid than in neutral soils; there are several (and complicated) reasons. The micronutrient molybdenum is less available in acid than in neutral soils. A lack of molybdenum especially affects *Rhizobium* bacteria, used in symbiotic nitrogen fixation by legumes.

Soil acidity may also influence the rate of mineralization of nutrients from soil organic matter. The literature is divided on whether nitrogen is mineralized more slowly from the organic matter of acid soils, but in the case of phosphorus and sulfur (Jackson, 1967) mineralization is decreased in acid soil. It has long been known that the formation of nitrates from ammonium in soils (nitrification) does not take place as quickly in acid as in nearly neutral soils. As a generality, the formation of nitrates slows markedly at pH 5, and ceases below pH 4.5 to 4.0. This means that, whereas plants growing in nearly neutral soils feed on nitrates as their source of nitrogen, plants growing in very acid soils feed on ammonium. In acid soils some species of plants do not feed on ammonium, as nitrate is their source of nitrogen, and aluminum toxicity can be accentuated and the uptake of calcium and magnesium decreased (Jackson, 1967).

Hence, there are a number of reasons why acid soils are markedly less fertile, and sulfur may produce soil acidity.

In agriculture, acid soils, or acidified soils, are corrected by treating with ground limestone and working the limestone into the cultivated layer of soil (to a depth of about 15 cm). Ground limestone used in agriculture is usually of such a fineness that at least 50% passes through a screen with 60 meshes/in. Fineness of the limestone and thorough incorporation into the cultivated layer are needed to react the limestone with the soil. The $CaCO_3$ is very low in solubility, 0.001 g 100 ml water, but at the same time this characteristic slows any leaching of the $CaCO_3$. Limestone is either $CaCO_3$, $CaCO_3 \cdot MgCO_3$, or intergrades. Also used as a liming substance in agriculture is $Ca(OH)_2$. It has a solubility 100 times that of $CaCO_3$ but can raise soil pH to 12 (if it is applied at too high a rate), whereas $CaCO_3$ will not raise soil pH above approximately 8.5, regardless of rate.

The amount of ground limestone needed varies greatly with the texture and the soil matter content of agricultural soil. Generally loam-textured soils need approximately 2 tonnes $CaCO_3$/ha to raise the pH by 1 unit. In the range of pH 4 to 7 (or a slightly wider range), similar amounts of lime, or acids, will change the pH as much near neutrality as further down the pH scale. Limestone is usually applied at rates of 1 to several tonnes per hectare, and the application is made only periodically (perhaps every 5 years at the most). The practice of liming is well understood in agriculture (e.g., Pearson and Adams, 1967) and is feasible whether the soil acidity arises naturally or is due to cropping, use of nitrogen and sulfur fertilizer, or soil pollution. However, there is the matter of the cost of liming.

Liming is not practiced in forests or on wild land. Were liming used in forests, the lime would, of necessity, be topdressed and not incorporated. There was some diffusion of topdressed $CaCO_3$ to a depth of 5 or 7 cm in two extremely acidified soils (pH 2 to 3) so that the pH was raised to 5.0 during a year in an aspen forest (Nyborg, 1974) and in a coniferous forest (Nyborg, 1976, unpublished study). In the laboratory, soils from another coniferous forest required only approximately 450 kg $CaCO_3$/ha to raise the pH of a 15-cm depth of the topsoil by 1.5 units; or conversely, only 150 kg S/ha was needed to lower the pH by a similar amount. The application of liming materials in forests, however, involves practical questions about the rates and types of liming materials and the effects on plants.

6. SULFUR DEFICIENCY OF SOILS AND USE OF SULFUR FERTILIZERS

Although plants need nitrogen and sulfur in a ratio of approximately 10:1, about 30 million tonnes of nitrogen is used as fertilizer annually, compared to approximately 10 million tonnes of sulfur. This could indicate that far more sulfur is being used, relative to nitrogen to overcome sulfur deficiency, but this in fact is not so. Most of the sulfur applied in fertilizer is incidental to sulfur needed by crops. The fertilizer $(NH_4)_2SO_4$, a by-product of H_2SO_4, contains 24% sulfur, but most of the 5 million tonnes applied yearly is used simply as nitrogen fertilizer; likewise, much of the 5 million tonnes of single or double superphosphate (which contains calcium phosphate and calcium sulfate) applied annually is used simply as phosphorus fertilizer. At the same time, superphosphate is applied both as phosphorus and as sulfur fertilizer in Australasia, where soils are frequently sulfur deficient. However, there is a continuing trend the world over toward the manufacture of more concentrated phosphate fertilizers (e.g., triple superphosphate and diammonium phosphate), and in the process the H_2SO_4 used to manufacture the phosphates is removed so that little sulfur is left in the fertilizer. Some $(NH_4)_2SO_4$ is added in small amounts to other nitrogen fertilizers to overcome sulfur deficiency (Bixby and Beaton, 1970).

The actual area of agricultural land that is sulfur deficient is not known precisely, but the largest proportion of sulfur deficient soils occurs in Australasia, western Canada, and the Pacific Northwest and California in the United States. The proportion of sulfur-deficient soils in Alberta has been estimated at more than 10% or 2.5 million ha (Beaton et al., 1971). Parts of Australasia and New Zealand apparently have higher proportions of sulfur-deficient soils (Jones et al., 1975; Blair and Nicolson, 1975; Walker and Gregg, 1975), used especially for forage crops for pasture (grass and clover). In addition, sulfur deficiency has been demonstrated in most countries of the world.

Legumes (clovers and alfalfa) do not need nitrogen fertilizer but require sulfur fertilizer on deficient soils. Nonlegumes usually need sulfur fertilizer only when an increase in growth from the application of nitrogen fertilizer causes sulfur deficiency. Sulfur deficiencies can be very severe in the field, essentially causing crop failure, of, for example, alfalfa and cereal grain (Nyborg, 1968) and rapeseed (Nyborg et al., 1974). Sulfur-deficient crops are low in sulfur and protein content (e.g., Martin and Walker, 1966; Stewart, 1969) and high in nonprotein content (Baker, et al., 1973), that is, the crops are of poor quality as foods or feeds. For efficient feeding, livestock need feeds with an N:S ratio of 15:1 (or less) to supply sulfur and the two amino acids cystine and methionine (e.g., Allaway and Thompson, 1966; Garrigus, 1969). Even when the growth of a crop does not increase from sulfur fertilizer (i.e., the crop is marginally sufficient in sulfur), addition of fertilizer sulfur may improve the quality of the crop as food or feed. Work by Stewart (1969) showed that an increment of sulfur fertilizer that hardly increased the growth of wheat markedly decreased the plant content of nonprotein nitrogen (amino acid-N, amide-N, and nitrate-N) and increased the content of true protein. Likewise, Baker et al. (1973) showed that even when sulfur fertilizer did not increase the yield of orchard grass it decreased the plant content of nonprotein nitrogen.

Sulfur deficiency is most likely to be caused by fast-growing agricultural plants, either legumes or nonlegumes fertilized with nitrogen; that is, in some soils not enough sulfur is available to keep pace with the nitrogen supply. Since about 30 million tonnes of fertilizer nitrogen is used yearly, and since about 13 million tonnes of nitrogen is fixed by agricultural legumes annually, a tenth of this amount (4.3 million tonnes) per year of fertilizer sulfur would be needed at the very maximum. Here the assumption was made that all soils that received nitrogen fertilizer or grew legumes were deficient in sulfur, and it would be more reasonable to assume that only 10% of the land is now sulfur-deficient; then the amount of fertilizer sulfur needed would be approximately 0.4 million tonnes. Regardless of what is now the actual figure, as the amount of nitrogen fertilizer used continues to increase year by year, the need for sulfur fertilizer also continues to increase (except for areas receiving heavy depositions of atmospheric sulfur). More soils will be depleted as the rates of nitrogen fertilization are increased and crop yields rise.

The various commercial sulfur fertilizers are outlined by Bixby and Beaton (1970), and most of them are acidifying. Elemental sulfur is also used to constitute a slow-release nitrogen fertilizer, where urea is coated with elemental sulfur in amounts of about 10 to 20%. Under specialized conditions this slow-release fertilizer will give increased efficiency of urea (e.g., Allen and Mays, 1971).

A number of the nitrification inhibitors of ammonium-based nitrogen fertilizer contain sulfur. Examples are CH_3SSCH_3 (Bremner and Bundy, 1974), CS_2 (Ashworth et al., 1975), and $(NH_4)_2CS_3$ (Malhi, 1977). All these compounds are acidifying in soils, but rates of application are small. The purpose of the inhibitors is to decrease losses of available fertilizer nitrogen by keeping the element in the ammonium rather than the nitrate form. In addition, the inhibition of nitrate formation lessens the acidification associated with ammonium-based fertilizers. The use of inhibitors and the placing of nitrogen fertilizer in rows in the soil essentially stop acidification of soils by the nitrogen fertilizer (e.g., Malhi, 1977). This approach to stopping soil acidity from nitrogen fertilizers is still in the experimental stage, but could be put into practice very quickly. If this technique involving sulfur-bearing inhibitors, and others, is used in practice, acidification of soils by nitrogen fertilizers will be greatly reduced, and will be less than the acidification by SO_2 deposition on agricultural land.

From time to time the acidification of agricultural land by SO_2 is compared to acidification by agricultural fertilizers, especially nitrogen. This comparison is made as a justification for SO_2 emissions because soil acidification due to these emissions is not as great as acidification caused by the use of nitrogen fertilizers and SO_2 emissions are beneficial to sulfur-deficient soils. Whereas acidification of agricultural soil by nitrogen fertilizers can be avoided, SO_2 emissions do not fall uniformly and so cannot overcome sulfur-deficiency in all cultivated land. If man-made SO_2 emissions fell uniformly on agricultural land the world over, the rate of deposition would be about 40 kg/ha, but since only 11% of the land surface is cultivated, there would be only 4 kg of sulfur allocated her hectare. This calculation is rather nonsensical, since in most of these areas where SO_2 emissions are heavy enough to overcome sulfur deficiency, more sulfur will be deposited than is needed, whereas in other areas with sulfur deficiency deposits will be inadequate or entirely lacking. In other words, the sulfur depositions are variable in amounts with distance (and direction) from SO_2 sources. This was shown clearly in Johansson's work in Sweden (1959), near an oil-burning plant that was a large SO_2 source. At five of the stations, set out at locations from less than 1 km (station K1) to 15 km (station KV111), the average concentrations of SO_2-S in the air ranged from 78 ug/m^3 at station K1 to 11 ug/m^3 at station KV111. Uncropped soils set out in pots for a period of 6 years accumulated 7 g/pot at K1, and 2 g at KV111. Pots cropped to barley showed sharp decreases in concentration of sulfur in going from K1 to KV111. Although Johansson's work may be somewhat atypical, it demonstrated that SO_2 emissions can be sharply reduced within short distances from the SO_2 source.

At least some crops can take in nearly sufficient SO_2 to meet their sulfur requirements through their leaves, and at the same time are not greatly harmed by the SO_2 concentration high enough to supply these needs (Coleman, 1966; Evans, 1975). Faller (1971) found that tobacco plants receiving an SO_2 content in the air of about 1350 ug/m^3 were nourished as well as tobacco plants fed on a root solution of sulfate, with no harm from the SO_2. Cowling et al. (1973) fed ryegrass plants grown in sulfur-deficient soil with SO_2 applied only to the leaves at a concentration of 0.045 ppm, and found that the plants were no longer deficient in sulfur as long as they were kept on the SO_2, and yielded almost as high as when the sulfur-deficient soils were fertilized with 10 ppm of sulfate. However, the concentrations of SO_2 used in these experiments were much above those usually found in the field, even in "polluted" areas. More important than leaf feeding with SO_2 may be soil adsorption of SO_2 from very low concentrations in the air. For example, Cox (1975) exposed three soils, wet and dry, for 2 days to 0.03 ppm of SO_2 and found that the soils gained from 3 to about 15 ppm of sulfate-S. Sulfur "pollution" can apparently nourish plants grown in soils not too well stocked with sulfur through the two lesser known mechanisms of leaf feeding by SO_2 and adsorption of SO_2 by soils, in addition to precipitation and fallout of particulates. Application of sulfur by these mechanisms, however, cannot usually overcome sulfur deficiency over great distances from the source of sulfur.

Sulfur deficiency in agricultural soils probably causes more lowering of crop yield or quality than does sulfur pollution of soils, simply because acidified soils are usually limed as needed. Sulfur deficiency in soils is overcome by sulfur pollution, but only very heavy sulfur deposits will satisfy the requirements of crops that have high needs for the element, such as rapeseed and alfalfa (Spencer, 1975). As a source of sulfur for crops, pollution sulfur has some disadvantages: the unreliability of the amount of sulfur from year to year, losses through leaching if the sulfur is deposited when the crop is not growing, and greater combination with soil organic matter. Agronomically, sulfur is most efficient when applied at the start of the growing season and only in the amount needed by the crop.

Correction of sulfur deficiency and of sulfur pollution can take place at the same time. Consider sulfur fertilizer as a "pollution" that slowly acidifies the soil, but initially promotes good growth of alfalfa. McCoy and Webster (1977) report that in long-term plots a treatment received $(NH_4)_2SO_4$ at a rate of 9 kg S/ha annually since 1939. Large increases were obtained in hay yield and quality. By 1968 the treatment was no longer able to grow alfalfa, and soil pH had dropped from 5.8 to 5.3. However, when the treatment was limed in 1972, along with receiving a sulfur fertilizer application, the alfalfa grew well.

7. AMENDMENT OF SODIUM-AFFECTED SOILS BY SULFUR COMPOUNDS

Soils that are high in sodium, and usually high in salts as well, have a poor structure and admit water only very slowly. Depending on their specific characteristics, the soils can be treated with gypsum ($CaSO_4 \cdot 2H_2O$) or the acidifying compounds, elemental sulfur and H_2SO_4 (and sometimes CaS_5 or $FeSO_4$). The acidifying compounds are added to reduce pH values when they are too high (because of sodium) and to increase the solubility of any $CaCO_3$ that is present. Gypsum may be added with or without the acidifiers.

The "reclamation" of these natural soils involves replacing the sodium with calcium and will result in better, friable soil structure and much faster water infiltration. Details on the principles of reclamation can be found in Richards (1969) and Loveday (1975). The amounts of amendments needed are determined by chemical tests on the soil, and requirements are quite variable from soil to soil. However, rates of gypsum addition might be from 5 to 10 tonnes/ha. The reclamation of these soils can produce very large increases in yield (Tisdale, 1970), and the benefit will last for many years.

Sodium-affected soils are a serious problem to agriculture, but, in practice, reclamation has not proceeded far as yet. In western Canada, for example, there are approximately 5 million ha of sodium-affected land where cultivation is difficult or impossible. Apart from correcting structure and water movement, reclaimed soils are no longer saline, and the soil pH is now nearly neutral. The sulfur added may be precipitated as $CaSO_4$ or be moved as Na_2SO_4 into the subsoil.

Miyamoto et al. (1975) noted that in the southwestern United States copper smelters may soon produce a yearly surplus of about 2 million tonnes of H_2SO_4, which could be used to reclaim the millions of hectares of calcareous soils high in sodium. The rates of application of H_2SO_4 might be 4 to 12 tonnes/ha on the calcareous sodium-affected soils, and excess H_2SO_4 (resulting from reducing SO_2 emissions) would find extensive areas where its addition to alkaline soil would increase crop yields.

8. SULFUR DIOXIDE DEPOSITION

While some 60 million tonnes of man-made SO_2 emissions are given off each year, and the majority comes down on land, some questions about the fate of the sulfur and its effect on soils and plants have not been answered with certainty. The amounts of sulfur in rain falling in remote rural areas and in industrial areas have been measured for many years, and the contribution of sulfur to the land through rainfall has been found to range from as little as 1 kg/ha in remote areas to sometimes over 100 kg/ha near cities or industry. Rains high in sulfur content are usually acid and contribute significant amounts of H_2SO_4 to soils, at least in western Europe (Anon., 1972). The questions are, how much of the SO_2 comes down as particulates, through intercepted rain, and as SO_2 adsorbed directly by soils; and to what degree does each of these mechanisms contribute sulfur and add to the acidification of soils? These three mechanisms in total probably far exceed rainfall in contributing sulfur to soils (Johansson, 1959; Coleman, 1966; Fox, 1976; Nyborg et al., 1977). Other questions concern the reaction rate of the acidifying sulfur in the fallout with soils, and the effect of soil acidification on plants.

Often, a minority of the sulfur in rain comes down as H_2SO_4, and calculations on data presented for northwestern Europe in 1965 show that about 15 to 40% of the sulfur in the precipitation was in acid form (Anon., 1972). Oden (1976) gives pH and sulfur values of precipitation at three locations in southern Sweden and Norway and one location in Denmark: the pH averaged about 4.5, the sulfur equaled 10 to 20 kg/ha, and a portion of this sulfur came down as H_2SO_4. Likewise, in the northeastern United States there is acid precipitation with a significant portion of sulfur as H_2SO_4 (Johnson et al., 1972). Tabatabai and Laflen (1976) found an average of 16 kg S/ha for precipitation at five rural sites in Iowa. However, the pH of the precipitation averaged about 6.2, indicating no significant H_2SO_4. In other words, acid precipitation will slowly acidify soils in some regions, but in other regions the precipitation is not acid, even though it may be high in sulfur content.

Tabatabai and Laflen (1976) point out that the amounts of sulfur found in precipitation in Iowa (13 to 17 kg/ha) may explain the sufficiency of sulfur for crops, even though many Iowa soils do not contain enough plant-available sulfur (Tabatabai and Bremner, 1972). Tabatabai and Laflen (1976) also state that direct adsorption of SO_2 from the atmosphere by soils and plants occurs in Iowa. An average figure for the yearly deposition of sulfur by precipitation in rural Wisconsin is 16 kg/ha (Hoeft et al., 1972). Furthermore, alfalfa grown in a greenhouse in a suburban area derived 44% of its sulfur uptake by direct absorption of SO_2 from the atmosphere.

In many parts of the world the amounts of sulfur deposited by precipitation are small (Coleman, 1966; Blair and Nicolson, 1975). For example, in Alberta sulfur deposited with precipitation was only 2 to 4 kg/ha yearly in the central part of the province, where there are SO_2-emitting plants (Walker, 1969). In other parts of Alberta the amount of sulfur in precipitation is usually 1 to 2 kg/ha. Deficiencies of sulfur occur in central Alberta for alfalfa and red clover (Walker, 1969), and in northern Alberta (Nyborg, 1968) for grain crops as well.

Emitted SO_2 also forms sulfates with cations in the atmosphere, and these settle out as particulates (e.g., $CaSO_4$). When the atmosphere has significant amounts of ammonia, the particles can also contain $(NH_4)_2SO_4$ or related ammonium compounds (Brosset, 1976). Brosset (1976) has results that indicate acid particles become more or less neutralized by ammonia with formation of $(NH_4)_2SO_4$, $(NH_4)_3H(SO_4)_2$, and NH_4HSO_4, and these particles will then acidify land. Reuss (1977) also points out that ammonium sulfate salts in particles that fall out because of SO_2 emissions will cause acidification through oxidation of ammonium to nitrate. It is well to remember that $(NH_4)_2SO_4$ in particles or in precipitation will react with soils in the same way as $(NH_4)_2SO_4$ used as a fertilizer, with concomitant soil acidification. The question remains of the frequency and extent of fallout of $(NH_4)_2SO_4$.

Rain intercepted by trees becomes more acid in areas with SO_2 emissions than in "clean" areas. Baker et al. (1977) found that the "throughfall" and "stemflow" of coniferous trees were markedly depressed in pH at exposed sites as compared to the control site. Although the rain not intercepted was only slightly acid, the stemflow from trees exposed to SO_2 was quite acid (pH 3 to 4 at one exposed site). Brosset (1976) suggests that deposition of ammonium sulfate particles on coniferous trees will give rise to intercepted rain that will acidify the soils beneath the trees. Abrahamsen et al. (1976) have shown that rain running over the crowns of spruce trees is depressed in pH and increased greatly in sulfur content in exposed areas. Therefore, although the exact mechanisms are not known by which intercepted rain is made higher in sulfur and lower in pH for coniferous trees exposed to SO_2 emission, this process apparently can acidify soils under forest trees at a substantial rate.

In some areas, at least, the majority of SO_2 emissions are apparently deposited by direct adsorption of the SO_2 by soils. Johansson (1959) found that in potted soils, set out downwind from an SO_2 source at distances ranging from less than 1 to 7 km, the ratio of sulfur gained through direct adsorption from the air exceeded the amount of sulfur gained through precipitation by factors of 7 to 2. In a similar experiment conducted by Nyborg et al. (1977) in Canada, the amount of sulfur added to soil directly from the air exceeded the sulfur added in rain by factors of 9 to 7, downwind from an SO_2 source. Cox (1975) set pots of three soils under canopies at distances of approximately 15 to 80 km downwind from two large SO_2-emitting smelters in Ontario. After 19 weeks of exposure, the pH of the 0- to 1-cm layer of a silty clay loam soil had been depressed by 0.2 and 0.4 unit at the stations 15 and 25 km from the plants, respectively, as compared to a control station. The other stations, up to 80 km away, showed smaller but statistically significant lowering of pH values. At the two stations at 15 and 25 km, the soluble sulfate-S content of soil increased by 15 and 70 ppm, respectively. In a similar experiment conducted by Nyborg et al. (1977), a lowering of pH values by 0.2 unit was obtained on two soils set out for 3 months directly downwind at distances up to 30 km from an SO_2-emitting plant. These experiments conducted in the field have the disadvantage of differences in soil temperature and moisture from station to station, and consequent differences in rate of mineralization and nitrification in experimental soils, that is, the amounts and ratios of ammonium and nitrate in soils, as well as their natural fluctuations, tend to dominate soil pH in the short term. However, exposure of soils to SO_2 under controlled conditions, but at fieldlike concentrations, has shown that depression of soil pH and gain in extractable sulfate-S occur at a surprisingly high rate (Cox, 1975). The rate of sulfur deposition could be calculated from Cox's results; when the average of three moistened soils treated with an atmosphere of 0.03 ppm of SO_2 for 48 hr was used, the gain in soluble sulfate-S was found to be approximately 0.6 kg/ha. At the same time, Ghiorse and Alexander (1976) have demonstrated that part of the SO_2 reacts with soil to form insoluble sulfur.

The rates and amounts of SO_2 adsorbed by soils are not known to any accuracy under field conditions, but adsorption seems an important, or even the most important, mechanism for depositing SO_2. In addition, this mechanism is the most acidifying to soils, in that the SO_2 deposited is soon turned into H_2SO_4 and no neutralizing substance is added from the SO_2.

The distance over which SO_2 emissions can travel before becoming completely oxidized is a matter of some hundreds of kilometers or more (e.g., Fisher, Part I; Ottar, 1976). A large portion of the SO_2, however, may be deposited within some tens of kilometers from a source (e.g., Benarie, 1976).

The acidification of soils has taken place rather definitely near large smelters. Cox (1975), working in the Sudbury area of Ontario, found that the soils lying up to 40 km to the northeast had been lowered by about 1 pH unit, and had increased in sulfate-S in the upper layer of their profiles, as compared to control profiles. In the same area, Hutchinson (1977) found that at distances of 14 km or less from a particular smelter pH of the top layer (0 to 5 cm) of soils was less than 3.0 in most cases. Soil samples taken from 5 to 10 cm and 10 to 15 cm were only slightly greater in pH. As compared to soil samples taken at greater distances, these soils were elevated in soluble sulfate, aluminum, and nickel. The toxicity of aluminum and nickel to vegetation was the probable cause of the inability of natural or cultivated plants to grow, but application of lime (which reduces the solubility of aluminum and nickel) made possible the growth of cultivated plants in field experiments. Hutchinson (1977) reports that the Sudbury area, where three large smelters were established at the turn of the century, now has at least 400 km^2 that grow little vegetation and where erosion has removed most of the soil. Hutchinson (1977) also reports that "sheet erosion of thousands of acres has already occurred," apparently because of SO_2 emissions and the subsequent acidification of soils and the increased solubility of metals (deposited by smelters) in acid soils. Katz et al. (1939), in a classical study of the effect of SO_2 emitted by a smelter in British Columbia, demonstrated that soil pH had been depressed to approximately 3.5 to a distance of at least 20 km from the smelter, with an accompanying increase in soil content of sulfate, compared to a natural soil pH of about 5 to 6 at greater distances. Likewise, the percent base saturation had been depressed to less than 20%, whereas more distant soils had about 50%. Bohn (1973) reported that a smelter in operation for 65 years in Arizona had changed a calcareous soil to one of pH 5.5 Oden (1976) found that the soils were acidified in a locality (Falun, Sweden) where alum shale had burned for decades, giving off SO_2. Outside the locality the degree of base saturation in the soil is about 25%, but in Falun the value had dropped to 10%. In agreement, the soil pH was lower in the locality than outside it. This evidence of soil acidification by smelters, gathered by various researchers in several countries, shows without question that soil acidification proceeds where SO_2 emissions are large and soil measurements are made locally.

In view of the fact that soil acidification by SO_2 emissions has been of much concern only within the last 10 years, measurements of the rate of soil acidification on a larger local or a regional basis cannot be made with much precision. Soil acidity will naturally fluctuate with season and year, and what is needed to isolate long-term changes in soils is sampling and analysis (by the same methods) from year to year. Unfortunately, by the time precise rates are established on some soils, acidification in forests may be well under way.

Oden (1976) reported that Swedish rivers discharge three to five times as much anthropogenic sulfate since the period 1909 to 1923. In southeastern Sweden the increase in sulfur discharged from rivers was 20 kg S/ha of land annually. In other words, the emissions of SO_2 reaching the land surface in Sweden have greatly increased in the last half of the century. Maps prepared for recent results of pH and base saturation for selected soil profiles (Anon., 1972; Oden, 1976) showed that soils of southwestern Scandanavia had the lowest values, and apparently these podozolic soils were lowered in content of bases in comparison to normal podzols. Oden (1976) suggested that this change in soils was caused by atmospheric acidification during the past hundred years. Jonsson (1977) obtained results suggesting that growth of Scots pine and Norway spruce has been reduced by acidification of forest soils in parts of Sweden. This work in Sweden indicates a regional effect of SO_2 emissions in acidifying soils.

Malmer (1976) is of the opinion that severe changes in soils could not yet be expected in Sweden because the acid precipitation phenomenon has taken place only since about 1950. He also points out that to detect acidification in its early stages is difficult because of the natural variation in soil types, and the variation of these types in resistance to acidification. Wiklander (1975), in laboratory work, has shown that addition of neutral salts along with H_2SO_4 reduces the acidifying effect of H_2SO_4. This "salt effect" in reducing the rate of acidification takes place in very acid podzolic and some brown earth soils, but not in neutral or slightly acid soils. Frink and Voigt (1977) are of the opinion that acid precipitation is not a new phenomenon in the northeastern United States. They state that additions of acid precipitation are insignificant in comparison to the acidity produced by fertilizers on agricultural soil, and that, although the effect of acid precipitation is greater on unmanaged land, it is probably exceeded at present by the effect of biological cycling of nitrogen and sulfur. On the other hand, Baker et al. (1977) found that seven soils taken at 18 km or less from two SO_2-emitting plants had lower pH and increased exchangeable aluminum values than did four soils 32 km or more from the plants. The mechanism of SO_2 deposition on the soil, in the work of Baker et al. (1977), was not acid precipitation; precipitation was nearly-neutral and low in sulfur. Halstead and Rennie (1977) reviewed the literature on effects of SO_2 on soils in Canada, despite the limited published work on the subject. They concluded that some of the effects of SO_2 on soils were an increase in acidity, sulfate content, and potentially toxic ions (aluminum, manganese); a decrease in bases and in the activity of microorganisms (nitrifiers and nitrogen-fixers in particular); and, in forests, soil erosion after the death of ground-cover vegetation. Hence authors vary in opinion on the amount of damage that has occurred from SO_2 and the rate at which such damage proceeds.

The complexity of the effect of acid precipitation on forest soils is emphasized by Tamm (1976, 1977). The sensitivity to pH of nitrogen fixation, mineralization, and nitrification is cited. Tamm reports an experiment in which small additions of acid sulfur decreased the evolution of CO_2 from the soil, but increased the rate of mineralization of nitrogen. An increase of ammonium and nitrate in the sulfur fallout plays a role in acidification, in that nitrification is acidifying and so is plant uptake of ammonium (Oden, 1976), and these processes are affected by acidification due to sulfur.

In work with SO_2 emissions, acidified soils are usually dealt with in terms of base saturation and pH, and apparently one of the fundamental causes of plant damage is lack of calcium. Base saturation of less than 20% and pH of less than 4.0 are reported in the literature (e.g., Anon., 1972; Oden, 1976). Agricultural crops were harmed by aluminum in soils with base saturation of more than 60% and pH slightly higher than 5.0 (Hoyt and Nyborg, 1971, 1972). It is therefore probable that some vegetation in forests or on wild land will be hurt by depressed pH values in the range of 5.0 and the concomitant increase in extractable aluminum.

9. ELEMENTAL SULFUR

Windblown elemental sulfur from storage piles can result in heavy local deposits (1 to 100 tonnes/ha, or more). Soils become completely barren, with pH values of 1 to 2. Reclamation by adding $CaCO_3$ in large amounts makes it possible for these soils to grow vegetation immediately (Bertrand, 1973; Nyborg, 1974). These results show that, at least in the short term, acidified soils can be restored without damage, as measured by plant growth.

The quick oxidation of fine windblown sulfur in the field stopped at about pH 2, and thereafter the deposited elemental sulfur accumulated (Nyborg, 1974). On liming, the elemental sulfur was oxidized at a rate of at least 10 tonnes/ha yearly. The application of $CaCO_3$ at a rate of 600 tonnes/ha in the field, and 1200 tonnes/ha in the greenhouse, gave normal yields and growth of barley or grasses. When soils that had a high content of elemental sulfur were limed, crops initially required heavy applications of nitrogen fertilizers to prevent extreme nitrogen deficiency. The nitrogen fertilizer also accelerated the oxidation of elemental sulfur.

Bertrand (1973) applied 224 tonnes lime/ha to two acidified soils (pH 1.4 and 2.1) that contained elemental sulfur, and sowed the soils with oats. With pure $CaCO_3$, or $CaCO_3$ with 5% $MgCO_3$, the oats grew well; but when larger portions of $MgCO_3$ were added, the oats scarcely grew because of elevated soluble salt content ($MgSO_4$). In other words, limestone needs to be almost free of $MgCO_3$; otherwise soils will become saline. In addition, Bertrand (1973) demonstrated that the crops grown after liming at the two locations had normal contents of N, P, K, S, Ca, Mg, Na, Al, Mn, Zn and Fe. It seems that severely sulfur-acidified soils can be restored to grow normal plants again.

10. CONCLUSIONS AND SPECULATIONS

The main way that sulfur "pollutes" soils is simply by making them acid. Soils become acid from sulfur only when acid-generating forms of sulfur are added to soils; for example, elemental sulfur or $(NH_4)_2SO_4$ is used as fertilizer, or a portion of the SO_2 emissions from industry reaches the soil as H_2SO_4, $(NH_4)_2SO_4$, or as SO_2 itself. When soils become acid, there are several ways by which their fertility is substantially reduced: the nodulation and the fixation of nitrogen legumes fail because of the high hydronium-ion concentration (this mechanism takes place at a pH of 6.0 and lower, with alfalfa); the solubility of aluminum and manganese begins to increase with decreases in pH below 5.5, and the toxicity of these two elements affects some species of plants to the extent that they will not grow; the plant availability of the most important nutrients, nitrogen, phosphorus, and potassium, is lowered; and calcium and magnesium are leached from soils that are naturally acid and are already low in these two nutrients. Acid soils do not recover when emissions of sulfur cease. The only method of rectifying these soils is by adding ground limestone. Also, if emissions are lowered to a level more tolerable to human beings and plants, soils will continue to take in sulfur, although at a lower rate.

Two obvious distinctions should be made. Acidifying fertilizers which contain sulfur or nitrogen or both can be replaced by neutral fertilizers that supply the same nutrients, but there is no known way (short of liming) of preventing acidification of soils by SO_2 emissions. Agricultural soils, when acid, can be easily limed (although a cost is involved), but with forest land that has grown very acid the application of lime would be very costly and the reaction of forest soils to liming is not well known. The scientific literature on the topic of sulfur acidification of soils by anthropogenic SO_2 emissions is in general agreement with few exceptions, that this process takes place in the industrial regions of the world. The following bold calculation may demonstrate the order of magnitude of soil acidification. Consider that man-made SO_2 emissions are about 60 million tonnes annually; and if only 20% of these emissions come down as strong acid [as H_2SO_4, as $(NH_4)_2SO_4$ in precipitation and particulates, and as SO_2 adsorbed by soils], this is still 12 million tonnes annually. If this amount were concentrated over 12 million ha, there would be 1 tonne acid S/ha·year, soil pH would be lowered by approximately 1 unit, and the fertility of most of the soils would be seriously impaired. Of course, SO_2 emissions are much more widely spread, and not in a uniform manner; also, they do not come down in the amounts per hectare used in the calculation. Nevertheless, accumulative damage from SO_2 to soils is alarming when viewed in the long run, especially for forest soils, where amelioration is not ensured.

Deficiency of sulfur in crops is of major importance to agriculture. Sulfur comes after the other fertilizer nutrients, nitrogen, phosphorus, and potassium, in frequency of deficiency. Possibly some 10% of cropped soil is deficient in sulfur for yield or crop quality. When a soil is very sulfur-deficient, there is almost a failure of the crop, and the crop will be of poor quality as a food or feed (low in protein, high in nonprotein nitrogen, including nitrates, and low in cystine and methionine). Even when fertilizer sulfur does not increase crop yield, it may narrow the N:S ratio toward 10:1, thereby increasing feed quality. In areas low in SO_2 emissions the need for sulfur fertilizers will probably increase with increased use of nitrogen fertilizers and the resulting heavier crops that need more sulfur. Sulfur deficiency in crops can be rectified by fertilizers (supplying approximately 10 to 20 kg S/ha) or by SO_2 emissions. When SO_2 emissions are expected to meet the sulfur needs of crops, more will be needed per hectare from this source than from fertilizer sulfur because the latter is applied in such a manner that losses are small. Fertilizer sulfur, when compared to fertilizer nitrogen, is inexpensive by weight and little is needed. These facts are presented to make the point that the presence of SO_2 emissions is not a great advantage for farmers as a means of overcoming sulfur deficiency in crops.

The use of neutral salts ($CaSO_4$, Na_2SO_4, K_2SO_4) as sulfur fertilizers will not harm the soil. The acidifying sulfur fertilizers [e.g., elemental sulfur, $(NH_4)_2SO_4$] will not greatly change calcareous soils. Soils high in organic matter and having a fine texture will acidify slowly, but more lime is required to neutralize such soils. Soils low in organic matter and having a coarse texture will acidify quickly, but require relatively small amounts of lime for neutralization. Whether to use lime, or to use neutral salts for a sulfur fertilizer instead, is a decision made by farmers on the basis of cost.

The deposition of SO_2 on agricultural soils is similar in action to the use of acidifying sulfur fertilizers. Some soils (calcareous) are immune to acidification, but those soils are in the minority. Most soils will need liming after a time. Depending on the climate, liming may be a regular practice with agricultural soils. The use of ammonium-based nitrogen fertilizers acidifies soils so that lime is needed periodically for that reason alone. All nitrogen fertilizers, however, do not acidify soils; neutral salts such as $Ca(NO_3)_2$ and KNO_3 do not have this effect, but are more expensive than ammonium-based fertilizers. Even acidification by ammonium-based fertilizers can be eliminated by placement or use of nitrification inhibitors; these two methods are mostly in the research stage, but will probably be in practice soon. The role of SO_2 emissions, as compared to nitrogen (and sulfur) fertilizers, in making agricultural soils more acid is difficult to determine when considering different regions and areas. Fortunately, agricultural soils can be rectified easily, but there is the cost to consider.

Acidification of forests by SO_2 emissions was brought to light only about 10 years ago, and consequently much is yet to be learned about the subject. Drastic acidification of soils is evident adjacent to very large SO_2-emitting plants that have been in operation for some time. The amount of soil acidification that may have occurred in areas more distant from the SO_2 sources, where depositions are light, cannot be estimated with any precision, because soil acidity in forests was not extensively recorded until the past decade. The effect of H_2SO_4 from SO_2 emissions will vary from one type of soil to another, and the influence of acidified soils on vegetation will vary from species to species. Putting aside these important complexities, we can say that forest soils in areas or regions with SO_2 emissions have probably increased in acidity, whether measured as pH, base saturation, exchangeable calcium, or extractable aluminum. The effect on growth of vegetation will probably have begun in some regions (the growth rate of pine and spruce has apparently decreased in Sweden), and will become apparent with time. It will require many years of research to determine the rates of deposition of acid sulfur in various areas, the rates of acidification of various soils, and the effects on different species of vegetation. By that time, the damage to soils and vegetation will have progressed. As this research advances, it would be well if the topic of liming of forest soils also received sufficient attention. Since liming can rectify extremely acid agricultural soils, there is good reason to think that it will also benefit acid forest soils, although there may be a slow reaction in neutralizing the entire soil profile. Even though spreading lime in forests would be costly, it might be a feasible way to maintain the productivity of forest soils, barring a reduction in SO_2 emissions.

REFERENCES

Abeles, F. B., Craker, L. E., Forrence, L. E., and Leather, G. R. (1971). *Science* **173,** 914-916.

Abrahamsen, G., Hornvedt, R., and Tveite, B. (1976). In L. S. Dochinger and T. A. Seliga, Eds., *Proceedings of the First International Symposium on Acid Precipitation and the Forest Ecosystem.* U. S. Department of Agriculture, Forest Service, Gen. Tech. Rept. NE-23, pp. 991-1009.

Allaway, W. H. and Thompson, J. F. (1966). *Soil Sci.* **101,** 240-246.

Allen, S. E. and Mays, D. A. (1971). *Agr. Food Chem.* **19,** 809-812.

Anon. (1972). *AMBIO* **1,** 15-20.

Ashworth, J., Briggs, G. G., and Evans, A. A. (1975). *Chem. Ind.,* pp. 749-750.

Baker, A. S., Mortensen, W. P., and Dermanis, P. (1973). *Sulphur Inst. J.* **9,** 14-16.

Baker, J., Hocking, D., and Nyborg, M. (1973). In D. Hocking and D. Reiter, Eds., *Proceedings of the Workshop on Sulphur Gas Research in Alberta.* Northern Forest Research Centre, Inf. Rept. NOR-X-72, Edmonton, pp. 98-102.

Baker, J., Hocking, D., and Nyborg, M. (1977). *Water Air Soil Pollut.* **7,** 449-460.

Barrett, E. and Brodin, G. (1955). *Tellus* **7,** 251-257.

Barrow, N. J. (1975). In K. D. McLachlan, Ed., *Sulphur in Australasian Agriculture.* Sydney University Press, pp. 50-57.

Beaton, J. D., Tisdale, S. L., and Platou, J. (1971). *Crop Responses to Sulphur in North America.* Sulphur Institute, Tech. Bull. 18, Washington, D. C., 29 pp.

Benarie, M. (1976). In L. S. Dochinger and T. A. Seliga, Eds., *Proceedings of the First International Symposium on Acid Precipitation and the Forest Ecosystem.* U. S. Department of Agriculture, Forest Service, Gen. Tech. Rept. NE-23, pp. 251-263.

Bertrand, R. A. (1973). M. Sc. thesis, University of Alberta.

Bettany, J. R., Stewart, J. W. B., and Halstead, E. H. (1973). *Soil Sci. Soc. Am. Proc.* **37,** 915-918.

Bixby, D. W., and Beaton, J. D. (1970). *Sulphur-Containing Fertilizers and Applications.* Sulphur Institute, Tech. Bull. 17, Washington, D. C. 27 pp.

Blair, G. J., and Nicolson, A. J. (1975). In K. D. McLachlan, Ed., *Sulphur in Australasian Agriculture.* Sydney University Press, pp. 137-143.

Bohn, H. L. (1973). *J. Environ. Qual.* **1,** 372-377.

Bremner, J. M., and Bundy, L. G. (1974). *Soil Sci. Biochem.* **6,** 161-165.

Brosset, C. (1973). *AMBIO* **2,** 2-8.

Brosset, C. (1976). In L. S. Dochinger and T. A. Seliga Eds., *Proceedings of the First International Symposium on Acid Precipitation and the Forest Ecosystem.* U. S. Department of Agriculture, Forest Service, Gen. Tech. Rept. NE-23, pp. 159-179.

Campbell, C. A., Paul, E. A., Rennie, D. A., and McCallum, K. J. (1967). *Soil Sci.* **104,** 217-224.

Chase, F. E., Corke, C. T., and Robinson, J. B. (1968). In T. R. G. Gray and D. Parker, Eds., *The Ecology of Soil Bacteria.* University of Liverpool Press, pp. 593-610.

Cogbill, C. V. (1976). In L. S. Dochinger and T. A. Seliga, Eds., *Proceedings of the First International Symposium on Acid Precipitation and the Forest Ecosystem.* U. S. Department of Agriculture, Forest Service, Gen. Tech. Rept. NE-23, pp. 363-370.

Coleman, R. (1966). *Soil Sci.* **101,** 230-239.

Cowling, D. W., Jones, L. P. H., and Lockyer, D. R. (1973). *Nature* **234,** 479-480.

Cox, G. L. (1975). M. Sc. thesis, University of Guelph.

Evans, L. T. (1975). In K. D. McLachlan, Ed., *Sulphur in Australasian Agriculture.* Sydney University Press, pp. 3-8.

Faller, N. (1971). *Sulphur Inst. J.* **7,** 5-6.

Ford, H. W. (1973). *J. Am. Hort. Sci.* **98,** 66-68.

Fox, D. G. (1976). In L. S. Dochinger and T. A. Seliga, Eds., *Proceedings of the First International Symposium on Acid Precipitation and the Forest Ecosystem.* U. S. Department of Agriculture, Forest Service, Gen. Tech. Rept. NE-23, pp. 57-85.

Freney, J. R., and Stevenson, F. J. (1966). *Soil Sci.* **101,** 307-313.

Freney, J. R., Barrow, N. J., and Spencer, K. (1962). *Plant Soil* **17,** 295-308.

Freney, J. R., Melville, G. E., and Williams, C. H. (1970). *Soil Sci.* **109,** 310-318.

Frink, C. R., and Voigt, G. K. (1977). *Water Air Soil Pollut.* **7,** 371-388.

Garrigus, U. S. (1969). In O. H. Muth, Ed., *Symposium: Sulfur in Nutrition.* Oregon State University, pp. 127-142.

Ghiorse, W. C. and Alexander, M. (1976). *J. Environ. Qual.* **5,** 227-230.

Halstead, R. L., and Rennie, P. J. (1977). In *Sulphur and Its Inorganic Derivatives in the Canadian Environment.* Environmental Secretariat, National Research Council, Publ. NRCC 15015, Ottawa, pp. 181-219.

Harward, M. E., and Reisenauer, H. M. (1966). *Soil Sci.* **101,** 326-333.

Harward, M. E., Chao, T. T., and Fang, S. C. (1962). *Agron. J.* **54,** 101-106.

Hoeft, R. G., Keeney, D. R., and Walsh, L. M. (1972). *J. Environ. Qual.* **1,** 203-208.

Holding, A. J., and Lowe, J. F. (1971). In T. A. Lie and E. G. Mulder, Eds., *Some Effects of Acidity and Heavy Metals on the* Rhizobium-*Leguminous Plant Association. Plant Soil,* Special Volume, pp. 153-166.

Hoyt, P. B., and Nyborg, M. (1971). *Soil Sci. Am. Proc.* **35,** 236-240.

Hoyt, P. B., and Nyborg, M. (1972). *Can. J. Soil Sci.* **52,** 163-167.

Humphreys, F. R., Lambert, M. J., and Kelly, J. (1975). In K. D. McLachlan, Ed., *Sulphur in Australasian Agriculture.* Sydney University Press, pp. 154-162.

Hutchinson, T. C., and Whitby, L. M. (1977). *Water Air Soil Pollut.* **7,** 421-438.

Jackson, M. L. (1962). In M. L. Jackson, Ed. *Soil Chemical Analysis.* Prentice-Hall, Inc., pp. 362–363.

Jacq, V., and Dommerques, V. (1970). *Zentralbl. Bakteriol. Parasitenkd. Infektionskr. Hyg.,* Abt. 2, **125,** 661-669.

Johansson, O. (1959). *Kgl. Landbr. Ann.* **25,** 57-161.

Johnson, N. M., Reynolds, R. C., and Likens, G. E. (1972). *Science* **177,** 514-516.

Jones, R. K., Probert, M. E., and Crack, B. J. (1975). In K. D. McLachlan, Ed. *Sulphur in Australasian Agriculture.* Sydney University Press, pp. 127-135.

Jonsson, B. (1977). *Water Soil Air Pollut.* **7,** 497-502.

Kamprath, E. J. (1970). *Soil Sci. Am. Proc.* **34,** 252-254.

Katz, M., Wyatt, F. A., and Atkinson, H. J. (1939). In *Effect of Sulphur Dioxide on Vegetation.* National Research Council of Canada, **815,** pp. 131-164.

Loveday, J. (1975). In K. D. McLachlan, Ed., *Sulphur in Australasian Agriculture.* Sydney University Press, pp. 163-171.

Lowe, L. E. (1964). *Can. J. Soil Sci.* **44,** 176-179.

Malhi, S. S. (1977). Ph.D. thesis, University of Alberta.

Malmer, N. (1976). *AMBIO* **5,** 231-233.

Martin, W. E. and Walker, T. W. (1966). *Soil Sci.* **101,** 248-255.

McCoy, D. A. and Webster, G. R. (1977). *Can. J. Soil Sci.* **57,** 119-127.

Miyamoto, S., Ryan, J., and Stroehlein, J. L. (1975). *J. Environ. Qual.* **4,** 431-437.

Neptune, A. M. L., Tabatabai, M. A., and Hanway, J. J. (1975). *Soil Sci. Am. Proc.* **39,** 51-55.

Nyborg, M. (1968). *Can. J. Soil Sci.* **48,** 37-40.

Nyborg, M. (1974). In D. Hocking and W. R. MacDonald, Eds., *Proceedings of the Workshop on Reclamation of Disturbed Lands in Alberta*. Northern Forest Research Centre, Inf. Rept. NOR-X-116, Edmonton, pp. 55-70.

Nyborg, M., Bentley, C. F., and Hoyt, P. B. (1974). *Sulphur Inst. J.* **10,** 14-15.

Nyborg, M., Crepin, J., Hocking, D., and Baker, J. (1977). *Water Air Soil Pollut.* **7,** 439-448.

Oden, S. (1976). In L. S. Dochinger and T. A. Seliga, Eds., *Proceedings of the First International Symposium on Acid Precipitation and the Forest Ecosystem*. U. S. Department of Agriculture, Forest Service, Gen. Tech. Rept. NE-23, pp. 1-36.

Ottar, B. (1976). In L. S. Dochinger and T. A. Seliga, Eds., *Proceedings of the First International Symposium on Acid Precipitation and the Forest Ecosystem*. U. S. Department of Agriculture, Forest Service, Gen. Tech. Rept. NE-23, pp. 105-116.

Overrein, L. N. (1972). *AMBIO* **1,** 145-147.

Overrein, L. N. (1976). In L. S. Dochinger and T. A. Seliga, Eds., *Proceedings of the First International Symposium on Acid Precipitation and the Forest Ecosystem*. U. S. Department of Agriculture, Forest Service, Gen. Tech. Rept. NE-23, pp. 37-42.

Pearson, R. W. and Adams, F. (1967). In R. W. Pearson and F. Adams, Eds., *Soil Acidity and Liming*. American Society of Agronomy, Madison, Wis. 274 pp.

Reuss, J. O. (1977). *Water Air Soil Pollut.* **7,** 461-478.

Rice, W. A., Penney, D. C., and Nyborg, M. (1977). *Can. J. Soil Sci.* **57,** 197-203.

Richards, L. A. (1969). In L. A. Richards, Ed., *Diagnosis and Improvement of Saline and Alkaline Soils*. U. S. Department of Agriculture, Agriculture Handbook 60, p. 160.

Spencer, K. (1975). In K. D. McLachlan, Ed., *Sulphur in Australasian Agriculture*. Sydney University Press, pp. 98-108.

Starkey, R. L. (1966). *Soil Sci.* **101,** 297-306.

Stewart, B. A. (1969). *Sulphur Inst. J.* **5,** 12-15.

Tabatabai, M. A. and Bremner, J. M. (1972). *Argon. J.* **64,** 40-43.

Tabatabai, M. A. and Laflen, J. M. (1976). In L. S. Dochinger and T. A. Seliga, Eds., *Proceedings of the First International Symposium on Acid Precipitation and the Forest Ecosystem*. U. S. Department of Agriculture, Forest Service, Gen. Tech. Rept. NE-23, pp. 293-308.

Tamm, C. O. (1976). *AMBIO* **5,** 235-238.

Tamm, C. O. (1977). *Water Air Soil Pollut.* **7,** 367-369.

Terraglio, F. P. and Manganelli, M. (1966). *Int. J. Air Water Pollut.* **10,** 783-791.

Tisdale, S. L. (1970). *Sulphur Inst. J.* **6,** 2-7.

Tisdale, S. L. and Nelson, W. L. (1975). *Soil Fertility and Fertilizers*. Macmillan Publishing Co., New York, pp. 278-301.

Vamos, R. (1964). *J. Soil Sci.* **15,** 103-109.

Walker, D. R. (1969). *Can. J. Soil Sci.* **49,** 409-410.

Walker, T. W. and Adams, A. F. R. (1958). *Soil Sci.* **85,** 307-318.

Walker, T. W. and Gregg, P. E. H. (1975). In K. D. McLachlan, Ed., *Sulphur in Australasian Agriculture*. Sydney University Press, pp. 145-153.

Wiklander, L. (1975). *Geoderma* **14,** 93-105.

Williams, C. H. (1967). *Plant Soil* **16,** 205-223.

Williams, C. H. (1975). In K. D. McLachlan, Ed., *Sulphur in Australasian Agriculture*. Sydney University Press, pp. 21-30.

Williams, C. H., Williams, E. G., and Scott, N. M. (1960). *J. Soil Sci.* **11,** 334-346.

Woodruff, C. A. (1967). In R. W. Pearson and F. Adams, Eds., *Soil Acidity and Liming*. American Society of Agronomy, Madison, Wis., pp. 207-231.

10

NATURALLY OCCURRING ORGANOSULFUR COMPOUNDS IN SOIL

John W. Fitzgerald

Department of Microbiology,
University of Georgia,
Athens, Georgia

1. INTRODUCTION

Organic forms of sulfur that occur naturally constitute a diverse group of compounds. This group includes the sulfur-containing amino acids and the sulfonates in which sulfur is directly bonded to carbon (C-bonded sulfur), as well as the true organic esters of sulfuric acid, in which sulfur is bonded to oxygen in the form of a C—O—SO_3^- linkage (organic sulfate). Examples of the latter type of sulfoconjugate include the sulfate esters of simple phenols (the arylsulfates), of aliphatic alcohols (the alkylsulfates and choline sulfate), or carbohydrates (i.e., chondroitin sulfate), and of amino acids (i.e., tyrosine sulfate). Less well studied, yet potentially important, organic sulfates are the sulfated thioglycosides (oxime-*o*-sulfate esters) and the sulfamates in which sulfur occurs in the form of an N—O—SO_3^- and an N—SO_3^- linkage, respectively. Although these sulfoconjugates and the sulfonates (C—SO_3^- linkage) have been considered elsewhere under the heading "ester sulfate," the term "ester sulfate" will be used here in reference only to compounds possessing the C—O—SO_3^- linkage.

The natural occurrence of sulfate esters (in urine) was first noted by Baumann in 1876, but it was primarily the pioneering and continuing efforts over the past 30 years of R. H. De Meio, K. S. Dodgson, J. R. Freney, and A. B. Roy that established the importance of these esters. In consequence, there is an extensive literature on sulfate ester metabolism that cannot be considered here. The reader's attention is therefore drawn to reviews by Dodgson and Rose (1970) and De Meio (1975) dealing with the occurrence and formation of these and other sulfoconjugates, as well as to articles on their hydrolysis (Dodgson and Rose, 1975) and soil biochemistry (Williams, 1975; Freney and Swaby, 1975). In addition, an excellent textbook on the entire field, which also includes valuable information on the chemistry of these esters, is available (Roy and Trudinger, 1970).

This chapter represents an extension and updating of a recent review (Fitzgerald, 1976) and deals primarily with the relationship of organosulfur compounds to the sulfur cycle. An attempt will be made here to include material relating to aerobic as well as to anaerobic environments, although information on the occurrence and participation of ester sulfate in the latter type of environment is meager. For an earlier account that is related to this general theme, see Dodgson and Rose (1966).

2. THE SULFUR STATUS OF SOIL

The close relationship between organic carbon, total nitrogen, and the total sulfur content of soil was considered by Freney and Stevenson (1966) to mean that most of the sulfur in soils throughout the world occurs in the organic rather than the inorganic state. This generalization has been substantiated by the results of soil sulfur analyses, which show (Table 1) that very little sulfur is found as elemental sulfur (S^0), sulfide (S^{2-}), or inorganic sulfate (SO_4^{2-}). Of 208 different soils investigated, only 5.2% (average value) of the total sulfur was present as SO_4^{2-}, and little, if any, sulfur was found in the elemental or sulfide form ($S^0 + S^{2-}$ represented 2.8% of the total sulfur in 14 soils). Indeed, Tabatabai and Bremner (1972b) were unable to detect sulfur in the latter two forms in an investigation of the sulfur contents of 37 representative samples of the major soil series in Iowa (Table 1).

Although Lowe (1964) and Kowalenko and Lowe (1975b) stressed that a third fraction may exist in some soils, soil organic sulfur is generally believed to consist of two major fractions called hydriodic acid-reducible sulfur (HI-reducible sulfur) and carbon-bonded sulfur. The latter fraction is tentatively defined as soil sulfur which is reduced to S^{2-} by Raney nickel (Papa et al., 1949; Lowe and De Long, 1963; Freney et al., 1970). This catalyst will reduce cystine and methionine but not ester sulfate (Freney, 1967; Freney et al., 1970). A close correlation was found between the Raney Nickel-reducible sulfur and the sum of the concentrations of these two amino acids in two Australian soils, and it was suggested (Freney et al., 1975) that the Raney nickel-reducible fraction consisted mainly of amino acid sulfur, although this may not be true for soils in general. Thus, using the Raney nickel-reduction method, Lowe (1969) determined the carbon-bonded sulfur content of the humic acid fraction of nine major soils in Alberta and found that only 39% of this fraction consisted of amino acid sulfur. Reviews dealing with the nature of the fulvic and humic acid components of soil are available (Hurst and Burges, 1967; Felbeck, 1971; Schnitzer, 1971; Van Dijk, 1971).

There is also evidence suggesting that some soils may contain carbon-bonded sulfur that is not reduced to S^{2-} by Raney nickel. In an investigation dealing with 17 Australian soils, Freney and co-workers (1970, 1975) found that, even under optimal conditions, the amount of

Table 1. Sulfur Status of Soils from Various Geographical Locations

Number of Samples	Location and Depth	Type of Sulfur (%) SO_4^{2-}	$S^\circ + S^{-2}$	C-Bonded S[a]	HI-Reducible S	Ester Sulfate[b]	Source
10	Eastern Australia, depth not reported	4.6	ND[c]	ND	48.3	43.7	Williams and Steinbergs (1959)
1	Solonized soil	17.3	ND	ND	59.1	41.8	
24	Eastern Australia, 0–6 in.	6.0	1.0	41.0	59.0	52.0	Freney (1961, 1967)
3	New South Wales, Australia, 0–4 in.	1.4	ND	ND	41.4	40.0	Freney et al. (1969), Melville et al. (1969)
15	Australia, 0–10 cm	ND	ND	30.0	50.0		Freney et al. (1970)
3	Australia, 0–10 cm	1.3	ND	68.7[d]	31.2[d]	31.2	Freney et al. (1971)
2	New South Wales, Australia, Crookwell regions, 0–10 cm						Freney et al. (1972)
	Virgin podzolic soil	ND	ND	46.2	53.8		
	Fertilized podzolic soil	ND	ND	64.7	35.3		
2	Same as above, different samples (?)						Freney et al. (1975)
	Virgin Podzolic soil	0.0	ND	56.6	43.4	43.4	
	Fertilized Podzolic soil	0.0	ND	71.8	28.2	28.2	
11	New South Wales, 0–10 cm	1.9	28	ND	ND		Melville et al. (1971)
20	Northern Nigeria, 0–15 cm	9.8	ND	ND	52.0	42.2	Cooper (1972)

14	England and Wales, U.K., 0–15 cm	4.4	ND	ND	ND		Jones et al. (1972)
2	Calcareous soils	16.9	ND	ND	ND		
3	Wales, U.K., 0–15 cm	0.6	ND	ND	36.2	35.6	Houghton and Rose (1976)
3	Quebec, Canada, depth not reported	1.7	ND	ND	ND		Lowe and DeLong (1961)
7	Quebec, 0–6 in.	ND	ND	38.4	ND		Lowe and DeLong (1963)
5	Quebec, 0–6 in.	6.6	ND	35.2	59.2	52.6	Lowe (1964)
3	British Columbia, Canada, depth not reported	2.1	ND	43.3	27.4	25.3	Kowalenko and Lowe (1975a)
1	Saskatchewan, Canada, depth not reported	0.7	ND	26.8	44.7	44.0	
64	Iowa, 0–15 cm	3.1	ND	ND	ND		Tabatabai and Brenner (1972a)
37	Iowa, 0–15 cm	2.6	0.0	10.7	52.8	50.2	Tabatabai and Bremner (1972b)

[a] See text for a description of C-bonded S (carbon-bonded sulfur) and HI-reducible S (hydriodic acid-reducible sulfur) and methods for determining them.

[b] Calculated by subtracting the percentage of SO_4^{2-} from the percentage of HI-reducible S. Values also corrected by subtracting the percentage of S° and the percentage of S^{2-} when determinations for S° and S^{2-} were reported.

[c] No determination reported.

[d] Values calculated from 1-day incubation data (Table 5; Freney et al., 1971). Samples treated to remove SO_4^{2-} before reduction.

Raney nickel-reducible sulfur was 50% less than the theoretical quantity of carbon-bonded sulfur (calculated by subtracting the HI-reducible sulfur from the total sulfur determined for these soils). Similar observations were reported by Lowe and De Long (1963) and by Kowalenko and Lowe (1975b) for Canadian soils. These results are not surprising since Raney nickel will not reduce other compounds containing the C—S linkage [such as the aliphatic sulfones or the sulfonic acids (Arkley, 1961)], and these compounds may be present in soil (see p. 426). Although our understanding of the precise nature of carbon-bonded sulfur is not clear, a consideration of this fraction is pertinent to the objectives of this chapter, since the sulfur-containing amino acids and the sulfonates can act as sources of SO_2^{2-} for these soils (see p. 427). In addition, the amino acids may act to regulate the synthesis of enzymes concerned with the release of SO_4^{2-} from ester sulfate (see p. 411). Of 99 soils that were tested (Table 1), carbon-bonded sulfur (Raney nickel-reducible sulfur) represented, on the average, 37.4% of the total sulfur present.

Hydriodic acid-reducible sulfur is more precisely defined as the fraction of organic sulfur that is reduced to H_2S by a mixture of hydriodic, formic, and hypophosphorous acids (Freney, 1958, 1961; Spencer and Freney, 1960). This mixture produces H_2S only from compounds containing the C—O—S linkage (ester sulfate) or the C—N—S linkage (sulfamate) and from some organic sulfites such as dimethyl sulfite or diethyl sulfite (Arkley, 1961; Freney, 1961). The C—S bond is not ruptured under these conditions (Johnson and Nishita, 1952; Freney, 1958). Lowe and De Long (1961) found that roughly 33% of the total sulfur of an alkaline soil extract was released as SO_4^{2-} after treatment of the dialyzed extract with 6 *N* HCl at 90°C for 12 hr. In a more detailed study, Freney (1961) quantitatively recovered the HI-reducible sulfur present in the fulvic acid component of soil as SO_4^{2-} under similar conditions. The liberation of SO_4^{2-} after acid hydrolysis is a unique characteristic of compounds possessing either *O*-sulfate or *N*-sulfate groups (see, e.g., Dodgson and Rose, 1970; Roy and Trudinger, 1970). Moreover, Freney (1961) demonstrated that the sulfur present in heparin and agar (which possess these ester linkages) was quantitatively recovered as H_2S by the HI-reduction method.

As further evidence for the existence of ester sulfate in soil, Freney (1961) extracted a substantial quantity (45 to 81%) of the total HI-reducible sulfur from five different soils with methanolic hydrogen chloride. This reagent is known to remove ester sulfate as methyl sulfate from chondroitin sulfate, a mucopolysaccharide (Kantor and Schubert, 1957). Although pure methyl sulfate was not isolated from the extracts,

Freney (1961) showed (*a*) that the sulfur present was not precipitated by the addition of Ba^{2+} (which normally distinguishes ester sulfate from SO_4^{2-}), (*b*) that infrared spectra of the extracts exhibited absorptions characteristic of covalently bound sulfate, and (*c*) that all of the sulfur present in these extracts was reducible to H_2S by the HI-reduction method. According to Freney (1961), these results suggested that methyl sulfate was extracted from soil treated with methanolic hydrogen chloride. It appears that compounds bearing the C—O—S or C—N—S linkages are confined to the HI-reducible sulfur fraction of total soil sulfur; and the general consensus (Freney, 1967; Lowe, 1969; Cooper, 1972; Tabatabai and Bremner, 1972b) is that this fraction is largely, if not entirely, composed of these sulfoconjugates. Although sulfamates are not known to occur widely in nature (Dodgson and Rose, 1970), there is no evidence that such compounds are absent from soils, and Tabatabai and Bremner (1972b) suggested that the term "organic sulfate" be used in place of "ester sulfate" to describe the sulfur linkages present in this fraction.

Since the HI-reduction method will also convert S^0 and SO_4^{2-} to H_2S (Freney, 1958), the true percentage of ester sulfate in soil is usually obtained by subtracting the percentage of total sulfur as S^0, S^{2-}, and SO_4^{2-} from the percentage of HI-reducible sulfur [unless SO_4^{2-}, e.g., (Freney et al., 1971), was extracted before reduction]. With two exceptions (Freney, 1961; Tabatabai and Bremner, 1972b), values for S^0 and S^{2-} are lacking from soil sulfur analyses (Table 1). Thus most ester sulfate values reported here (Table 1) are not corrected for these forms of inorganic sulfur. Because of the limited concentrations of S^0 and S^{2-} in soil (Table 1), it is felt that HI-reducible sulfur values, uncorrected for S^0 and S^{2-}, still represent a reasonable estimate of the ester sulfate content of soil. Despite these limitations the data presented in Table 1 show that ester sulfate represents a substantial proportion of the total sulfur in soils from various geographical locations. Of 112 different soils that were analyzed for SO_4^{2-} and HI-reducible sulfur, ester sulfate represented 40.8% (average) of the total sulfur present. The lowest value reported (25.3%) was that for a soil belonging to the black chernozemic group, taken from British Columbia (Kowalenko and Lowe, 1975a, 1975b), whereas Tabatabai and Bremner (1972b) reported values as high as 93.1% for subsoils from Iowa. The latter workers further showed that the percentage of total sulfur as ester sulfate increased with increases in sample depth (Table 2), thus demonstrating that this parameter should be considered in soil sulfur analysis in general. Available sample depth values for other studies are included in Table 1.

Although Freney (1961) found in a study of several Australian soils

Table 2. The Sulfur Status of Some Iowa Soils in Relation to Sample Depth[a]

		Type of Sulfur (%)			
Sample	Depth (cm)	SO_4^{2-}	HI-Reducible S	C-Bonded S	Ester Sulfate
Sharpsburg	0–15	1.8	50.9	6.3	49.1
	15–30	1.3	62.5	4.9	61.2
	30–60	2.0	68.0	3.4	66.0
	60–90	1.9	71.3	1.1	69.4
	90–120	1.1	—	0.9	—
Fayette	0–15	2.1	54.9	9.7	52.8
	15–30	2.7	66.4	9.4	63.7
	30–60	4.0	69.8	5.0	65.8
	60–90	3.7	77.4	4.9	70.7
	90–120	4.9	—	1.6	—
Webster	0–15	3.0	46.2	8.3	43.2
	15–30	2.3	59.9	8.3	47.6
	30–60	4.6	73.4	8.3	68.8
	60–90	1.1	—	4.7	—
	90–120	1.4	94.5	1.9	93.1
Clarion	0–15	2.4	55.9	7.6	53.5
	15–30	2.3	66.0	7.9	63.7
	30–60	2.4	70.2	5.8	67.8
	60–90	9.4	88.2	5.5	78.8
	90–120	6.6	—	1.3	—
Hagener	0–15	2.7	58.7	20.0	56.0
	15–30	2.3	60.2	17.0	57.9
	30–60	3.6	71.4	14.3	67.8
	60–90	7.9	—	7.9	—
	90–120	7.5	—	2.5	—
Hamburg	0–15	1.9	57.9	3.0	56.0
	15–30	1.8	65.1	1.4	63.3
	30–60	5.4	86.5	1.3	81.1
	60–90	9.4	87.5	0.6	78.1
	90–120	15.5	90.0	0.4	74.5

[a] Data taken in part from Tabatabai and Bremner (1972b), Table 4.

that most of the HI-reducible sulfur was associated with fulvic acid, it is now apparent (Freney et al., 1969) that this result was due to the extraction procedure. Thus, using a variety of mild extractants (Melville et al., 1969) unlikely to degrade humic acid, Freney et al. (1969) found that more HI-reducible sulfur was associated with the latter soil colloid than with fulvic acid. This result is consistent with those of Lowe (1969) and Houghton and Rose (1976), who found that HI-reducible sulfur represented 39% and 51% of the total sulfur present in humic acid isolated from Alberta and Welsh soils, respectively. In the latter study the presence of sulfate ester groups in the humic acid fraction was confirmed by extraction with methanolic hydrogen chloride and by release of SO_4^{2-} after hydrolysis in 6 *N* HCl at 100°C. The collective results of these studies (Lowe, 1964; Freney et al., 1971, 1975) suggest that no one procedure is entirely suitable for assessing the heterogeneity of compounds comprising the HI-reducible sulfur fraction of soil.

It is likely that a significant proportion of this fraction occurs as a nonintegral part of soil colloids such as the humic and fulvic acids. Using extractants designed to remove SO_4^{2-}, Lowe (1964) extracted from soil substances that he considered to be sulfated polysaccharides. Similarly, Freney and co-workers (1961, 1971) found that HI-reducible sulfur could be extracted from soil by using a dilute potassium phosphate solution at pH 7. Although some soils have little capacity to absorb SO_4^{2-} (Freney et al., 1971; Tabatabai and Bremner, 1972a), other soils absorb substantial quantities of this anion (Kamprath et al., 1956; Harward and Reisenauer, 1966; Barrow et al., 1969), and there is evidence that HI-reducible sulfur may also be adsorbed to soil particles. Thus Houghton and Rose (1976) found that a wide variety of different ^{35}S-labeled sulfate esters were adsorbed to Welsh soils to the extent of 67% (average) of the total concentration of the ester that was initially added to these soils. Furthermore, procedures that do not release SO_4^{2-} from humic acid [such as heating at neutral pH (Williams and Steinbergs, 1959; Spencer and Freney, 1960; Barrow, 1961) or simple grinding of soil before extraction (Freney, 1961)] caused a substantial increase in the SO_4^{2-} content of some soils. This observation suggests that sulfate ester linkages of differing labilities are present but are not firmly bound to soil colloids.

In terms of the possible occurrence of sulfamate linkages in soil, it may be pertinent to mention that the *N*-sulfate group is more unstable than the *O*-sulfate linkage to acid hydrolysis. Thus *N*-desulfated heparin is prepared under milk conditions that do not cause the hydrolysis of *O*-sulfate groups in this polysaccharide (Dodgson and Rose, 1970). The instability of the *N*-sulfate linkage was also demonstrated for simpler sulfamates which are excreted after the administration of arylamines to mammals (Boyland et al., 1957; Parke, 1960) and spiders (Smith, 1962). Authentic sulfate esters are known to differ in stability toward nonenzymic hydrolysis [i.e., the arylsulfates are easily hydrolyzed, whereas esters such as choline *O*-sulfate (Spencer and Harada, 1960) and keratan sulfate (Kitamikado and Ueno, 1970) are reportedly resistant to autoclaving]. Indeed, Segel and Johnson (1963) reported that choline sulfate underwent only one-half hydrolysis after 30 min at 100°C in 1 *N* HCl.

3. FORMATION AND HYDROLYSIS OF SOIL ESTER SULFATE

Freney and co-workers (1971, 1975) demonstrated that the HI-reducible sulfur fraction is not an inert or stable end product of sulfur metabolism in soil. Thus, when fallow soils or soils awaiting planting were incubated in the presence of ^{35}S-labeled SO_4^{2-}, ^{35}S was incorporated into both the HI-reducible and the carbon-bonded sulfur fractions. The HI-reducible sulfur fraction exhibited greater specific radioactivity; and, using a mild extractant, Freney et al. (1971) found that 75% of the ^{35}S was present in the fulvic acid component of these soils. Approximately 90% of the fulvic acid sulfur was reduced to $^{35}S^{2-}$ by hydriodic acid. Soil sterilized by autoclaving failed to incorporate ^{35}S, suggesting that isotopic exchange did not occur in unsterilized samples and that microbial activity was responsible for the incorporation of the isotope.

Plants (*Sorghum vulgare*) were found to utilize sulfur from both the HI-reducible and the carbon-bonded sulfur fractions (Freney et al., 1975). Results obtained by incubating initially SO_4^{2-} -free soils, after incorporation with $^{35}SO_4^{2-}$, in the presence of *S. vulgare* suggested that immediate utilization of sulfur for growth involved the HI-reducible fraction. Thus the radioactivity present in this fraction decreased considerably during plant growth, whereas no measurable decrease occurred in the ^{35}S-labeled carbon-bonded fraction. In fact, the radioactivity in the latter fraction tended to increase, suggesting that some of the HI-reducible sulfur was converted to carbon-bonded sulfur during plant growth. Incubation of soils in a parallel experiment in the absence of *S. vulgare* resulted in the release of SO_4^{2-} (labeled and unlabeled), and most of this anion was derived from the HI-reducible sulfur fraction. Indeed, many of the changes observed in the carbon-bonded fraction were considered not to be significant by the authors (Freney et al., 1975). Changes in the indigenous (nonradioactive) sulfur present in these fractions were also followed during plant growth. The findings suggested that *S. vulgare* obtained 40 and 60% of its sulfur requirement (for a 36-week growing period) from the HI-reducible and carbon-bonded sulfur fractions, respectively. The authors stressed that this experiment did not take into account interconversion and exchange of sulfur between the two fractions. Thus the quantity of HI-reducible sulfur that may have been converted to carbon-bonded sulfur is not known.

It is unfortunate that the already detailed study by Freney and coworkers (1975) was not extended to include changes in the HI-reducible sulfur present in the humic and fulvic acid components of these soils. These soil colloids were considered by Dodgson and Rose (1975) to be resistant to SO_4^{2-} release by microorganisms, but the available evidence for this view is not convincing. In the only reported study of the degradability of humic acid sulfur, Houghton and Rose (1976) found that the sulfate ester groups present in this colloid were resistant to hydrolysis by extracts possessing alkylsulfatase, arylsulfatase, and glycosulfatase activities. It is well established that depolymerization occurs before the desulfation of other sulfated macromolecules (Dodgson and Rose, 1970, 1975); and, as Houghton and Rose (1976) pointed out, the desulfation of humic acid may result from a sequential attack by depolymerizing and desulfating enzymes.

4. ORIGINS OF SOIL ESTER SULFATE

The presence of high concentrations of ester sulfate in soil is not surprising in view of numerous reports dealing with the natural occurrence of these esters. Mammalian connective tissue consists of keratan sulfate, dermatan sulfate, and chondroitin sulfate, as well as heparin and heparan sulfate. The last two polysaccharides possess the C—O—S and the C—N—S (sulfamate) linkages and represent the only known naturally occurring substances with the latter sulfur linkage (Dodgson and Lloyd, 1968). Rahemtulla and Lovtrup (1974a, 1974b) demonstrated the presence of chondroitin sulfate in many invertebrates that inhabit soils. Collectively, these polysaccharides are released into soil from decaying animal matter. A number of sulfate esters are returned to soil in animal excreta. These include the arylsulfates (Heacock and Mahon, 1964; Davis et al., 1966; Dziewiatkowski, 1970; Curtis et al., 1974), sulfate esters of steroids (Schachter and Marrian, 1938; Roy and Trudinger, 1970; De Meio, 1975), the amino acid *O*-sulfate esters (Tallan et al., 1955; John et al., 1966; Hext et al., 1973), and ascorbic acid 2-*O*-sulfate (Baker et al., 1971; Mumma and Verlangiera, 1972). To emphasize the magnitude of the latter contribution, Dodgson and Rose (1975) stated that "a rough estimate suggests that in terms of human excreta alone almost 50 tons of sulfur are daily returned to the sulfur cycle in the form of sulfate esters."

Although the esters mentioned above probably represent the major contribution of mammals to the organic sulfate content of soil, other esters of animal origin may be returned to soil less frequently. These include the sulfate esters of glycoproteins (Embery and Whitehead, 1975) and of bile alcohols (Boström and Vestermark, 1959; Spencer, 1960; Haslewood, 1967), lactose 6-*O*-sulfate and neuramin lactose 6-*O*-sulfate of mammary glands (Barra and Caputto, 1965; Ryan et al., 1965), heparin from mast cells (Dodgson and Rose, 1975), the mammalian sulfated glycolipids (Svennerholm and Stalberg-Stenhagen, 1968; Ishizuka et al., 1973; Fluharty et al., 1974), the polyhexose sulfate esters of primitive animals (Inoue and Egami, 1963; Hunt and Jevons, 1966; Doyle, 1967; Katzman and Jeanloz, 1973), and UDP-*N*-acetyl-D-galactosamine sulfate and isopropyl sulfate from the hen's oviduct (Strominger, 1955) and egg (Yagi, 1966), respectively.

To these animal sources of ester sulfate can be added a growing list of esters that are synthesized by other forms of life. Taylor and Novelli (1961) reported that an unidentified bacterial isolate from soil was capable of synthesizing an extracellular polysaccharide possessing ester sulfate groups. Similarly, *Pseudomonas multiphilia* synthesizes a complex mixture of sulfated polysaccharides which are released into the medium during the growth of this bacterium (R. Tornabene, unpublished observation). Sulfate esters of short-carbon-chain monocarboxylic acids similar to those formed and released into the medium by *Pseudomonas fluorescens* (Fitzgerald and Dodgson, 1970, 1971a, 1971b) were also detected in soils incubated with ^{35}S-labeled D-glucose 6-*O*-sulfate (C. Houghton, personal communication). These esters may also occur in soils as a consequence of the mammalian (Denner et al., 1969; Ottery et al., 1970) and microbial (Dronkers and Von der Vet, 1964) degradation of the primary alkylsulfate detergent, sodium dodecyl sulfate. Both primary and secondary alkylsulfate esters are employed as components of commercial surface-active agents (Bogan and Sawyer, 1954; Allen, 1962); and, in at least one case, an alkylsulfate ester was used as a herbicide (Vlitos, 1953). Other alkylsulfate esters were detected in a variety of microorganisms (Haines and Block, 1962; Mayers and Haines, 1967; Mayers et al., 1969), and the possibility that naturally occurring alkylsulfates are widely distributed is strengthened by the observations that mammals (Vestermark and Bostrom, 1959; Spencer, 1960) and lower vertebrates (Scully et al., 1970) are capable of sulfating a wide variety of alcohols. Burns and Wynn (1975) demonstrated that extracts of *Aspergillus oryzae* can synthesize tyrosine *O*-sulfate as well as a number of different arylsulfate esters. Heretofore, phenol sulfotransferase activity was thought to be restricted to mammals (Grimes, 1959; Segal and Mologne, 1959; Mattock and Jones, 1970), but it appears now that the fungi and possibly the bacteria as a group may also possess this type of activity. In this regard, Burns and Wynn (1975) stressed that the presence of phenol sulfotransferase in *A. oryzae* suggests that sulfate esters of phenols may be of more widespread occurrence in microorganisms than has hitherto been thought.

The point that should be borne in mind when assessing the natural occurrence of any sulfate ester is that the demonstration of sulfotransferase activity depends, in many cases, upon the fortuitous choice of a suitable sulfate donor and alcohol acceptor. We may, therefore, be underestimating the number and variety of organisms contributing to the production of these esters. The additional occurrence of bacterial lipids bearing the C—O—SO_3^- linkage (Kates et al., 1967; Goren, 1970, 1971; Hancock and Kates, 1973; Kates and Deroo, 1973; Langworthy et al., 1974) suggests that nonmammalian sources of ester sulfate may be far more prevalent in soil. The real magnitude of this contribution will not be appreciated, however, until investigations are conducted to assess the ability of various microorganisms to synthesize these esters. In many cases the observation of the biological occurrence of a sulfate ester is accidental, arising as an interesting sideline from a sometimes totally unrelated study. This author's observation (1973) that bacteria can synthesize the *O*-sulfate ester of choline (see p. 406) represents a good case in point.

5. OCCURRENCE AND FORMATION OF CHOLINE SULFATE

Choline *O*-sulfate (choline sulfate) is returned to soil from a variety of sources. This ester occurs in high concentration in lichens (Lindberg, 1955a; Harper and Letcher, 1966; Feige and Simonis, 1969; Solberg, 1971), in algae (Lindberg, 1955b; Ikawa and Taylor, 1973; Taylor et al., 1974), and in plants (Nissen and Benson, 1961; Benson and Atkinson, 1967; Thompson et al., 1970).

6. FUNGAL CHOLINE SULFATE

This ester is also widely distributed in fungi that inhabit soils (Ballio et al., 1959; Harada and Spencer, 1960) as well as waters (Catalfomo et al., 1973), and it is from these sources that the ester has been most extensively investigated with respect to its formation, utilization, and biological function. Since choline sulfate-producing fungi are able to transport the ester even under conditions that permit intracellular synthesis, the ester is considered as an important source of sulfur for microbial growth (Bellenger et al., 1968) and, by virtue of its resistance to nonenzymic hydrolysis (Spencer and Harada, 1960; Segel and Johnson, 1963; Fitzgerald, 1973) as an important nonacid storage form of soil SO_4^{2-}. Sulfate esters can occur in the free acid form, but choline sulfate is an internally compensated salt. As such, its presence in soil should not alter the pH.

In agreement with the results of Ballio et al. (1959), choline sulfate was found to be synthesized by most of the higher fungi but was absent in members of the orders Mucorales and Endomycetales when these fungi were grown in enriched media supplemented with sodium sulfate (Harada and Spencer, 1960). Even when grown under conditions in which the sulfate for choline sulfate must be derived endogenously from taurine, the ester can accumulate in the mycelia of *Aspergillus nidulans* to a concentration of 0.6% of the dry weight (Spencer et al., 1968). Similar results (0.2 to 0.3% of the dry weight) were reported for *A. sydowi* (Woolley and Peterson, 1937) and *Penicillium chrysogenum* (De Flines, 1955; Stevens and Vohra, 1955). In contrast to various *Pseudomonas* species that form this ester (see p. 406), choline sulfate was not detected in culture filtrates of the choline sulfate-producing fungi (Harada and Spencer, 1960).

Choline sulfate is also a major component of the conidiospores of several species of *Aspergillus,* accounting for as much as 1.5% of the dry weight and 40% of the total sulfur present in the spores (Takebe, 1960). The SO_4^{2-} concentration of the sporulation medium was not critical for ester formation since spores containing similar quantities of choline sulfate were formed when either an enriched but SO_4^{2-} -unsupplemented medium or a synthetic medium containing Na_2SO_4 was employed (Takebe, 1960). In contrast, the formation of this ester by mycelia and conidia of *Neurospora crassa* is related to the SO_4^{2-} content of the environment. Thus a 98% increase in the choline sulfate content of both developmental stages was observed after growth on 2.0 m*M* Na_2SO_4, as opposed to growth in the presence of 0.1 m*M* Na_2SO_4 (McGuire and Marzluf, 1974a). No similar increase occurred when conidia were formed in media containing increasing concentrations of L-methionine. Ascospores of *N. crassa* possessed the highest levels of choline sulfate (80% of the total soluble sulfur), but the formation of the ester by this developmental stage was independent of the external SO_4^{2-} content of the medium before and for periods up to 40 hr after germination (McGuire and Marzluf, 1974a). When various choline sulfate-producing fungi were grown under conditions that permit choline sulfate formation (0.05% w/v $MgSO_4$) and then transferred to a sulfur-deficient medium, the fungi continued to grow; whereas, in a parallel experiment, fungi incapable of forming the ester failed to grow further in the sulfur-deficient medium (Spencer and Harada, 1960). Similarly, ^{35}S-labeled conidiospores of *Aspergillus niger* were observed to undergo complete germination in a sulfur-free medium (Yanagita, 1957; Takebe and Yanagita, 1959). Under these conditions about 50% of the total radioactivity present in the choline sulfate of the spores was recovered in various sulfur-containing amino acids after germination (Yanagita, 1957). These findings led Spencer and Harada (1960) and Takebe (1960) to conclude, independently, that choline sulfate acts as a reserve source of sulfur for fungal growth and conidiospore germination, respectively.

Kaji and co-workers (Kaji and McElroy, 1958; Kaji and Gregory, 1959) were the first to demonstrate that the formation of choline sulfate in the fungus *Aspergillus sydowi* required the initial formation of 3′-phosphoadenosine 5′-phosphosulfate (PAPS), a sulfated nucleotide now known to act as the immediate sulfur donor in the formation of numerous other sulfate esters (Dodgson and Rose, 1970; De Meio, 1975). The formation of PAPS is a two step process in which adenosine 5′-phosphosulfate (APS) is synthesized first from adenosine 5′-triphosphate (ATP) by ATP-sulfurylase (Robbins and Lipmann, 1958; Tweedie and Segel, 1971), the product of this reaction being subsequently converted to PAPS in the presence of ATP and APS-kinase (Robbins, 1962). A sulfotransferase catalyzes the transfer of SO_4^{2-} from PAPS to the appropriate alcohol acceptor—in this case, choline chloride. Spencer and Harada (1960) demonstrated the involvement of PAPS in choline sulfate formation by a variety of other fungi and further showed that lack of ability in some fungi to synthesize the ester was due to a concomitant lack of ability to synthesize choline sulfotransferase. Thus fungi that failed to form the ester were nevertheless capable of synthesizing PAPS from SO_4^{2-} and ATP. Despite the pronounced instability of the enzyme, Orsi and Spencer (1964) and more recently Renosto and Segel (1977) were able to purify the choline sulfotransferase from *Aspergillus nidulans* and *Penicillium chrysogenum*, respectively, thus proving that this enzyme does indeed catalyze the transfer of SO_4^{2-} from PAPS to choline in these fungi. A study of the enzyme (purified 31-fold) from *P. chrysogenum* revealed a high specificity for choline ($K\hat{m}$ = 17 μM) and PAPS ($K\hat{m}$ = 12 μM). Renosto and Segel (1977) also managed to stabilize the enzyme in 25% sucrose as a frozen solution, a noteworthy achievement since instability has represented an unsolved problem with other *in vitro* choline sulfate-forming systems [Orsi and Spencer, 1964; Fitzgerald and Luschinski, 1977 (see also below)].

7. BACTERIAL CHOLINE SULFATE

Choline sulfate may also occur in soils as a result of bacterial synthesis. All of eight randomly selected *Pseudomonas* species formed the ester when cultured on growth-limiting concentrations of SO_4^{2-} (Fitzgerald, 1973). Unlike fungi, which retain most of the choline sulfate they synthesize, these bacteria released a large proportion of the ester into the culture medium. This work dealt with the synthesis only of ^{35}S-labeled choline sulfate by intact bacterial cells and indicated that growth on Na_2SO_4 (> 0.01 m*M*) resulted in decreased synthesis of the ester. This result appears to be due to the dilution of $^{35}SO_4^{2-}$ with unlabeled SO_4^{2-} present in the medium at concentrations above 0.01 m*M*. Assays monitoring total (labeled + unlabeled) choline sulfate production revealed (Table 3) that the formation of the ester was largely independent of the Na_2SO_4 concentration (0.01 to 1.0 m*M*) of the culture medium not only for *Pseudomonas* $C_{12}B$ but also for *Comamonas terrigena* and *Escherichia coli*. The synthesis of choline sulfate was previously not known to occur with the latter bacteria, and these results show that the ability to synthesize the ester is not confined to the genus *Pseudomonas* (cf. Fitzgerald, 1973). The synthesis and release of this ester by all three bacteria was accompanied by the release into the culture medium of two additional ^{35}S-labeled metabolites (Table 3). The electrophoretic mobilities (for conditions see Fitzgerald, 1973) of these metabolites relative to SO_4^{2-} (0.47 and 0.68) suggest their nonidentity with APS or PAPS. The method for determining total choline sulfate formation (see footnote *b*, Table 3) has not been reported elsewhere and should facilitate studies designed to screen other bacteria for the ability to form and release this ester into soils and other environments.

Factors regulating the formation of choline sulfate were investigated further (Fitzgerald, 1973) using a *Pseudomonas* species isolated from soil (Payne, 1963; Payne and Feisal, 1963), and with this isolate (designated as *Pseudomonas* $C_{12}B$) the ester was found in the culture medium at all stages of the culture cycle. Maximum quantities were discerned in stationary phase culture supernatants. Adenosine 5′-triphosphate and Mg^{2+} were required for the formation of the ester by cell extracts. In these respects the choline sulfate-synthesizing system in *Pseudomonas* $C_{12}B$ (Fitzgerald, 1973) is similar to that present in various fungi. The exact mode of regulation of the formation of the ester by *Pseudomonas* $C_{12}B$ is presently undefined. It is known (Fitzgerald, 1973) that this system is not repressed by the presence during growth of SO_4^{2-} or L-methionine but is inhibited *in vitro* by L-cysteine. These results were interpreted (Fitzgerald, 1973) to mean that, at growth-limiting

Table 3. Release of Choline Sulfate into the Culture Medium by Bacteria during Growth in the Presence of Varying Concentrations of Sodium Sulfate[a]

Bacterium	Concentration (m*M*)	Total Choline Sulfate[b] (μmol choline/100 ml hydrolyzed medium)	Metabolites (% total radioactivity)	
			D	E
Pseudomonas $C_{12}B$	0.01	147	30.0	0
	0.1	97	3.1	1.7
	1.0	140	9.5	27.4
Comamonas terrigena	0.01	74	24.3	23.8
	0.1	86	7.6	34.0
	1.0	62	0	0
Escherichia coli	0.01	177	6.1	6.1
	0.1	180	0	0
	1.0	183	0	0

[a] Bacteria were grown to the stationary phase in a basal-salts medium (Fitzgerald, 1973) containing $Na_2{}^{35}SO_4$ (0.5 μl/ml medium; sp. radioactivity, 716 mCi/μmol) and 1% w/v sodium citrate, sodium pyruvate, or D-glucose, respectively. For concentrations of $Na_2SO_4 > 0.01$ m*M*, unlabeled Na_2SO_4 was added to give the final concentration stated above.

[b] The quantity of choline sulfate was determined as free choline after hydrolysis of the ester as follows. An aliquot of the culture was centrifuged, and 0.7 ml of the clear supernatant was mixed with an equal volume of 2 *N* HCl. This solution was heated for 3 hr at 121°C in an autoclave. The resulting hydrolyzate was neutralized with KOH, and a 1.0-ml aliquot was assayed for choline chloride by the method of Hayashi et al. (1962).

concentrations of SO_4^{2-}, the endogenous concentration of cysteine was not sufficient to inhibit choline sulfate formation.

Work dealing with the *in vitro* synthesis of the ester revealed that, although cell extracts of *Pseudomonas* $C_{12}B$ form choline sulfate (cf. Fitzgerald, 1973), most of the activity was associated with the cell-free culture fluid (Fitzgerald and Luschinski, 1977). Moreover, results of localization studies showed that the responsible enzymes were associated with the cell envelope of this bacterium and that these enzymes were easily released from intact cells during growth. It is noteworthy that the choline sulfotransferase of the fungus *Aspergillus nidulans* was not inhibited by cysteine; in fact, this amino acid was employed as a reducing agent in an effort to stabilize the enzyme (Orsi and Spencer,

1964). A detailed study (Fitzgerald and Luschinski, 1977) of the effect of cysteine on the *in vitro* synthesis of the ester by culture fluids of *Pseudomonas* $C_{12}B$ showed that this amino acid acted as an inhibitor at concentrations as low as 10^{-3} m*M*. The formation of choline sulfate from SO_4^{2-} and choline chloride was stimulated three- and fivefold by the additional presence of APS and PAPS, respectively. Radioactively-unlabeled APS (10 m*M*) was also found to completely inhibit the incorporation of $^{35}SO_4^{2-}$ into choline chloride (Fitzgerald and Luschinski, 1977). These results suggest that APS and PAPS can serve as sulfate donors for the formation of choline sulfate in a manner that may be analogous to the role of these sulfated nucleotides in the formation of this ester by fungi (see also p. 405).

Choline sulfate was also detected as an intracellular component of *Lactobacillus plantarum* when this bacterium was grown in an enriched medium in the absence of added sulfur-containing compounds (Ikawa et al., 1972). The authors extracted their lyophilized cells twice with 95% (v/v) aqueous ethanol, resuspended the dry extract in water, and partitioned the suspension against *n*-butanol. Chromatography of the aqueous partition phase in 16 different solvents revealed the presence of choline sulfate in substantial quantity in these cells. This extraction procedure was reportedly at least 96% effective, but the efficiency of the procedure for the extraction of authentic choline sulfate was not reported.

8. MINERALIZATION OF SOIL ESTER SULFATE

The reaccession to the sulfur cycle of SO_4^{2-} immobilized in organic sulfate esters is dependent upon the ability of microorganisms, plants, and mammals to produce enzymes (sulfohydrolases) that hydrolyze these esters. Bacteria and fungi appear to be the major sources of these enzymes in soil. However, plant roots are known to hydrolyze choline sulfate (Nissen, 1968), and the possibility exists that mammalian urine may contain such enzymes. It is known that arylsulfatase is present in human urine in quantities that permit the purification of the enzyme from this source (Breslow and Sloan, 1972; Stevens et al., 1975). The almost ubiquitous occurrence of this enzyme in the organs of other mammals (see, e.g., Roy and Trudinger, 1970) suggests the likelihood of its presence, and eventual discovery, in the urine of other animals.

It has been known for some time that soils contain a number of different classes of enzymes that originate from but exist outside of living tissue (for a review, see Skujins, 1967). More recently, it was demonstrated that sulfohydrolases, present in soils, are capable of desulfating esters of many different types. Cooper (1971) found that wetting Nigerian soils caused the release of SO_4^{2-} and that SO_4^{2-} liberation was associated with a corresponding decrease in the size of the HI-reducible sulfur fraction of these soils. Inorganic sulfate release was inhibited completely only in the presence of compounds that both suppress microbial growth and inhibit the activity of extracellular enzymes. Twenty different soils were found to contain arylsulfatase (arylsulfate sulfohydrolase, EC 3.1.6.1) activity, and in all cases the activity was positively correlated with the HI-reducible sulfur fraction of each soil (Cooper, 1972).

Although arylsulfatase plays a major role in mineralizing ester sulfate for the sulfur cycle in Nigerian soils, there are suggestions that this particular sulfohydrolase may not be involved to the same extent in soils from other parts of the world. Thus Tabatabai and Bremner (1970a, 1970b, 1971) assayed 21 different Iowa soils and found that all samples possessed appreciable arylsulfatase activity. Soils sterilized by gamma irradiation still possessed the enzyme but at reduced levels, suggesting that a large proportion of the activity was due to arylsulfatase present in or released by nonviable microorganisms (Tabatabai and Bremner, 1970a). This result implies a high degree of stability for the enzyme, and Tabatabai and Bremner (1970b) reported only an 18% decrease in arylsulfatase activity for field-moist soils after storage at 22 to 24°C for 3 months. This enzyme also appears to function well under adverse conditions of temperature; optimal activity for arylsulfatase in six different soils was detected at an incubation temperature of roughly 67°C (Tabatabai and Bremner, 1970a). Results with the purified enzyme from microbial sources complement these findings (see p. 000).

Factors other than arylsulfatase content appear to be involved in mediating SO_4^{2-} release from Iowa soils. Although arylsulfatase activity was significantly correlated with the organic carbon content of each soil examined (Tabatabai and Bremner, 1970b), this activity was not significantly correlated with the total amount of SO_4^{2-} released after incubation of these soils in the absence of added substrate. For example, only 0.6% of the total sulfur was mineralized in the soil with the highest initial level of arylsulfatase after aerobic incubation at 30°C for 10 weeks. However, 3.0% of the total sulfur was converted to SO_4^{2-} in the soil with the lowest initial level of the enzyme (Tabatabai and Bremner, 1972a). Unfortunately, the HI-reducible sulfur contents of these particular samples were not reported. Kowalenko and Lowe (1975b) found that a Prest soil from British Columbia possessed the highest initial level of arylsulfatase of four soils examined and demonstrated that this same soil also released the greatest amount of SO_4^{2-} during a 14-week incubation period in the absence of added substrate. Maximum SO_4^{2-} release occurred during the first 2 weeks of incubation, after which a leveling off occurred. Moreover, it was also shown that the arylsulfatase activity was just significantly correlated with SO_4^{2-} release over the entire incubation period. However, the arylsulfatase activity of this soil decreased sharply over the first 4 weeks, and the authors concluded that the presence of arylsulfatase was not a major factor controlling the release of SO_4^{2-} in this soil. Similar results were obtained with the other three soils examined, but supporting data were not presented (Kowalenko and Lowe, 1975b). As will become apparent (below), microbial arylsulfatase synthesis is subject to end product regulation by SO_4^{2-}. If microbial activity is responsible for SO_4^{2-} release under these conditions, a decrease in the intracellular as well as the extracellular levels of the enzyme would be expected once maximum SO_4^{2-} release was achieved.

Arylsulfates are not the only sulfate esters present in soils (see p. 401), and it follows from this fact that, irrespective of the involvement of arylsulfatase, other sulfohydrolases may also be responsible for sulfur mineralization in soil. Such a possibility is currently being investigated in the laboratories of F. A. Rose, and this work to date shows that Welsh soils (Houghton and Rose, 1976) possess enzymes capable of hydrolyzing ^{35}S-labeled choline sulfate, dodecyl sulfate, glucose 6-*O*-sulfate, and tyrosine *O*-sulfate, as well as arylsulfate esters such as 2-hydroxy-5-nitrophenyl sulfate and phenyl sulfate.

It can also be demonstrated that, for every sulfate ester which could be found in soil, a corresponding soil microorganism has been isolated which releases SO_4^{2-} from that ester. By far the greatest effort has been directed toward a study of arylsulfatase, and there are numerous detailed reports on the occurrence of this enzyme in bacteria, fungi, and algae (Table 4). This table does not include a complete listing; further references will be given elsewhere in this chapter. The information contained in this table further shows that enzymes of microbial origin have been reported that desulfate mono-, di- and tetrasaccharide sulfate esters, carbohydrate-related sulfate esters, polysaccharide sulfate esters, amino acid sulfate esters, alkylsulfate esters, and choline sulfate. A detailed consideration of this area is clearly beyond the scope of this chapter. The reader's attention is also directed to articles (Roy and Trudinger, 1970; Dodgson and Rose, 1975) dealing with the properties and general occurrence of these enzymes.

9. FUNCTION OF SULFOHYDROLASES IN ESTER SULFATE MINERALIZATION

To assess effectively the involvement of microorganisms in the conversion of ester sulfate to SO_4^{2-}, it is necessary to know (*a*) the physiological factors that regulate the synthesis of the enzymes concerned with SO_4^{2-} release, (*b*) the location of these enzymes within the cells that produce them, and (*c*) the stability characteristics of the enzymes when they exist outside the cell.

10. REGULATION OF SULFOHYDROLASE FORMATION

With some exceptions (Yamagata et al., 1966; Fitzgerald and Laslie, 1975; Fitzgerald et al., 1975), the sulfohydrolases (sulfatases) are not constitutive in bacteria or fungi. The synthesis of these enzymes is controlled by either the sulfur or the carbon content of the environment. *Pseudomonas* $C_{12}B$ (a soil isolate) synthesizes two substrate-inducible primary alkylsulfatases, P1 and P2 (Williams and Payne, 1964; Fitzgerald et al., 1974), and a secondary alkylsulfatase (S3) whose formation is induced by sulfate esters of secondary alcohols in the presence of the corresponding parent alcohol (Dodgson et al., 1974). The induction of none of these enzymes is sulfate ester specific, and the synthesis of the P2 form of primary alkylsulfatase, for example, is induced by secondary as well as primary alkylsulfates (Dodgson et al., 1974) and to a limited extent by a number of primary alkylsulfonates (Cloves, Dodgson, and Fitzgerald, unpublished observations).

Knowledge of the factors controlling sulfatase induction is necessary for an assessment of the relative importance of these enzymes in sulfate ester mineralization. For example, sulfatase induction, which is regulated by the carbon or by the sulfur status of soil, would be expected to play a greater role in the microbial mineralization process than would an enzyme whose synthesis was inhibited in the presence of both carbon- and sulfur-containing compounds. An example of the latter circumstance is the formation of choline sulfatase in *Pseudomonas* 5-A (a sewage isolate). The induction of this enzyme by choline sulfate is inhibited by SO_4^{2-}, cysteine, methionine, and choline chloride (Fitzgerald and Scott, 1974). Microorganisms having sulfohydrolase enzymes regulated in this way will synthesize only enough enzyme to satisfy their growth requirements.

A less rigidly controlled system is the induction of both primary and secondary alkylsulfatases in *Pseudomonas* $C_{12}B$. The formation of these

Table 4. Occurrence of Enzymes from Microbial Sources that Release Inorganic Sulfate from Various Sulfate Esters

Sulfate Ester	Enzyme Nomenclature	Enzyme Found in:	References[a]
Arylsulfates, e.g., *p*-nitrophenyl sulfate, *p*-nitrocatechol sulfate	Arylsulfatase, arylsulfohydrolase, aryl-sulfate sulfohydrolase (EC 3.1.6.1)	Bacteria	Dodgson (1959), Harada and Kamogawa (1963), Fowler and Rammler (1964), Harada and Spencer (1964), Delisle and Milazzo (1970), Fitzgerald and Milazzo (1970, 1975), Okamura et al. (1976b)
		Fungi	Harada and Spencer (1962), Lougheed and Milazzo (1965), Milazzo and Lougheed (1967), Apte and Siddigi (1971), Benkovic et al. (1971), Scott et al. (1971), Sampson et al. (1975)
		Algae	Lien and Schreiner (1975)
Alkylsulfates			
Choline *O*-sulfate	Choline sulfatase, choline sulfohydrolase, choline-sulfate sulfohydrolase (EC 3.1.6.6)	Bacteria	Takebe (1961), Harada (1964), Ikawa et al. (1972), Lucas et al. (1972), Fitzgerald and Scott (1974), Sukhumavasi et al. (1975)
		Fungi	Segel and Johnson (1963), Hussey et al. (1965)
Sodium dodecan-1-yl sulfate (sodium dodecyl sulfate)	Primary alkylsulfatase	Bacteria	Dronkers and Von der Vet (1964), Hsu (1965), Payne et al. (1965), Uesugi et al. (1967), Payne et al. (1974), Sukhumavasi et al. (1975)

Sodium hexan-1-yl sulfate			Fitzgerald (1974), Fitzgerald et al. (1974)
Potassium decan-5-yl sulfate, potassium pentan-3-yl sulfate	Secondary alkylsulfatase	Bacteria	Payne et al. (1967), Payne and Painter (1971), Dodgson et al. (1974), Fitzgerald et al. (1975)
Amino acid-*O*-sulfate, i.e., Tyrosine sulfate, Serine sulfate, Threonine sulfate	Not established. Probably an α,β-elimination reaction for serine sulfate	Bacteria	Harada (1964), Fitzgerald and Payne (1972a), O'Neill (1973), Sukhumavasi et al. (1975), Tudball and O'Neill (1975, 1976), Okamura et al. (1976b)
		Fungi	Burns and Wynn (1975)
Polysaccharide sulfate esters	Not established	Bacteria	Rosen et al. (1960), Dietrich (1969b), Kitamikado et al. (1970), Nakazawa et al. (1975)
Polysaccharide-related sulfate esters			
Monosaccharide sulfates	Glycosulfatase, glycosulfohydrolase, sugar-sulfate sulfohydrolase (EC 3.1.6.3)	Bacteria	Large et al. (1964), Weigl and Yaphe (1966), Fitzgerald and Dodgson (1971b), Fitzgerald (1975)
		Fungi	Lloyd et al. (1968), Reinert and Marzluf (1974)
Tetrasaccharide sulfates derived from chondroitin sulfate	Chondrosulfatase, chondrosulfohydrolase	Bacteria	Dodgson et al. (1957), Dodgson and Lloyd (1957), Linker et al. (1960), Yamagata et al. (1966), Michelacci and Dietrich (1973), Seno et al. (1974)
D-Glycerate 3-*O*-sulfate	Not established	Bacteria	Fitzgerald and Dodgson (1970, 1971a), Dodgson and Rose (1975)

[a] Space does not permit a complete listing; see text for further references, especially those dealing with arylsulfatase.

enzymes was unaffected by the presence or absence of SO_4^{2-} or the sulfur-containing amino acids but was inhibited by a variety of primary and secondary alcohols and by some Krebs cycle intermediates. The latter results prompted Fitzgerald and Payne (1972a) to suggest that the main function of these enzymes was to obtain carbon and energy for the growth of *Pseudomonas* $C_{12}B$. It is likely that the availability of energy regulates alkylsulfatase formation in *Pseudomonas* $C_{12}B$ since the Krebs cycle intermediates act as growth-supporting carbon sources, and this bacterium can oxidize primary alcohols (Payne and Painter, 1971), as well as some secondary alcohols (Fitzgerald and Payne, 1972b). Pentan-3-ol was an effective inhibitor of secondary alkylsulfatase (S-3) formation, and this alcohol, as well as its immediate oxidation product (pentan-3-one), was rapidly oxidized by cell extracts of this bacterium. Lijmbach and Brinkhuis (1973) demonstrated that the formation of the ketone (namely, pentan-3-one) represents the second step in the bacterial utilization of secondary alcohols for energy purposes.

This possibility may also apply to other alkylsulfatase-producing bacteria of soil origin. Thus the synthesis of secondary alkylsulfatase by *Comamonas terrigena* is unaffected by the sulfur status of the culture and is confined exclusively to the stationary phase of the culture cycle (Fitzgerald et al., 1975). Since the energy charge (see Atkinson, 1971) of the bacterial cell is very low at this stage of the culture cycle (Chapman et al., 1971; Wiebe and Bancroft, 1975), results with *C. terrigena* suggest that a depletion of carbon and/or energy triggers the formation of the enzyme. Similarly, primary alkylsulfatase induction in a recently isolated soil bacterium (identified as *Pseudomonas aeruginosa*) is subject to inhibition by glucose, catabolites of glucose, and adenosine 5′-triphosphate (Fitzgerald and Kight, manuscript in preparation).

The same type of sulfatase may be regulated differently, depending upon whether or not it is synthesized by bacteria or fungi. Thus choline sulfatase is substrate inducible in bacteria (Takebe, 1960; Harada, 1964; Lucas et al., 1972; Fitzgerald and Scott, 1974), but its synthesis in *Aspergillus nidulans* (Scott and Spencer, 1968; Spencer et al., 1968) and *Neurospora crassa* (Metzenberg and Parson, 1966; Marzluf and Metzenberg, 1968; McGuire and Marzluf, 1974b) is regulated by a sulfur-mediated de-repression mechanism. Although the enzyme was synthesized only when these fungi (mycelial stage) were cultivated in sulfur-deficient media, the same was not true for the choline sulfatase of *Penicillium chrysogenum*. Lucas and co-workers (1972) found that the formation of this enzyme by mycelia was repressed to only 20% of the maximum level by excess SO_4^{2-} or methionine. The work of McGuire and Marzluf (1974b) demonstrates that factors regulating the synthesis of choline sulfatase may vary, depending upon the fungal developmental stage that is under consideration. Whereas choline sulfatase formation by *N. crassa* during the mycelial stage was repressed by 2 m*M* methionine or SO_4^{2-}, the same concentration of SO_4^{2-} was ineffective in repressing the synthesis of this enzyme by *N. crassa* conidia. Choline sulfatase formation by conidia was regulated only by the methionine content of the medium, with full repression and de-repression taking place at 5 m*M* and 0.25 m*M*, respectively (McGuire and Marzluf, 1974b).

The sulfohydrolases that are required for the degradation of heparin (Dietrich, 1969a) and of keratan sulfate (Kitamikado et al., 1970; Nakazawa et al., 1975) are induced by these mucopolysaccharides, but reports on the factors regulating the induction process are unavailable.

Arylsulfatase formation by many bacteria and fungi is repressed by SO_4^{2-} and, in most cases, by other components of the cysteine biosynthetic pathway. The actual corepressor(s) is (are) not known with certainty for any one system, and there is evidence suggesting that different effectors may be involved. Thus work with cysteine auxotrophs suggested that SO_4^{2-} was most directly involved in repressing arylsulfatase formation by *Aerobacter aerogenes* (Rammler et al., 1964). However, cysteine was much more effective than SO_4^{2-} as an effector regulating the synthesis of this enzyme in *Pseudomonas* $C_{12}B$ (Fitzgerald and Payne, 1972c); and, on the basis of work with a cysteine auxotroph (Adachi and Harada, 1972) and a number of other mutants of *Klebsiella aerogenes,* Adachi et al. (1975) suggested that both SO_4^{2-} and cysteine acted independently to repress arylsulfatase formation in this bacterium. Harada and Spencer (1962) suggested that SO_4^{2-} was the corepressor of arylsulfatase synthesis by a number of fungi, but Metzenberg and Parson (1966) suggested that S^{2-} was most directly involved in repressing the synthesis of this enzyme by *N. crassa*.

De-repression of arylsulfatase formation occurs when various bacteria are grown on methionine as the sole source of sulfur (Harada and Spencer, 1964; Milazzo and Fitzgerald, 1967; Fitzgerald and Payne, 1972c; Adachi et al., 1975). For the synthesis of this enzyme by *A. aerogenes,* Rammler et al. (1964) interpreted this result to mean that methionine is not converted directly (or rapidly) to cysteine in this bacterium. In these studies the various bacteria responded to increasing concentrations of methionine by synthesizing increasing levels of the enzyme. Methionine does not exert a similar effect on arylsulfatase formation by fungi. This amino acid acted as a repressor at high (5 m*M*) concentrations (Harada and Spencer, 1962; Metzenberg and Parson, 1966; Marzluf and Metzenberg, 1968), and de-repression occurred when *N. crassa* was grown on low (0.25 m*M*) concentrations (Marzluf and

Metzenberg, 1968; McGuire and Marzluf, 1974b). Metzenberg and Parson (1966) suggested that methionine acted independently of S^{2-} as an additional corepressor of arylsulfatase formation in the latter fungus. In light of results obtained by Benko et al. (1967), it would not be surprising that methionine and components of the cysteine pathway might act independently in this way. Thus, in a study of the transport of ^{35}S-labeled methionine by various fungi, these workers found that most of the isotope was associated with cystathionine and little, if any, ^{35}S-labeled cysteine or cystine was detected in mycelial extracts.

11. REGULATION OF ARYLSULFATASE SYNTHESIS AND ACTIVITY: THE TYRAMINE EFFECT

The regulation of arylsulfatase synthesis is equally complex in species of the taxanomically closely related genera *Aerobacter (Enterobacter)* and *Klebsiella*. The formation of the enzyme in these bacteria is controlled not only by the sulfur status of the environment (Harada and Spencer, 1964; Rammler et al., 1964; Adachi et al., 1975), but also by glucose-mediated repression (Harada, 1959) and de-repression by tyramine [*p*-(2-aminoethyl)phenol], a substance that is related structurally to the phenolic product of the action of this enzyme on its substrate. The early work of Harada and colleagues identified tyramine as a factor present in enriched media which elicited arylsulfatase formation (Harada and Kono, 1955; Harada and Hattori, 1956; Harada, 1957; Harada and Spencer, 1964). Although tyramine oxidation occurs in cells synthesizing arylsulfatase (Harada, 1959), results of studies with mutants incapable of degrading tyramine suggested that the action of this monoamine was independent of its subsequent metabolism (Adachi et al., 1973; Okamura et al., 1976a). In these mutants, tyramine was shown by Adachi and coworkers (1973) to overcome the repression of arylsulfatase formation due to SO_4^{2-}, an effect that may occur as a consequence of de-repression rather than induction (cf. Harada and Spencer, 1964; Roy and Trudinger, 1970). Further work by Harada's group (Adachi et al., 1974) indicates that a factor (believed to be protein in nature) may mediate the de-repression of arylsulfatase synthesis in the presence of tyramine. Although this factor has not yet been isolated and identified, results of the study by Adachi et al. (1974) clearly show that its synthesis is regulated by catabolite repression in a tyramine oxidase-deficient mutant. The induction in bacteria of the synthesis of many proteins, especially those catalyzing catabolic pathways, is regulated by catabolite repression (for a current review of this phenomenon, see Pastan and Adhya, 1976).

The effect of tyramine on arylsulfatase formation by other microorganisms is not as pronounced. Thus *Pseudomonas aeruginosa* failed to form the enzyme in response to this compound (Harada, 1964), and a similar observation was made with *Proteus rettgeri* (J. W. Fitzgerald and F. H. Milazzo, unpublished observations). Fitzgerald and Payne (1972c) found that tyramine partially relieved full repression due to L-cysteine of arylsulfatase formation in *Pseudomonas* $C_{12}B$. Tyramine produced only a slight stimulation of arylsulfatase synthesis in SO_4^{2-}-repressed cultures of this bacterium. The latter results, representing a ten- and a two-fold increase in synthesis, respectively, should be compared with those obtained by Harada and Spencer (1964) with *Aerobacter aerogenes*. In this case, tyramine caused a 230-fold and a 450-fold increase in enzyme synthesis in the presence of SO_4^{2-} and L-cysteine, respectively. It is unlikely that the differential response of arylsulfatase-forming systems to tyramine is due to the ability to degrade this compound (or the lack of it) among the bacteria that have been examined (cf. Adachi et al., 1974). Even when substantial tyramine utilization occurred with wild-type cultures, the rates of arylsulfatase synthesis were essentially the same in these cells and in mutants that were incapable of degrading this monoamine (Adachi et al., 1973). It is possible that the determining factor may be the rate at which tyramine is transported by various bacteria.

Hussey and Spencer (1967) found that the de-repression of arylsulfatase formation achieved by transfer of the fungus *Aspergillus nidulans* from a cysteine-containing medium to a sulfur-deficient one was increased by 50% if the latter medium contained tyramine. However, retransfer of de-repressed cells (mycelia) to a medium containing cysteine was observed to result in full repression of arylsulfatase synthesis, which could not be prevented by tryamine. The possibility that tyramine exerted an effect on the arylsulfatase-forming system in this fungus should be reinvestigated since cell extracts were apparently not dialyzed before arylsulfatase assay, and it is possible, in light of recent results with another species of *Aspergillus,* that tyramine produced its effect by interacting directly with the action of the enzyme on its substrate. Thus Burns and Wynn (1975) found that tyramine caused an 18-fold activation per se of one of the arylsulfatase isozymes present in *A. oryzae*. For earlier work relating to the occurrence of multiple forms (isozymes) of arylsulfatase in this and other microbial species, see, for example, Cherayil and Van Kley (1962), Drnec and Van Kley (1968), Vessell and Weyer (1968), Rasburn and Wynn (1973), Apte et al. (1974), and Fitzgerald and Milazzo (1975). In *A. oryzae,* "arylsulfatase II" exhibited sulfotransferase activity for tyramine, thus accounting for the apparent activation of this isozyme with respect to its sulfohydrolase activity (Burns and Wynn, 1975). This study also emphasizes that the inclusion of a hydroxyl-containing compound in sulfohydrolase assay media should be avoided unless it is known that the enzyme does not possess a sulfotransferase function for that compound. This caution may partially apply to the use of Tris [Tris(hydroxymethyl)aminomethane] as a buffer component in the assay of some arylsulfatases (see Dennen and Carver, 1969) and certainly applies to assay media containing 3′-phosphoadenosine 5′-phosphosulfate (see Adams, 1962).

On the basis of the different substrate specificities exhibited by the three arylsulfatase isozymes present in *A. oryzae,* Burns and Wynn (manuscripts in preparation) have developed methods for the independent assay of each enzyme in crude cell extracts. This novel approach has been applied to a study of the factors regulating the formation of each isozyme separately, and their results indicate that these enzymes differ not only in their substrate specificities but also in the sensitivities of the genetic loci responsible for their synthesis. Unfortunately, tyramine was not tested as a potential effector regulating arylsulfatase synthesis; thus it is not known whether the monoamine exerts a dual control, influencing the synthesis as well as the activity of the arylsulfatase isozymes in this fungus. Recent work by Harada's group (Okamura et al., 1976b) has established that tyramine does exert a dual function in the bacterium *Klebsiella aerogenes*. Thus tyramine functions *in vivo* as a de-repressor of arylsulfatase synthesis (see above) but *in vitro* as a potent competitive inhibitor of the enzyme purified to the single-protein stage from this source (Okamura et al., 1976*b*). In the latter study it was also confirmed that the arylsulfatase synthesized in response to tryamine was identical to the enzyme with the same activity synthesized in the presence of a nonrepressor source of sulfur (methionine). Neither arylsulfatase possessed sulfotransferase activity for tryamine, tyrosine, or pheonl; indeed, the enzymes hydrolyzed the sulfate esters of these compounds (Okamura et al., 1976b). In contrast, tyramine enhanced the arylsulfatase activity of crude cell extracts of *Pseudomonas* $C_{12}B$ (J. R. George and J. W. Fitzgerald, unpublished observations), and studies are currently under way to determine the mechanism of this effect with a purified source of the enzyme.

12. RELATIONSHIP BETWEEN ARYLSULFATASE FORMATION AND FUNCTION IN INORGANIC SULFATE-DEFICIENT AND SUFFICIENT SOILS

Arylsulfatase formation may also be de-repressed when bacteria (see, e.g., Rammler et al., 1964), fungi (see, e.g., Hussey and Spencer, 1967;

Dennen and Carver, 1969; Apte et al., 1974), and algae (Schreiner et al., 1975) are cultivated in sulfur-deficient media. Consequently, it has been suggested (Dodgson and Rose, 1975) that the function of this enzyme is to provide SO_4^{2-} for microbial growth in SO_4^{2-}-deficient environments containing arylsulfate esters. These results may explain why SO_4^{2-}-deficient soils (see p. 408 and Table 1) possess appreciable levels of the enzyme. However, small differences in arylsulfatase activity among the soils assayed were not related to either the total sulfur content or the SO_4^{2-} content of these samples (Tabatabai and Bremner, 1970b, 1972a; Cooper, 1972; Kowalenko and Lowe, 1975b). Since the soils contained high levels of carbon-bonded sulfur (Table 1), it is likely that cysteine and methionine may also have been present. These amino acids may act to regulate sulfate ester mineralization in soil mediated by microorganisms that are able to synthesize either arylsulfatase or choline sulfatase. The concentrations of cysteine and/or methionine required for complete repression or de-repression of arylsulfatase formation by a bacterial culture can be as low as 10^{-2} m*M* (for repression by cysteine) or 10^{-4} m*M* (for de-repression by methionine) (Fitzgerald and Payne, 1972c). Unfortunately, the ability of soils to hydrolyze choline sulfate has been tested only once (Houghton and Rose, 1976), and data correlating soil arylsulfatase activity with the cysteine and/or methionine content of the same soils is unavailable. In addition, it is not known to what extent these amino acids occur free of peptide bond linkage to other amino acids in the soils where they have been detected (Freney et al., 1972, 1975).

The oceans are rich in SO_4^{2-} (see e.g., Kellog et al., 1972), and there are indications of the existence in marine soils of arylsulfatase-forming systems of microbial origin that are refractive to repression by SO_4^{2-} or cysteine. Thus Chandramohan and co-workers (1974) showed that soils of mangrove swamps and marine sediments of the Indian Ocean possess arylsulfatase activity. Dodgson and co-workers (1954) isolated a strain of *Alcaligenes metalcaligenes* from intertidal mud that was capable of synthesizing arylsulfatase when grown in nutrient broth in quantities suitable for its isolation (Dodgson et al., 1955). This enriched medium generally contains enough sulfur to repress the synthesis of arylsulfatase by other bacteria. Oshrain and Wiebe (manuscript in preparation) have investigated the arylsulfatase content of anaerobic soils in the salt marshes on Sapelo Island (off the coast of Georgia). Significant enzyme activity was detected in core samples taken from depths as great as 30 cm when an area having a high rate of primary production (vegetative growth) was considered. As expected from the fact that these soils are flooded periodically with seawater, the SO_4^{2-} content in the interstitial water of the core samples was a high as 17 m*M*. Arylsulfatase activity was not associated with plant roots, suggesting the existence (for the first time) of anaerobic microorganisms that are also capable of synthesizing the enzyme in SO_4^{2-}-sufficient soils. Although these soils have not been assayed for their ester sulfate contents, data from other studies (see Table 2) suggest that this enzyme may function in anaerobic environments as a means of enabling the microbial community to utilize arylsulfates as sources of carbon.

13. LOCALIZATION OF ALKYLSULFATASE AND "POLYSACCHARIDE" SULFATASE

A knowledge of the location of the sulfohydrolases in microorganisms is essential in order to evaluate the contribution made by viable cells to sulfate ester mineralization in soil. If a particular sulfohydrolase is not found somewhere on the cell periphery, its action on a sulfate ester will be more likely to yield SO_4^{2-} for microbial growth than for plant growth. For example, primary alkylsulfatase is cell wall associated in *Pseudomonas* $C_{12}B$, and high concentrations of the enzyme (Fitzgerald and Laslie, 1975) and SO_4^{2-} (Fitzgerald et al., 1974) were released into the culture medium when this soil isolate was grown in the presence of a primary alkylsulfate ester. Similar considerations apply to the secondary alkylsulfatases of this isolate (Fitzgerald and Laslie, 1974, 1975) and to the choline sulfatase present in another *Pseudomonas* isolate (Fitzgerald and Scott, 1974). Inorganic SO_4^{2-} accumulated in the medium during the growth of the latter bacterium on choline sulfate. The sulfohydrolases involved with heparin (Dietrich, 1969a) and keratan sulfate degradation (Kitamikado and Ueno, 1970) may also be cell wall associated since roughly 50% of the total activity after growth was found in the culture medium.

Unlike many other enzymes (see e.g., Tanford, 1968, 1970; Marsden, 1975), the alkylsulfatases are resistant to denaturation in the presence of sodium dodecyl sulfate (K. S. Dodgson, unpublished results, this laboratory). These enzymes share this unique characteristic with alkaline phosphatase (Mather and Keenan, 1974), and there are good theoretical reasons for expecting that enzymes which deal with protein-dissociating agents such as sodium dodecyl sulfate (Waehneldt, 1975) would be located on the cell periphery, as are other hydrolases (see, e.g., Neu and Heppel, 1965; Neu and Chou, 1967; Nisonson et al., 1969; Wetzel et al., 1970). It may be unwise, however, to generalize at this stage, especially since the secondary alkylsulfatases of *Comamonas terrigena* were not released by osmotic shock or during spheroplast formation (Fitzgerald et al., 1975). Recent unpublished data obtained from membrane vesicles by G. Matcham in this laboratory suggest that these enzymes are associated with the cytoplasmic (inner) membrane in this bacterium.

14. LOCALIZATION OF ARYLSULFATASE

Although concrete visual evidence is lacking, the fact that arylsulfatase can be assayed using intact cells or mycelial pellets (Lougheed and Milazzo, 1967) suggests that this sulfohydrolase also occupies an exocytoplasmic location within the bacterial cell (Adachi et al., 1975; Sukhamavasi et al., 1975), the algal cell (Lien and Schreiner, 1975), the fungal mycelium (Dennen and Carver, 1969), and the fungal conidium (Scott and Metzenberg, 1970). Rammler et al. (1964) used intact cell suspensions to measure arylsulfatase activity in *Aerobacter aerogenes* and reported a complete recovery of activity in the supernatant after cell rupture and centrifugation. With the exception of the temperature optimum, Lein and Schreiner (1975) found that arylsulfatase, present in whole cells of the alga *Chlamydomonas reinhardti,* was similar in a number of properties to the same enzyme purified to the single-protein stage from this source. No increase in activity accompanied cell rupture, and the authors suggested that the enzyme was attached to the cell surface of this alga. Preliminary results of an investigation of the cytochemical location of the enzyme (referred to by Lien and Schreiner, 1975) supported this conclusion, but these results have not yet been published. Rammler et al (1964) concluded that arylsulfatase was also located on the surface of *A. aerogenes,* and a similarity in properties of the enzyme present in intact cells, as opposed to cell extracts of *Proteus rettgeri,* was also noted. In the latter studies the same substrate-dependent activation of arylsulfatase by PO_4^{3-} was observed with intact cells (Milazzo and Fitzgerald, 1966) and with partially purified cell extracts (Fitzgerald and Milazzo, 1975).

Results of work by Metzenberg's group indicate that the location of this enzyme in *Neurospora crassa* is dependent upon the developmental stage that is examined. Thus in young mycelia arylsulfatase is associated with particles resembling the lysosomes of mammalian cells (Scott et al., 1971), but in conidia the enzyme was found to be surface associated as well as intracellular (Scott and Metzenberg, 1970). Approximately 70% of the arylsulfatase present in *Pseudomonas* $C_{12}B$ was released either after osmotic shock or during spheroplast formation. This suggests that the enzyme is cell wall associated in this soil isolate (unpublished results, this laboratory).

15. LOCALIZATION OF CHOLINE SULFATASE

Data from studies of the transport of choline sulfate by fungi (Spencer et al., 1968; Marzluf, 1972a; Cuppoletti and Segel, 1974) suggest that choline sulfatase is not cell surface associated in these microorganisms. Segel's group (Bellenger et al., 1968) found that a mutant of *Penicillium notatum,* deficient in SO_4^{2-} activation, accumulated choline sulfate to an intracellular concentration of 0.075 *M* after suspension for 3 hr in a medium containing 5 m*M* ester. The intracellular concentration of SO_4^{2-} (from the subsequent hydrolysis of the ester) was 0.035 *M*, and no SO_4^{2-} was detected in the assay medium. Choline sulfate was transported unchanged by mycelia when short (2-min) incubations were employed. The transport of this ester by *P. notatum,* as well as by a number of other fungi (Bellenger et al., 1968), is due to a highly specific permease whose synthesis in *P. notatum* and *P. chysogenum* is regulated by a sulfur-mediated repression/de-repression type of control similar to that regulating the synthesis of some sulfohydrolases (see p. 411).

In a similar study, Marzluf (1972a) used a mutant of *Neurospora crassa,* deficient in SO_4^{2-} transport, to confirm that the intact ester (and not its hydrolysis products) was taken up by mycelia. Inorganic sulfate was not detected in the assay medium when this mutant was incubated with choline sulfate for 2 hr. It was concluded that extracellular hydrolysis did not take place. Using a mutant incapable of activating SO_4^{2-}, Marzluf (1972a) found that 70% of the transported choline sulfate was hydrolyzed by *N. crassa* intracellularly during this time interval. *Neurospora crassa* also has a specific choline sulfate permease, and the synthesis of this transport protein too is regulated by the sulfur status of the culture, being repressed by methionine or SO_4^{2-}. These effectors did not inhibit transport activity (Marzluf, 1972a, 1972b).

It is important, in terms of assessing the contribution to the SO_4^{2-} content of soils made by viable fungi, to point out that liberated SO_4^{2-} was observed to be retained by mycelia against a concentration gradient in both investigations (Bellenger et al., 1968; Marzluf, 1972a). These results disagree with those obtained with bacteria in which excess SO_4^{2-} originating from the hydrolysis of a primary alkylsulfate (Fitzgerald et al., 1974) or choline sulfate (Fitzgerald and Scott, 1974) was found in the culture medium. Although choline sulfatase may have a surface location in the latter isolate, Nissen (1968) reported that a wide variety of other soil bacteria were capable of transporting choline sulfate. The ability or inability of these bacteria to hydrolyze the ester was not reported.

16. LOCALIZATION OF GLYCOSULFATASE

Potassium D-glucose 6-*O*-sulfate (glucose sulfate), an ester that is analogous to those comprising some sulfated polysaccharides (Dodgson and Lloyd, 1968), is utilized by *Neurospora crassa* via a mechanism that also involves the transport of the ester followed by its hydrolysis within the mycelium (Reinert and Marzluf, 1974). A mutant of this fungus that is deficient in SO_4^{2-} transport was nevertheless capable of transporting glucose sulfate and of growing on this ester as a sulfur source. The observation that SO_4^{2-} could not be detected in the transport assay medium even after a 6-hr incubation was interpreted to mean that the SO_4^{2-} required for growth was obtained by intracellular hydrolysis (Reinert and Marzluf, 1974). Although SO_4^{2-} was found to inhibit the transport of glucose sulfate, the uptake system for this ester is distinct from those that act to transport SO_4^{2-} (Marzluf, 1970), choline sulfate (Marzluf, 1972a), or glucose (Schneider and Wiley, 1971). The glucose sulfate system is subject to a repression/de-repression type of control mediated primarily by methionine (Reinert and Marzluf, 1974). It is fortunate that Tris [Tris(hydroxy methyl)amino methane] was not used as a buffer component for transport studies with glucose sulfate since this primary amine ($\geqq$ 25 m*M*) can catalyze the isomerization of the glucose ester, as well as the partial hydrolysis of the resulting fructose sulfate (Fitzgerald, 1975).

In work that established for the first time the existence of a glycosulfatase in fungi, Lloyd and co-workers (1968) found that *Trichoderma viride* released SO_4^{2-} into the culture medium when grown on the 6-*O*-sulfate esters of either glucose or galactose. Since no attempt was made to localize the sulfohydrolase in this study, it is not known whether these results indicate a surface location for the enzyme or whether this fungus differs from *N. crassa* (Reinert and Marzluf, 1974) and various *Penicillium* spp. (Bellenger et al., 1968) in being unable to retain the SO_4^{2-} that was generated by intracellular hydrolysis.

17. STABILITY OF SULFOHYDROLASES

Obviously, a soil microorganism having a sulfohydrolase that lacks thermal and/or pH stability will make little contribution to the sulfur cycle in terms of sulfate ester mineralization. Although published data are unavailable on the stability of these enzymes in a crude state, it has been noted in this laboratory and elsewhere that the sulfohydrolases as a class appear to be very stable to elevated temperature when present in crude cell extracts. Presumably, in this state the enzyme is protected from denaturation by the presence of extraneous proteins and polysaccharides. Soils are known to contain these substances in addition to a highly complex mixture of other organic compounds of unknown origin (see, e.g., Bremner, 1967; Finch et al., 1971), and the protective environment found in soil may be similar to that existing in a crude cell extract. Previous reference has been made to the stability of arylsulfatase in soil samples (see p. 409).

Results of stability studies on arylsulfatase, purified to the single-protein stage, attest to the inherent durability of these proteins. Thus no appreciable loss in activity was noted when the enzyme (purified from *Chlamydomonas reinhardti*) was incubated in the absence of substrate for 16 min at a temperature as high as 60°C. The optimum temperature for activity of arylsulfatase in intact cells was also 60°C (Lien and Schreiner, 1975). The α- and β-isoenzymes of arylsulfatase from *Pseudomonas aeruginosa* were stable to the same temperature for 5 min, and Delisle and Milazzo (1972) also observed these enzymes to be equally stable to pH over the range 6.5 to 10.0 for periods of storage at 37.5°C up to 24 hr. The enzymes also exhibited appreciable activity over this pH range. Similar results were obtained with a partially purified source of arylsulfatase from *Aerobacter (Enterobacter) aerogenes*. In this study Fowler and Rammler (1964) found, in addition, that this enzyme was maximally active over a wide temperature range extending from 28 to 60°C.

With the exception of work by Payne et al. (1965), stability studies of other purified sulfohydrolases of microbial origin have not been reported. In the study by Payne and co-workers it was demonstrated that primary alkylsulfatase, present in $(NH_4)_2SO_4$-fractionated extracts of *Pseudomonas* $C_{12}B$, hydrolyzed sodium dodecyl sulfate best at 70°C. This enzyme was stable in the presence of its substrate for 1 hr at 60°C. Primary alkylsulfatase was active over a pH range of 5 to 9, but its stability over this range was not reported. It may be anticipated from observations made with crude cell extracts that similar studies with other purified sulfohydrolases will reveal these enzymes to be inherently stable as well.

18. SOURCES OF INORGANIC SULFATE IN AEROBIC SOILS

Although the oxidation of reduced forms of sulfur (S^0 and S^{2-}) represents a major mechanism for the generation of SO_4^{2-} in anaerobic environments (see, e.g., Postgate, 1968; Peck, 1970; Le Gall, 1972; Jorgensen and Fenchel, 1974), there is evidence suggesting that this is not necessarily the case for well-drained terrestrial soil. Thus, whereas many bacterial photoautotrophs are capable of these oxidations in aquatic environments, soil is not a suitable medium for their survival. Light and anaerobiosis cannot occur together in this kind of environment. According to Ehrlich (1971), the aerobic oxidation of S^0 is carried out principally by three species of the genus *Thiobacillus*. The ability to oxidize S^{2-} appears to be even further restricted, being confined to a single bacterial species, *Thiobacillus thioparus* (Ehrlich, 1971). Indeed, over 50% of the 329 Australian surface soils investigated by Swaby and co-workers oxidized added S^0 either very slowly or not at all (Swaby and Vitolins-Maija, 1968; Vitolins-Maija and Swaby, 1968; Swaby and Fedel, 1973). The inability to oxidize this form of sulfur was correlated with the absence of thiobacilli from these soils (Swaby and Fedel, 1973).

The occurrence of sulfate esters in aerobic soils and the additional presence of microorganisms capable of liberating SO_4^{2-} from these esters suggest that sulfate ester hydrolysis represents a source of this anion for the sulfur cycle. This possibility is further substantiated by the stability characteristics of the sulfohydrolases which catalyze this reaction and by the cell surface location of many of these enzymes. However, because of the limited data available, it is not possible at present to evaluate the extent of this contribution to the sulfur cycle. Certainly the hydrolysis of these esters may represent a major means of generating SO_4^{2-} in Australian soils that lack thiobacilli (Swaby and Fedel, 1973) but contain high concentrations of ester sulfate (Table 1). Detailed studies on sulfate ester mineralization by microorganisms in these particular soils are clearly warranted.

In addition to mammalian formation and excretion of ester sulfates, soils are also capable of synthesizing these esters; and sulfate ester formation by soil microorganisms may represent an important mechanism for regulating the availability of SO_4^{2-} for the cycle. Again, however, insufficient data obviate an evaluation of the general applicability of this possibility, since much of the work (see p. 403) was confined to studies of the generation of a single sulfate ester, namely, choline sulfate. Although fungi are capable of forming this ester in large quantities, almost all of the ester is retained intracellularly and thus is released only after cell lysis. On the other hand, bacteria of the genus *Pseudomonas* form and release choline sulfate when SO_4^{2-} is growth limiting (Fitzgerald, 1973), and this may represent a mechanism for storing readily utilizable SO_4^{2-} in a form that will not alter the existing soil pH. The release of SO_4^{2-} from this ester need not be mediated only by microorganisms since plant roots (Nissen and Benson, 1964; Nissen,

1974) and leaves (Nissen, 1974) are capable of transporting and hydrolyzing choline sulfate (Nissen, 1968).

Since commonly used fertilizers are free of sulfur (Bixby et al., 1964; Tisdale, 1966), atmospheric pollution represents a major source of SO_4^{2-} for inland aerobic soils (Kellog et al., 1972; Likens and Bormann, 1974), and seawater aerosols (Kellog et al., 1972) contribute some of the SO_4^{2-} entering coastal soils. There is evidence that highly industrialized areas receive proportionally more sulfur from pollution than do rural areas of low industrialization. For example, in coastal areas such as Norfolk, Virginia, 33.5 lb S/acre·year was deposited in rainfall, whereas a less industrialized area such as Halifax, Nova Scotia, received only 12.5 lb (Le Gall, 1972). The pollution-derived SO_4^{2-} in rain and snow originates as a consequence of the oxidation (by a number of different mechanisms; see Kellog et al., 1972) of SO_2 and SO_3 which occur in air as pollutants. The results of studies by Freney and co-workers (1971, 1975) suggest that this SO_4^{2-} will be either used directly for plant growth or converted to ester sulfate and later utilized for plant growth. In fallow soils or soils weeded before planting, exogenously derived SO_4^{2-} does not accumulate as such before utilization by growing plants. Freney et al. (1971) found that $^{35}SO_4^{2-}$ was readily incorporated into HI-reducible sulfur (ester sulfate; see p. 396) in soils of this type. Sulfur dioxide can also enter soils directly (Dodgson and Rose, 1966). Smith et al. (1973) found that many agricultural soils sorb substantial amounts of this gas, but it is uncertain whether this SO_2 can be subsequently oxidized to SO_4^{2-}.

Apart from SO_4^{2-} returned to soils from atmospheric pollution, a review of factors generating SO_4^{2-} for the sulfur cycle would not be complete without a consideration of the metabolism of compounds that possess the C—SO_3^- (sulfonate), N—SO_3^- (sulfamate), and N—O—SO_3^- linkages. Sulfolipids and sulfocarbohydrates possessing the sulfonate linkage are widely distributed in green plants (Wintermans, 1969; Lepage et al., 1961; Benson, 1963; Kates, 1970; Harwood, 1975). Sulfonates also occur to a lesser extent in algae (Shibuya et al., 1963), protozoans (Davies et al., 1965, 1966), bacterial membranes (Benson et al., 1959; Langworthy et al., 1976), and bacterial spores (Bonsen et al., 1969). Although a sulfonate has never been isolated from soil, it is obvious that these compounds (like compounds possessing the C—O—S linkage) may be easily released and possibly mineralized in this environment. The latter consideration is given some support by the isolation from soil of several *Flavobacterium* species which liberated SO_4^{2-} from a lipid sulfonate found in plant tissue (Martelli and Benson, 1964). The existence of an analogous enzyme, capable of hydrolyzing a phosphonate linkage, has been demonstrated in plants, insects, animals, and microorganisms (Kelly and Butler, 1975). Moreover, many of the alkylbenzene sulfonates, which arise in soils and waters as pollutants, are readily degraded by microorganisms (Payne, 1963; Payne and Feisal, 1963; Prochazka and Payne, 1965; Cardini et al., 1966; Heyman and Molof, 1967; Huddleston and Allred, 1967).

The sulfated thioglycosides (oxime *O*-sulfate esters) which possess the N—O—SO_3^- linkage are synthesized and released into soil by most, if not by all, members of the Cruciferae family of plants (Kjaer, 1961; Virtanen, 1965). Once acted upon by a thioglycosidase of widespread occurrence in plants and fungi (Dodgson and Rose, 1975), these esters can serve as sources of SO_4^{2-}. At present a great deal of confusion exists as to whether the subsequent release of SO_4^{2-} occurs enzymically or by a nonenzymic rearrangement, and evidence for and against the involvement of a sulfohydrolase was considered in reviews by Roy and Trudinger (1970) and by Dodgson and Rose (1975). Irrespective of the mechanism involved, it is clear that these esters may represent a major source of SO_4^{2-} for soils possessing cruciferous plants. Korn and Payza (1956) and Dietrich (1969a, 1969b) demonstrated the existence of an enzyme in *Flavobacterium heparinum* [a soil isolate (Payza and Korn, 1956)] which liberated SO_4^{2-} from the sulfamate groups present in heparin. The occurrence of this sulfated polysaccharide in soil was discussed on p. 401.

A consideration of cyst(e)ine degradation as a source of SO_4^{2-} is necessary since this may represent the major means of completing the sulfur cycle in aerobic soils that cannot oxidize S^0. Various fungi and bacteria of soil origin (Mothes, 1938; Stahl et al., 1949; Starkey, 1950, 1956; Frederick et al., 1957) converted cysteine and cystine to SO_4^{2-} aerobically, and Freney (1958b, 1960) demonstrated that aerated soils were also capable of degrading these amino acids directly to SO_4^{2-}. In addition, the sulfur present in methionine was converted aerobically to this anion by a mixed population of soil microorganisms (Hesse, 1957) and by *Aspergillus niger* (Garreau, 1941). Possible pathways for the direct aerobic conversion of cysteine to SO_4^{2-} were considered in a review by Freney (1967). The sulfur in this amino acid was also degraded aerobically to S^{2-} by *Escherichia coli* and by *Proteus vulgaris* (Kondo, 1923; Almy and James, 1926; Starkey, 1950), but the mechanism for this transformation is not well understood. In more recent work, Swaby and Fedel (1973) found that 41% of 56 Australian soils converted cystine to S^{2-}, but results of experiments designed to test the further oxidation of S^{2-} were not reported. Although S^{2-} can be oxidized nonenzymically to S^0, only 14% of these soils rapidly oxidized S^0 to

SO_4^{2-}, demonstrating that S^0 and cystine are not suitable precursors of SO_4^{2-} for these soils. With the exception of a brief report (Hagedorn, 1975), similar published accounts of the ability of other soils to oxidize S^0 are unavailable. Although the application of S^0 is sometimes used to lower soil pH, it is not known to what extent S^0 oxidation acts as a source of SO_4^{2-} for soils in general. For lack of a suitable alternative, it is equally possible that the Australian soils which are deficient in the ability to oxidize S^0 may depend largely on their reserve of ester sulfate (see Table 1) as a major source of SO_4^{2-}. Possible sources of this anion for well-drained, aerobic soils in general are summarized in Figure 1.

19. PRACTICAL AND FUTURE CONSIDERATIONS

Whereas a great deal is known about the parameters regulating the supply of SO_4^{2-} in anaerobic environments, much less is known concerning the analogous processes in aerobic soils, yet aerobic soils are more extensively used for agriculture. Indeed, the question may be asked, does a cyclic transformation of sulfur always occur in aerobic soils? At present, there is insufficient information to answer this question, even though an affirmative response has been obtained for partially anaerobic environments (see e.g., Butlin and Postgate, 1954; Peck, 1974; Jorgensen and Fenchel, 1974).

Since aerobic soils represent the major source of the world's food, the factors that generate SO_4^{2-} for these soils not only are related to an understanding of the sulfur cycle but also determine how much SO_4^{2-} will be available for plant growth and hence animal survival. Toward this end, Tisdale (1966) stressed that a "greater knowledge of the factors affecting the release of sulfur (from organically bound ester sulfate) would be of immense value in predicting the supply of this element available to growing crops." Despite the SO_4^{2-} that is returned in the form of air pollution and ocean aerosols, sulfur deficiencies have occurred in inland aerobic soils (Bixby et al., 1964; Jordan, 1964; Anderson and Futral, 1966; Tisdale, 1966; Swaby and Fedel, 1973). The belief held by some ecologists that sulfur is nonlimiting in this environment may now need to be reevaluated, especially if the proposed transfer and use of nitrogen-fixing genes to improve plant growth becomes a reality. In this connection Postgate (1974) stressed that, in soils where nitrogen is not the nutrient limiting plant growth, a limitation of sulfur as well as of other elements could be expected if plant growth is intensified. In view of the continued use of fertilizers of limited sulfur content, existing and proposed restrictions on SO_2-SO_3 pollution may

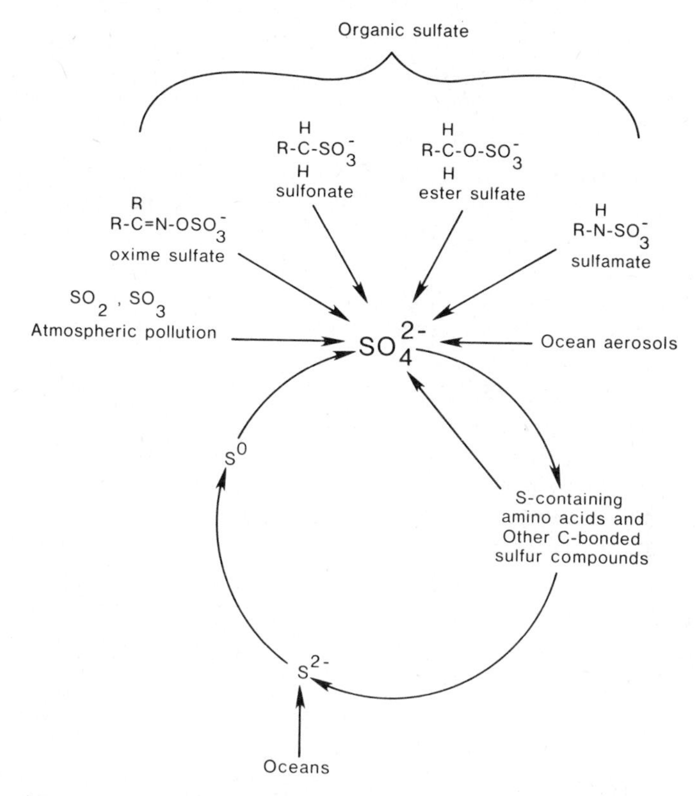

Figure 1. Sources of inorganic sulfate for aerobic, well-drained soils: SO_4^{2-}, inorganic sulfate; sulfur-containing amino acids cysteine, cystine, and methionine; S^{2-}, sulfide; S^0, elemental sulfur. Taken in part from Fitzgerald (1976), Fig. 1.

mean that in the future a greater demand will be made by plants on the soil's reserves of ester sulfate than on free SO_4^{2-}. This possibility is particularly pertinent in underdeveloped countries in which land is still largely fertilized with animal excreta.

It is clear that sulfate ester hydrolysis represents a source of SO_4^{2-} for soil, but at this stage the magnitude of this contribution cannot be evaluated. In addition to the elegant studies of Freney and co-workers, further investigations of the nature, formation, and release of SO_4^{2-} from organic sulfate need to be carried out not only for Australian soils but also for aerobic soils throughout the world. Although it is clear that sulfur is present in this fraction as ester sulfate, the kind(s) of ester linkage(s) involved and other properties of the macromolecules comprising this fraction remain to be determined. Soil is decidely the most complex mixture on this planet, not only from a microbiological but also from a physical and a chemical viewpoint. Since it is impossible to control all the variables involved, it is believed that questions regarding the generation of SO_4^{2-} in this environment will not be totally answered by studies dealing entirely with the natural system. These investigations should be complemented by further studies of the localization of enzymes catalyzing the release of SO_4^{2-} and of factors governing the synthesis of these enzymes in microorganisms and plants grown or maintained in pure culture. The latter investigations should be directed toward determining the properties of the appropriate enzymes in microorganisms, isolated from soils known to contain high concentrations of ester sulfate. To date this has not been done, and yet the exhaustive studies by Tabatabai and Bremner of the sulfur status of Iowa soils represent an ideal foundation for such an investigation in the United States. Considering the direction of future research on sulfur in agriculture, Tisdale (1966) listed identification of the factors influencing the mineralization and immobilization of sulfur in soils as having high priority, second only to definition of the pathways of sulfur metabolism in cells. Tisdale further stated that "when this knowledge is translated into agricultural practice, it will contribute immeasurably to more efficient crop and animal production."

ACKNOWLEDGMENTS

The author is grateful for assistance provided by the University of Georgia's Computer Center for Information Retrieval. He also wishes to thank I. H. Segel, W. J. Wiebe, and C. H. Wynn for access to unpublished information.

REFERENCES

Acheson, R. M. and Hands, A. R. (1961). *J. Chem. Soc.*, p. 746.

Adachi, T. and Harada, T. (1972). *Seikagaku* **44**, 382.

Adachi, T., Murooka, Y., and Harada, T. (1973). *J. Bacteriol.* **116**, 19-24.

Adachi, T., Okamura, H., Murooka, Y., and Harada, T. (1974). *J. Bacteriol.* **120**, 880-885.

Adachi, T., Murooka, Y., and Harada, T. (1975). *J. Bacteriol.* **121**, 29-35.

Adams, J. B. (1962). *Biochim. Biophys. Acta* **62**, 17-26.

Allen, B. F. (1962). *Md. Pharm.* **37**, 671-675.

Almy, L. H. and James, L. H. (1926). *J. Bacteriol.* **12,** 319-331.

Anderson, O. E. and Futral, J. G. (1966). Georgia Agricultural Experiment Stations, N. S., Bull. 167.

Apte, B. N. and Siddigi, O. (1971). *Biochim. Biophys. Acta* **242,** 129-140.

Apte, B. N., Bhavsar, P. N. and Siddigi, O. (1974). *J. Mol. Biol.* **86,** 637-648.

Arkley, T. H. (1961). Ph.D. thesis, University of California, Berkeley.

Atkinson, D. E. (1971). *Ann. Rev. Microbiol.* **23,** 47-68.

Baker, E. M., Hammer, D. C., March, S. C., Tolbert, B. M., and Canham, J. E. (1971). *Science* **173,** 826-827.

Baker, E. M., Knight, M. K., Tillotson, J. A., Johnsen, D. O., and Tolbert, B. M. (1974). *Fed. Am. Soc. Exp. Biol.* **33,** 665.

Ballio, A., Chain, E. B., Dentice di Accadia, N. F., Rossi, C., and Ventura, M. T. (1959). "The Distribution of Sulphurylcholine in Fungi." Reprinted from *Selected Scientific Papers from the Instituto de Superiore di Sanita,* Vol. 2, Part 2, pp. 343-353.

Barra, H. S. and Caputto, R. (1965). *Biochim. Biophys. Acta* **101,** 367-369.

Barrow, N. J. (1961). *Aust. J. Agric. Res.* **12,** 306-319.

Barrow, N. J., Spencer, K., and McArthur, W. M. (1969). *Soil Sci.* **108,** 120-126.

Bellenger, N., Nissen, P., Wood, T. C., and Segel, I. H. (1968). *J. Bacteriol.* **96,** 1574-1585.

Benko, P. V., Wood, T., and Segel, I. H. (1967). *Arch. Biochem. Biophys.* **122,** 783-804.

Benkovic, S. J., Vergara, E. V., and Hevey, R. C. (1971). *J. Biol. Chem.* **246,** 4926-4933.

Benson, A. A. (1963). Adv. *Lipid Res.* **1,** 387-394.

Benson, A. A. and Atkinson, M. R. (1967). *Fed. Am. Soc. Exp. Biol.* **26,** 394.

Benson, A. A., Daniel, H., and Wiser, R. (1959). *Proc. Natl. Acad. Sci. U.S.* **45,** 1582-1587.

Bixby, D. W., Tisdale, S. L., and Rucker, D. L. (1964). The Sulphur Institute, Tech. Bull. 10.

Bogan, R. H. and Sawyer, C. N. (1954). *Sewage Ind. Wastes* **26,** 1069-1080.

Bonsen, P. P. M., Spudich, J. A., Nelson, D. L., and Kornberg, A. (1969). *J. Bacteriol.* **98,** 62-68.

Boström, H. and Vestermark, A. (1959). *Nature* **183,** 1593-1594.

Boyland, E., Manson, D., and Orr, S. F. D. (1957). *Biochem. J.* **65,** 417-423.

Bremner, J. M. (1967). "Nitrogenous Compounds," In A. D. McLaren and G. H. Peterson, Eds., *Soil Biochemistry,* Vol. 1, pp. 19-66.

Breslow, J. L. and Sloan, H. R. (1972). *Biochem. Biophys. Res. Commun.* **46,** 919-925.

Burns, G. R. J. and Wynn, C. H. (1975). *Biochem. J.* **149,** 697-705.

Burton, E. G. and Metzenberg, R. L. (1973). *J. Bacteriol.* **113,** 519-520.

Butlin, K. R. and Postgate, J. R. (1954). "The Microbial Formation of Sulfur in Cryenaican Lakes." In *Biology of Deserts,* Symposium Inst. Biology, London, pp. 112-122.

Catalfomo, P., Block, J. H., Constantine, G. H., and Kirk, P. W. (1973). *Mar. Chem.* **1,** 157-162.

Cardini, G., Catelani, D., Sorlini, C., and Treccani, V. (1966). *Ann. Microbiol. Enzymol.* **16,** 217-223.

Chandramohan, D., Devendran, K., and Natarajan, R. (1974). *Marine Biol.* **27,** 89–92.

Chapman, A., Fall, L., and Atkinson, D. E. (1971). *J. Bacteriol.* **108,** 1072-1086.

Cherayil, J. D. and Van Kley, H. (1962). *Fed. Proc. Fed. Am. Soc. Exp. Biol.* **21,** 230.

Cooper, P. J. M. (1971). Ph.D. thesis, University of Reading, England.

Cooper, P. J. M. (1972). *Soil Biol. Biochem.* **4,** 333-337.

Cuppoletti, J. and Segel, I. H. (1974). *J. Membrane Biol.* **17,** 239-252.

Curtis, C. G., Hearse, D., and Powell, G. M. (1974). *Xenobiotica* **4,** 595-600.

Dalgliesh, C. E. (1955). *Biochem. J.* **61,** 334-337.

Davies, W. H., Mercer, E. T., and Goodwin, T. W. (1965). *Phytochemistry* **4,** 741–749.

Davies, W. H., Mercer, E. T., and Goodwin, T. W. (1966). *Biochem. J.* **98,** 369–373.

Davis, V. E., Huff, J. A., and Brown, H. (1966). *Clin. Chim. Acta* **13,** 380-382.

De Flines, J. (1955). *J. Am. Chem. Soc.* **77,** 1676-1677.

Delisle, G. and Milazzo, F. H. (1970). *Biochim. Biophys. Acta* **212,** 505-508.

Delisle, G. J. and Milazzo, F. H. (1972). *Can. J. Microbiol.* **18,** 561-568.

DeLong, W. A. and Lowe, L. E. (1962). *Can. J. Soil Sci.* **42,** 223.

De Meio, R. H. (1975). "Sulfate Activation and Transfer." In D. M. Greenberg, Ed., *Metabolic Pathways,* Vol. 7, pp. 287-358.

Dennen, D. W. and Carver, D. D. (1969). *Can. J. Microbiol.* **15,** 175-181.

Denner, W. H. B., Olavesen, A. H., Powell, G. M., and Dodgson, K. S. (1969). *Biochem. J.* **111,** 43-51.

Dietrich, C. P. (1969a). *Biochemistry* **8,** 3342-3347.

Dietrich, C. P. (1969b). *Biochem. J.* **111,** 91-95.

Dodgson, K. S. (1959). *Enzymologia* **20,** 301-312.

Dodgson, K. S. and Lloyd, A. G. (1957). *Biochem. J.* **66,** 532-539.

Dodgson, K. S. and Lloyd, A. G. (1968) "Metabolism of Acidic Glycosaminoglycans (Mucopolyaccharides)." In F. Dickens, P. J. Randle, and W. J. Whelan, Eds., *Carbohydrate Metabolism and Its Disorders,* Vol. 1, pp. 169-212.

Dodgson, K. S. and Rose, F. A. (1966). *Nutr. Abstr. Rev.* **36,** 327-342.

Dodgson, K. S. and Rose, F. A. (1970). "*Sulfoconjugation and Sulfohydro-*

lysis." In W. H. Fishman, Ed., *Metabolic Conjugation and Metabolic Hydrolysis,* Vol. 1, pp. 239-325.

Dodgson, K. S. and Rose, F. A. (1975) "Sulfohydrolases." In D. M. Greenberg, Ed., *Metabolic Pathways,* Vol. 7, pp. 359-431.

Dodgson, K. S., Lloyd, A. G., and Spencer, B. (1957). *Biochem. J.* **65,** 131–137.

Dodgson, K. S., Melville, T. H., Spencer, B., and Williams, K. (1954). *Biochem. J.* **58,** 182-187.

Dodgson, K. S., Spencer, B., and Williams, K. (1955). *Biochem. J.* **61,** 374–380.

Dodgson, K. S., Fitzgerald, J. W., and Payne, W. J. (1974). *Biochem. J.* **138,** 53–62.

Doyle, J. (1967). *Biochem. J.* **103,** 325–330.

Drnec, J. and Van Kley, H. (1968). *Fed. Proc. Fed. Am. Soc. Exp. Biol.* **27,** 590.

Dronkers, H. and Von Der Vet, A. P. (1964). *Investigations into the Mechanisms of Microbial Breakdown of Dodecyl Sulfate.* IVth International Congress on Surface Active Agents, Brussels.

Dziewiatkowski, D. D. (1970) In: P. H. Muth and J. E. Oldfield, Eds., *Symposium: Sulfur in Nutrition: Metabolism of Sulfate Esters.* AVI Publishing Co., Westport, Conn., pp. 97–125.

Ehrlich, H. L. (1971) "Biogeochemistry of the Minor Elements in Soil." In A. D. McLaren and J. Skujins, Eds., *Soil Biochemistry,* Vol. 2, pp. 361–384.

Embery, G. and Whitehead, E. (1975). *Biochem. Soc. Trans.* **3,** 661.

Feige, B. and Simonis, W. (1969). *Planta* **86,** 202-204.

Felbeck, G. T. (1971). "Chemical and Biological Characterization of Humic Matter." In A. D. McLaren and J. Skujins, Eds., *Soil Biochemistry,* Vol. 2, pp. 36–59.

Fenchel, J. M. and Rield, R. J. (1970). Mar. Biol. **7,** 255–268.

Finch, P., Hayes, M. H. B., and Stacey, M. (1971). "The Biochemistry of Soil Polysaccharides." In A. D. McLaren and J. Skujins, Eds., *Soil Biochemistry,* Vol. 2, pp. 257–319.

Fitzgerald, J. W. (1973). *Biochem. J.* **136,** 361–369.

Fitzgerald, J. W. (1974). *Microbios* **11,** 153–158.

Fitzgerald, J. W. (1975). *Can. J. Biochem.* **53,** 906–910.

Fitzgerald, J. W. (1976). *Bacteriol. Rev.* **40,** 698–721.

Fitzgerald, J. W. and Dodgson, K. S. (1970). *Biochem. J.* **119,** 30 p.

Fitzgerald, J. W. and Dodgson, K. S. (1971a). *Biochem. J.* **121,** 521–528.

Fitzgerald, J. W. and Dodgson, K. S. (1971b). *Biochem. J.* **122,** 277–283.

Fitzgerald, J. W. and Laslie, W. W. (1974). *Biochem. Soc. Trans.* **2,** 1072–1073.

Fitzgerald, J. W. and Laslie, W. W. (1975). *Can. J. Microbiol.* **21,** 59–68.

Fitzgerald, J. W. and Luschinski, P. C. (1977). *Can. J. Microbiol.* **23,** 483–490.

Fitzgerald, J. W. and Milazzo, F. H. (1970). *Can. J. Microbiol.* **16,** 1109–1115.

Fitzgerald, J. W. and Milazzo, F. H. (1975). *Int. J. Biochem.* **6,** 769–774.

Fitzgerald, J. W. and Payne, W. J. (1972a). *Microbios* **5,** 87–100.

Fitzgerald, J. W. and Payne, W. J. (1972b). *Microbios* **6,** 55–67.

Fitzgerald, J. W. and Payne, W. J. (1972c). *Microbios* **6,** 147–156.

Fitzgerald, J. W. and Scott, C. L. (1974). *Microbios* **10,** 121–131.

Fitzgerald, J. W., Dodgson, K. S., and Payne, W. J. (1974). *Biochem. J.* **138,** 63–69.

Fitzgerald, J. W., Dodgson, K. S., and Matchem, G. (1975). *Biochem. J.* **149,** 477–480.

Fluharty, A. L., Stevens, Miller, R. T., and Kihara, H. (1974). *Biochem. Biophys. Res. Commun.* **61,** 348–354.

Fowler, L. R. and Rammler, D. H. (1964). *Biochemistry* **3,** 230–237.

Frederick, L. R., Starkey, R. L., and Segal, W. (1957). *Soil Sci. Soc. Am. Proc.* **21,** 287–292.

Freney, J. R. (1958a). *Soil Sci.* **86,** 241–244.

Freney, J. R. (1958b). *Nature* **182,** 1318–1319.

Freney, J. R. (1960). *Aust. J. Biol. Sci.* **13,** 387–392.

Freney, J. R. (1961). *Aust. J. Agr. Res.* **12,** 424–432.

Freney, J. R. (1967). "Sulphur-Containing Organics." In A. D. McLaren and G. H. Peterson, Eds., *Soil Biochemistry,* Vol. 1, pp. 220–259.

Freney, J. R. and Stevenson, F. J. (1966). *Soil Sci.* **101,** 307–316.

Freney, J. R. and Swaby, R. J. (1975). In K. D. McLachlan, Ed., *Sulphur in Australasian Agriculture: Sulphur Transformations in Soils.* Prentice-Hall, Hemel Hempstead, England, pp. 31–39.

Freney, J. R., Melville, G. E., and Williams, C. H. (1969). *J. Sci. Food Agric.* **20,** 440–445.

Freney, J. R., Melville, G. E., and Williams, C. H. (1970). *Soil Sci.* **109,** 310–318.

Freney, J. R., Melville, G. E., and Williams, C. H. (1971). *Soil Biol. Biochem.* **3,** 133–141.

Freney, J. R., Stevenson, F. J., and Beavers, A. H. (1972). *Soil Sci.* **114,** 468–476.

Freney, J. R., Melville, G. E., and Williams, C. H. (1975). *Soil Biol. Biochem.* **7,** 217–221.

Garreau, Y. (1941). *C. R. Acad. Sci.* **135,** 508–510.

Goren, M. B. (1970). *Biochim. Biophys. Acta* **210,** 127–138.

Goren, M. B. (1971). *Lipids* **6,** 40–46.

Graef, V. and Fuchs, M. (1975). *Z. Klin. Chem. Klin. Biochem.* **13,** 163–167.

Grimes, A. J. (1959). *Biochem. J.* **73,** 723–729.

Grisley, D. W. and McComb, R. B. (1974). *Clin. Chem.* **20,** 892.

Hagedorn, C. (1975). *Bacteriol. Proc.* **75,** 137.

Haines, T. and Block, R. J. (1962). *J. Protozool.* **9,** 33–38.

Hanahan, D. J., Everett, N. B., and Davis, C. D. (1949). *Arch. Biochem. Biophys.* **23,** 501–503.

Hancock, A. J. and Kates, M. (1973). *J. Lipid Res.* **14,** 422–429.

Harada, T. (1957). *Bull. Agr. Chem. Soc. Jap.* **21,** 267–273.

Harada, T. (1959). *Bull. Agr. Chem. Soc. Jap.* **23,** 222–230.

Harada, T. (1964). *Biochim. Biophys. Acta* **81,** 193–196.

Harada, T. and Hattori, F. (1956). *Bull. Agr. Chem. Soc. Jap.* **20,** 110–115.

Harada, T. and Kamogawa, A. (1963). *J. Ferment. Technol.* **41,** 132–136.

Harada, T. and Kono, K. (1955). *Mem. Inst. Sci. Ind. Res. Osaka Univ.* **12,** 183.

Harada, T. and Spencer, B. (1960). *J. Gen. Microbiol.* **22,** 520–527.

Harada, T. and Spencer, B. (1962). *Biochem. J.* **82,** 148–156.

Harada, T. and Spencer, B. (1964). *Biochem. J.* **93,** 373–378.

Harper, S. H. and Letcher, R. M. (1966). *Chem. Ind.,* p. 419.

Harward, M. E. and Reisenauer, H. M. (1966). *Soil Sci.* **101,** 326–335.

Harwood, J. L. (1975). *Biochim. Biophys. Acta* **398,** 224–230.

Haslewood, G. A. D. (1967). *J. Lipid Res.* **8,** 535–550.

Hayashi, M., Unemoto, T., and Miyaki, K. (1962). *Chem. Pharm. Bull. Tokyo* **10,** 533–535.

Heacock, R. A. and Mahon, M. E. (1964). *Can. J. Biochem.* **42,** 813–819.

Hesse, P. R. (1957). *Plant Soil* **9,** 86–96.

Hext, P. M., Thomas, S., Rose, F. A., and Dodgson, K. S. (1973). *Biochem. J.* **134,** 629–635.

Heyman, J. J. and Molof, A. H. (1967). *J. Water Pollut. Control Fed.* January, pp. 50–62.

Houghton, C. and Rose, F. A. (1976). *Appl. Environ. Microbiol.* **31,** 969–976.

Hsu, Y. (1965). *Nature* **207,** 385–388.

Huddleston, R. L. and Allred, R. C. (1967). "Surface-Active Agents: Biodegradability of Detergents." In A. D. McLaren and G. H. Peterson, Eds., *Soil Biochemistry,* Vol. 1, pp. 343–370.

Hunt, S. and Jevons, F. R. (1966). *Biochem. J.* **98,** 522–529.

Hurst, H. M. and Burges, N. A. (1967). "Lignin and Humic Acids." In A. D. McLaren and G. H. Peterson, Eds., *Soil Biochemistry,* Vol. 1, pp. 260–286.

Hussey, C. and Spencer, B. (1967). *Biochem. J.* **103,** 56–57.

Hussey, C. B., Orsi, B. A. Scott, J., and Spencer, B. (1965). *Nature* **207,** 632.

Ikawa, M. and Taylor, F. R. (1973). In D. F. Martin and G. M. Padilla, Eds., *Marine Pharmacognosy: Choline and Related Substances in Algae.* Academic Press, New York, pp. 203–240.

Ikawa, M., Chakravarti, A., and Taylor, R. F. (1972). *Can. J. Microbiol.* **18,** 1241–1245.

Inoue, S. and Egami, F. (1963). *J. Biochem. Tokyo* **54,** 557–558.

Ishihara, K. (1968). *Exp. Cell Res.* **51,** 473–484.

Ishizuka, I., Suzuki, M., and Yamakawa, T. (1973). *J. Biochem. Tokyo* **73,** 77–87.

John, R. A., Rose, F. A., Wusteman, F. S., and Dodgson, K. S. (1966). *Biochem. J.* **100,** 278–281.

Johnson, C. M. and Nishita, H. (1952). *Anal. Chem.* **24,** 736–742.

Jones, L. H. P., Cowling, D. W., and Lockyer, D. R. (1972). *Soil Sci.* **114,** 104–114.

Jordan, H. V. (1964). *Sulfur as a Plant Nutrient in the Southern United States.* Agricultural Research Service., U.S. Department of Agriculture, Tech. Bull. 1287.

Kaji, A. and Gregory, J. D. (1959). *J. Biol. Chem.* **234,** 3007–3009.

Jorgensen, B. B. and Fenchel, T. (1974). *Mar. Biol.* **24,** 189–201.

Kaji, A. and McElroy, W. D. (1958). *Biochim. Biophys. Acta* **30,** 190–191.

Kamprath, E. J., Nelson, W. L., and Fitts, J. W. (1956). *Soil Sci. Soc. Am. Proc.* **20,** 463–466.

Kantor, T. G. and Schubert, M. (1957). *J. Am. Chem. Soc.* **79,** 152–153.

Kates, M. (1970). *Adv. Lipid Res.* **8,** 225–265.

Kates, M. and Deroo, P. W. (1973). *J. Lipid Res.* **14,** 438–445.

Kates, M., Palameta, B., Perry, M. P., and Adams, G. A. (1967). *Biochim. Biophys. Acta* **137,** 213–216.

Katzman, R. L. and Jeanloz, R. W. (1973). *J. Biol. Chem.* **248,** 50–55.

Kellog, W. W., Cadle, R. D., Allen, E. R., Lazrus, A. L., and Martell, E. A. (1972). *Science* **175,** 587–596.

Kelly, S. J. and Butler, L. G. (1975). *Biochem. Biophys. Res. Commun.* **66,** 316–321.

Kitamikado, M. and Ueno, R. (1970). *Bull. Jap. Soc. Sci. Fish.* **36,** 1175–1180.

Kitamikado, M., Ueno, R., and Nakamura, T. (1970). *Bull. Jap. Soc. Sci. Fish.* **36,** 592–596.

Kjaer, A. (1961). In N. Karasch, Ed., *Organic Sulfur Compounds.* Pergamon Press, Oxford, p. 409.

Kondo, M. (1923). *Biochem. Z.* **136,** 198–202.

Korn, E. D. and Payza, A. N. (1956). *J. Biol. Chem.* **223,** 859–864.

Kowalenko, C. G. and Lowe, L. E. (1975a). *Can. J. Soil Sci.* **55,** 1–8.

Kowalenko, C. G. and Lowe, L. E. (1975b). *Can. J. Soil Sci.* **55,** 9–14.

Langworthy, T. A., Mayberry, W. R., and Smith, P. F. (1974). *J. Bacteriol.* **119,** 106–116.

Langworthy, T. A., Mayberry, W. R., and Smith, P. F. (1976). *Biochim. Biophys. Acta* **431,** 550–569.

Large, P. J., Lloyd, A. G., and Dodgson, K. S. (1964). *Biochem. J.* **90,** 12 p.

Le Gall, J. (1972). In L. J. Guarraia and R. K. Balletine, Eds., *The Aquatic Environment: Microbial Transformations and Water Management Implications: The Sulfur Cycle*. Environmental Protection Agency Symposium, pp. 75–85.

Lepage, M., Daniel, H., and Benson, A. A. (1961). *J. Am. Chem. Soc.* **83,** 157–159.

Lien, T. and Schreiner, Ø. (1975). *Biochim. Biophys. Acta* **384,** 168–179.

Lijmbach, G. W. M. and Brinkhuis, E. (1973). *Antonie van Leeuwenhoek* **39,** 415–423.

Likens, G. E. and Bormann, F. H. (1974). *Science* **184,** 1176–1179.

Lindberg, B. (1955a) *Acta Chem. Scand.* **9,** 917–919.

Lindberg, B. (1955b). *Acta Chem. Scand.* **9,** 1323–1326.

Linker, A., Hoffman, P., Meyer, K., Sampson, P., and Korn, E. D. (1960). *J. Biol. Chem.* **265,** 3061–3067.

Lloyd, A. G., Large, P. J., Davies, M., Olavesen, A. H., and Dodgson, K. S. (1968). *Biochem. J.* **108,** 393–399.

Lougheed, G. and Milazzo, F. H. (1965). *Can. J. Microbiol.* **11,** 959–966.

Lowe, L. E. (1964). *Can. J. Soil Sci.* **44,** 176–179.

Lowe, L. E. (1969). *Can. J. Soil Sci.* **49,** 129–141.

Lowe, L. E. and De Long, W. A. (1961). *Can. J. Soil Sci.* **41,** 141–146.

Lowe, L. E. and De Long W. A. (1963). *Can. J. Soil Sci.* **43,** 151–155.

Lucas, J. J., Burchiel, S. W., and Segel, I. H. (1972). *Arch. Biochem. Biophys.* **153,** 644–672.

Marsden, J. C. (1975). *Biochem. Soc. Trans.* **3,** 314–316.

Martelli, H. L. and Benson, A. A. (1964). *Biochim. Biophys. Acta* **93,** 169–171.

Marzluf, G. A. (1970). *Arch. Biochem. Biophys.* **138,** 254–263.

Marzluf, G. A. (1972a). *Biochem. Genetics* **7,** 219–233.

Marzluf, G. A. (1972b). *Arch. Biochem. Biophys.* **150,** 714–724.

Marzluf, G. A. and Metzenberg, R. L. (1968). *J. Mol. Biol.* **33,** 423–437.

Mather, I. H. and Keenan, T. W. (1974). *FEBS Lett.* **44,** 79–82.

Mattock, P. and Jones, J. G. (1970). *Biochem. J.* **116,** 797–803.

Mayers, G. L. and Haines, T. H. (1967). *Biochemistry* **6,** 1665–1671.

Mayers, G. L., Pousada, M., and Haines, T. H. (1969). *Biochemistry* **8,** 2981–2986.

McGuire, W. G. and Marzluf, G. A. (1974a). *Arch. Biochem. Biophys.* **161,** 570–580.

McGuire, W. G. and Marzluf, G. A. (1974b). *Arch. Biochem. Biophys.* **161,** 360–368.

McKenna, J., Menini, E., and Norymberski, J. K. (1961). *Biochem. J.* **79,** 11 p.

Melville, G. E., Freney, J. R., and Williams, C. H. (1969). *J. Sci. Food Agric.* **20,** 203–206.

Melville, G. E., Freney, J. R., and Williams, C. H. (1971). *Soil Sci.* **112,** 245–248.

Metzenberg, R. L. and Parson, J. W. (1966). *Proc. Natl. Acad. Sci. U.S.* **55,** 629–635.

Michelacci, Y. M. and Dietrich, C. P. (1973). *Biochimie* **55,** 893–898.

Milazzo, F. H. and Fitzgerald, J. W. (1966). *Can. J. Microbiol.* **12,** 735–744.

Milazzo, F. H. and Fitzgerald, J. W. (1967). *Can. J. Microbiol.* **13,** 659–664.

Milazzo, F. H. and Lougheed, G. J. (1967). *Can. J. Bot.* **45,** 532–534.

Mothes, K. (1938). *Planta* **29,** 67–109.

Mumma, R. O. and Verlangiera, J. (1972). *Biochem. Biophys. Acta* **273,** 349–253.

Nakazawa, K., Suzuki, N., and Suzuki, S. (1975). *J. Biol. Chem.* **250,** 905–911.

Neu, H. C. and Chou, J. (1967). *J. Bacteriol.* **94,** 1934–1945.

Neu, H. C. and Heppel, L. A. (1965). *J. Biol. Chem.* **240,** 3685–3692.

Nisonson, I., Tannenbaum, M., and Neu, H. C. (1969). *J. Bacteriol.* **100,** 1083–1090.

Nissen, P. (1968). *Biochem. Biophys. Res. Commun.* **32,** 696–703.

Nissen, P. (1974). *Plant Physiol.* **30,** 307–316.

Nissen, P. and Benson, A. A. (1961). *Science* **134,** 1759.

Nissen, P. and Benson, A. A. (1964). *Plant Physiol.* **39,** 586–589.

Nose, Y. and Lipmann, F. (1958). *J. Biol. Chem.* **233,** 1348–1351.

Okamura, H., Murooka, Y., and Harada, T. (1976a). *J. Bacteriol.* **127,** 24–31.

Okamura, H., Yamada, T., Murooka, Y., and Harada, T. (1976b). *Agric. Biol. Chem.* **40,** 2071–2076.

O'Neill, J. G. (1973). Ph.D. thesis, University of Wales, United Kingdom.

Orsi, B. A. and Spencer, B. (1964). *J. Biochem.* **56,** 81–91.

Ottery, J., Olavesen, A. H., and Dodgson, K. S. (1970). *Life Sci.* **9,** 1335–1340.

Papa, D., Schwenk, K. E., and Ginsberg, H. F. (1949). *J. Org. Chem.* **14,** 723–731.

Parke, D. V. (1960). *Biochem. J.* **77,** 493–503.

Pastan, I. and Adhya, S. (1976). *Bacteriol. Rev.* **40,** 527–551.

Payne, W. J. (1963). *Biotechnol. Bioeng.* **5,** 355–365.

Payne, W. J. and Feisal, V. E. (1963). *Appl. Microbiol.* **11,** 239–344.

Payne, W. J. and Painter, B. G. (1971). *Microbios* **3,** 199–206.

Payne, W. J., Williams, J. P., and Mayberry, W. R. (1965). *Appl. Microbiol.* **13,** 698–701.

Payne, W. J., Williams, J. P., and Mayberry, W. R. (1967). *Nature* **214,** 623–624.

Payne, W. J., Fitzgerald, J. W., and Dodgson, K. S. (1974). *Appl. Microbiol.* **27,** 154–158.

Payza, A. N. and Korn, E. D. (1956). *J. Biol. Chem.* **223,** 853–858.

Peck, H. D. (1970). In O. H. Muth and J. E. Oldfield, Eds., *Symposium: Sulfur in Nutrition: Sulfur Requirements and Metabolism of Microorganisms.* AVI Publishing Co., Westport, Conn., pp. 61–79.

Peck, H. D. (1974). In *Evolution in the Microbial World: The Evolutionary Significance of Inorganic Sulfur Metabolism.* Symposium, Society for General Microbiology, No. 24, pp. 241–262.

Postgate, J. R. (1968). In G. Nickless Ed., *Inorganic Sulphur Chemistry: The Sulphur Cycle.* Elsevier Publishing Co., New York, pp. 259–272.

Postgate, J. R. (1974). *J. Appl. Bact.* **37,** 185–202.

Rahemtulla, F. and Lovtrup, S. (1974a). *Comp. Biochem. Physiol.* **49B,** 631–637.

Rahemtulla, F. and Lovtrup, S. (1974b). *Comp. Biochem. Physiol.* **49B,** 639–646.

Rammler, D. H., Grado, C., and Fowler, L. R. (1964). *Biochemistry* **3,** 224–230.

Rasburn, M. and Wynn, C. H. (1973). *Biochim. Biophys. Acta* **293,** 191–196.

Reinert, W. R. and Marzluf, G. A. (1974). *Biochem. Genetics* **12,** 97–108.

Renosto, F. and Segel, I. H. (1977). *Arch. Biochem. Biophys.* **180,** 416–428.

Robbins, P. W. (1962). "Sulfate Activating Enzymes." In S. P. Colowick and No. O. Kaplan, Eds., *Methods in Enzymology,* Vol. 5, pp. 964–977.

Robbins, P. W. and Lipmann, F. (1958). *J. Biol. Chem.* **233,** 686–690.

Rosen, O., Hoffman, P., and Meyer, K. (1960). *Fed. Proc. Fed. Am. Soc. Exp. Biol.* **19,** 147.

Roy, A. B. and Trudinger, P. A. (1970). *The Biochemistry of Inorganic Compounds of Sulphur.* Cambridge University Press.

Ryan, L. C., Carubelli, R., Caputto, R., and Trucco, R. E. (1965). *Biochim. Biophys. Acta* **101,** 252–258.

Sampson, E. J., Vergara, E. V., Fedor, J. M., Funk, M. O., and Benkovic, S. J. (1975). *Arch. Biochem. Biophys.* **169,** 372–383.

Schachter, B. and Marrian, G. F. (1938). *J. Biol. Chem.* **126,** 663–669.

Schneider, R. P. and Wiley, W. R. (1971). *J. Bacteriol.* **106,** 479–486.

Schnitzer, M. (1971). "Characterization of Humic Constituents by Spectroscopy." In A. D. McLaren and J. Skujins, Eds., *Soil Biochemistry,* Vol. 2, pp. 60–95.

Schreiner, Ø., Lien, T., and Knutsen, G. (1975). *Biochim. Biophys. Acta* **384,** 180–193.

Scott, J. M. and Spencer, B. (1968). *Biochem. J.* **106,** 471–477.

Scott, W. A. and Metzenberg, R. L. (1970). *J. Bacteriol.* **104,** 1254–1265.

Scott, W. A., Munkres, K. D., and Metzenberg, R. L. (1971). *Arch. Biochem. Biophys.* **142,** 623–632.

Scully, M. F., Dodgson, K. S., and Rose, F. A. (1970). *Biochem. J.* **119,** 20 p.

Segal, H. L. and Mologne, L. A. (1959). *J. Biol. Chem.* **234,** 909–911.

Segel, I. H. and Johnson, M. J. (1963). *Biochim. Biophys. Acta* **69,** 433–434.

Seno, N., Akiyama, F., and Anno, K. (1974). *Biochim. Biophys. Acta* **362,** 290–298.

Shibuya, I., Yagi, T., and Benson, A. A. (1963). In *Japanese Society of Plant Physiologists' Studies on Microalgae and Photosynthetic Bacteria: Sulfonic Acids in Algae.* University of Tokyo Press, Japan, pp. 627–636.

Skujins, J. J. (1967). "Enzymes in Soil." In A. D. McLaren and G. H. Peterson, Eds., *Soil Biochemistry,* Vol. 1, pp. 371–414.

Smith, J. N. (1962). *Nature* **195,** 399–400.

Smith, K. A., Bremner, J. M., and Tabatabai, M. A. (1973). *Soil Sci.* **116,** 313–319.

Solberg, Y. J. (1971). *Bryologist,* **74,** 144–150.

Spencer, B. (1960). *Biochem. J.* **77,** 294–304.

Spencer, B. and Harada, T. (1960). *Biochem. J.* **77,** 305–315.

Spencer, B., Hussey, E. C., Orsi, B. A., and Scott, J. M. (1968). *Biochem. J.* **106,** 461–469.

Spencer, K. and Freney, J. R. (1960). *Aust. J. Agr. Res.* **11,** 948–959.

Stahl, W. H., McQue, B., Mandels, G. R., and Siu, R. G. H. (1949). *Arch. Biochem. Biophys.* **20,** 422–432.

Starkey, R. L. (1950). *Soil Sci.* **70,** 55–65.

Starkey, R. L. (1956). *Ind. Eng. Chem.* **48,** 1429–1437.

Stevens, C. H. and Vohra, P. (1955). *J. Am. Chem. Soc.* **77,** 4935–4936.

Stevens, R. L., Fluharty, A. L., Skokut, M. H., and Kihara, H. (1975). *J. Biol. Chem.* **250,** 2495–2501.

Strominger, J. L. (1955). *Biochim. Biophys. Acta* **17,** 283–285.

Sukhumavasi, J., Murooka, Y., and Harada, T. (1975). *J. Ferment. Technol.* **53,** 62–66.

Svennerholm, L. and Stalberg-Stenhagen, S. (1968). *J. Lipid Res.* **9,** 215–225.

Swaby, R. J. and Fedel, R. (1973). *Soil Biol. Biochem.* **5,** 773–781.

Swaby, R. J. and Vitolins-Maija, I. (1968). *Int. Congr. Soil Sci. Trans.* **4,** 673–681.

Tabatabai, M. A. and Bremner, J. M. (1970a). *Soil Sci. Soc. Am. Proc.* **34,** 225–229.

Tabatabai, M. A. and Bremner, J. M. (1970b). *Soil Sci. Soc. Am. Proc.* **34,** 427–429.

Tabatabai, M. A. and Bremner, J. M. (1971). *Soil Biol. Biochem.* **3,** 317–323.

Tabatabai, M. A. and Bremner, J. M. (1972a). *Agron. J.* **64,** 40–44.

Tabatabai, M. A. and Bremner, J. M. (1972b). *Soil Sci.* **114,** 380–386.

Takebe, I. (1960). *J. Gen. Appl. Microbiol.* **6,** 83–89.

Takebe, I. (1961). *J. Biochem. (Tokyo)* **50,** 245–255.

Takebe, I. and Yanagita, T. (1959). *Plant and Cell Physiol.* **1,** 17–28.

Tallan, H. H., Bella, S. T., Stein, W. H., and Moore, S. (1955). *J. Biol. Chem.* **217,** 703–708.

Tamiya, N. (1951). *J. Chem. Soc. Jap.* **72,** 118–124.

Tanford, C. (1968). *Adv. Protein Chem.* **23,** 121–282.

Tanford, C. (1970). *Adv. Protein Chem.* **24,** 2–95.

Taylor, M. B. and Novelli, G. D. (1961). *Bacteriol. Proc.,* p. 190.

Taylor, R. F., Ikawa, M., Sasner, J. J., Thurberg, F. P., and Andersen, K. K. (1974). *J. Phycol.* **10,** 279–283.

Thompson, J. F., Smith, I. K., and Moore, D. P. (1970). In O. H. Muth and J. E. Oldfield Eds., *Symposium: Sulfur in Nutrition: Sulfur in Plant Nutrition.* AVI Publishing Co., Westport, Conn., pp. 80–96.

Tisdale, S. L. (1966). *Sulphur Inst. J.* **2,** 15–18.

Troen, P., Nilsson, B., Wiqvist, N., and Diczfalusy, E. (1961). *Acta Endocrinol.* **38,** 361–382.

Tudball, N. and O'Neill, J. G. (1975). *Microbios* **13,** 217–224.

Tudball, N. and O'Neill, J. G. (1976). *Biochim. Biophys. Acta* **429,** 616–623.

Tweedie, J. W. and Segel, I. H. (1971). *J. Biol. Chem.* **246,** 2438–2446.

Uesugi, Y., Hashimoto, H., and Harada, T. (1967). *Abstract,* Agricultural Chemical Society of Japan, p. 165.

Van Dijk, H. (1971). "Colloid Chemical Properties of Humic Matter." In A. D. McLaren and J. Skujins, Eds., *Soil Biochemistry,* Vol. 2, pp. 16–35.

Vesell, E. S. and Weyer, E. M. (1968). *Ann. N. Y. Acad. Sci.* **151,** 1–689.

Vestermark, A. and Boström, H. (1959). *Exp. Cell Res.* **18,** 174–177.

Virtanen, A. I. (1965). *Phytochemistry* **4,** 207–228.

Vitolins-Maija, I. and Swaby, R. J. (1968). *Aust. J. Soil Res.* **7,** 171–183.

Vlitos, A. J. (1953). *Contrib. Boyce Thompson Inst.* **17,** 127–149.

Waehneldt, T. V. (1975). *Bio Systems* **6,** 176–187.

Weigl, J. and Yaphe, W. (1966). *Can. J. Microbiol.* **12,** 847–856.

Wetzel, B. K., Spicer, S. S., Dvorak, H. F., and Heppel, L. A. (1970). *J. Bacteriol.* **104,** 529–542.

Weibe, W. J. and Bancroft, K. (1975). *Proc. Natl. Acad. Sci. U.S.* **72,** 2112–2115.

Williams, C. H. (1975). In K. D. McLachlan, Ed., *Sulphur in Australasian Agriculture: The Chemical Nature of Sulphur Compounds in Soils.* Prentice-Hall, Hemel Hempstead, England, pp. 21–31.

Williams, C. H. and Steinbergs, A. (1959). *Aust. J. Agric. Res.* **10,** 340–352.

Williams, J. and Payne, W. J. (1964). *Appl. Microbiol.* **12,** 360–362.

Wintermans, J. F. G. M. (1960). *Biochim. Biophys. Acta* **44,** 49–54.

Woolley, D. W. and Peterson, W. H. (1937). *J. Biol. Chem.* **122,** 213–218.

Yagi, T. (1966). *J. Biochem.* (Tokyo) **59,** 495–500.

Yamagata, T., Kawamura, Y., and Suzuki, S. (1966). *Biochim. Biophys. Acta* **115,** 250–252.

Yanagita, T. (1957). *Arch. Mikrobiol.* **26,** 329–344.

NOTES ADDED IN PROOF

Since the writing of this chapter a number of papers have been published or have been brought to the authors attention, the contents of which should be of some benefit to the reader. These are summarized below.

1. Formation of sulfate esters primarily by mammals.

   Dodgson, K. S. (1976). "Conjugation with Sulphate." In *Drug Metabolism—From Microbe to Man.* Taylor and Francis Ltd., London, pp. 91–104.

2. A sensitive method for ester sulfate determination in biological fluids.

   Ginsberg, L. C. and Di Ferrante, N. (1977). *Biochem. Med.* **17,** 80–86.

3. Occurrence of ester sulfate in soils of high organic content: the importance of ester sulfate in precursors of coal.

   Casagrande, D. and Siefert, K. (1977). *Science* **195,** 675–676.

4. Regulation of arylsulfatase formation.

   Burns, G. R. J. and Wynn, C. H. (1977). *Biochem. J.* **166,** 415–420.

   Jacobson, E. S. and Metzenberg, R. L. (1977). *J. Bacteriol.* **130,** 1397–1398.

   Okamura, H., Murooka, Y., and Harada, T. (1977). *J. Bacteriol.* **129,** 59–65.

   Murooka, Y., Adachi, T., Okamura, H., and Harada, T. (1977). *J. Bacteriol.* **130,** 74–81.

   Murooka, Y., Yamada, T., Tanabe, S., and Harada, T. (1977). *J. Bacteriol.* **132,** 247–253.

   Fitzgerald, J. W. and Cline, M. E. (1977). *FEMS Microbiol. Lett.* **2,** 221–224.

   Fitzgerald, J. W., Cline, M. E., and Rose, F. A. (1977). *FEMS Microbiol. Lett.* **2,** 217–220

   Yamada, T., Murooka, Y., and Harada, T. (1978). *J. Bacteriol.* **133,** 536–541.

5. Independent assays for arylsulfatase isoenzymes and kinetics of the phenol sulfotransferase activity associated with isoenzyme 2 of *Aspergillus oryzae*.

Burns, G. R. J. and Wynn, C. H. (1977). *Biochem. J.* **166,** 411–413.

Burns, G. R. J., Galanopoulou, E., and Wynn, C. H. (1977). *Biochem. J.* **167,** 223–227.

6. Localization of arylsulfatase in bacteria.

Fitzgerald, J. W. and George, J. R. (1977). *Appl. Environ. Microbiol.* **34,** 107–108.

7. Substrate specificity, mechanism of action, and cellular localization of enzymes catalyzing the biodegradation of primary and secondary alkylsulfate ester detergents.

Matcham, G. W. J., Bartholomew, B., Dodgson, K. S., Fitzgerald, J. W., and Payne, W. J. (1977). *FEMS Microbiol. Lett.* **1,** 197–200.

Bartholomew, B., Dodgson, K. S., Matcham, G. W. J., Shaw, D. J., and White, G. F. (1977). *Biochem. J.* **165,** 575–580.

Matcham, G. W. J. and Dodgson, K. S. (1977). *Biochem. J.* **167,** 717–722.

Matcham, G. W. J., Dodgson, K. S., and Fitzgerald, J. W. (1977). *Biochem. J.* **167,** 723–729.

Cloves, J. M., Dodgson, K. S., Games, D. E., Shaw, D. J., and White, G. F. (1977). *Biochem. J.* 843–846.

8. Occurrence and metabolism of the sulfonates. These are important publications that were inadvertently overlooked in the writing of this chapter. Although the natural occurrence of these sulfoconjugates is well established (especially in plants and algae), a reading of these papers reveals that little is yet known regarding their biodegradation especially with reference to the scission of the C-S bond.

Busby, W. F. (1966). *Biochim. Biophys. Acta* **121,** 160–161.

Martelli, H. L. (1967). *Nature* **216,** 1238–1239.

Lee, R. F. and Benson, A. A. (1972). *Biochim. Biophys. Acta.* **261,** 35–37.

Willetts, A. J. and Cain, R. B. (1972a). *Biochem. J.* **129,** 389–402.

Willetts, A. J. and Cain, R. B. (1972b). *A. van Leeuwenhoek J. Microbiol. Serol.* **38,** 543–555.

Busby, W. F. and Benson, A. A. (1973). *Plant Cell Physiol.* **14,** 1123–1132.

Taylor, C. D. and Wolfe, R. S. (1974). *J. Biol. Chem.* **249,** 4879–4885.

Bird, J. A. and Cain, R. B. (1974). *Biochem. J.* **140,** 121–134.

Swisher, R. D. (1976). In J. M. Sharpley and A. M. Kaplan, Eds., *Proc. 3rd Int. Biodegradation Symp.* Applied Science Publishers Ltd., London, pp. 853–865.

Haraguchi, T. and Morohoshi, N. (1976). *Ibid.,* pp. 719–729.

Cook, A. M., Daughton, C. G., and Alexander, M. (1978). *J. Bacteriol.* **133,** 85–90.

11

MICROBIAL TRANSFORMATIONS OF SULFUR IN THE ENVIRONMENT

Stephen. H. Zinder
Thomas D. Brock

Department of Bacteriology,
University of Wisconsin,
Madison, Wisconsin

1. INTRODUCTION

Sulfur in the environment can exist in a variety of oxidation states, depending on the *Eh* and pH of the environment. However, many thermodynamically feasiule reactions of sulfur compounds are relatively slow, and microorganisms play an important role in the sulfur cycle as catalysts of these reactions.

Traditionally, reviews on the microbial contribution to the sulfur cycle deal almost exclusively with the roles of the sulfate-reducing bacteria and the chemolithotrophic and photolithotrophic sulfur oxidizers. Although these organisms are of great importance to the sulfur cycle, this chapter will attempt to be more comprehensive and to cover also other transformations carried out by microorganisms. As can be seen in Figure 1, microorganisms are capable of a wide variety of sulfur transformations, including oxidations, reductions, incorporations into organic material, mineralizations, and production of volatile products. This chapter discusses these conversions and the environments in which they will be important.

2. ASSIMILATION AND REDUCTION OF SULFUR COMPOUNDS

2.1. Assimilation

Sulfur usually accounts for 0.4 to 0.8% of microbial dry weight (1). Almost all of this sulfur is in proteins and polypeptides as the amino acids cysteine and methionine, and is in the 2—oxidation state. Small amounts of sulfur-containing vitamins and cofactors, such as thiamin,

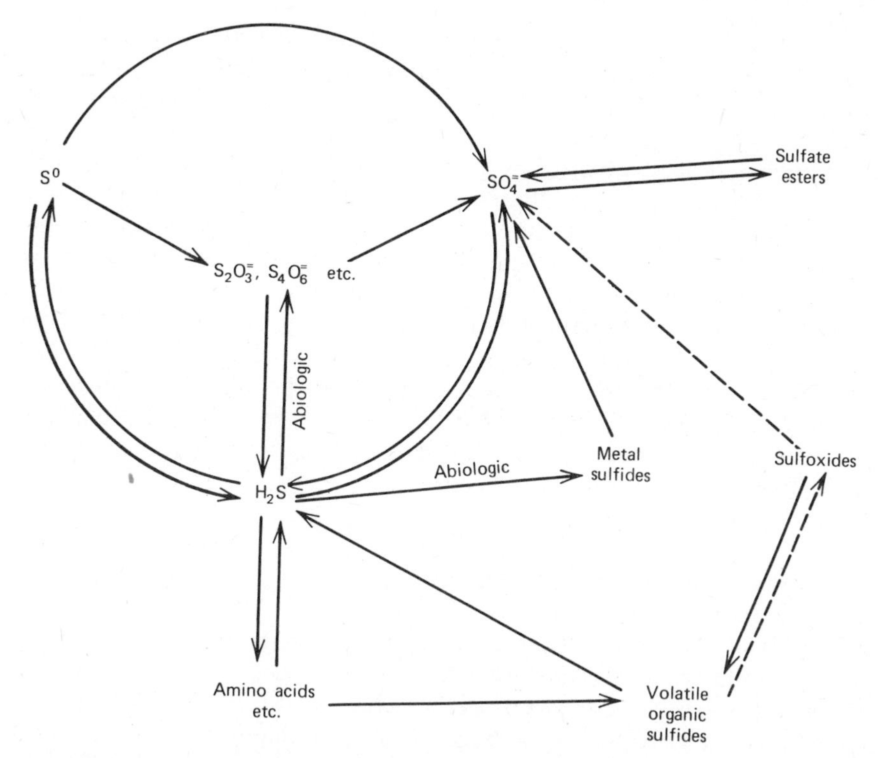

Figure 1. The microbial sulfur cycle. All arrows with solid lines denote known biological reactions, except the two marked "abiologic." Arrows with dashed lines denote possible reactions.

biotin, and coenzyme M, are also found, and a few organisms produce sulfate esters of polysaccharides as structural material.

Most microorganisms are capable of using sulfate as a sole sulfur source for growth; in doing so, they reduce it to hydrogen sulfide intracellularly and then replace the hydroxyl groups on serine and homoserine with sulfhydryl groups (2). Some microorganisms cannot reduce sulfate, but can reduce forms of intermediate oxidation state such as thiosulfate or sulfite. These reductions are termed assimilatory since virtually all the hydrogen sulfide produced is incorporated into amino acids. None is released, as is the case in dissimilatory reduction, which is discussed in the next section. Other microbes may require hydrogen sulfide as a sulfur source, or even have an outright requirement for a sulfur-containing amino acid or cofactor. Often, a sulfur-containing amino acid will be assimilated in preference to sulfate and will repress assimilatory sulfate reduction (3).

There have been very few studies of microbial sulfur assimilation in nature. Monheimer (4) studied the uptake of ^{35}S-labeled sulfate by freshwater phytoplankton, and found that sulfate uptake correlated well with carbon dioxide uptake and that the ratio between them was close to the predicted 500: 1. From this he concluded that sulfate uptake may be a method for measuring productivity in aquatic systems. Jassby (5) used ^{35}S-sulfate uptake to measure bacterial growth and obtained similar results. More studies are required before the general utility of this method for measuring productivity can be assessed.

Freney et al. (110) studied the fate of ^{35}S-sulfate added to soils. About 18 to 25% of the sulfur added ended up as insoluble carbon-bonded sulfur (assumed to be amino acids, and perhaps some humic material) after a long-term incubation (168 days). Another 10 to 20% went to soluble carbon-bonded sulfur, and a third, sizable fraction to sulfate esters, which are discussed in Section 6.

2.2. Dissimilatory Sulfate Reduction

There is no known chemical mechanism for the reduction of sulfate by organic matter at normal earth temperatures and pressures, so that microorganisms are completely responsible for this process (6). Reduction of sulfate to sulfide by microorganisms can be found in virtually any anoxic habitat containing organic matter and sulfate. These include ocean and lake sediments, flooded soils, sewage digester sludge, and the rumen (1,6,7).

Dissimilatory sulfate reduction is the process in which sulfate is used as the electron acceptor for the oxidation of organic matter, much as oxygen is used by aerobes. The end product of this oxidation is hydrogen sulfide, which is excreted by the cells (as opposed to being assimilated). As shown in Table 1, the energy obtained from using sulfate as an electron acceptor is much less than that from oxygen or nitrate, and is slightly greater than that from bicarbonate reduction to methane.

Three genera of sulfate-reducing bacteria have been described. *Desulfovibrio* is a small, motile, heterotrophic, obligatorily anaerobic, spiral-shaped organism (8). *Desulfomonas* is a recently described short, nonmotile rod which is physiologically similar to *Desulfovibrio* (9). *Desulfotomaculum* is a larger, motile, heterotrophic rod that forms heat-resistant spores (10) and at one time was considered to belong to the genus *Clostridium*. The strains described use a limited range of carbon sources, usually fermentation products such as alcohols and organic

Table 1. Free Energies for the Reduction of Various Electron Acceptors by Hydrogen[a]

Electron Acceptor	Product	$\Delta G'_0$ (kcal/mol H_2)
O_2	H_2O	−56.7
NO_3^-	N_2	−53.6
$(CH_3)_2SO$	$(CH_3)_2S$	−29.6
SO_3^{2-}	H_2S	−12.9
$S_2O_3^{2-}$	H_2S	−10.3
Pyruvate	Lactate	−10.3
SO_4^{2-}	H_2S	−9.2
HCO_3^-	CH_4	−8.1
S°	H_2S	−6.5

[a] Data from Decker et al. (112), except for the reduction of dimethyl sulfoxide, which was calculated by the authors. Reduction of pyruvate to lactate is representative of enzymatic reduction during fermentation.

acids. Molecular hydrogen can serve as an electron donor, but an organic source of carbon is required (11). Since virtually all strains described have been isolated using lactate as the electron donor, it is quite possible that the metabolic diversity of the sulfate reducers is greater than is presently thought. It had long been believed that sulfate-reducing bacteria could not use acetate as an electron donor, but in 1977 Widdel and Pfennig (13) isolated *Desulfotomaculum acetoxidans* from muds. This organism oxidizes acetate to carbon dioxide and reduces sulfate to hydrogen sulfide.

Sulfate-reducing bacteria, as a group, can tolerate a wide variety of environmental extremes, including temperature, salinity, and pressure (12). One extreme that can limit sulfate reduction, however, is low pH. Tuttle et al. (14) studied the feasibility of using sulfate reducers to raise the pH of a stream receiving acid mine drainage. They developed a mixed culture system, using sawdust as a carbon source, which, at pH 2.8, could reduce sulfate. However, they could not isolate any sulfate-reducing bacteria that grew below pH 5 and suggest that the sawdust may provide a microenvironment of higher pH.

Although much work has been done with sulfate reducers in pure culture, there have been few studies of sulfate reduction rates in natural environments. Some rate measurements for sediments are shown in Table 2. The results of Ivanov (16), Sorokin (16a), and Lien et al. (16b) were obtained by incubating sediments in the laboratory with ^{35}S-sulfate.

Table 2. Sulfate Reduction Rates in Various Sediments[a]

Locality	Salinity	SO_4^{2-} Concentration (mg/l)	Rate (mg H_2S/1·day)	Reference
Lake Beloe	Fresh	48–205	$2–6 \times 10^{-4}$	16
Gor'kii Resevoir	Fresh	32–132	0.04–2.94	16
Lake Belovod	Fresh	428–625	0.067–0.124	16
Linsley Pond	Fresh	7.5–17.5	0.8–1.5	17
Lake Solenoe	Saline	Up to 4200	0.388–19	16
Black Sea	Saline	Up to 2200	0.027–2.31	16a
Santa Barbara Basin	Saline	—	0.17	6
Pacific Ocean core	Saline			
35–50 cm		2440	0.10	16b
120–135 cm		1827	0.053	16b
310 cm		1086	0.0098	16b

[a] From Trudinger et al. (15) and Lien et al. (16b).

Stuiver (17) added 110 mCi (millicuries) of ^{35}S-sulfate directly to the hypolimnion of Linsley Pond and followed the fate of the label. Goldhaber and Kaplan (6) based their calculation of the rate of sulfate reduction on the rate of sedimentation and the increase in sulfide concentration with depth in sediment cores. As can be readily seen in Table 2, sulfate reduction rates in different environments are highly variable. Note the decrease in sulfate reduction rate with depth in the Pacific Ocean sediment core described by Lien et al. (16b), probably because of the depletion of easily degradable carbon compounds.

A further indication of the activity of sulfate-reducing bacteria can be found in stable isotope studies. An enrichment of ^{32}S over ^{34}S is found in ocean sediment sulfide, as compared with sulfate. This isotope enrichment is similar to that found in cultures of sulfate-reducing bacteria when growth is not limited by sulfate (6). Other evidence of sulfate reduction in ocean sediments is the depletion in pore waters of sulfate relative to chloride, a conservative ion. Although chloride concentrations are not given in Table 2, note the depletion of sulfate with depth in the Pacific Ocean sediment core.

The activities of sulfate-reducing bacteria are geochemically and environmentally important. Their most important role in the global sulfur cycle is in oceanic sediments, because of the large areas involved, and the high concentration of sulfate in seawater (28 m*M*). Sulfate reduction affects the carbon cycle in ocean sediments in that a significant portion of the carbon compounds metabolized in sediments are oxidized using sulfate as the electron acceptor (6). The hydrogen sulfide produced in sediments rapidly combines with metal ions, when they are present, and forms metal sulfides, considered to be the first step in the formation of pyrite and other minerals (6,15,18). In organic-rich or polluted waters, hydrogen sulfide, when produced in excess of the iron and other metal ions present, can diffuse out of the sediments and cause fish kills and odor problems. Sulfate reduction can inhibit methanogenesis in sediments (19,20,21), probably by competition for hydrogen and acetate (21). In waterlogged soils, hydrogen sulfide can be toxic to plants (22) and cause corrosion of metal pipes (23).

2.3. Reduction of Compounds Other than Sulfate

Thiosulfate, Polythionates, and Sulfite

These compounds are common intermediates in the abiological, and perhaps biological, oxidation of hydrogen sulfide (24) and may find their way back to an anaerobic environment. Many microorganisms, both procaryotic and eucaryotic, are capable of reducing them to hydrogen sulfide, which they may either assimilate or excrete. They can be reduced in a dissimilatory manner for energy production by sulfate-reducing bacteria (25), *Proteus mirabilis* (26), some marine heterotrophs (27), and a spiral-shaped organism from lake sediments (28,28a), which also reduced elemental sulfur.

Elemental Sulfur

Elemental sulfur is found in many sedimentary environments and may result from the oxidation of sulfide by ferric iron (6); it also occurs in sulfur deposits. A wide variety of bacteria and fungi readily reduce it to hydrogen sulfide. Very little is known about energy generation from the reduction of sulfur. Postgate (25) found that six strains of sulfate-reducing bacteria did not reduce sulfur, but the more recent work of Biebl and Pfennig (29) has shown that some strains can. Pfennig and Biebel (30) isolated a rod-shaped anaerobic bacterium, *Desulfuromonas acetoxidans,* which used sulfur as an electron acceptor and acetate as the donor, and did not reduce other sulfur compounds tested, including sulfate, sulfite, and thiosulfate.

Sulfoxides and Sulfones

Dimethyl sulfoxide is found in paper mill wastes and may be a product of the atmospheric oxidation of dimethyl sulfide (31). Methionine and biotin sulfoxides are oxidation products of methionine and biotin. We have surveyed both procaryotic and eucaryotic microorganisms and have found that many are capable of reducing dimethyl sulfoxide to dimethyl sulfide (28). Reduction of biotin sulfoxide (32) and methionine sulfoxide (33) has also been reported. *Rhodopseudomonas spheroides* (34), *Proteus vulgaris* (28), and a spiral organism from lake sediments (28) are capable of growth using dimethyl sulfoxide as an electron acceptor, producing dimethyl sulfide. Certain pesticides containing thionophosphorus and thioether groups are susceptible to oxidation of sulfoxides (35) which can be highly toxic to man and animals. Microbial reduction may serve to detoxify these sulfoxides.

Sulfones are very difficult to reduce chemically (36) and, apparently, are also difficult to reduce biologically. None of the organisms we surveyed was capable of reducing dimethyl sulfone (28), and methionine sulfone is degraded much more slowly than methionine sulfoxide (37,38).

3. OXIDATION OF SULFUR COMPOUNDS

A great variety of microorganisms are capable of oxidizing inorganic sulfur compounds. The spontaneous oxidation of sulfide by oxygen proceeds fairly rapidly in neutral pH environments, but sulfide minerals and elemental sulfur are relatively stable. The sulfur-oxidizing bacteria are usually divided into two groups, the colorless and the colored bacteria. The colorless sulfur bacteria are so called because they lack photosynthetic pigments, whereas the phototrophic or colored sulfur bacteria use the energy derived from photosynthesis to drive the oxidation of sulfur compounds in anaerobic environments, where oxidation would not otherwise occur. Because the diversity of the sulfur bacteria is so great, oxidation of sulfur compounds will be discussed in terms of groups of microorganisms. Sulfur oxidation by heterotrophic bacteria and fungi will also be discussed.

3.1. Colorless Sulfur Bacteria

The most thoroughly studied colorless sulfur bacteria are in the genus *Thiobacillus*. The thiobacilli are rod-shaped cells that can obtain energy from the oxidation of inorganic sulfur compounds to sulfate, using oxygen as an electron acceptor. In addition, *T. denitrificans* can use nitrate as an electron acceptor, and *T. ferroxidans* and *T. thiooxidans* have been shown to utilize ferric iron for this purpose (38a). All the thiobacilli, with the exception of *T. perometabolis,* are capable of autotrophic growth. The various members of the thiobacilli can oxidize hydrogen sulfide, elemental sulfur, thiosulfate, and polythionates (39, 40). The current status of *Thiobacillus* taxonomy is presented in Bergey's manual (41). *Thiomicrospira pelophila,* a spiral-shaped organism physiologically similar to the nonacidophilic thiobacilli, is present in tidal flats (42).

The nonacidophilic thiobacilli can be found in some soils, fresh and salt waters, and muds. Plate counts of thiobacilli in soils are usually low (43,44), but this may be a reflection of the unsuitability of the plate count method. Their activity *in situ* in these environments is difficult to assess, since often many other organisms are present that are capable of either oxidizing sulfur compounds or fixing carbon dioxide, thereby interfering with measurements (43).

The acidophilic thiobacilli, *T. thiooxidans* and *T. ferrooxidans,* can be found in neutral pH soils, probably in acid microenvironments, but their most important activity, at least from an economic standpoint, is their participation in the oxidation of sulfide minerals leading to the formation of acid mine drainage (45). At acid pH values the chemical oxidation of ferrous iron by oxygen takes place slowly. *Thiobacillus ferrooxidans* catalyzes this oxidation, using the energy derived from it for autotrophic growth (45). Singer and Stumm (46) have postulated that pyrite in coal refuse piles is attacked by ferric iron, and the oxidation of the ferrous iron produced back to ferric iron is the rate-limiting step. Since *T. ferrooxidans* can rapidly oxidize ferrous iron, the overall rate of reaction is greatly enhanced, as shown in Figure 2. *Thiobacillus thiooxidans* may also participate by oxidizing intermediates such as elemental sulfur. The net result is the oxidation of pyrite sulfur to sulfuric acid, thereby greatly decreasing the pH of the water percolating through the coal refuse pile.

The role of *T. ferrooxidans* in acid mine drainage has been confirmed by *in situ* studies. Belly and Brock (48), studying coal refuse piles, found high rates of carbon dioxide fixation, which could be stimulated by the addition of ferrous iron, and which correlated with high numbers of *T. ferrooxidans* as measured by the most-probable-number technique. Apel et al. (49) have made fluorescent antibodies to *T. ferrooxidans* and have found cells that react with the antibody in samples from coal refuse piles. In laboratory studies they found a correlation between acid production and an increase in cell numbers. Although the use of various biological, chemical, and physical agents to inhibit these bacteria has been suggested, no practical solutions have yet been found (50).

Thiobacillus ferrooxidans is capable of leaching other sulfide minerals, such as covellite (CuS), chalcopyrite ($CuFeS_2$), galena (PbS), sphalerite

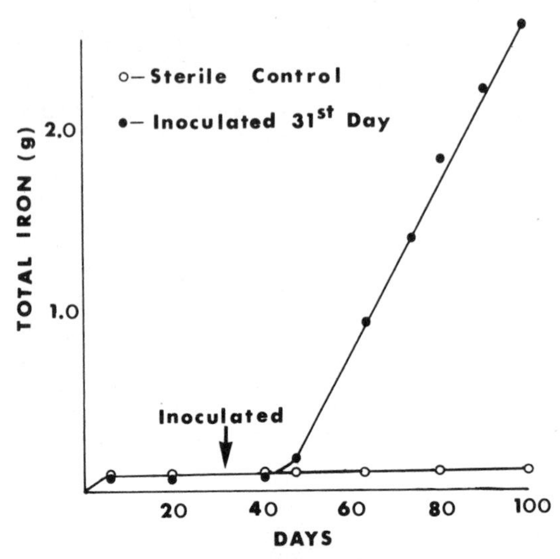

Figure 2. Oxidation of iron pyrite by *Thiobacillus ferrooxidans* as measured by the appearance of soluble ferric iron. The initial pH was 2.6. Data from Bryner et al. (47).

(ZnS), and orpiment (As_2S_3) (18). This ability is taken advantage of in using *T. ferrooxidans* to leach out metals, such as copper, from low-grade ores.

Sulfolobus acidocaldarius, described by Brock et al. (51), is a thermophilic, acidophilic, lobed sphere capable of autotrophic growth oxidizing elemental sulfur. It can also oxidize ferrous iron (52) and hydrogen sulfide (53) and can grow heterotrophically. *Sulfolobus* can be found in concentrations up to 10^8 cells/ml in acid hot springs all over the world. Mosser et al. (54) showed, using ^{35}S-labeled elemental sulfur, that natural populations of *Sulfolobus* were responsible for the oxidation of elemental sulfur to sulfuric acid in these springs.

Beggiatoa is a filamentous, gliding bacterium, usually containing intracellular sulfur globules, with a remarkable morphological resemblance to the cyanobacterium (blue-green alga) *Oscillatoria*. Although the oxidation of hydrogen sulfide to intracellular sulfur granules and its subsequent oxidation to sulfate led Winogradsky (55) to develop the concept of energy generation via inorganic oxidations, the function of sulfur oxidation in these heterotrophic organisms is still unclear. Burton and Morita (56) found that the heterotrophically grown strains they studied lack catalase and proposed that the hydrogen sulfide serves to scavenge peroxides which could otherwise cause damage to the cell. *Beggiatoa* forms tufts or mats at sulfide-oxygen interfaces, such as those where sediments meet water. An interesting habitat where it can be found is the rhizospheres of plants growing in flooded soils, such as rice (57) and cattails (*Typhalatafolia*) (58). It is possible that *Beggiatoa* reduces hydrogen sulfide levels near the roots, thereby alleviating toxicity to the plant (57).

Thiothrix, a filamentous, colorless sulfur bacterium which also forms intracellular sulfur granules and grows in habitats similar to those of *Beggiatoa,* is morphologically similar to the marine bacterium *Leucothrix.* Since it never has been grown in culture, its physiological properties are obscure. An adverse economic activity of *Thiothrix* occurs when activated sludge digesters receive sewage containing hydrogen sulfide. Filamentous masses of *Thiothrix* can form, causing poor settling of the sludge, termed bulking. For greater detail on the diversity, physiology, and ecology of the colorless sulfur bacteria, the review by Kuenen (43) is recommended.

3.2. Heterotrophic Sulfur-Oxidizing Bacteria

If one inoculates a soil sample into thiosulfate-mineral salts medium, to enrich for autotrophic thiobacilli, sometimes one finds, instead of the characteristic pH drop accompanying sulfuric acid production, a slight rise in pH due to the production of tetrathionate. If this enrichment is then streaked out on thiosulfate-mineral salts agar, large colonies develop within 24 hr instead of the usual 2 to 4 days for thiobacilli. These organisms are often fluorescent pseudomonads, heterotrophs that grow on trace organic materials contaminating the media. These organisms had been classified as *Thiobacillus trautweinii,* but the most recent classification of the genus (41) has justifiably eliminated the species (39,59). Such heterotrophs can be isolated from river waters and soils (60,61), as well as the thermocline of the Black Sea and other oceanic environments (62). In the Black Sea they may be totally responsible for the oxidation of thiosulfate, formed via abiologic oxidation of hydrogen sulfide (24), since they are present in large numbers and no thiobacilli can be found (63). Caldwell et al. (64) incubated thiosulfate gradients *in situ* in a freshwater environment and found only colonies of sulfur-encrusted fluorescent pseudomonads.

Both Guittonneau (65), who wa studying vineyard soil that was dusted with sulfur as a fungicide, and Viltolins and Swaby (44), who were studying sulfur oxidation in Australian soils, found a wide variety of heterotrophic organisms that could oxidize sulfur compounds, including members of the genera *Arthrobacter, Bacillus, Micrococcus, Pseudo-*

monas, Mycobacterium, some actinomycetes, and the fungi *Saccharomyces* and *Debaryomyces*. Guittonneau (66) could set up two-membered cultures, one of which oxidized sulfur to thiosulfate, and a second that oxidized thiosulfate to sulfate. This is similar to what he (65) and Nor and Tabatabai (67) found to occur in soils, suggesting that heterotrophs take an active part in sulfur oxidation in some soils.

The role of heterotrophic sulfur oxidizers in various ecosystems remains obscure, for the same reasons that it is difficult to determine the role of thiobacilli in complex environments. What advantage is there to these heterotrophs in oxidizing sulfur compounds? Tuttle et al. (59) demonstrated that thiosulfate stimulated the growth of a marine pseudomonad when it was growing on low concentrations of acetate. Whether the growth of other heterotrophs can also be so stimulated is unknown. Perhaps the autotrophic thiobacilli, because they are physiologically interesting, have been given a greater role than they deserve in the sulfur cycle in certain habitats. In any event the heterotrophic sulfur oxidizers merit further attention.

3.3. Photosynthetic Bacteria

The photosynthetic bacteria, which can be found in anaerobic microcosms where light and hydrogen sulfide are present, are responsible for oxidation of sulfur compounds in the absence of oxygen. They use sunlight as an energy source and the electrons derived from the oxidation of sulfur compounds to reduce carbon dioxide to the level of cell material. The green sulfur bacteria (Chlorobiniaceae) are generally found in areas of high hydrogen sulfide (4 to 8 m*M*), while the purple sulfur bacteria (Chromatiaceae) are usually in areas of intermediate hydrogen sulfide (0.8 to 4 m*M*) (68). The so-called purple nonsulfur bacteria (Rhodospirillaceae) are found at the lowest hydrogen sulfide concentrations (0.4 to 2 m*M*). Whereas the first two groups can grow autotrophically, the Rhodospirillaceae grow primarily by using organic matter as the electron donor. It was thought that they were incapable of growth on hydrogen sulfide, but the work of Hansen and Van Gemerden (69) has shown that very low concentrations of hydrogen sulfide can be used for growth. When oxygen is present, the Rhodospirillaceae stop photosynthesizing and grow as aerobic heterotrophs. A new family, the Chloroflexaceae, has been proposed (70) to include the genus *Chloroflexus,* a filamentous thermophile, which has pigments like those of the green sulfur bacteria but is physiologically similar to the Rhodospirillaceae. *Chloroflexus* is common in hot springs in mats, along with cyanobacteria and other bacteria (71,72). A nonthermophilic bacterium, similar to *Chloroflexus,* has been isolated from lake water by Gorlenko (72a).

The green sulfur bacteria generally oxidize hydrogen sulfide to sulfate, forming extracellular sulfur as an intermediate at high hydrogen sulfide concentrations. Elemental sulfur and thiosulfate may also be oxidized. The purple sulfur bacteria (with one exception, *Ectothiorhodospira)* oxidize hydrogen sulfide, forming intracellular sulfur globules, which may then be oxidized to sulfate. The Rhodospirillaceae may oxidize hydrogen sulfide to elemental sulfur or sulfate. The sulfur relations of these bacteria are discussed more completely by Pfennig (68). It should be added that some cyanobacteria are capable of growing phototrophically by oxidizing hydrogen sulfide (73) and thiosulfate (74).

Photosynthetic bacteria can be found in large numbers in stagnant waters over sulfide-containing muds. One of the most interesting habitats where they may be found is the hypolimnia of lakes. If the hypolimnion is anaerobic and light reaches it (as is the case in some meromictic and dimictic lakes), a bloom of photosynthetic bacteria can form (75,76). This bloom can be quite thick, since nutrient concentrations are often high in the hypolimnion, and may actually account for a large portion of the primary productivity of a lake [83% in one case (76)]. Zooplankton may migrate down to the hypolimnion to feed on the photosynthetic bacteria, making them an important part of the food web (76). The potential importance of these bacteria to the sulfur cycle in some lakes may be demonstrated by Pfennig's calculation (68) from the data of Culver and Brunskill (76) that in Fayetteville Green Lake (about 0.25 km^2) the photosynthetic bacteria oxidize 84 tons of hydrogen sulfide to sulfate a year.

Photosynthetic bacteria may play an important part, in concert with sulfate reducers, in the formation of sulfur deposits. The sulfate reducers produce hydrogen sulfide, which can then be photosynthetically oxidized to elemental sulfur. Butlin and Postgate (77) have postulated that this was the mechanism of formation of sulfur deposits at the bottom of certain African lakes. Van Gemerden (78) has shown that purple sulfur bacteria and sulfate reducers can grow together tightly coupled, so that only a small amount of sulfur, to be cycled many times from one to another, is needed for optimal growth of the culture. Often in natural samples one sees a consortium (bacterial association) in which a large, colorless central bacterium is surrounded by attached green sulfur bacteria. Pfennig and Biebel (30) have isolated from such a consortium a colorless organism, *Desulfuromonas,* which oxidizes acetate and reduces elemental sulfur. The review of Pfennig (68) gives further information on the ecology of the photosynthetic bacteria.

One can imagine a whole stable ecosystem based on photosynthetic bacteria as producers and bacteria that reduce sulfate or sulfur as consumers. Sulfur compounds would replace oxygen compounds (oxygen and water) as the primary movers of the carbon cycle. This sort of cycle has been termed a sulfuretum by Baas Becking (79). The earth may have been a sulfuretum before oxygen-producing phototrophs evolved (1).

4. DEGRADATION OF AMINO ACIDS AND OTHER SULFUR COMPOUNDS

The degradation of organic sulfur to sulfate or hydrogen sulfide is termed mineralization. Protein breakdown is an important part of the carbon cycle; and although some of the amino acids in a protein are assimilated directly, others are degradated by microorganisms as carbon and energy sources. Depending on the environment, sulfur amino acids can be broken down to sulfate, hydrogen sulfide, and volatile organic sulfur compounds. The last two have a variety of biological effects and are highly odorous. The recent development of the flame photometric detector for gas chromatography and the development of inert GC columns (80,81) and inert sample enclosures (81) now allow the quantification of nanogram quantities of volatile sulfur compounds.

4.1. Sulfate Production from Amino Acids

When cystine is added to soils, it is rapidly degraded to sulfate (82,83), sometimes via the formation of cysteic acid (84). Frederick et al. (82) found significant sulfate production in soils amended with cystine, taurine, sodium taurocholate, and thiamin. A variety of other sulfur compounds, including methionine, were not metabolized to sulfate. However, Hesse (83) found quantitative conversion of methionine to sulfate in the soil he studied. A note of caution should be added here in that the addition of large amounts of a sulfur compound, as in these studies and in others, may not accurately simulate the response of the environment to the small amounts of the compound actually present, although such studies may be useful in outlining possible pathways of metabolism.

4.2. Hydrogen Sulfide Production

The anaerobic production of hydrogen sulfide from organic sulfur compounds, especially cyst(e)ine (84), is well known and can be very important in freshwater environments where the sulfate concentration is low. Nriagu (86) calculated that half of the 5 mg/g dry weight of sulfide found in Lake Mendota sediments was derived from organic sulfur, while Stuiver (17), from *in situ* reduction rates of $^{35}SO_4^{2-}$, calculated that less than 10% of the sulfide in Linsley Pond sediments was from organic sources. Hydrogen sulfide production from methionine is less well characterized. Most studies find that methyl mercaptan is the main product of methionine degradation (see the next section). In our own studies (87), using nanogram quantities of ^{35}S-labeled methionine, we found that lake Mendota sediments readily degraded methionine sulfur to hydrogen sulfide.

4.3. Methyl Mercaptan Production

Methyl mercaptan (also called methane thiol) and its oxidized form, dimethyl disulfide, are commonly formed via cleavage to methionine. Methyl mercaptan can be found in paper mill effluents, river water (88), rice paddies (89), soils (90), spoiling foods (85), and manures (91), and as the product of the decomposition of cruciferous plants (92) and other organic materials (93). In general, it is present in environments where a large amount of protein decomposition is occurring. Methyl mercaptan is phytotoxic (22), although some plants use thiols and sulfides with longer carbon chains to discourage fungal pathogens (92). Methyl mercaptan and other volatile sulfur compounds are potent inhibitors of nitrification (94).

Methyl mercaptan is produced when methionine is added to a variety of environments, including aerobic and anaerobic pure cultures of bacteria and fungi (85), and from such natural habitats as soils (37,82,-95), manures (91), sewage sludge (96), lake sediments (96), and the bovine rumen (97).

4.4. Dimethyl Sulfide Production

Dimethyl sulfide (DMS) is often formed as a breakdown product of sulfonium compounds, and usually as a minor product in the breakdown of methionine. It can be found in a variety of environments, including ocean water (31), paper mill effluents, soils (31, 93,95), river water (88), lake water (98), Great Salt Lake water (96), and spoiled foods (85).

Algae and seaweeds produce sulfonium compounds [e.g., thetins (99) and sulfonium lipids (100)] which may be readily degraded to form DMS; in fact the smell of DMS at low concentrations is reminiscent of the seashore. Although little is known about microbial metabolism of sulfonium compounds, a *Clostridium* that ferments dimethyl propethetin (101) and a *Pseudomonas* that metabolizes trimethyl sulfonium (102), both of which produce DMS, have been described. Lovelock et al. (31) have found small quantities (approximately 10 ng/1) of DMS in open ocean water and in the atmosphere over soils, and have suggested that DMS plays an important role in the global transfer of sulfur. Certain fungi also produce DMS from sulfate in the growth medium (99). Probably, where a large amount of protein degradation, marked by methyl mercaptan production, is occurring, DMS is derived from methionine, whereas in other environments the source is likely to be sulfonium compounds.

4.5. Carbon Disulfide and Carbonyl Sulfide Production

Much less is known about these compounds than the ones previously discussed. Carbon disulfide was found by Banwart and Bremner (37) to be a major product of the decomposition of cysteine in soil, and also occurs in manures (91) and in very small amounts in seawater (31). Care should be taken when studying carbon disulfide, since it often emanates from rubber fittings. Carbon disulfide is extremely effective at inhibiting nitrification (94) and is very toxic in general. Carbonyl sulfide has been detected from manures (91,103) and soils (90). Although its production in soils was stimulated by djenkolic acid (37), its natural source remains unclear.

5. METABOLISM OF VOLATILE ORGANIC SULFUR COMPOUNDS

Very little is known about the transformation of volatile sulfur compounds by microorganisms. Sivela and Sundman (105) reported the isolation, from a filter used to treat paper mill wastes, of a *Thiobacillus* capable of oxidizing dimethyl sulfide and dimethyl disulfide to an unknown product. Indirect evidence for the microbial metabolism of these gases was obtained by Bremner and Banwart (106), who found that soil sorbtion of dimethyl sulfide, dimethyl disulfide, carbonyl sulfide, and carbon disulfide was decreased by autoclaving. Our own studies (107) have shown that ^{14}C-labeled methyl mercaptan and dimethyl sulfide are anaerobically metabolized to methane and carbon dioxide, and presumably hydrogen sulfide, by lake sediments and sewage sludge. Kurita et al. (108) isolated an anaerobic, rod-shaped bacterium that produced hydrogen sulfide from dimethyl sulfide, but did not characterize it further.

6. SULFATE ESTER FORMATION AND HYDROLYSIS IN SOIL

When one analyzes soil sulfur, only a small fraction, frequently less than 10% of the total sulfur in soils, is found as inorganic sulfate (109). Usually about half of what is left is carbon-bonded sulfur, and the remainder is the so-called hydroiodic acid-reducible fraction, consisting of sulfate esters. Although the exact form of sulfate esters in soil is unclear, they are probably polysaccharide sulfates, such as condroitin sulfate, keratin sulfates, choline sulfate, and arylsulfates (109). Sources of these include animal bodies and excreta, plant materials, and formation from sulfate by the indigenous microflora. Freney et al. (110) incubated soils with ^{35}S-labeled sulfate for 168 days and found most of the label as sulfate esters. This incorporation was inhibited by autoclaving the soils, indicating microbial activity. In addition, microorganisms are capable of hydrolyzing these esters and may be instrumental in providing available sulfate for plant growth (109). It has been hypothesized (111) that sulfate esters in peat bogs are the precursors of organic sulfur in coals. Understanding of the factors controlling the formation and hydrolysis of sulfate esters may be useful in predicting the effect of sulfate-containing acid rains on the sulfur budgets of soils.

7. DISCUSSION

Microbial communities are capable of a wide variety of sulfur transformations and are the prime movers of the sulfur cycle. Every reaction denoted by an arrow in Figure 1 has been discussed. This chapter has served more to outline possibilities than to make predictions about the behavior of microbial communities toward certain sulfur compounds. One can predict that over geological time the net flow of sulfur will be to the 2— oxidation state in dark anaerobic habitats and to the 6+ oxidation state in aerobic habitats. Thermodynamics predicts this, and microorganisms can only serve as catalysts. However, on the human time scale, reaction rates become critical, and our knowledge of these rates in the many environments affected by human activity is small.

Knowledge is needed as to what conditions will prevent certain reactions that we find undesirable, such as the production of odorous volatile organic sulfur compounds or the leaching of coal refuse. On the other hand, we would like to promote and improve other reactions, such as the leaching of low-grade ores. Our understanding of some aspects of the cycle, such as the metabolism of volatile organic sulfur compounds, is poor.

Studies of pure cultures of microorganisms have been very useful in making predictions and should continue, but more needs to be known about their activities in their natural habitats. Analytical methods with great sensitivity have been developed in recent years, especially ^{35}S radiotracer methods and gas chromatography, which can help us. Our understanding of the interactions between microorganisms and sulfur compounds in the environment is still rudimentary, and must improve if we are to predict the effects of human activity on the sulfur cycle.

ACKNOWLEDGMENT

The preparation of this chapter was aided in part by research contract COO-2161-38 from the Energy Research and Development Administration.

REFERENCES

1. Postgate, J. R. (1968). "The Sulfur Cycle." In G. Nickless, Ed., *Inorganic Sulfur Chemistry,* Elsevier Publishing Co., New York.
2. Roy, A. B. and Trudinger, P. A. (1970). *The Biochemistry of Inorganic Compounds of Sulfur.* Cambridge University Press, New York.
3. Roberts, R. B., Abelson, P. H., Cowie, D. B., Bolton, E. T., and Britten, R. J. (1955). Carnagie Institute of Washington, Publ. 607.
4. Monheimer, R. H. (1974). *Can. J. Microbiol.* **20,** 825.
5. Jassby, A. D. (1975). *Ecology* **56,** 627.
6. Goldhaber, M. S., and Kaplan, I. R. (1974). "The Sulfur Cycle." In E. G. Goldberg, Ed., *The Sea,* Vol. 5. John Wiley and Sons, New York, pp. 569–655.
7. Howard, B. H. and Hungate, R. E. (1976). *Appl. Environ. Microbiol.* **32,** 598.

8. Postgate, J. R. and Campbell, L. L. (1966). *Bacteriol. Rev.* **30,** 732.
9. Moore, W. E. C., Johnson, J. L., and Holdman, L. V. (1976). *Int. J. Syst. Bacteriol.* **26,** 236.
10. Campbell, L. L. and Postgate, J. R. (1966). *Bacteriol. Rev.* **29,** 359.
11 Mechalas, B. J. and Rittenberg, S. C. (1960). *J. Bacteriol.* **80,** 501.
12. Zobell, C. E. (1958). *Producer's Mon.,* **22,** 12.
13. Widdel, F. and Pfennig, N. (1977). *Arch. Microbiol.* **112,** 119.
14. Tuttle, J. H., Dugan, P. R., Macmillan, C. B., and Randalls, C. I. (1969). *J. Bacteriol.* **97,**·594.
15. Trudinger, P. A., Lambert, I. B., and Skyring, G. W. (1972). *Econ. Geol.* **67,** 1114.
16. Ivanov, M. V. (1968). *Microbial Processes in the Formation of Sulfur Deposits.* Israel Programme for Scientific Translations, Jerusalem.
16a. Sorokin, Y. I. (1970). *Oceanology* **10,** 329.
16b. Lien, A. J., Kudriavzeva, A. I., Natrosov, A. G., and Ziakun, A. M. (1976). "Isotope Composition of Sulfur Compounds in Pacific Ocean Sediments." In I. I. Volkov, Ed., *Biogeochemistry of Diagenisis in Ocean Sediments.* Academy of Science, USSR, Moscow, pp. 179–185.
17. Stuiver, M. (1967). *Geochim. Cosmochim. Acta* **31,** 2152.
18. Silverman, M. P. and Ehrlich, H. L. (1964). *Adv. Appl. Microbiol.* **6,** 153.
19. Martens, C. S. and Berner, R. A. (1974). *Science* **185,** 1167.
20. Cappenberg, T. (1974). *Antonie van Leeuwenhoek* **40,** 285.
21. Winfrey, M. R. and Zeikus, J. G. (1977). *Appl. Environ. Microbiol.* **33,** 275.
22. Alexander, M. (1974). *Adv. Appl. Microbiol.* **18,** 1.
23. Starkey, R. L. and Wight, K. M. (1945). *Anaerobic Corrosion of Iron in Soil.* American Gas Association, Tech. Rept., Distribution Committee, New York.
24. Chen, K. Y. and Morris, J. C. (1972). *Adv. Water Pollut. Res.,* **III (32),** 1.
25. Postgate, J. R. (1951). *J. Gen. Microbiol.* **5,** 725.
26. Oltmann, L. F., Van der Beck, E. G., and Stouthamer, A. H. (1975). *Plant Soil* **43,** 153.
27. Tuttle, J. H. and Jannasch, H. W. (1973). *J. Bacteriol.* **115,** 732.
28. Zinder, S. H. and Brock, T. D. (1978). *Arch. Microbiol. 116,* 35.
28a. Wolfe, R. S. and Pfennig, N. (1977). *Appl. Environ. Microbiol.* **33,** 427.
29. Biebl, H. and Pfennig, N. (1977). *Arch. Microbiol.* **112,** 115.
30. Pfennig, N. and Biebl, H. (1976). *Arch. Microbiol.* **110,** 3.
31. Lovelock, J. E., Maggs, R. J., and Rasmussen, R. A. (1974). *Nature* **237,** 452.
32. Cleary, P. P. and Dykhuizen, D. (1974). *Biochem. Biophys. Res. Commun.* **56,** 629.
33. Sourkes, T. L. and Trano, Y. (1953). *Arch. Biochem. Biophys.* **42,** 321.

34. Yen, H. C. and Marrs, B. (1976). *Abstracts for the Annual Meeting of the American Society for Microbiology*. American Society for Microbiology, Washington, D.C., p. 124.
35. De Baun, J. R. and Menn, J. J. (1976). *Science* **191,** 187.
36. Vogel, A. I. (1956). *Practical Organic Chemistry,* 3rd ed. Longmans, Green and Co., London, p. 1078.
37. Banwart, W. J. and Bremner, J. M. (1974). *Soil Biol. Biochem.* **7,** 359.
38. Tonzetich, J. (1976). *Anal. Biochem.* **73,** 290.
38a. Brock, T. D. and Gustafson, J. (1976). *Appl. Environ. Microbiol.* **32,** 567.
39. Vishniac, W. and Santer, M. (1957). *Bacteriol. Rev.* **21,** 195.
40. Suzuki, I. (1974). *Ann. Rev. Microbiol.* **28,** 85.
41. Buchanan, R. E. and Gibbons, M. E., Eds. (1974). *Bergey's Manual of Determinative Bacteriology,* 8th ed. Williams and Wilkins Co., Baltimore.
42. Kuenen, J. G. and Veldkamp, H. (1972). *Antonie van Leeuwenhoek* **38,** 241.
43. Kuenen, J. G. (1975). *Plant Soil* **43,** 49.
44. Vitolins, M. I. and Swaby R. J. (1969). *Aust. J. Soil Res.* **7,** 171.
45. Lundgren, D. G., Vestal, J. R., and Tabita, F. R. (1972). "The Microbiology of Mine Drainage Pollution." In R. Mitchell, Ed., *Water Pollution Microbiology*. Wiley-Interscience, New York pp. 69–88.
46. Singer, P. G., and Stumm, W. (1970). *Science* **167,** 1121.
47. Bryner, L. C., Beck, J. V., Davis, D. B., and Wilson, D. G. (1954). *Ind. Eng. Chem.* **46,** 2587.
48. Belly, R. T. and Brock, T. D. (1974). *J. Bacteriol.* **117,** 726.
49. Apel, W. A., Dugan, P. R., Filippi, J. A., and Rheins, R. S. (1976). *Appl. Environ. Microbiol.* **32,** 159.
50. Dugan, P. R. (1975). *Ohio J. Sci.* **75,** 266.
51. Brock, T. D., Brock, K. M., Belly, T. T., and Weiss, R. L. (1972). *Arch. Mikrobiol.* **84,** 54.
52. Brock, T. D., Cook, S., Petersen, S., and Mosser, J. L. (1976). *Geochim. Cosmochim. Acta* **40,** 493.
53. Zinder, S. and Brock, T. D. unpublished results.
54. Mosser, J. L., Mosser, A. G., and Brock, T. D. (1973). *Science* **179,** 1075.
55. Winogradsky, S. (1976). In T. D. Brock, Ed., *Milestones in Microbiology*. American Society for Microbiology, Washington, D.C., p. 227.
56. Burton, S. D. and Morita, R. Y. (1964). *J. Bacteriol.* **88,** 1755.
57. Joshi, M. M. and Hollis, J. P. (1977). *Science* **195,** 179.
58. Brock, T. D. personal observation.
59. Tuttle, J. H., Holmes, P. E., and Jannasch, H. W. (1974). *Arch. Mikrobiol.* **99,** 5.
60. Starkey, R. L. (1935). *Soil Sci.* **39,** 197.

61. Zinder, S. H. personal observation.
62. Tuttle, J. H. and Jannasch, H. W. (1972). *Limnol. Oceanogr.* **17,** 532.
63. Kuenen, J. G. personal communication.
64. Caldwell, D. E., Caldwell, S. J., and Tiedje, J. M. (1975). *Plant Soil* **43,** 101.
65. Guittonneau, G. and Keiling, J. (1932). *Ann. Agron.,* N. S., **2,** 690.
66. Guittonneau, G. (1927). *C. R. Acad. Sci.* **184,** 45.
67. Nor, Y. M. and Tabatabai, M. A. (1976). *Soil Sci.* **122,** 171.
68. Pfennig, N. (1975). *Plant Soil,* **43,** 1.
69. Hansen, T. A. and Van Gemerden, H. (1972). *Arch. Microbiol.,* **86,** 49.
70. Truper, H. G. (1976). *Int. J. System. Bacteriol.* **26,** 74.
71. Pierson, P. K. and Castenholz R. W. (1974). *Arch. Microbiol.* **100,** 5.
72. Bauld, J. and Brock, T. D. (1973). *Arch. Mikrobiol.* **92,** 267.
72a. Gorlenko, V. M. (1975). *Microbiology* **44,** 682.
73. Cohen, Y., Padan, E., and Shilo, M. (1975). *J. Bacteriol.* **123,** 855.
74. Utkilen, H. C. (1976). *J. Gen. Microbiol.* **95,** 177.
75. Takahashi, M. and Ichimura, S. (1968). *Limnol. Oceanogr.* **13,** 644.
76. Culver, D. A. and Brunskill, G. J. (1969). *Limnol. Oceanogr.* **14,** 862.
77. Butlin, K. R. and Postgate, J. R. (1954). In *Biology of Deserts.* Institute of Biology, London, p. 112.
78. Van Gemerden, H. (1967). Ph.D. thesis, Leiden, Holland.
79. Baas Becking, L. G. M. (1925). *Ann. Bot.* **39,** 613.
80. Stevens, R. K., Mulik, J. D., O'Keefe, A. E., and Krost, K. J. (1971). *Anal. Chem.* **43,** 827.
81. Banwart, W. L. and Bremner, J. M. (1974). *Soil Biol. Biochem.* **6,** 113.
82. Frederick, L. R., Starkey, R. L., and Segal, W. (1957). *Proc. Soil Sci. Soc. Am.* **21,** 287.
83. Hesse, P. R. (1957). *Plant Soil* **9,** 86.
84. Freney, J. R. (1976). *Aust. J. Biol. Sci.* **13,** 387.
85. Kadota, H. and Ishida, Y. (1972). *Ann. Rev. Microbiol.* **26,** 127.
86. Nriagu, J. (1968). *Limnol. Oceanogr.* **13,** 431.
87. Zinder, S. H. and Brock, T. D. (1978). *Appl. Environ. Microbiol.* (in press).
88. Rasmussen, R. A. (1974). *Tellus* **26,** 254.
89. Asami, T. and Takai, Y. (1963). *Soil Sci. Plant Nutr.* **9,** 23.
90. Banwart, W. L. and Bremner, J. M. (1976). *Soil Biol. Biochem.* **8,** 19.
91. Banwart, W. L. and Bremner, J. M. (1975). *J. Environ. Qual.* **4,** 363.
92. Lewis, J. A. and Papavizas, G. C. (1970). *Soil Biol. Biochem.* **2,** 239.
93. Banwart, W. L. and Bremner, J. M. (1976). *Soil Biol. Biochem.* **8,** 439.
94. Bremner, J. M. and Bundy, L. G. (1974). *Soil Biol. Biochem.* **6,** 161.

95. Francis, A. J., Duxbury, J. M., and Alexander, M. (1975). *Soil Biol. Biochem.,* **7,** 51.
96. Zinder, S. H. and Brock, T. D. unpublished results.
97. Salsbury, T. L. and Merricks, D. L. (1975). *Plant Soil* **43,** 191.
98. Zinder, S. H. and Brock, T. D. article to be published.
99. Challenger, F. (1951). *Adv. Enzymol.* **12,** 429.
100. Anderson, T., Kates, M., and Volcani, B. E. (1976). *Nature* 263, 51.
101. Wagner, C. and Stadtman, E. R. (1976). *Arch. Biochem. Biophys.* **98,** 331.
102. Wagner, C., Lusty, S. M., Jr., Kung, H. F., and Rogers, J. R. (1966). *J. Biol. Chem.* **241,** 1923.
103. Elliot, L. F. and Travis, T. A. (1973). *Soil Sci. Soc. Am. Proc.* **37,** 700.
104. Lovelock, J. E. (1974). *Nature* **248,** 625.
105. Sivela, S. and Sundman, V. (1975). *Arch. Microbiol.* **103,** 303.
106. Bremner, J. M. and Banwart, W. L. (1976). *Soil Biol. Biochem.* **8,** 79.
107. Zinder, S. H. and Brock, T. D. (1978). *Nature 273,* 226.
108. Kurita, S., Endo, T., and Nakamura, H. (1971). *J. Gen. Appl. Microbiol.* **17,** 185.
109. Fitzgerald, J. W. (1976). *Bacteriol. Rev.* **40,** 698.
110. Freney, J. R., Melville, G. E., and Williams, C. H. (1971). *Soil Biol. Biochem.* **3,** 133.
111. Casagrande, D. and Siefert, K. (1977). *Science* **195,** 675.
112. Decker, K., Jungermann, K., and Thauer, R. K. (1970). *Angew. Chem.,* Int. Ed. **9,** 138.

INDEX